Lizard Ecology

Lizard Ecology

Studies of a Model Organism

Edited by
Raymond B. Huey,
Eric R. Pianka,
and Thomas W. Schoener

Harvard University Press
Cambridge, Massachusetts
and London, England
1983

Printed in the United States of America
10 9 8 7 6 5 4 3 2 1

This book is printed on acid-free paper, and its binding materials have been chosen for strength and durability.

Library of Congress Cataloging in Publication Data

Main entry under title:

Lizard ecology.

"Most of the chapters were originally presented in December 1980 at a symposium held during the annual meetings of the American Society of Zoologists in Seattle, Washington"—Pref.
Bibliography: p.
Includes index.
1. Lizards—Ecology—Congresses. 2. Reptiles—Ecology—Congresses. I. Huey, Raymond B. II. Pianka, Eric R. III. Schoener, Thomas W., 1943–
IV. American Society of Zoologists.
QL666.L2L58 1983 597.95'045 82-15686
ISBN 0-674-53673-8

Preface

PROGRESS IN SCIENCE is achieved by an understanding of the past as well as of the present. An awareness of historical trends leading to contemporary generalizations inevitably deepens one's appreciation of the nature and limits of those generalizations.

This book addresses the current status of studies on lizard ecology but has a historical basis traceable to the publication in 1967 of a volume edited by the late W. W. Milstead, *Lizard Ecology: A Symposium* (Columbia: University of Missouri Press). That volume, which was based on a symposium held in 1965, demonstrated to a broad audience that lizards are often model organisms for ecological studies: many species are abundant, easily observed, low in mobility, easy to capture, and hardy in captivity. Moreover, several key papers provided an intellectual standard and a framework that have helped to guide and stimulate subsequent research.

During the past 15 years studies of lizard ecology have advanced so dramatically that a new critical synthesis of the field is required. In response, we organized this book to summarize advances in several areas of active research on lizards and to evaluate progress made during those 15 years. Most of the chapters were originally presented in December 1980 at a symposium held during the annual meetings of the American Society of Zoologists in Seattle, Washington.

The authors have diverse backgrounds and approaches, and they include younger as well as established researchers. Authors were selected by the editors and by three participants in the 1965 symposium, William R. Dawson, Rodolfo Ruibal, and Donald W. Tinkle.

We dedicate this book to the memory of three great lizard ecologists: Raymond B. Cowles, William W. Milstead, and Donald W. Tinkle. Cowles's insights on behavioral thermoregulation of ectotherms have had a fundamental impact on physiological ecology—indeed, Cowles and Bogert's 1944 paper is a landmark in lizard ecology. Milstead's pioneering research on competition in *Cnemidophorus* came in the early 1960s, a time when studies on competition were not yet a dominant theme in population ecology, and he organized and edited the 1967 volume. Tinkle's monumental studies on life-history phenomena and his advocacy of long-term studies have left a lasting imprint on population biology and ecology. Don Tinkle planned to contribute to this book his analysis of a 10-year study of a lizard population, but his tragic death in early 1980 deprived us of this contribution.

Each chapter of this book was critically reviewed by at least two extramural reviewers. We sincerely thank the following colleagues as well as several anonymous reviewers for their efforts: George A. Bartholomew, James P. Collins, Robert K. Colwell, Justin D. Congdon, David Duvall, Paul W. Ewald, Joseph Felsenstein, John H. Gillespie, Neil Greenberg, Paul E. Hertz, Robert Holt, A. Ross Kiester, Armand Kuris, Polley A. McClure, Timothy C. Moermond, F. Harvey Pough, Stuart L. Pimm, A. Stanley Rand, James N. M. Smith, James R. Spotila, Stephen C. Stearns, Christopher H. Stinson, Donald R. Strong, Jr., Laurie J. Vitt, R. Haven Wiley, and John W. Wright.

Kathryn Rahn was a great help with editorial matters. We wish to acknowledge George A. Bartholomew (President, American Society of Zoologists [A.S.Z.]), Malcolm S. Gordon (Chairperson, Division of Ecology, A.S.Z.), Alan J. Kohn (Local Meeting Chairman, A.S.Z.), and especially Mary Wiley (Business Manager, A.S.Z.) for their untiring support of the symposium. We also thank the Ecological Society of America (Robert T. Paine, President) for cosponsorship. A grant from the National Science Foundation (D. W. Kaufman, Associate Program Director) provided financial assistance.

R.B.H.
E.R.P.
T.W.S.

Contents

Lizard Ecology

Introduction

SINCE THE 1960s studies of the ecology of lizards have developed considerably in both number and importance, have added new dimensions, and are increasingly theoretical, experimental, and synthetic. Current research projects are not only generating new and better examples of established ecological phenomena but also developing novel or more general theories of ecology, physiology, and behavior. Some studies of these "low-energy" animals are challenging long-held precepts that were derived largely from studies of "high-energy" species such as birds. Thus, research on lizards is playing an increasingly creative role and is helping to establish a new and more general paradigm for ecology.

This book considers advances in several areas of active research on lizard ecology. Our principal objectives are to describe some current developments (empirical, theoretical, methodological), to promote debate and integration among areas, to evaluate recent progress, and to discuss when possible how ecological studies of lizards complement those of other animals. We are neither advocating a single approach to ecological studies nor surveying the entire field.

The book covers three broad topics. The first part, on physiological ecology, discusses the ecological and behavioral significance of activity metabolism, the dynamics of energy flow, how the biophysical environment influences activity and distribution, and how malaria affects physiology and ecology. A second part, on behavioral ecology, contains an analysis of the adaptive zone and behavior of lizards and examines the significance of chemoreception, the establishment and maintenance of territories, the interactions among sexual selection and territoriality, and the psychobiology of parthenogenesis. The third part, on population and community ecology, evaluates patterns of life history, the relationship between niche overlap and interspecific competition, whether temporal differences in activity reduce dietary overlap, the effects of sympatry on patterns of body size, the predictability of adaptive radiations, and the integration of coevolutionary theory with biogeography. Each part begins with an overview that provides historical background. The book concludes with a chapter by an avian ecologist critically assessing current studies of lizards. This volume is thus designed to address where we are, where we have come from, and where we should be going.

Changing Perspectives

Prior to the late 1950s, natural history dominated the focus of most ecological research on lizards. The pioneering work of R. B. Cowles, H. S. Fitch, G. K. Noble, A. M. Sergeyev, and R. C. Stebbins, among others, laid a solid foundation for the field.

Research became increasingly comparative during the late 1950s and early 1960s. Blair (1960) dynamically portrayed the population ecology of the rusty lizard (*Sceloporus olivaceus*). Milstead (1961) pioneered analyses of interspecific competition in lizards. Collette (1961), Ruibal (1961), and Rand (1964) documented an unexpected degree of interspecific diversity in the ecologies of *Anolis* lizards in the Caribbean. This diversity quickly attracted the attention of numerous scientists interested in evolutionary, physiological, behavioral, and community ecology.

This burst of research during the early sixties inspired Milstead's volume (1967). Several papers from that volume have had lasting impact. Tinkle's analysis of temporal variation in the population dynamics of *Uta stansburiana* (Tinkle, 1967) set a high standard for long-term studies, inspiring a generation of students and leading to some of the first empirical analyses of life-history patterns in vertebrates (Tinkle, 1969; Tinkle, Wilbur, and Tilley, 1970; Turner, 1977; Chapter 11 below). Rand's and Ruibal's chapters in Milstead on social and territorial behavior of West Indian *Anolis* (Rand, 1967; Ruibal, 1967) and Carpenter's intensive analysis of iguanid displays (Carpenter, 1967) presaged important work on communication, social behavior, and reproduction in lizards (Rand and Williams, 1971; Jenssen, 1977; Rand and Rand, 1976; Stamps, 1977a,b, and Chapter 9 below; Crews, 1979, and Chapter 10 below; Chapters 7 and 8 below). Norris's chapter on color adaptation and thermal relations (Norris, 1967) represented an early application of the biophysical principles and procedures that are playing increasingly important roles in population ecology and physiological ecology (Chapter 3 below). Dawson's synthesis of thermal physiology (Dawson, 1967) provided a physiological basis for interpreting behavioral temperature regulation by lizards—a subject that continues to be a central feature in many ecological studies (Dawson, 1975; Heatwole, 1976; Huey and Slatkin, 1976; Avery, 1982; Huey, 1982; Chapters 1 and 3 below).

Developments since 1967

Energetics and Activity. Demonstrations that lizards have strikingly lower energetic requirements (Bennett and Nagy, 1977; Turner, Medica, and Kowalewsky, 1976; Nagy, 1981, and Chapter 2 below) and endurance capacities (Moberly, 1968; Bennett, 1978, and Chapter 1 below) than do birds and mammals have led to an awareness of fundamental ecological and behavioral distinctions between ectotherms and endotherms (Case, 1978; Regal, 1978, and Chapter 5 below; Bennett, 1980, and Chapter 1 below; Pough, 1980) and have led to new ideas of the evolution of endothermy (Bennett and Ruben, 1979). Research on the physiological effects of malarial parasites on lizards (Schall, Chapter 4 below) is constructing links between physiological ecology, parasitology, and population biology.

Temperature Regulation. Our understanding of the mechanisms and the physiological significance of temperature regulation has increased enormously (Dawson, 1975; Avery, 1982; Bartholomew, 1982). Field research on thermoconformity in *Anolis* (Ruibal, 1961; Rand, 1964; Hertz, 1974) has led to some of the first theoretical models of the ecology of temperature regulation for ectotherms (Huey and Slatkin, 1976; Huey, 1982). Complex biophysical models of heat and mass flux, now greatly expanded, are being applied to diverse and complex ecological problems (Porter et al., 1973; Roughgarden, Porter, and Heckel, 1981; Tracy, 1982; Chapter 3 below). Discoveries revealing the medical and adaptive significance of fever have relied heavily on experiments with lizards (Kluger, 1979).

Social Behavior. Research on comparative analyses of display patterns (Carpenter, 1967) quickly led to more general studies of reproductive isolation and communication (Gorman, 1969; Rand and Williams, 1971; Ferguson, 1971; Carpenter and Ferguson, 1977; Crews and Williams, 1977; Jenssen, 1977; Kiester, 1977). Lizards were found to be ideal subjects on which to test ideas concerning sexual selection, mating strategies, and the differential behavior of males and females (Andrews, 1971; Stamps 1977a,b, and Chapter 9 below; Trivers, 1972). Interdisciplinary linkages were forged (1) between social behavior and physiology, particularly endocrinology (Crews, 1975; Crews and Williams, 1977; Greenberg and McLean, 1978; Chapters 8 and 10 below), and (2) between social behavior and population biology, especially as concerned spacing systems (Kiester and Slatkin, 1974; Simon, 1975; Stamps, 1977a,b; Stamps and Tanaka, 1981; Schoener and Schoener, 1982a; Chapters 7 to 9 below).

Foraging. Field observations of lizards helped to inspire early models of optimal foraging behavior (MacArthur and Pianka, 1966; Schoener, 1969, 1971). Recent field and experimental studies (Schoener, Huey, and Pianka, 1978; Huey and Pianka, 1981; Stamps, Tanaka, and Krishnan, 1981; Chapters 5 and 12 below) extend and modify certain aspects of these models. Other studies examine interactions among foraging, habitat structure, and risk of predation (Vitt and Congdon, 1978; Moermond, 1979b).

Reproductive Strategies. Studies of geographic and temporal variation in life-history determinants are now almost routine and form a basis for many comparative and theoretical analyses of life-history evolution (Tinkle, Wilbur, and Tilley, 1970; Hirshfield and Tinkle, 1975; Pianka and Parker, 1975; Turner, 1977; Ballinger, 1978; Vitt and Congdon, 1978; Andrews, 1979; Chapters 2 and 11 below).

Interspecific Interactions. Except for Milstead's research, few studies of lizards considered the possible role of competition until the mid-1960s. Intensive observational studies (Rand, 1964; Schoener, 1968, 1975; Schoener and Gorman, 1968; Sexton and Heatwole, 1968; Pianka, 1969; Jenssen, 1973; Huey and Pianka, 1974; Roughgarden, 1974; Schoener, 1974a; Lister, 1976; Roughgarden and Fuentes, 1977; Case, 1979; Huey, 1979; Moermond, 1979a; Chapters 12 to 14 below) as well as experimental work (Nevo et al., 1972; Dunham, 1980; Smith, 1981; Roughgarden, Rummel, and Pacala, 1982; Pacala and Roughgarden, 1982; Chapter 16 below) not only are providing definitive examples of resource partitioning as it relates to environmental availability and potential competitors but also are inspiring development of complex competition models (Roughgarden, 1972, 1979; Schoener, 1974b, 1977; Chapters 14 and 16 below). Although direct observation of predation on lizards is infrequent, indirect measures have suggested predation rates may differ between types of places, such as islands versus mainlands, and between kinds of lizards, such as males versus females (Vitt, Congdon, and Dickson, 1977; Andrews, 1979; Schoener, 1979; Schall and Pianka, 1980; Schoener and Schoener, 1980).

Community Ecology. At the time of the Milstead symposium, studies of community ecology were limited. MacArthur's insights on species diversity helped inspire considerable work on this topic in lizards (Pianka, 1967, 1973, 1974, 1975, 1977; Sage, 1973; Case, 1975; Fuentes, 1976; Inger and Colwell, 1977; Schall and Pianka, 1978). A unique development has been the attempt to deduce the evolutionary and ecological history of radiations of *Anolis* (Williams, 1972, and Chapter 15 below). Current experimental studies are testing many general hypotheses on community structure (Roughgarden, Rummel, and Pacala, 1982; Pacala and Roughgarden, 1982; Chapter 16 below).

Island Biogeography. MacArthur and Wilson's theories of island biogeography and ecology received some of their first tests and challenges from research with lizards (Grant, 1967; Williams, 1969; Case, 1975), and *Anolis* lizards have become prime subjects for experimental biogeography (Schoener and Schoener, 1982b).

The years since Milstead (1967) have been exceedingly productive for all areas of lizard ecology. As Fig. I.1 illustrates, while the number of new studies per year in some areas (sociobehavioral, community) is now

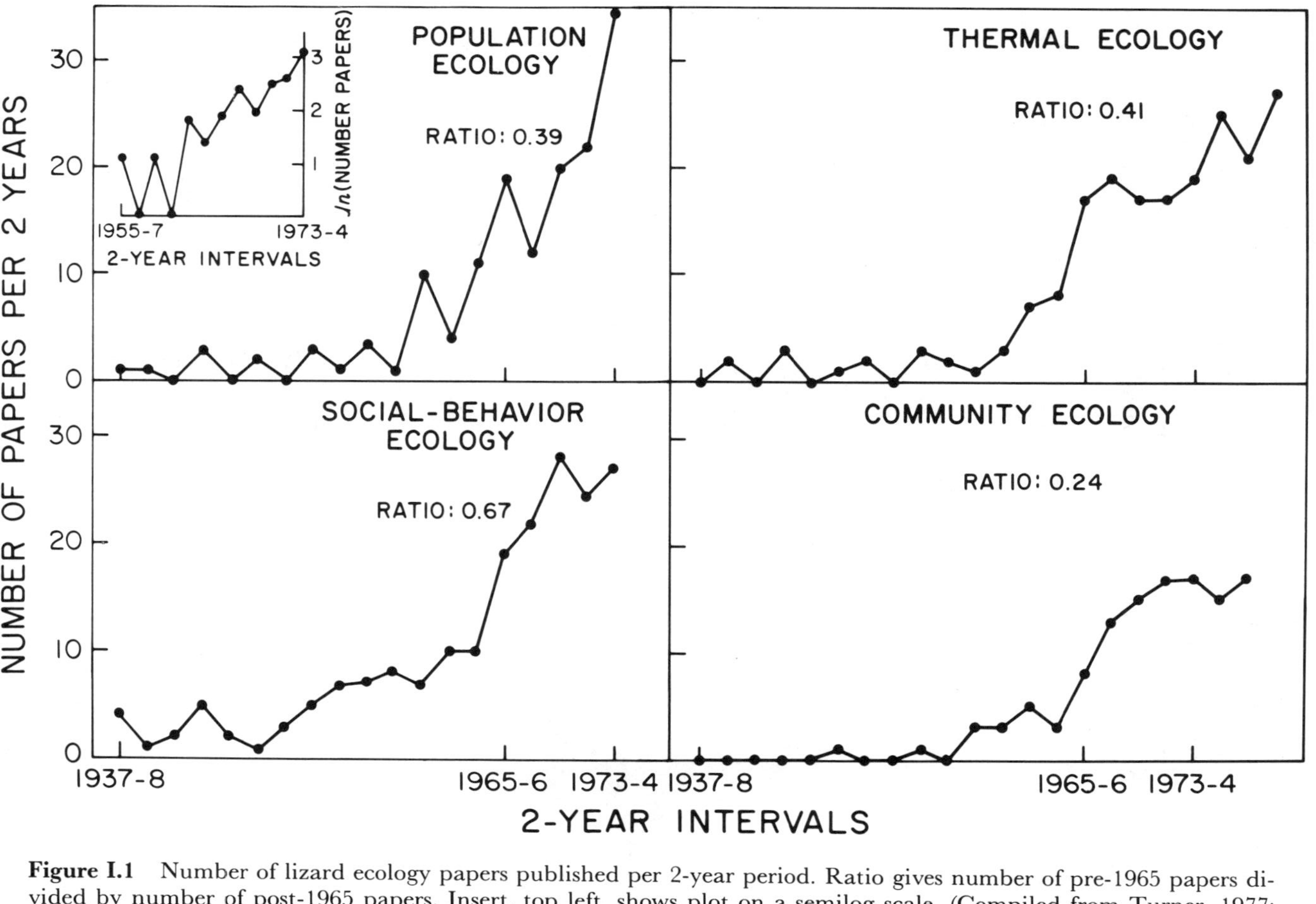

Figure I.1 Number of lizard ecology papers published per 2-year period. Ratio gives number of pre-1965 papers divided by number of post-1965 papers. Insert, top left, shows plot on a semilog scale. (Compiled from Turner, 1977; Pianka, 1977; Stamps, 1977a; Schoener, 1977; Avery, 1982; Huey, 1982; and Scott, 1982.)

beginning to level off, in one (population ecology—see insert) growth is still exponential. Indeed, the cumulative number of papers in the latter field is growing exponentially with $r \simeq 0.14$, a rate substantially higher than average for major scientific disciplines (Price, 1963). Clearly, for many types of ecological studies lizards are model organisms—moreover, they now challenge birds as the paradigmatic organism of ecology.

Raymond B. Huey
Eric R. Pianka
Thomas W. Schoener

PART I
PHYSIOLOGICAL ECOLOGY

PHYSIOLOGICAL ECOLOGY of animals occupies common ground between comparative physiology and ecology, and it has contributions to make to each field. Topics that appear prominent in this area include, according to my view, the following.

Energetics. An analysis of the capture and expenditure of free energy by animals is important not only to the understanding of their autecology but to the assessment of the dynamics of populations and communities as well. Such analysis provides us with a means for estimating in a universal currency the costs of particular activities and, in some cases, their benefits. Nagy (Chapter 2) skillfully applies a battery of techniques including one involving use of doubly labeled water to determine the energy budget of *Uta stansburiana.* His useful analysis documents lizards as "low-energy machines" (see Chapter 1). The provision of data on *U. stansburiana* is especially fortunate because of the prominent place this species occupies in lizard demography (see Turner, 1977). Bennett (Chapter 1) provides another view of energetics emphasizing metabolic support of activity. The limited aerobic powers of many lizards could force a heavy dependence on anaerobic metabolism during vigorous activity. Such a dependence would have profound biochemical and behavioral consequences (see below). Bennett's chapter reminds us that the pathways through which energy is made available may rival in ecological importance considerations of power input and total energy utilization. His work also shows us the substantial progress that has been made in characterizing the activity physiology of lizards since the beginning provided by Bartholomew and Tucker (1963, 1964), Bartholomew, Tucker, and Lee (1965), and Moberly (1966, 1968a, 1968b).

Effect of Physical and Biotic Factors. When coupled with adequate studies of the animals in nature, laboratory investigation of the physiological effects of physical factors upon these organisms has provided insight concerning patterns of adaptation to harsh environments, mechanisms of acclimatization, control of reproductive timing, and so on. Bennett illustrates the importance of these effects in his analysis of the thermal dependence of oxygen transport in lizards (Chapter 1). This dependence is an important determinant of a lizard's potential for activity.

Biotic factors undoubtedly play an important role in conditioning physiological response, yet we know relatively little about the details of their actions in wild populations of most organisms. The involvement of such things as nutritional state, parasite burden, and social interactions in affecting the physiological state and ultimately the fitness of organisms remains an important item on the agenda for physiological ecology. For this reason Schall's work (Chapter 4) on the effects of malarial parasites on the western fence lizard (*Sceloporus occidentalis*) is particularly welcome. It shows just how crucial biotic factors can be in the biology of free-living lizards. In addition to having ecological importance, Schall's study is valuable in providing a broadened framework for parasitological studies.

Distributional Limits. Except in the most obvious cases, knowledge of the manner in which myriad biotic and physical factors act in imposing limitations on the distribution of organisms is not well understood. This topic has been of interest to physiological ecologists since their early efforts at "tolerance physiology." Porter and Tracy (Chapter 3) show us how adequate information on the physiology and ecology of a lizard such as the desert iguana (*Dipsosaurus dorsalis*) can be used to explore situations in which physical factors appear limiting.

Model Species. Physiological ecology deals primarily with wild animals beyond the realms of agriculture, medicine, or anthropology. It therefore allows contact with a wide assortment of species which may depart conspicuously in physiological characteristics from more conventional laboratory subjects. This sometimes makes these species especially promising material for investigating the distribution and underlying mechanisms of particular physiological processes. Such has been the case with *D. dorsalis,* one of the subjects serving to establish the presence of febrile responses in ectothermic vertebrates (see Kluger, 1979). The lizards employed in Bennett's studies of activity physiology (Chapter 1) may be very useful subjects for investigations dealing with the mechanisms by which vertebrates cope with metabolic acidosis. The *Sceloporus* studies by Schall (Chapter 4) should also be valuable for further assessment of host-parasite interactions in natural populations.

Behavioral Interpretations. Behavior contributes importantly to maintenance of homeostasis in various animals. One aspect of this relates to these animals' positioning themselves in suitable microclimates. Appreciation of the significance of such behavior often depends upon analysis of the flow of energy and material between the animals and their environment, as well as upon an understanding of physiological requirements. Porter and Tracy (Chapter 3) provide helpful examples of this flow, for example, in showing just how movements of a land iguana (*Conolophus*) through its home range on Isla Sante Fe in the Galapagos affect thermal economy.

Bennett's analysis (Chapter 1) of activity metabolism in lizards gives us an opportunity to interpret behavior through understanding of functional capacities. It is much easier to appreciate the predilection of many of these animals for sit-and-wait patterns of predation or low-speed patrol when one is aware of their low aerobic capacities (Regal 1978, and Chapter 5 below).

Evolutionary Interpretations. Using results on contemporary species to develop inferences concerning evolution of physiological functions has considerable risk attached to it. Nevertheless, such information does allow fruitful speculation about certain topics and identification of key evolutionary steps. The characterization of reptiles and other ectothermic tetrapods as low-energy machines (Pough, 1980; Bennett, Chapter 1) emphasizes anew the importance of enhancement of aerobic power in the evolution of endothermy and ultimately homeothermy. Bennett also notes that the limitations on aerobic input effectively bar many lizards from exploiting niches requiring sustained vigorous activity. This has important implications not only for foraging and predator avoidance but also for the feasibility of terrestrial migrations. In the present context, we can better appreciate the evolutionary possibilities of lizards if we understand the functional substrate represented in a saurian grade of organization (Bennett, Chapter 1; Regal, Chapter 5).

William R. Dawson

1 Ecological Consequences of Activity Metabolism

Albert F. Bennett

LIZARDS EXHIBIT LOW RESTING levels of oxygen consumption and energy utilization compared to mammals and birds. Their rate of resting metabolism is similar to that of other reptiles and other ectothermic vertebrates and invertebrates of similar temperature and size (Hemmingsen, 1960; Bennett and Dawson, 1976). This low rate of energy turnover is insufficient to provide substantial physiological thermoregulation, and lizards must consequently thermoregulate behaviorally or assume the temperature of their environment. Such a low rate of energy demand has obvious advantages in permitting lizards and other ectotherms to survive on very little food, to tolerate periods of limited food availability or unpredictability of food resources, and to convert a large fraction of ingested food into biomass. The thermal and energetic consequences of low metabolic rate in lizards are fairly familiar to ecologists and have been summarized elsewhere (Dawson, 1967; Tucker, 1967; Templeton, 1970; Bennett and Dawson, 1976; Bennett, 1978, 1982; Pough, 1980). However, there are apparently other consequences associated with this low level of energy metabolism and oxygen consumption, among which is the fact that maximal levels of oxygen consumption for lizards are also relatively low compared to those of endotherms. Since sustained activity requires increased oxygen utilization, behavioral capacities of lizards are curtailed by low absolute levels of available aerobic power. These behavioral constraints structure interactions between lizards and both their biotic and abiotic environments. Thus lizards' physiological capacities for activity have direct effects on both their behavior and ecology.

Aerobic Power Input

Power inputs from aerobic metabolism available to lizards during activity have been examined in different ways. One method is the determination of maximal oxygen consumption and calculation of maximal aerobic power input on that basis. The implicit assumption that all this oxygen consumption can be diverted to physical activity is incorrect since many processes besides contraction of skeletal muscle are maintained during exercise. Although this method yields an overestimate, it is at least relatively simple to perform. Another metric, the scope for activity, was proposed by Fry (1947) as a measure of work capacity and was defined as the difference between maximal and resting rates of oxygen consumption. But use of aerobic metabolic scope assumes that physiological processes continue at the resting level even during activity. This assumption is also almost certainly incorrect since some "maintenance" functions will be augmented and some will decrease during activity. This measure has the advantage, however, of recognizing that a portion of aerobically derived energy is not available for physical activity.

Aerobic power capacities of lizards have been summarized in a series of mass-dependent power expressions. For lizards, the following relationships describe aerobic metabolic power input at 35 °C (Bennett, 1982):

$$\text{resting oxygen consumption: } y = 2.51\, m^{0.78}, \qquad (1.1)$$
$$\text{maximal oxygen consumption: } y = 16.3\, m^{0.76}, \qquad (1.2)$$
$$\text{aerobic metabolic scope: } y = 13.8\, m^{0.76}, \qquad (1.3)$$

where y is aerobic power input in mW (divide by 5.6 to obtain values in ml $O_2\ h^{-1}$) and m is body mass in g. Thus, a 100-g lizard would have a maximal aerobic power input of 540 mW and an aerobic metabolic scope of 460 mW at 35 °C. Similar relationships have been derived for lizards at other body temperatures as well (Wilson, 1974; Bennett and Dawson, 1976; Bennett, 1982). These levels of resting and active metabolism are similar to those of fish and amphibians but are substantially less than those of mammals or birds of equal size (Table 1.1).

The maximal aerobic metabolic power input of which lizards are capable is quite low. Additionally only a small portion of this power input appears as power output (work involved in activity), because organisms are not completely efficient in energetic conversions. As lizards rarely do work in the sense of moving objects from one place to another, their low work capacities may not appear to be important. However, low aerobic power inputs have severe effects on locomotor capacities, and these effects underlie all of the behavioral repertoire of these animals. The consequences of such limited capabilities may be best envisioned in an examination of locomotor energetics.

Locomotor Energetics and Limitations

The energetics of quadrupedal terrestrial locomotion have been measured in a variety of animals, including lizards, and theoretical discussions and summaries are available elsewhere (Taylor, Schmidt-Nielsen,

Table 1.1 Maximal aerobic power input during activity in small vertebrates (data summarized in Bennett and Ruben, 1979).

Species	Mass (g)	Aerobic power input (mW/g)
Lizards		
Dipsosaurus dorsalis	35	11
Iguana iguana	800	3
Varanus gouldii	674	6
Other reptiles, amphibians, and fish		
Bufo cognatus	40	9
Carassius auratus	66	2
Pituophis catenifer	548	3
Pseudemys scripta	305	6
Rana pipiens	38	3
Salmo gairdneri	66	3
Mammals and birds		
Larus atricilla	322	71
Melopsittacus undulatus	35	171
Mus musculus	34	52
Pteropus gouldii	779	60

and Raab, 1970; Tucker, 1970; Schmidt-Nielsen, 1972; Taylor, 1973). Aerobic power input (oxygen consumption) is measured as a function of locomotor speed, which is set by a motor-driven treadmill. As speed increases, oxygen consumption increases approximately linearly. The slope of the line is designated the net cost of transport, is expressed in units of ml $O_2/(g \cdot km)$ or $J/(g \cdot km)$, and is independent of speed. From one point of view, this number may be interpreted as the amount of work required to move a unit of body mass over a unit distance. However, since this function does not have an intercept of zero, there are other costs in addition to those represented by the net cost of transport. These are accounted for by dividing total oxygen consumption by speed, yielding a value termed the cost of transport (Tucker, 1970) or the total cost of locomotion (Schmidt-Nielsen, 1972). This value is expressed in identical units, $J/(g \cdot km)$, but is dependent on speed and decreases as speed increases. Consequently, the total cost of moving a unit of mass over a unit distance decreases as speed increases. This relationship has obvious energetic consequences for locomotor activity. Both net and total cost of transport are useful in assessing locomotor costs in energy budgets, but it is important to distinguish between them because they are easily confused. The intercept of the metabolism-speed curve also merits some comment since it has been commonly observed that it exceeds the value of resting metabolic rate, often by a factor of 1.5 to 2.0. The significance of this relation is unclear. However, it does indicate that locomotion and activity, even at very slow speeds, almost always involve substantial increments in metabolic rate. Locomotion and activity are expensive and can represent a major portion of the energy budget of an animal, even if activity is infrequent or low level.

The relationship of aerobic metabolic rate to speed does not increase indefinitely. Oxygen consumption increases linearly with speed up to the point at which maximal oxygen consumption is attained (the maximal aerobic speed). Maximal oxygen consumption is then maintained as speed increases further. Power input requirements, of course, continue to increase and must be met with other, anaerobic sources of energy provision. Behavior and exertion in excess of maximal aerobic speed are not sustainable, and fatigue begins to limit performance. This transition velocity between sustainable and nonsustainable locomotion is surprisingly low in lizards. Maximal aerobic speeds for a variety of lizards are given in Table 1.2. With the exception of *Varanus,* all these are less than 1 km/h. These represent very slow velocities, an order of magnitude below those of which these same lizards are capable during burst activity.

An additional factor influencing oxygen consumption and aerobic power input is the effect of body temperature. Most aerobic processes in lizards, including maximal aerobic power input, are strongly temperature dependent. Maximal oxygen consumption generally has a temperature coefficient (Q_{10}) of 2 to 3, that is, aerobic capacity is reduced by one-half to two-thirds by every 10° C decrement in body temperature. In view of the limiting relationship of oxygen consumption and speed, we might anticipate that low temperature would similarly curtail sustainable activity capacity and behavior. The limited data available suggest that this is so. Walking ability in *Iguana iguana* is strongly influenced by temperature (Moberly, 1968). The greatest walking speeds sustainable for 10 to 20 minutes by *Iguana* have the following thermal dependence: 20° C, 0.17 km/h; 25° C, 0.23 km/h; 30° C, 0.45 km/h; 35° C, 0.52 km/h; 40° C, 0.45 km/h. Low body temperature strongly curtails stamina; note again how low these velocities are. A similar temperature-dependent decrement in performance is found in the desert iguana, *Dipsosaurus dorsalis* (Fig. 1.1) (John-Alder and Bennett, 1981). Maximal aerobic speed declines from 0.8 km/h at 40° C to 0.3 km/h at 25° C. Body temperature obviously exerts an important influence on activity and stamina, and low or even moderate temperatures may greatly restrict behavioral capacity. It should be remembered that there are definite energetic benefits to be gained from the overall metabolic depression associated with low body temperatures. The energetic saving accrued is considered to be one of the major advantages of the poikilothermic condition (Pough, 1980).

Body temperature may also influence locomotor costs directly, beyond placing limitations on maximal aerobic speed. At present, data are too few to permit confident generalizations. In *Iguana* (Moberly, 1968)

Table 1.2 Maximal aerobic speeds and burst speeds in lizards.

Species	Maximal aerobic speed (km/h)	Burst speed	T_B	Reference
Amblyrhynchus cristatus	1.0	9.0	35	Gleeson, 1979
Cnemidophorus murinus	0.3	8.8	40	Bennett and Gleeson, 1979; Bennett, 1980
Dipsosaurus dorsalis	0.8	7.8	40	John-Alder and Bennett, 1981; Bennett, 1980
Gerrhonotus multicarinatus	0.3	3.9	35	John-Alder, unpublished observations; Bennett, 1980
Iguana iguana	0.5	16.4	35	Gleeson, Mitchell, and Bennett, 1980; unpublished observations
Tupinambis nigropunctatus	0.6	14.2	35	Unpublished observations
Varanus exanthematicus	1.2	12.5	35	Gleeson, Mitchell, and Bennett, 1980; unpublished observations

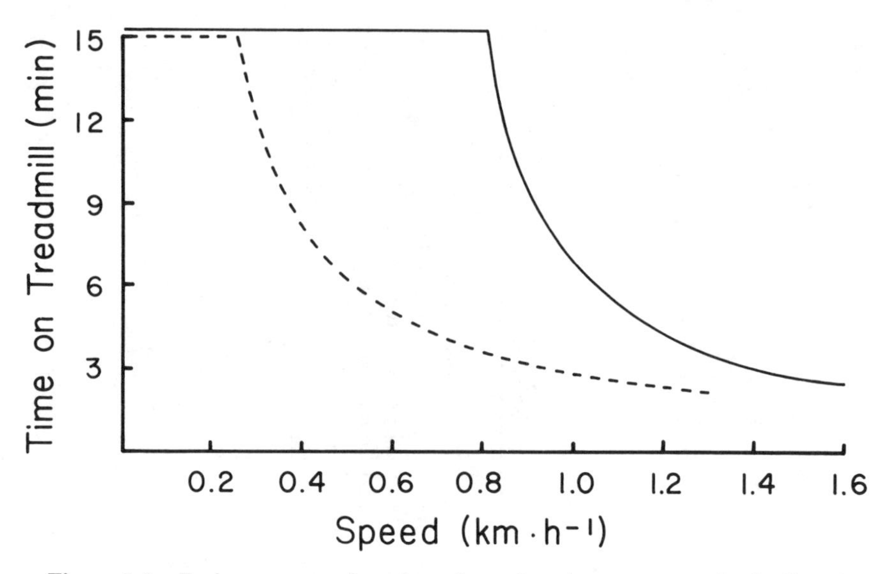

Figure 1.1 Endurance as a function of speed and temperature in the lizard *Dipsosaurus dorsalis.* Lines indicate the average duration of sustained walking behavior at each speed. The broken line indicates a body temperature of 25 ° C; the solid line, 40 ° C. Experiments were terminated after 15 min at both temperatures. (After John-Alder and Bennett, 1981.)

and *Dipsosaurus* (John-Alder and Bennett, 1981) increasing body temperature increases the metabolic rate of a lizard walking at a given sustainable speed but does not alter the net cost of transport, that is, the intercept but not the slope of the speed-metabolism curve is increased. Thus, walking at 35 ° C is more expensive than walking at 25 ° C, but a further increase in speed results in equal metabolic increments at both temperatures. Locomotion is consequently more expensive at higher temperatures in these lizards. This is reported not to be the case for the agamid *Uromastix aegyptius* (Dmi'el and Rappeport, 1976), in which metabolic rate while walking at a given speed is independent of temperature. Only further observations will make clear which of these patterns, if either, predominates.

It is now easier to appreciate the conflicting pressures of temperature on locomotor ability in lizards. On the one hand, low body temperatures reduce metabolic and locomotor costs but result in very low stamina. On the other hand, high temperatures greatly expand the range of sustainable behaviors but increase energy demand. The cost of activity at high temperatures may be balanced somewhat by increasing speed, since the total cost of locomotion decreases because maintenance costs form a smaller proportion of total metabolic expenditure. These relationships are given in Figure 1.2 for *D. dorsalis.* At any speed sustainable at both

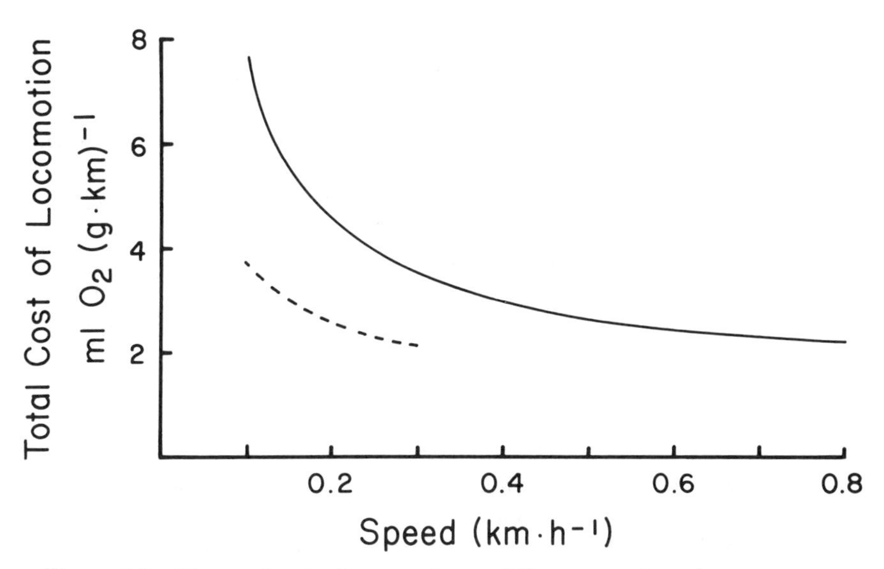

Figure 1.2 The total cost of locomotion at different speeds and temperatures in the lizard *Dipsosaurus dorsalis.* The broken line indicates a body temperature of 25 ° C; the solid line, 40 ° C. (After John-Alder and Bennett, 1981.)

body temperatures, energy expenditure is less at the lower temperature. However, the total cost of locomotion decreases as speed increases, and these values are nearly equal at maximal aerobic speeds at each temperature. That is, the cost differential associated with the difference in body temperature can be offset by walking faster at the higher temperature. The greater scope for activity and stamina at high temperatures permits behavior which may ameliorate the increased costs incurred.

These low aerobically supportable velocities set limits on the behavioral capacities of lizards. A variety of behavioral and ecological roles are not available to animals with this limited aerobic metabolic scope. For instance, certain lizards may be specialized for gliding, but the power requirements for flapping flight are far beyond the capacities of modern reptiles. The expenses cannot be met regardless of body design. Likewise, we do not anticipate herds of herbivorous lizards grazing in open fields, galloping off and outdistancing mammalian predators such as dogs. Nor do we have dog-like lizard predators, chasing down their prey. Short, intense bursts of exertion, similar to those of cats, are of course possible by carnivorous lizards, but these can last for only 1 to 2 minutes. The capacity for terrestrial migrations of any distance are also greatly limited, since they would have to be undertaken at such slow speeds that they would result in long periods of exposure to predators. Nocturnal behavior is also restricted by aerobic power input, since body temperatures are necessarily low and maximal oxygen consumption is restricted by thermal effects.

The relations between oxygen consumption, locomotion capacity, and behavior may be better understood with reference to a specific example. Locomotor and field energetics have been examined in *Cnemidophorus murinus*, a medium-sized teiid lizard found on a few Caribbean islands (Bennett and Gleeson, 1979; Bennett and Gorman, 1979). *C. murinus* is a relatively active lizard, similar to its congeners in North America or lacertids in Europe. It is a highly alert, curious, and voracious lizard. It spends about 75 percent of its emergent time in active foraging, walking more or less continuously at approximately 0.13 km/h, searching for plant or insect material. Even when moving so slowly, its oxygen consumption is four times resting values. This activity makes a substantial impact on the energy budget of this lizard, representing nearly 70 percent of the total daily energy expenditure. *C. murinus* may be unusual in devoting so much time and energy to foraging; foraging has been estimated to represent only about 10 percent of daily energy budgets of *Egernia cunninghami* (Wilson and Lee, 1974) and *Amblyrhynchus cristatus* (Gleeson, 1979). *Cnemidophorus* forages well within its aerobic capacity at about half maximal aerobically sustainable speed. The latter is attained at only 0.3 km/h, and exhaustion occurs in less than 5 minutes at 0.5 km/h. There is considerable reserve burst capacity, however, and *C. murinus* can run at speeds of about 9 km/h for short distances (Bennett, 1980). Thus, aerobically supportable behavior is only a small fraction of the performance capabilities of this lizard, even though it is apparently so active and invests so much energy in foraging activity. Normally, foraging is restricted within these aerobic limits, but additional capacity for escape or pursuit is maintained.

Estimates of locomotor costs for use in the calculation of energy budgets may be obtained from summaries of similarly derived data for lizards exercised on treadmills. The net cost of transport has been calculated to be

$$y = 75.8\ m^{-0.25}, \tag{1.4}$$

where y is net cost of transport in J/(g • km) and m is mass in g (Gleeson, 1979). As is clear from the equation, it is relatively less expensive for a large lizard to move a unit mass than for a small lizard, or, stated another way, total locomotor costs increase as size increases, but they do not grow in direct proportion to body mass. For example, it costs a 1-kg lizard only one-third as much to move a unit of its weight as it costs a 0.01-kg lizard. The underlying reasons for this mass-dependent relationship are unknown. They may partially reflect the mass dependence of resting metabolic rate ($b = 0.80$), the basis of which also has not been satisfactorily explained. The utility of these allometric relationships to ecologists is that they permit the estimation of foraging costs to a lizard if body size, temperature, speed, and duration of locomotion are known. As a rough approximation, the intercept value of the speed-metabolism curve can be estimated as 1.7 times resting metabolic rate (observed values often range between 1.5 and 2.0). To this is added the product of speed and the net cost of transport. Such an equation might take the following form for a lizard at 35 °C:

$$y = 1.7(2.51\ m^{0.78}) + s(21.2\ m^{0.75}), \tag{1.5}$$

where y equals metabolic rate in mW, m is mass in g, and s is speed in km/h. Thus, a 100-g lizard travelling at 1.0 km/h would have a power input of approximately 820 mW. These relationships permit very general estimates of foraging costs in regard to total energy budget or in regard to a cost-benefit analysis of return gained for energy expended. Such estimates would also fit comfortably with a time-energy budget and provide more realistic estimates of activity costs than are currently available.

The components of Eq. 1.5 contain considerable variability: the net cost of transport, for example, may in fact be double or only half that predicted by Eq. 1.4. If greater precision is required, one should make direct observations on the lizard in question. Moreover, these values represent steady-state determinations which are achieved only after 5 to 10 minutes of constant-level activity. Initial oxygen consumption during the first 1 to 2 minutes of activity is considerably less than the level anticipated by Eq. 1.5, and there is often an overshoot during the next several minutes of activity. The predictive value of these relations in estimating the cost of short-term activity, such as walking between two bushes, is less than certain.

The previous discussions have treated lizards as interchangeable animals in a relatively undifferentiated group. The submergence of interspecific differences is in many ways the desired function of these analyses. However, given the very different phylogenetic histories of modern lizards and their great diversity of body form, behavioral type, and ecological role, we certainly would anticipate a degree of diversity in aerobic capacity, behavioral performance capacity, and stamina. The one group of lizards which is clearly exceptional in these regards is the varanids. They possess substantially greater aerobic capacities than do other lizards investigated so far. Maximal levels of oxygen consumption (Bartholomew and Tucker, 1964; Bennett, 1972; Wood et al., 1978; Gleeson, Mitchell, and Bennett, 1980) and maximal aerobic speeds (Bakker, 1972; Gleeson, Mitchell, and Bennett, 1980) are nearly double those of iguanid lizards of similar size. There are obvious correlates with the predatory ability and aggressive nature of these carnivorous lizards. Apart from the varanids, however, it is much more difficult to differentiate among saurian groups on the basis of aerobic ability. Maximal oxygen consumption is very similar in iguanids, agamids, skinks, and teiids, the only saurian families examined in any detail. The teiids are particularly surprising in this regard since they are considered highly active and comparatively intelligent animals (Regal, 1978), fleet, and able to subdue large prey. However, aerobic capacity of *Cnemidophorus* (Asplund, 1970; Bennett and Gleeson, 1979) is no greater than that of the similarly-sized iguanid *D. dorsalis,* and the latter has a substantially greater maximal aerobic speed (Bennett and Gleeson, 1979; John-Alder and Bennett, 1981). The aerobic ability of the large teiid *Tupinambis nigropunctatus* is very similar to that of *Iguana iguana* and maximal aerobic speeds are almost identical (Bennett, unpublished observations). Consequently, we may not be able to draw easy generalizations between apparent activity level and oxygen consumption on an interfamilial or interspecific basis. With the exception of the varanids, there appears to be little differentiation between the saurian groups in maximal oxygen consumption. Aerobic support in lizards may be sufficient for and permit a variety of different types of low-level behavior, ranging from sit-and-wait predation to herbivorous foraging or low-velocity patrol.

Supplemental Anaerobic Energy Metabolism

When the demand for energy utilization exceeds the supply capacity of aerobic metabolism, other sources of energy generation are activated. These are termed anaerobic, a catchall term signifying that their only common element is the lack of oxygen in their function. Anaerobic metabolism may take place in lizards under some anoxic circumstances such as breath holding or diving. However, a more common circumstance when anaerobic metabolism takes place is during *physical activity.* Anaerobiosis plays a significant role under two circumstances: during the initial stages of activity and during intense exercise. In the former case, oxygen consumption cannot be increased instantaneously, as can muscular movements during pursuit or escape. Until aerobic supply can catch up with demand, anaerobiosis provides supplemental energy. Sustained exercise may also create demands for energy utilization that are beyond the maximal levels which oxygen consumption can support. These exercise levels are fairly low in lizards. The more intense the activity, the greater the level of anaerobic metabolism.

The principal anaerobic pathway during activity in lizards is the breakdown of glycogen within the skeletal muscles to lactic acid via glycolysis. Endogenous stores of adenosine triphosphate (ATP) and creatine phosphate stored in the muscle can be catabolized for fueling some muscle contraction. These compounds are used relatively early during activity over a period of a few seconds, whereas lactic acid generation provides energy for 1 to 2 minutes of maximal levels of activity in small lizards. After this time, lizards are visibly fatigued and can be driven to exhaustion shortly thereafter. The proximate causes of this fatigue after intense exercise are unknown, even in mammals.

The rate of anaerobic energy generation in lizards can be high in comparison to aerobic energy generation: maximal anaerobic power input rates range from 20 to 36 mW/g for small lizards during 30 seconds of activity (see Table 1.1) (Bennett and Licht, 1972). Rates of ATP formation via anaerobic metabolism may over short time periods greatly exceed those of aerobic metabolism: 60 to 80 percent of ATP production during 2 minutes of burst activity is produced by lactic acid formation in *D. dorsalis* and *Sceloporus occidentalis* (Bennett and Dawson, 1972; Bennett and Gleeson, 1976). The anaerobic contribution is even greater when shorter time periods are considered. This differential is still more impressive since anaerobic glycolysis is only 10 percent as efficient as aerobiosis at producing energy equivalents (ATP) from equal amounts of carbohydrate fuel. Consequently, the flow of carbohydrate into the anaerobic pathway is necessarily very much greater than that entering aerobic metabolic pathways. This great capacity for anaerobic energy production is responsible for the impressive capabilities for burst activity possessed by most lizards, far beyond those which aerobic systems can support. Performance capabilities are nearly tenfold greater than can be accounted for by aerobic metabolism alone (Table 1.2). Clearly anaerobiosis can provide a very significant component to the behavioral repertoire of lizards.

Another important feature of anaerobic metabolism in lizards besides its absolute magnitude is its very limited dependence on body temperature. The capacity of a variety of lizards to form lactic acid during activity is nearly as great at body temperatures of 20 ° C as at 40 ° C (Bennett and Licht, 1972) (Fig. 1.3). Temperature coefficients of anaerobic metabolism over this range are typically 1.1 to 1.3, in comparison with Q_{10}'s of 2 to 3 for aerobic metabolism. Energy mobilization for rapid activity is thus possible over a broad range of temperatures, a very important feature for rapid escape by a lizard even when it is far from its thermal preferendum. This anaerobic independence is particularly significant in view of the strong thermal dependence of aerobic metabolism. Low body temperature retards not only maximal levels of aerobically supported exertion but also the rate of development of maximal levels. Evasive behavior must be fueled almost totally anaerobically in lizards with low body temperatures. Burst speeds of lizards, as anticipated, have a fairly low thermal dependence over significant spans of body temperature (Bennett, 1980).

Although anaerobiosis provides a greatly expanded behavioral repertoire for lizards, its prolonged use may be very debilitating. One or two

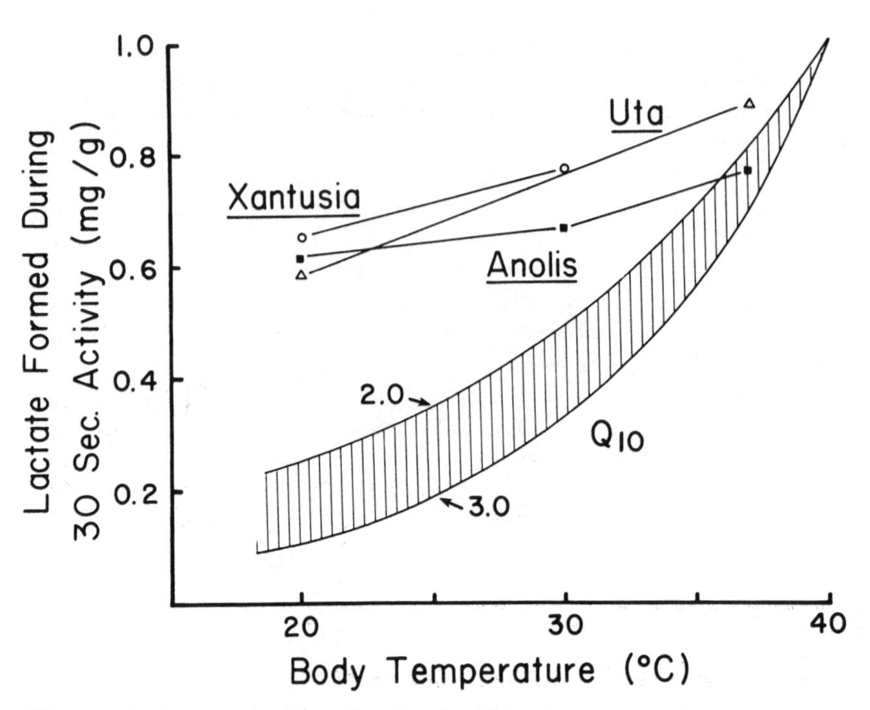

Figure 1.3 Lactate formation [mg lactate/(g body mass • 30 s)] in the lizards *Anolis carolinensis, Uta stansburiana,* and *Xantusia vigilis.* The hatched area indicates the anticipated thermal dependence of a biochemical reaction rate of a process with a rate of 1.0 mg lactate/(g • 30 s) at 40 ° C. (Based on data from Bennett and Licht, 1972.)

minutes of intense activity can leave a lizard exhausted, even to the point of abolishing its righting reflex, and the consequent physiological disruptions may persist for hours. Consequently, we might expect an avoidance of prolonged anaerobically supported activity and its use only under infrequent conditions that demand high levels of power output. An additional factor that argues against frequent, extensive anaerobic metabolism is its inefficiency: compared to aerobic activity, 10 times as much foodstuff is required for comparable levels of exertion during anaerobic activity. Although the lactic acid formed may later be catabolized further or reconverted to glycogen, the immediate demand for ATP during activity and this inefficiency of anaerobic metabolism may severely deplete fuel reserves in muscle. If anaerobiosis occurs only infrequently, these inefficiencies would have relatively little impact on the ecological energetics of lizards and the behavioral consequences would undoubtedly be more severe than the energetic ones. The extent of utilization of anaerobic metabolism under field conditions has not received much examination, primarily because of technical difficulties in securing measurements. Some field-caught specimens of the lizard *Anolis bon-*

airensis possess high lactate contents (Bennett, Gleeson, and Gorman, 1981). The intensity of territorial defense also parallels lactate accumulation in these lizards. In contrast, little evidence of anaerobiosis during either basking behavior or diving was found in the marine iguanas *Amblyrhynchus cristatus,* even though anaerobic capabilities are well developed in this lizard (Gleeson, 1980). We may certainly anticipate further research into the extent of utilization of anaerobic metabolism in free-living lizards, including the circumstances of its activation and the extent of its use.

We have come to understand lizards as low-energy machines, with low maintenance costs and low capacities for sustained power output. Much of their behavior takes place within the constraints imposed by oxygen transport capacities. Anaerobic metabolism provides additional power output for emergency situations. It can greatly extend the performance capacities of lizards but has detrimental effects on subsequent behavioral capacity. Escape capabilities (burst performance) are retained over a broad thermal span, although low body temperature tends to retard sustainable behavior. This appears to be a particularly useful behavioral system, operating at low cost yet retaining high capacities for short-term performance, even under environmentally unfavorable circumstances.

All of this information, with its implications and interpretations, has been gathered in the relatively recent past. Not all the directions of future research are clear, but we can make some educated guesses as to the ways in which this area will continue to develop. First and foremost, we anticipate and require data on metabolic expenditure, both aerobic and anaerobic, under field conditions. Laboratory approaches to these problems have been very fruitful, but their interpretation and ecological significance rest ultimately upon behavior under natural circumstances. We may also anticipate on the one hand a broadening and on the other a narrowing of the phylogenetic scope of these sorts of investigations. Only two or three species have received a truly adequate examination of both their metabolic and locomotor capacities. Data are scattered for other species, and whole groups remain uninvestigated. In contrast, other studies may attempt to focus more finely on interspecific differences between closely related lizards to determine the latitude in behavioral performance which common history and similar structure and function can provide. Finally, we may expect a continuing interest in locomotor energetics per se. Interspecific differences, mass-dependent phenomena, and the cost of other types of locomotor patterns (swimming, crawling, climbing) will all undoubtedly receive attention. Lizards have in the past served as excellent models for exploring the interrelations of physiology, behavior, and ecology. We can anticipate that they still have much to provide us in the future.

2 Ecological Energetics

Kenneth A. Nagy

THE ACTUAL AND RELATIVE rates at which energy flows through animals are of central and unifying importance in many areas of contemporary biology. The trophic dynamic concept of Lindemann (1942) stimulated suggestions that energy flow is the major process underlying and driving ecosystem function and that obtaining adequate energy may be the most important challenge facing wild animals. This has led to detailed questions and hypotheses concerning feeding rates and energetic costs of obtaining food (foraging cost-benefit analysis or optimal foraging theory), the ability of various animals to extract energy from different kinds of food (assimilation efficiency; optimal diet), and how an animal apportions its assimilated energy into respiration, growth, and reproduction (reproductive effort hypotheses; *r* versus *K* selection; sexual dimorphism in size). Because food energy may be a major limiting resource for animals, knowledge of the relationships between food supply and demand could provide keys to understanding intraspecific and interspecific competition and natural selection, and perhaps many other related properties of animals (life-history strategy; territoriality; temperature regulation; morphology). Other questions concern differences in energetics due to sex, age, size, altitude, latitude, habitat productivity, predation, physiology (endothermy versus ectothermy), and season.

Despite the presumed pivotal position of energetics in ecology and evolution, little is known about the details of energy flow in free-living animals. Many studies have been done on the momentary state or standing crop of energy in wild animals, and much is known about rates of energy flow in captive animals. Until recently, measuring rates of energy flux in the field has been very difficult. However, such assessments can now be made by using doubly labeled water to measure energy metabolism (Lifson and McClintock, 1966; Nagy, 1975), along with determinations of the momentary status of energy in animals at successive intervals to measure rates of growth and reproduction.

This research was designed primarily to measure in as much detail as possible the ecological energetics of an animal living in its natural habitat. This study represents the first complete account of the energy economics of a free-ranging vertebrate. Secondary purposes were (1) to provide baseline (control) information needed for assessment of low-level radiation effects on animals in an adjacent experimental plot in Rock Valley, Nevada, (2) to demonstrate the potential of isotope tracer methodology in field studies of animal energetics, and (3) to provide information bearing on as many of the theoretical questions outlined above as possible, and perhaps (4) to stimulate generation of new ideas.

The lizard *Uta stansburiana* was chosen because it is small (2 to 5 g), abundant, and diurnal, and is already well known (see, for example, Carpenter, 1962; Tinkle, McGregor, and Dana, 1962; Irwin, 1965; Dixon and Medica, 1966; Dixon, 1967; Norris, 1967; Tinkle, 1967; Alexander and Whitford, 1968; Turner et al., 1969, 1970; Tanner and Hopkin, 1972; Parker, 1974; Turner, Medica, and Smith, 1974; Parker and Pianka, 1975; Turner, Medica, and Kowalewsky, 1976; Fox, 1978). Thus, this species is relatively easy to study in the field. These iguanid lizards are sit-and-wait predators, feeding primarily on insects. They are considered to be essentially an annual species in Rock Valley, although some individuals may live 3 or more years. They begin breeding when 6 to 8 months old. These lizards are active at almost any time of year, even during winter when they have been seen basking in sunny, bare patches between snow drifts at air temperatures near freezing. They are primarily ground dwellers, rarely climbing in shrubs, and they retreat to shallow burrows in the soil at night.

Rates of energy metabolism (using doubly labeled water) and body weight change were measured in the field at 2- to 4-week intervals throughout a year. Rates of change in body energy located in abdominal fat pads, somatic lipid, fat-free dry matter, and reproductive tissue were estimated in animals collected for autopsy. The stomach contents of autopsied lizards were analyzed for diet, and they were measured for body water. Determinations of energy digestion and assimilation were done in laboratory feeding trials, in conjunction with assessments of the accuracy of doubly labeled water for measuring metabolic rates. Feeding rates in the field were calculated from the sums of energy loss and production. Measurements were done separately for the cohorts of yearling females, older females, adult males, and juveniles. Itemized energy budgets were constructed for each cohort for each month throughout a year.

Materials and Methods

Study Site. Field work on *U. stansburiana* was done from January 1975 through March 1976 at Rock Valley (Nevada) Test Site, in the same areas used by Turner (Turner et al., 1970; Turner, Medica, and Kowalewsky, 1976). This creosote-scrub desert habitat is part of the Mojave

Desert and is described in detail by Beatley (1976). Air temperatures and rainfall were measured at a weather station in Rock Valley.

Marked Lizards. During the course of this study, 230 *Uta* were marked by toe clipping and by painting, measured for snout-to-vent and tail length, weighed, sexed, ages estimated (based on length, using the criteria of Turner et al., 1970), and released where captured in the study areas. Most lizards were injected with doubly labeled water (HTO-18) to measure field metabolic rates (CO_2 production). Lizards were given 10 to 50 microliters (depending on body weight) of HTO-18 (30 atom percent oxygen-18 and 0.7 millicurie tritium per milliliter) by intraperitoneal injection. At least an hour was allowed to pass for complete isotope mixing (a time previously determined in the laboratory), and 5 to 40 microliter blood samples were taken from an eye sinus, flame sealed in glass tubes, and refrigerated for later analysis. Marked animals were recaptured for weighing and blood sampling at intervals ranging between 3 and 130 days, but most intervals were between 7 and 14 days. Lizards were reinjected with HTO-18 as necessary. An effort was made to keep disturbance of the animals and the habitat to a minimum.

Autopsied Lizards. Late each month, unmarked lizards living between about 0.5 to 2 km from the study plots were collected for analysis of diet and body composition. We attempted to capture at least 5 animals in each cohort per month, but in late December inclement weather set in, and despite intensive searching for 3 days, we could find no unmarked adult females. Lizards were captured for autopsy around midday, so they had the opportunity to fill their stomachs before sampling. Animals were kept cool to inhibit gut function until they were weighed and killed that evening. Stomach contents were removed, sorted under a dissecting microscope, and dried to constant weight at 65° C. Average diets for each cohort each month were calculated on a dry-matter basis. Abdominal fat bodies and reproductive tissues (complete reproductive tracts) were carefully removed and dried, as was the rest of the carcass, to determine percent of live mass that was (1) water, (2) total dry matter (including gut contents), (3) dry abdominal fat bodies, and (4) dry reproductive tissues. Later, dry carcasses were pulverized in acetone using a kitchen blender, soaked in ether and acetone for several days, filtered, dried at 65° C, and reweighed to measure somatic lipid content by weight difference. The solvent was evaporated from the filtrate to obtain samples of somatic lipids, which were retained for measurement of energy content along with fat-free carcass, abdominal fat pads, and reproductive tissue.

Diet Samples. The day after stomach content analyses were done, we collected fresh samples of diet items near the study area for analysis of water and energy content. Arthropods were collected by hand from the soil surface or under surface objects, by pitfall traps, and by vacuuming shrubs (D-Vac). Vegetation samples were collected from the portions of plants that were accessible to *Uta*. We were not able to collect adequate samples of every single diet item every month, but we obtained enough information to estimate energy content of the monthly diet.

Laboratory Feeding Experiment. Feeding trials were done to measure assimilation efficiency (energy ingested minus energy defecated) and metabolic efficiency (energy ingested minus energy defecated and urinated) for *Uta* that were eating arthropods and to test the accuracy of HTO-18 measurements of metabolic rate. Five male and 5 female *Uta* captured near the field study area in June 1975 were housed individually in the laboratory in plastic shoeboxes fitted with 7.5-watt light bulbs to provide a photothermal gradient. These bulbs were on between 0800 and 1800 h; the ceiling lights in the animal room were on between 0700 and 1900 h daily. Air temperatures in the room ranged between 24° C (nighttime) and 33° C (daytime), and ambient relative humidity ranged from 35 to 60 percent.

We were unable to collect sufficient amounts of natural prey items for this experiment, despite our vacuuming and trapping efforts. Consequently, we used small larvae of *Tenebrio molitor* (mealworms, about 4th instar) as food. Most lizards would not voluntarily eat enough to maintain constant body masses, so they were force-fed. Each lizard was fed various amounts each morning to determine the consumption rate at which body mass was constant. For 12 days thereafter the lizards were weighed in the morning and force-fed just enough to increase their body masses to their steady-state values; the amounts of fresh food given were calculated from increases in body mass.

Starting one day after the measurements of food input began, all feces and urine voided during the subsequent 12 days were collected separately and dried. Samples of fresh mealworms were taken periodically for measurement of water and energy contents. Rates of CO_2 production during the trial were measured using HTO-18 as with field animals (sampling interval, 12 days). Conversion of CO_2 measurements to units of energy for comparison with metabolizable energy values required estimates of assimilable protein, fat, and carbohydrate in the diet. These were based on measurements of the energy and nitrogen content of samples of food and feces.

Sample Analyses. Body fluid samples were distilled under vacuum, and the resulting water was analyzed for oxygen-18 content by cyclotron-generated proton activation (Wood et al., 1975). Tritium was measured by liquid scintillation spectrometry (Nagy and Costa, 1980). Composition analysis of lizard, diet, and excreta samples was done as follows: water content by drying to constant weight at 65 ° C, lipid content by weight loss following acetone/ether extraction, energy content by Phillipson microbomb calorimetry, and nitrogen content by micro-Kjeldahl technique.

Calculations. Field metabolic rates were calculated from isotope data using equation 2 of Nagy (1980). This equation requires values for total volume of body water, *W*, at the beginning and end of each measurement period. It was not possible to measure *W* for each labeled lizard at each recapture. The value for *W* was measured by the isotope dilution principle (Nagy, 1980) when each lizard was injected or reinjected, and by drying lizards collected monthly for body composition diet analyses. The mean of all measurements of *W* during the appropriate time period was calculated (as a percent of live weight) and multiplied by individual body-mass values to estimate *W* for each injected lizard.

Average field metabolic rates were calculated for the approximately 30-day intervals shown in Table 2.3 later in this chapter. If the time period of a given measurement of field CO_2 production was not clearly assignable to one of these intervals, it was not used in calculating interval means. Results from marked and autopsied animals in January, February, and March of 1976 were not significantly different from those for the same months in 1975. Accordingly, results for the two years were combined.

Mean monthly body masses of the 4 cohorts of *Uta* in the field were calculated from measurements of marked animals. Care was taken to use data only from marked lizards that were recaptured many times during the study, in an attempt to avoid random errors due to small, independent samples.

Conversion of metabolic rate measurements from units of CO_2 to joules was done on the basis of estimates of carbohydrate, fat, and protein assimilated from the food. *Uta* eating small mealworms assimilated 0.88 g dry matter per gram of dry food, and the assimilated matter contained 26.1 kJ/g. Measured nitrogen fluxes indicated that each gram of assimilated matter contained 0.0785 g nitrogen, and, assuming 6.25 g protein per gram of nitrogen, 0.49 g protein per gram. As protein contains 17.8 metabolizable kJ/g for uricotelic animals (Schmidt-Nielsen, 1979), protein provided 8.7 kJ of the assimilated energy per gram, so the fat and carbohydrate in the assimilated matter must have accounted for the remaining 17.4 kJ and 0.51 g dry matter. Thus the nonprotein dry matter contained 34.1 kJ/g. Linear interpolation between the energy content of fat (39.34 kJ/g) and carbohydrate (17.57 kJ/g) indicated that the nonprotein matter was 76 percent fat and 24 percent carbohydrate, so one gram of assimilated dry matter contained 0.49 g protein, 0.39 g fat, and 0.12 g carbohydrate. Oxidation of this mixture should yield 26 J of heat per milliliter of CO_2 produced. This factor was used in comparing laboratory metabolic rates, simultaneously estimated with doubly labeled water and with the balance method.

For field measurements, conversion factors for CO_2 to joules were calculated on the basis of the field diet. The factor for field arthropods was based on values for *Tenebrio* larvae. Mealworms (even the early instar stages used here) probably contain more fat than do field arthropods, so this assumed factor is probably slightly higher (2 to 5 percent) than the true factor, because fat yields 27.7 J/ml CO_2, whereas protein (in uricotelic animals) yields 24.8 and carbohydrate yields 20.9. A factor for plants was calculated as above from feeding trial results for a herbivorous iguanid lizard (21.7 J/ml of CO_2 produced) (Nagy and Shoemaker, 1975), and a weighted factor was calculated from the proportions of dry arthropod and plant matter in the diet for each month for each cohort.

Rates of energy flow, in J per animal per day, for free-living *U. stansburiana* were calculated from isotope results and body composition analyses as follows. Respiration was estimated from values for mean monthly CO_2 production. Rates of chemical-potential-energy flows to or from abdominal fat bodies, fat-free somatic tissue, somatic lipids, and reproductive tissues were estimated as the differences from one month to the next between the products of mean cohort body mass of marked lizards, percent body composition of autopsied lizards, and the energy contents of the various tissues. The sums of the rates of energy flows to heat production and to the various components must equal the rate that metabolizable energy was obtained from ingested food. Ingestion rates were calculated from metabolizable energy influxes on the basis of relative amounts of dry arthropod and plant matter in the diets, along with estimated metabolic efficiencies for arthropods (0.81 for *Tenebrio* larvae) and plants (0.50) (Nagy and Shoemaker, 1975).

No adult females could be captured for autopsy on December 20, so mean changes in daily body composition between November 25 and January 24 were used for both monthly intervals. Also, no gravid 2-year+ females were captured on April 22 because they had already laid their eggs. However, prelaying and postlaying body masses were measured in marked lizards, and mass changes of these animals were assumed to be due only to eggs. This egg mass (1.20 g fresh mass, or dry mass of eggs equals 17.2 percent of live body mass) was combined with body composition results from postlaying females for purposes of energy-flux calculations.

Statistics. Results are expressed as means, ± standard errors; N represents the number of measurements. Differences were considered significant when $P \leqslant 0.05$. Lines were calculated by least-squares-regression procedures. For analyses of covariance, results were first tested for a significant difference between slopes. If none existed, the regressions were recalculated using the common slope, and the difference between intercepts (elevations) was tested for significance. Other statistical procedures used are identified in the text.

Results

Field CO_2 Production. Field metabolic rates (FMR) of adult *Uta* were highest in late spring and early summer and lowest in winter (Fig. 2.1). Males were establishing and maintaining territories in early spring (Feb-

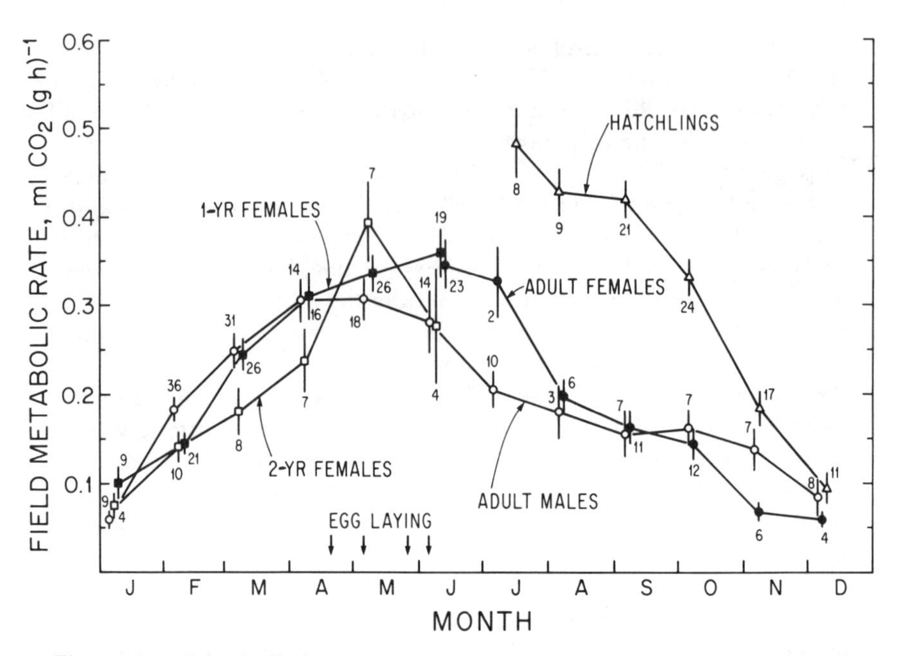

Figure 2.1 Metabolic rates (measured with HTO-18) of free-living *Uta stansburiana* in southern Nevada. Symbols indicate monthly means, vertical lines are plus or minus one standard error, and numbers show sample sizes. Means for yearling and 2-year+ females are shown separately from January through June, but are combined thereafter. Periods of egg laying, based on abrupt body mass declines (see Fig. 2.2), are indicated by arrows.

ruary, March, and April), and they had significantly higher mass-specific FMRs than did females (yearling and 2-year+ females combined) during that particular season—analysis of covariance, $F(1,166) = 4.12$, $0.05 > P > 0.025$. Further analysis of these results indicated that FMRs of 2-year+ females were significantly lower than in males—$F(1,103) = 8.35$, $P < 0.005$—but those of yearling females were not—$F(1,144) = 1.06$, $P > 0.25$. In late April, May, and early June, females were producing and laying eggs, and territorial behavior by males had become infrequent. FMRs of females (combined) were significantly higher than males during this period—analysis of covariance, $F(1,85) = 4.36$, $0.05 > P > 0.025$. Yearling females had higher FMRs during this interval than males—$F(1,74) = 4.08$, $0.05 > P > 0.025$—but the older females did not—$F(1,40) = 1.71$, $P > 0.10$.

Beginning in July the age of females was no longer reliably predictable from their body size, so results from all adult females were combined within monthly intervals through December. Females had significantly higher FMRs than males in July, a month after egg laying ceased—analysis of variance, $F(1,10) = 7.15$, $P < 0.025$. Feeding rates of females were about 25 percent greater than those of males, and growth rates of females were about twice those of males during that month (Table 2.3). Thus, the difference in FMRs in July may reflect a higher foraging effort in females than in males. In November, FMRs of females were significantly lower than those of males—analysis of variance, $F(1,11) = 6.78$ $P < 0.025$—but females were growing more rapidly than males during that month as well (Table 2.3). Males may have have had higher FMRs in November because they exhibited territorial behavior, but our field observations for that month are too scanty to confirm this.

Hatchlings had significantly higher mass-specific FMRs than adults (combined) during July, August, September, October, and November (analysis of variance within monthly intervals). Part of these variations are probably due to differences in body mass (see below and Nagy, 1981). Beginning on December 21, hatchlings were classed as adults.

Body Mass. Males were heavier than females except when females were gravid (Fig. 2.2). Body masses of adults increased during late winter, early spring, and autumn. Males lost some weight in June, whereas females lost much weight periodically between April and June while laying two clutches of eggs. Egg laying was synchronous within cohorts, but 2-year+ females laid their clutches 1 to 2 weeks earlier than the 1-year females. Hatchlings grew rapidly in summer and autumn. All cohorts lost weight in December.

Body Composition. Adult lizards were from 23 to 29 percent dry matter throughout the year (Table 2.1). Hatchlings were about 21 to 23 percent dry matter when they were 1 to 3 months old. Total fat contents in males varied from a high of about 5 percent of live body mass in August to a low near 1 percent in March. In females the peak in fat content

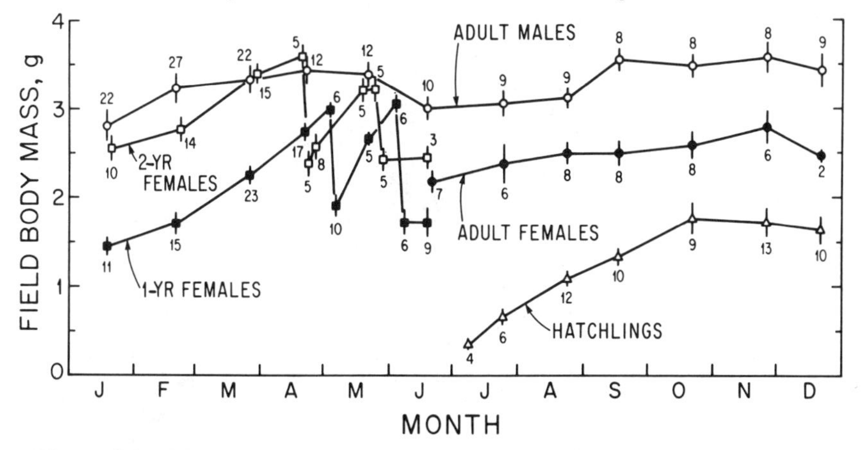

Figure 2.2 Mean body masses of successively recaptured, marked *Uta stansburiana* in southern Nevada. Symbols as in Fig. 2.1. Values for yearling and 2-year+ females were combined for summer and autumn.

Table 2.1 Seasonal changes in body composition of *Uta stansburiana.* Values are mean percent of live body mass; standard error in parentheses.

	Jan. 24	Feb. 21	Mar. 24	April 22		May 21		June 18	July 16	Aug. 22	Sept. 16	Oct. 21	Nov. 25	Dec. 20
Adult males														
Dry matter	26.1	23.6	23.3	24.4		24.3		26.8	28.1	27.8	26.3	26.7	26.2	26.6
	(0.4)	(0.3)	(0.3)	(0.3)		(0.1)		(0.3)	(0.3)	(1.2)	(0.1)	(0.3)	(0.6)	(0.4)
Dry abdominal fat bodies	0.14	0.06	<0.01	<0.01		0.20		0.52	0.89	1.41	0.51	0.43	0.54	0.61
	(0.06)	(0.03)	–	–		(0.09)		(0.26)	(0.11)	(0.76)	(0.10)	(0.13)	(0.17)	(0.07)
Extractable somatic lipid	2.61	1.73	1.05	1.52		2.03		2.80	2.74	3.86	2.84	3.26	3.80	3.23
	(0.21)	(0.22)	(0.13)	(0.07)		(0.28)		(0.19)	(0.12)	(0.63)	(0.12)	(0.29)	(0.54)	(0.16)
Dry testes	0.48	0.37	0.41	0.55		0.23		0.03	0.02	0.02	0.01	0.09	0.30	0.54
	(0.04)	(0.04)	(0.05)	(0.09)		(0.04)		(0.00)	(0.00)	(0.00)	(0.00)	(0.01)	(0.04)	(0.06)
N	10	15	10	5		4		5	5	5	5	5	5	8
	2-year + females							**Adult females**						
				A	B	C	D							
Dry matter	29.0	26.6	26.8	–	26.5	27.9	22.8	26.5	27.9	27.5	26.5	27.4	28.8	–
	(0.7)	(0.6)	(0.7)	–	(0.5)	(0.9)	–	(0.7)	(0.5)	(0.6)	(0.6)	(1.0)	(0.5)	–
Dry abdominal fat bodies	2.02	1.06	0.06	–	0.17	0.06	0.01	0.48	0.98	1.41	1.06	1.04	2.73	–
	(0.37)	(0.18)	(0.04)	–	(0.02)	(0.04)	–	(0.38)	(0.33)	(0.41)	(0.12)	(0.36)	(0.58)	–
Extractable somatic lipid	4.08	4.40	1.99	–	3.04	1.55	1.70	2.76	3.74	3.74	3.58	3.66	4.48	–
	(0.36)	(0.44)	(0.25)	–	(0.20)	(0.10)	–	(0.62)	(0.52)	(0.38)	(0.41)	(0.33)	(0.19)	–

	Jan. 24	Feb. 21	Mar. 24	April 22		May 21		June 18	July 16	Aug. 22	Sept. 16	Oct. 21	Nov. 25	Dec. 20
Dry ovaries and eggs	0.05	0.27	5.63	–	0.94	9.64	0.06	0.34	0.03	0.01	0.01	0.01	0.04	–
	(0.02)	(0.08)	(1.07)	–	(0.22)	(0.34)	–	(0.31)	(0.01)	–	–	–	(0.01)	–
N	4	9	8	0	5	4	1	5	5	5	5	5	4	0
	1-year females								Hatchlings					
				A	B	C	D							
Dry matter	27.7	25.6	25.1	27.6	26.1	27.3	25.3		22.6	21.1	23.6	23.6	25.1	25.0
	(0.8)	(0.6)	(0.9)	(0.6)	(0.0)	(0.6)	(1.1)		(0.8)	(0.8)	(0.4)	(0.5)	(0.4)	(0.6)
Dry abdominal fat bodies	1.24	0.64	0.28	0.05	0.21	0.06	0.09		<.01	<.01	0.09	0.01	0.39	0.17
	(0.37)	(0.15)	(0.15)	(0.03)	(0.19)	(0.04)	(0.12)		–	–	(0.04)	(0.01)	(0.14)	(0.09)
Extractable somatic lipid	3.80	3.39	2.68	2.13	2.70	1.73	1.50		3.72	1.68	1.90	2.04	3.37	2.75
	(0.30)	(0.35)	(0.46)	(0.30)	(0.90)	(0.18)	(0.10)		(0.35)	(0.24)	(0.09)	(0.40)	(0.23)	(0.28)
Dry reproductive tissue	0.05	0.10	1.37	6.77	0.39	8.40	1.49		<.01	<.01	<.01	<.01	<.01	0.07
	(0.01)	(0.04)	(0.51)	(1.36)	(0.27)	(1.22)	(1.41)		–	–	–	–	–	(0.02)
N	7	9	6	4	2	4	2		9	5	5	5	6	4

COLUMNS:
A: With large follicles or oviductal eggs on April 22.
B: Without large follicles or oviductal eggs on April 22.
C: With large follicles or oviductal eggs on May 21.
D: Without large follicles or oviductal eggs on May 21.

occurred in November (about 7 percent) and the lowest value was about 1.6 percent in April. The general body (tail included) appeared to be somewhat more important as a lipid storage depot than the abdominal fat pads, in terms of relative masses of lipid present, but these depots were about equally important in terms of actual energy fluxes (see below). In males, the testes were enlarged from December through April, but female reproductive tissues were large only during March, April, and May.

Body Energy. The energy densities of most body components did not vary with sex, age, or season, so mean values were calculated for these data (fat-free carcass: 18.5 kJ/g dry matter, S.E. = 0.11, N = 54; somatic lipid: 38.1 kJ/g, S.E. = 0.22, N = 14; eggs and ovaries: 26.4 kJ/g, S.E. = 0.47, N = 17). Abdominal fat bodies, however, had significantly higher energy contents between October and April (35.7 kJ/g, S.E. = 0.40, N = 14) than between May and September (33.8 kJ/g, S.E. = 0.26, N = 13).

Composition and Energy of Diet. The diet of *Uta* consisted mostly of arthropods. Beetles and ants were consumed throughout the year. Larvae of moths and butterflies were important in winter and spring. Grasshoppers, spiders, hemipterans, and flies were important in late spring and summer, and termites were important during autumn. Surprisingly (see Pough, 1973), much plant leaf tissue was eaten in late winter and early spring (up to 43 percent of the dry mass of stomach contents), when seedlings of annual plants were just beginning to grow rapidly. Hatchlings ingested almost no plant matter. Some herbivory by *Uta* has previously been reported (Hoff and Kay, 1970; Sanborn, 1977).

The energy contents of dietary items were relatively constant between seasons and for different plant or arthropod taxa, so means were calculated. Plants had consistently lower energy contents—mean = 17.2 kJ (gram of dry matter)$^{-1}$—than arthropods—mean = 22.8 kJ (gram of dry matter)$^{-1}$.

Energy Assimilation. The *Uta* fed small mealworms in the laboratory digested and assimilated 91 percent of the energy in their diet. Urinary losses amounted to about 10 percent of ingested energy, so metabolic efficiency was 81 percent (Table 2.2). The efficiency values from the 5 males and the 5 females did not differ significantly, so all the results were combined in Table 2.2. Laboratory metabolic rates (about 0.32 ml CO_2 per gram per hour, estimated from metabolizable energy influx) were close to metabolic rates in the field during May and June (Fig. 2.1).

HTO-18 Validation. Using calculated rates of metabolizable energy input, and assuming all metabolizable energy was in fact oxidized (balance method), comparison with simultaneous doubly labeled water measurements indicates that the HTO-18 method underestimated energy metabolism by about 7 percent—balance: 200 ± 10 kJ (kg day)$^{-1}$, HTO-18: 183 ± 7 kJ (kg day)$^{-1}$; difference: −16 ± 8 kJ (kg day)$^{-1}$;

Table 2.2 Assimilation of energy from an arthropod diet (*Tenebrio* larvae) by 10 *Uta stansburiana*.

Avenue	Dry mass flux g (kg day)$^{-1}$	Energy content kJ (g dry mass)$^{-1}$	Energy flux kJ (kg day)$^{-1}$
Influx			
Food	10.1	25.1	254
	(0.5)	(0.4)	(12)
Efflux			
Feces	1.22	18.2	21.5
	(0.13)	(0.4)	(1.8)
Urine	2.14	11.6	25.4
	(0.12)	(0.2)	(1.4)
Metabolism[a]			207
			(10)

Assimilative efficiency[b] = 0.909 (0.012)
Metabolic efficiency[c] = 0.812 (0.011)

a. Obtained by difference between food influx and feces and urine efflux.
b. Assimilative efficiency = (food energy influx − feces energy efflux) (food energy influx)$^{-1}$.
c. Metabolic efficiency = (food energy influx − feces and urine energy efflux) (food energy influx)$^{-1}$.

error, −7.3 ± 3.4 percent, N = 10 (5 males, 5 females). However, this difference is not significant ($P > 0.05$, paired t test). Laboratory validation studies in a variety of animals indicate that doubly labeled water measurements are accurate to within ± 8 percent (Nagy, 1980).

The apparent cause of the 7 percent discrepancy is the incorrect assumption that all metabolizable energy was oxidized by the females. For the males (mean body mass 3.74 g, S.E. = 0.19), metabolic rate estimates were 175 ± 7.5 kJ (kg day)$^{-1}$ by the balance method, versus 175 ± 14.2 kJ (kg day)$^{-1}$ by the HTO-18 method ($P > 0.60$, paired t test), but for females (body mass 2.44 g, S.E. = 0.05), these values were 224 ± 9.0 kJ (kg day)$^{-1}$ by the balance method, versus 191 ± 0.8 by the HTO-18 method ($0.05 > P > 0.025$, paired t test). This suggests that the females were storing some fat, despite our attempts to maintain their energy budget in a steady-state condition by feeding them just enough to maintain constant body mass. If the females had stored the difference as body fat (assumed to be accompanied by 15 percent water in vivo), their body weights over the 12-day trial would have increased only 0.03 g, an amount which would have gone unnoticed in the normal daily fluctuation in body mass. The difference in energy processing between males and females in this feeding trial may be related to reproductive physiology differences, as the measurements were done late in the breeding season (June and July).

Energy Flow. Rates of energy flux through individual, free-living *U. stansburiana* were highest in spring and lowest in December (Table 2.3). Feeding rates (ingestion) in spring were 5 to 15 times higher than in winter. Most metabolizable energy was oxidized in respiration by all cohorts in all monthly intervals. However, in yearling females more than 30 percent of their metabolizable energy was allocated to reproductive tissue in May. In general, energy was stored as fat in summer and fall, and fat reserves were used during winter and spring.

Discussion

The information contained in this detailed energy budget can be interpreted at many levels of biological organization, ranging from organ systems to ecosystems. One can examine seasonal energy storage and cycling of body fat (Hahn and Tinkle, 1965; Bustard, 1967; Hahn, 1967; Avery, 1970, 1974; Derickson, 1974, 1976a; Andrews, 1979; and many others), differences in reproductive organs by season, sex, and age (Ferguson, 1966; Mayhew, 1967; Fitch, 1970; Licht and Gorman, 1970, 1975; Licht, 1973; Derickson, 1976b; Medica and Turner, 1976; Goldberg, 1977), reproductive effort and reproductive strategy (Williams, 1966a, b; Tinkle, 1969; Tinkle, Wilbur, and Tilley, 1970; Tinkle and Hadley, 1975; Ballinger and Clark, 1973; Hirshfield and Tinkle, 1975; Pianka and Parker, 1975; Vitt and Ohmart, 1975; Derickson, 1976b; Pianka, 1976; Stearns, 1976; Licht, 1979; Shine, 1980; Nussbaum, 1981, among others), growth and annual net production (Brody, 1945; Golley, 1968; McNeill and Lawton, 1970; Turner, 1970; Humphreys, 1979; Banse and Mosher, 1980), daily time/energy budgets (Kitchell and Windell, 1972; Wilson and Lee, 1974; Huey and Slatkin, 1976; Bennett and Nagy, 1977; Anderson and Karasov, 1981), annual energetics of populations (Alexander and Whitford, 1968; Nagy and Shoemaker, 1975; Tinkle and Hadley, 1975; Turner, Medica, and Kowalewsky, 1976), and diet selection and the costs, benefits, and optimization of foraging (Schoener, 1971; Pough, 1973; Wilson and Lee, 1974; Charnov, 1976; Pyke, Pulliam, and Charnov, 1977; Pyke, 1978).

It is not possible to discuss all of these topics here. Readers are invited to apply the results herein to topics of interest to them. I have tried to provide as many details as possible to facilitate this.

Energetics of Individuals

Lipid Storage and Use. Most animals are able to ameliorate seasonal fluctuations in energy supply and demand somewhat by storing chemical energy in their bodies when energy input exceeds demands and using these reserves when energy needs exceed input. Many lizards store lipids when food is abundant and use these stores when breeding or overwintering, resulting in seasonal cycling of body fat (see review by Derickson, 1976a). In *Uta* abdominal fat bodies and somatic lipids (including the tail) were about equally important as storage sites for chemical potential energy, and nonlipid chemical energy in the body was important as well (Table 2.3). All of these reserves were built up or drawn upon at various times of the year. Conversion of body substances into egg material by females could account for at most about 25 percent of the energy contained in eggs, so most of the egg material must have come directly from the food. Derickson (1974) reached a similar conclusion regarding *Sceloporus graciosus.*

Stored energy provided *Uta* at most about 35 percent of the energy used in respiration during the winter, and these lizards must have fed to obtain the remaining energy. In *Sceloporus jarrovi* in Arizona, stored body fat accounted for less than half of the energy expended during the winter (Congdon, Ballinger, and Nagy, 1979). However, *S. graciosus* could overwinter on lipid from fat bodies alone (Derickson, 1974). There is probably much variation among different species or populations of lizards regarding their dependence upon food resources during winter. Lizards at high latitudes or altitudes may not be able to feed during winter because of temperature limitations on the activity of their prey or themselves or because of snow cover. Changes in temperature and in lizard energetics are probably parallel. If it is warm enough for an insectivorous lizard to be active, it may also be warm enough for its prey to be active and available as food. When temperature declines, food availability should decline, but so should the energy requirement of the lizard. At low body temperatures, even relatively small fat reserves could sustain a lizard for many weeks or months. There is some evidence that, during winter, some lizards can reduce their metabolic rates (and hence their energy requirements) below those expected from low body temperatures alone (Bennett and Dawson, 1976). More information about this capability is needed. Also, more observations are needed regarding the extent to which lizards can alter their energy expenditure during winter through behaviorial selection of body temperature. Is winter behavior influenced by autumn or winter food availability, or by levels of stored fat in the body, or is it determined solely by ambient thermal conditions?

Respiration: Season and Sex. Bennett and Nagy (1977) found no seasonal or sexual differences in field metabolic rates of *Sceloporus occidentalis* during 5- to 20-day measurement periods in the spring reproductive season and the fall nonreproductive season, but Merker and Nagy (1982) found significant seasonal and sexual differences during 7- to 14-day measurement periods in *Sceloporus virgatus.* Examination of continuous measurements of field metabolic rates during an entire year in *U. stansburiana*

Table 2.3 Energy flow in *Uta stansburiana*. Values are J (animal day)$^{-1}$.

	Time periods (days)											
	Dec. 21– Jan. 23 (35)	Jan. 24– Feb. 21 (28)	Feb.22– Mar. 24 (31)	Mar. 25– Apr. 22 (29)	Apr. 23– May 21 (29)	May 22– Jun. 18 (28)	Jun. 19– Jul. 16 (28)	Jul. 17– Aug. 22 (37)	Aug. 23– Sep. 16 (25)	Sep. 17– Oct. 21 (35)	Oct. 22– Nov. 25 (35)	Nov.26– Dec. 20 (25)
	Adult males											
Respiration	117	337	494	630	645	560	389	353	329	361	309	190
Somatic tissue	78	33	19	28	−17	−25	29	−19	80	−9	−14	−8
Somatic lipid	7	−24	−26	23	21	21	−1	32	−31	13	25	−39
Fat bodies	−5	−3	−2	0	8	11	14	16	−36	−2	5	2
Testes	3	−1	1	3	−7	−4	0	0	0	2	4	6
Metabolizable	200	340	486	684	650	563	431	382	342	365	329	151
Ingestion	249	471	697	961	884	723	545	474	424	449	405	186
	2-year+ females						Adult females					
Respiration	112	232	343	437	707	494	472	306	257	232	116	99
Somatic tissue	−6	−4	43	−49	−25	19	34	5	−8	21	8	−6
Somatic lipid	−14	24	−66	14	−37	57	39	5	−6	5	33	−14
Fat bodies	−15	−28	−31	3	−3	12	15	11	−12	2	50	−15
Ovaries and eggs	0	6	156	119	259	13	−6	0	0	0	1	0
Metabolizable	77	230	445	524	901	595	554	327	231	260	208	64
Ingestion	96	305	596	686	1167	742	681	402	284	329	263	80

	1-year females						Hatchlings				
Respiration	98	141	294	476	563	496	228	321	326	204	136
Somatic tissue	−18	26	59	52	−30	−71	58	58	50	−6	−4
Somatic lipid	11	4	3	4	−25	4	−1	11	12	24	−20
Fat bodies	19	−9	−5	−4	−1	8	1	1	−1	7	−6
Reproductive tissue	0	1	25	140	320	194	0	0	0	0	1
Metabolizable	110	163	376	668	827	631	286	391	387	229	107
Ingestion	137	223	533	894	1075	788	352	481	476	282	132

(Fig. 2.1) indicates that one may or may not find significant differences between sexes and seasons, depending on which calendar intervals one selects. Clearly, one should know as much as possible about the reproductive status and behavior of the animals being studied when interpreting results of short-term measurements of field metabolism.

The seasonal pattern of respiration that seems to emerge from the above studies of primarily insectivorous, arid-habitat, sceloporine lizards is one of rising metabolic rates in spring, high rates during summer, and declining rates in late summer and fall, roughly paralleling average air temperature. In contrast, field metabolic rates in a herbivorous desert lizard declined precipitously in late spring (Nagy and Shoemaker, 1975), in conjunction with the rapid dehydration of food plants and a reduction in feeding and other activity abroad (Nagy, 1972, 1973). Because many of the arthropods preyed upon by *Uta* also were plant eaters, one might expect the welfare and hence the availability of many species upon which *Uta* prey to decline in summer, as the plants became dehydrated. However, *Uta* apparently were able to find adequate food to fuel their high metabolic rates in summer by switching to arthropods such as beetles, ants, and termites, which can feed on dead or relatively dry plant parts. Thus, the adaptations of desert animals in the primary-consumer trophic level may inadvertently serve to smooth out or buffer seasonal variations in the food supply for secondary consumers.

Respiration Associated with Somatic Growth plus Reproduction. The metabolic rate of a lizard that is producing new biomass should be higher than if it were not growing or reproducing. The respiratory cost of production due to converting food into new tissue has been estimated to be about 0.33 J respired per joule of new biomass in laboratory mammals (Ricklefs, 1974) and birds' eggs (Vleck, Vleck, and Hoyt, 1980). For a free-ranging animal, the respiratory cost of production should be higher than this because of the added costs of obtaining the extra food needed for growth and reproduction in the field. For a lizard, this may involve lengthening its daily activity period or increasing the intensity of its activity while abroad, or both. Any increased metabolic costs due to increased social, thermoregulatory, or predator-avoidance behavior that is associated with increased foraging should also be included as respiratory costs of production.

The results of this study offer a way of estimating the respiratory costs of growth and reproduction in free-living *Uta* lizards. If these costs comprise a substantial proportion of total field metabolic rate, then there should be a positive correlation between field metabolic rate and production rate (sum of the rates of chemical potential energy appearing as new somatic tissue, somatic lipids, and reproductive tissue). Results from all cohorts, for the measurement periods between March 25 and October 21 (selected in an attempt to reduce variation due to seasonal thermal conditions), are shown in Figure 2.3. In spring, the field meta-

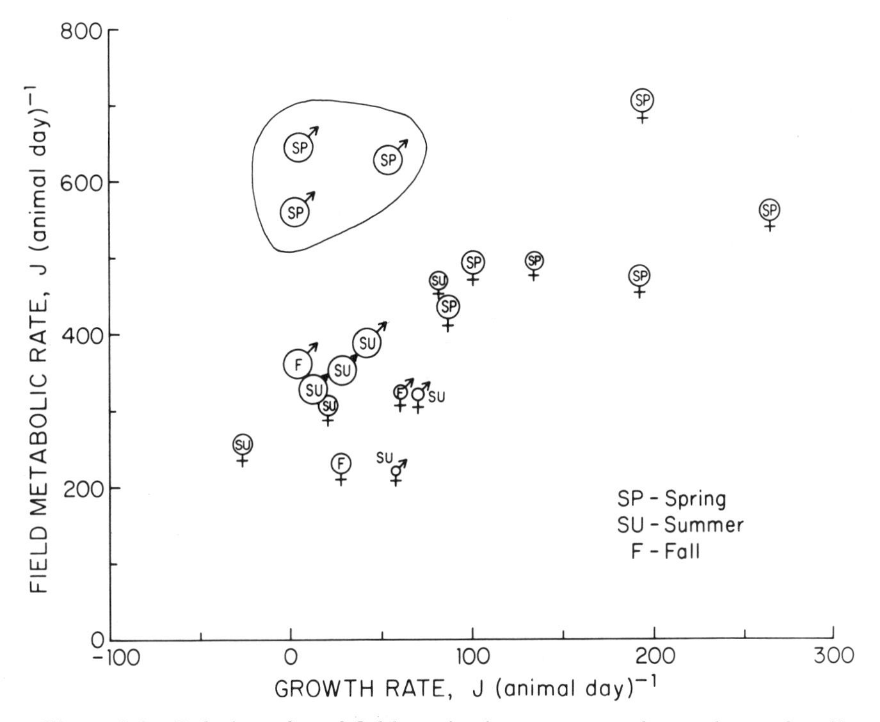

Figure 2.3 Relation of total field respiration rate to total growth rate for all cohorts of *Uta* during the warm part of the year. Symbol sizes indicate relative body sizes of the different groups. Data points for adult males during spring are circled. Combined male-female symbols indicate unsexed juveniles.

bolic rates of males, circled in the figure, were relatively high, probably because of the additional costs of their territorial behavior at that time. They grew very little during spring. The remaining results indicate that field metabolic rates increased with increasing growth rates, but this correlation is complicated by variation in field metabolic rates due to differences in body mass. Accordingly, the variation in field metabolic rate associated with variation in both growth rate and body mass was determined by multiple regression analysis of the $\log_{10}$ transformed data. Logarithms were used because metabolic rate and growth rate are exponentially related to body mass in endothermic vertebrates (Kleiber, 1975), as is field metabolic rate in iguanid lizards (Nagy, 1982). The three points for males in spring and the point for summer females having negative growth rates (Fig. 2.3) were excluded from this calculation. The least-squares regression is significant—$F(1,14) = 18.4$, $P < 0.001$; multiple $r = 0.860$; multiple $r^2 = 0.740$—and the equation for the line is

$$\text{Field metabolic rate in J per (animal day)} = 114 \times [\text{growth rate in J per (animal day)}]^{0.21} (\text{body mass in g})^{0.48}. \quad (2.1)$$

The proportion of variation in field metabolic rate accounted for by variation in growth rate is 0.366, and that accounted for by variation in body mass is 0.373. About one-fourth of the variation in field metabolic rate is not associated with variation in body mass and growth rate. When the effects of body-mass differences are removed, the partial correlation coefficient between field metabolism and growth rate equals 0.767; with the effects of differences in growth rate removed, the partial correlation coefficient between field metabolism and body mass is 0.818.

By assuming that most of the observed increase in field metabolic rate with increasing growth rate represents respiratory costs associated with production, one can estimate these costs using Eq. 2.1. (The exponential form of this equation precludes its use for situations where growth rate is zero or negative.) Equation 2.1 indicates that the metabolic costs associated with production are greater at the lower growth rates, and greater for larger lizards than smaller ones. For example, predicted field metabolic rate for a 1 g *Uta* increases from 111 to 180 J (animal day)$^{-1}$ as its growth rate increases from 1 to 10 J (animal day)$^{-1}$. The increase in respiration rate per unit increase in growth rate is 69 J (9 J)$^{-1}$ = 7.7 for this growth increment and body mass. However, for the same size lizard to increase its growth rate from 1 to 200 J (animal day)$^{-1}$, the ratio is 1.1 J increased respiration (J increased growth)$^{-1}$. For a 3 g *Uta,* comparable ratios for the same growth-rate intervals are 11.7 and 1.9, respectively. "Partial" ratios (Kleiber, 1975), which indicate the change in respiration associated with a single unit of change in growth rate (at some specified growth rate), emphasize these differences more strongly. There are some theoretical reasons why growth might be more expensive at low growth rates (if food is in short supply or if foraging time is limited), but consideration of these is beyond the scope of this chapter.

Respiration Associated with Reproduction. Two ways were used to estimate the rates of respiration associated with reproduction alone (R_R) in female *Uta.* First, Eq. 2.1 was used to predict field metabolic rates corresponding (1) to somatic growth rates alone (somatic tissue plus somatic lipid plus fat bodies; Table 2.3), and (2) to total observed production, including reproduction (metabolizable minus respiration; Table 2.3). The value R_R was estimated as the difference. Results of these calculations are shown in Table 2.4. Estimates of R_R for the first clutch of both female cohorts are less reliable than those for the second clutch, because Eq. 2.1 was also applied to the February 22 through March 22 portion of the first clutch, but this time period was not used in the derivation of the equation. When mean monthly somatic growth rates of females were zero or negative, the value of 1 was used in Eq. 2.1 (see explanation above). Some of the chemical energy in a clutch of eggs could have come from body chemicals stored prior to the breeding season (see below). The respiratory energy expenditures associated with the initial accumulation of those chemicals should be included as reproduction-associated

Table 2.4 Reproductive effort of free-living *Uta stansburiana* in southern Nevada during 1975–1976. Numerical values shown are either rates of energy flow, in kJ per animal per total time period[a] or ratios of various flow rates.

	First clutch	Second clutch	Breeding season	Annual
1-year females:				
Reproduction				
Chemical energy to eggs (R_c)	8.3	11.2	19.5	19.5
Respiration cost (R_R)	9.0	16.3	25.3	25.3
Total (R_T)	17.3	27.5	44.8	44.8
Ingestion (I)	54.3	41.4	95.7	170.5
Metabolizable (M)	40.1	32.6	72.7	132.5
Storage (S)	2.7	−2.7[b]	0.0	6.9
$R_c\ (M-S)^{-1}$	0.222	0.317	0.268	0.155
$R_R\ (M-S)^{-1}$	0.241	0.462	0.348	0.201
$R_T\ I^{-1}$	0.319	0.664	0.468	0.263
$R_T\ (M-S)^{-1}$	0.463	0.779	0.616	0.357
2-year+ females:				
R_c	8.3	7.9	16.2	16.2
R_R	18.1	11.0	29.2	29.2
R_T	26.4	18.9	45.4	45.4
I	38.4	54.6	93.0	168.7
M	29.0	42.8	71.8	132.3
S	−2.6	0.6	−2.0	2.4
$R_c\ (M-S)^{-1}$	0.263	0.187	0.220	0.125
$R_R\ (M-S)^{-1}$	0.573	0.261	0.396	0.225
$R_T\ I^{-1}$	0.688	0.346	0.488	0.269
$R_T\ (M-S)^{-1}$	0.835	0.448	0.615	0.349
Males:				
$R_R \cong R_T$	–	–	21.5	26.6
I	–	–	95.4	198.7
M	–	–	69.5	148.8
S	–	–	1.7	6.8
$R_T\ I^{-1}$	–	–	0.255	0.134
$R_T\ (M-S)^{-1}$	–	–	0.317	0.187

a. Time periods are first clutch, 1-year females, February 22 to May 3 (71 days); first clutch, 2-year+ females, February 22 to April 22 (60 days); second clutch, 1-year females, May 4 to June 18 (46 days); second clutch, 2-year+ females, April 23 to June 18 (57 days); breeding season, February 22 to June 18 (117 days); and annual, January 1 to December 31 (365 days).

b. Negative signs for storage values indicate that the sum of chemical energy in somatic lipid and other somatic tissues, and fat bodies (Table 2.3), declined during that time interval.

respiration, but they are ignored in the calculations of R_R herein. Considering all the assumptions involved, these R_R values should be viewed as crude estimates only.

A second, independent way of estimating R_R in female *Uta* is to compare field metabolic rates of reproductively active animals with those of nonreproducing animals. Field metabolic rates of radiation-sterilized 2-year+ females were available for this comparison (Nagy and Medica, unpublished results). Sterile females were studied in a fenced, 9-hectare plot that has received continuous, low-level gamma irradiation (^{137}Cs source, centrally located) since 1964. The irradiated plot is located in Rock Valley, about 2 km from the study area described above. The irradiated female *Uta* become sterile sometime in their second year (Turner and Medica, 1977). Their ovaries disappear, apparently from radiation destruction of the germ cells, rather than via radiation effects on pituitary function (Pearson et al., 1978). The irradiated population persists because of successful reproduction by yearling females, which are not yet sterile. Field metabolic rates (measured with doubly labeled water) of females lacking ovaries (determined in the field by palpation and later confirmed by autopsy in most cases) were measured between April 24 and June 19, 1975, coincident with the period of production of a second clutch by 2-year+ females in the control area (see note a in Table 2.4). Sterile, 2-year+ female *Uta* (mean mass = 3.48 g, S.E. = 0.12, N = 8) had field metabolic rates averaging 0.193 ml CO_2 $(g \cdot h)^{-1}$ (S.E. = 0.026, N = 8), which is equivalent to 422 J $(\text{animal day})^{-1}$, or 24.1 kJ $(\text{animal})^{-1}$ for the entire 57-day interval. The mean field metabolic rate of fertile females during this period was 34.3 kJ $(\text{animal})^{-1}$ (calculated from Table 2.4). The difference (an estimate of R_R) is 10.2 kJ $(\text{animal})^{-1}$, which is remarkably close to the value of 11.0 (Table 2.4) estimated using Eq. 2.1. This agreement lends some confidence in the R_R values shown in Table 2.4.

The value of R_R in male *Uta* during spring was estimated as the difference between observed R (Fig. 2.3) and R predicted by Eq. 2.1. Males grew slowly in spring, but their metabolic rates were high. The increment of observed R above predicted R was assumed to be due to territorial behavior, which was assumed to be associated with reproduction. Many previous studies of *Uta,* along with our field observations, lend support to these conjectures. Males also displayed territorial behavior during February and March. Energy expenditures associated with this were estimated as the difference between metabolic rates of males and 2-year+ females (which were not territorial) in February and March (see Fig. 2.1).

The estimated respiration costs associated with reproduction (Table 2.4) tended to be higher for the second clutch in yearling females but higher for the first clutch in 2-year+ females. Even though both young and old females invested about the same amount of chemical potential energy in their first clutches, 2-year+ females allocated about twice as much metabolic energy to their first clutch compared to the yearlings. Older females may have spent more energy locating and preparing nest sites than young females. If so, this could markedly affect relative hatching success, leading to a greater number of breeding offspring and greater reproductive success. This form of investment in reproduction (similar, one might suggest, to parental care) could have benefits that are quite disproportionate to its energetic cost, and it would not be detected by the traditional methods of measuring reproductive effort in terms of the energy content, dry mass, or number of eggs.

Chemical Energy in Eggs. The proportion of available energy flow allocated by females to chemical-potential-energy accumulation in eggs was calculated for each clutch, for the breeding season, and for an entire year (Table 2.4). In previous reports in the literature, "available" energy has been defined as ingested (I), metabolizable (M), or metabolizable energy corrected for energy flow to or from somatic energy depots ($M - S$). The latter definition is probably the most realistic estimate of energy available for reproduction, but the others are included in Table 2.4 to facilitate comparison with other studies.

The fraction of available energy invested in eggs as chemical energy, $R_c\,(M - S)^{-1}$, over the entire breeding season tended to be higher in yearling females than in older females (Table 2.4). Young females tended to allocate a greater fraction of available energy to eggs in their second clutch than in their first, but the reverse occurred in older females.

Are the differences in these fractions statistically significant? Formal calculation of probability values is difficult, because these ratios are several calculation steps from the measured parameters of body composition and mass, tissue energy contents, and field metabolic rates, and because the various measurements are not completely independent. However, the variances of these ratios can be roughly estimated (Mood, Graybill, and Boes, 1974, p. 180). For first year females over the breeding season, estimated standard deviation (S.D.) of the $R_c\ (M - S)^{-1}$ value (0.268 in Table 2.4) is S.D. = 0.156, and for the 2-year+ females, $R_c\ (M - S)^{-1} = 0.220$, S.D. = 0.108. Thus, these ratios are not significantly different: z statistic = 0.254, $P = 0.6$ (Dixon and Massey, 1969). These high ratio variances result in part from the methods of data gathering and the calculation procedures necessitated in this study. More direct measurements of rates of energy ingestion and energy flow to reproductive tissue would facilitate statistical analyses, but such measurements are not yet technically feasible.

Reproductive Effort. Lizard reproductive effort has been assessed in a variety of ways, including static measures like ratios of the mass or the energy contents of eggs to bodies of females (see references in Pianka and

Parker, 1975), as well as dynamic (rate-process) measures and their ratios (Tinkle and Hadley, 1975; Andrews, 1979). In this chapter reproductive effort is expressed as the fraction of available energy flow (metabolizable energy, M, corrected for energy flow to or from body storage depots, S) used for reproductive purposes (chemical potential energy in eggs, R_c, and associated respiratory costs, R_R). The correction for stored energy accounts for previously accumulated body fat being used to form eggs, as well as accounting for simultaneous growth of somatic and reproductive tissues. These results permit analysis of reproductive effort in relation to sex, age, and clutch sequence, in a natural field population.

Female *Uta* channeled between 45 and 84 percent of their available energy flow into reproduction during the breeding season (Table 2.4). The highest reproductive effort (0.835) occurred in 2-year+ females during their first clutch of eggs. The highest absolute rate of energy flow to reproduction, 598 J $(\text{animal day})^{-1}$, occurred during production of the second clutch of eggs by yearling females. First-year females were smaller than 2-year+ females (Fig. 2.2), so the reproductive power output per gram of body mass was quite high in yearlings, as compared with older females.

Pianka and Parker (1975) have theorized that annual reproductive effort in an individual female *U. stansburiana* should stay the same or increase slightly from its first to second breeding season, because its expectation of future reproductive success declines with age. By the same logic, they predict that reproductive effort within a breeding season should increase in successive clutches. (See Nussbaum, 1981, for an alternative hypothesis.) The values in Table 2.4 partly support and partly refute these hypotheses. Annual reproductive effort was about the same in first- and second-year females, as predicted, and reproductive effort did tend to increase from first to second clutches in yearling females. However, reproductive effort for the second clutch of 2-year+ females was lower than for their first clutch. A limitation in food availability does not account for this decline, because 2-year+ females had a higher feeding rate during their second clutch than during their first (Table 2.3). Moreover, their feeding rate during the second clutch was about the same as that of yearling females (who exhibited a higher reproductive effort) during the same time of year in the same habitat. A parsimonious interpretation of these results is that reproduction by older females was reduced because of some debilitating physiological effects of age or previous reproductive effort.

Pianka and Parker (1975) pointed out that it is difficult to assess the risk of mortality as an additional energetic cost of reproduction. The R_R values in Table 2.4 should include any additional metabolic costs of predator avoidance and other actions that reduce harm and death, but they may still exclude some risks, such as disease and old age, that are not readily measurable in the currency of energetics.

Males spent about as much energy as did females for reproduction-related respiration during the breeding season (Table 2.4). Chemical energy investment in sperm production was very small (Table 2.3), so the total energy expenditure in reproductive effort of males was lower than females.

Annual and Lifetime Energy Budgets. Itemized energy budgets for one year (January 1 to December 31) and for a lifetime (from hatching in July to death in June two years later) are shown for male and female *Uta* in Table 2.5. Tinkle and Hadley (1975) estimated annual metabolic rates in 3.5-g *Uta* in Texas to be about 130 kJ in 1-year females and 145 kJ in 2-year females. Alexander and Whitford (1968) estimated that annual metabolic rate of 4-g *Uta* in New Mexico was about 243 kJ. Both of these studies involved estimation of field metabolic rates by extrapolation of laboratory measurements of oxygen consumption on the basis of field observations of behavior and body temperatures. Correcting for body mass differences (Nagy, 1982), these values become about 48 kJ $(g^{0.80}yr)^{-1}$ for yearling females in Texas, 53 for Texas second-year females, and 80 for adults in New Mexico. These are similar to the rates of 54 kJ $(g^{0.80}yr)^{-1}$ in yearling females (2.33 g), 52 for 2-year+ females (2.65 g), and 51 for adult males (3.28 g) in Nevada (Table 2.5), despite the differences in methods and habitats among the three studies.

Annual production efficiencies in Nevada *Uta* during this study (P/M = 0.20, for first-year females, and 0.14 for older females) were generally similar to those in Texas *Uta* (0.22 for first-year and 0.21 for older females, calculated from table 7 in Tinkle and Hadley, 1975). Andrews (1979) found a somewhat higher production efficiency (0.25) in a small tropical lizard (*Anolis limifrons*), which also had a higher annual respiration cost—about 67 kJ $(g^{0.80}yr)^{-1}$—in its less thermally diverse habitat.

Comparison with Endotherms. Field metabolic rates of iguanid lizards, including *U. stansburiana,* during their activity seasons are closely correlated ($r = 0.99$) with body mass, and are described by the equation

$$\text{kJ (animal day)}^{-1} = 0.224\ \text{g}^{0.80} \qquad (2.2)$$

(Nagy, 1982). Comparison of this relationship with similar equations for rodents and birds in the field (King, 1974) reveals that the predicted field metabolic rate of a 100-g rodent or bird is about 18 times and 30 times, respectively, that of a 100-g iguanid lizard. For a rodent and a bird as small as a typical *Uta* (2.5 g in this study), predicted field respiration rates are 29 and 42 times that of *Uta.* This means that a rodent or bird needs 20 to 40 times as much food as a similarly sized lizard to meet its metabolic demands during a spring or summer day.

Annual respiration rate of one wild mammal, the blacktailed jackrabbit; *Lepus californicus,* 1.8 kg (Shoemaker, Nagy, and Costa, 1976) is available for comparison with the annual metabolic rate of *Uta.* Using

Table 2.5 Annual and lifetime (2-year) energy budgets of *Uta stansburiana* in Nevada. Values are kJ $(\text{animal})^{-1}$.

	Annual			Lifetime	
	1-year female	2-year female	Male	Female	Male
Ingestion (I)	171	169	199	347	383
Feces and urine ($F + U$)	−38	−37	−50	−76	−90
Metabolizable (M)	133	132	149	271	293
Total respiration	106	113	142	225	281
Associated with reproduction (R_R)	25	29	27	54	54
Remaining metabolic costs	81	84	115	171	227
Total production (P)	27	19	7	46	12
Reproduction (R_c)	20	16	≈0	36	≈0
Growth and storage	7	3	7	10	12
(Production efficiency)					
($P(M)^{-1}$)	(0.203)	(0.144)	(0.047)	(0.170)	(0.041)
($P(I)^{-1}$)	(0.158)	(0.112)	(0.035)	(0.133)	(0.031)

the body mass correction for rodents (King, 1974), the weight-corrected annual field respiration of jackrabbits is 2740 kJ $(g^{0.67}yr)^{-1}$, or about 53 times the *Uta* respiration rate of 52 kJ $(g^{0.80}yr)^{-1}$. These large differences in field metabolic rates of lizards and endotherms are partly due to the 5 to 10 times higher resting metabolic rates of endotherms, and partly because, when the environment cools at night and during the winter, the metabolic rate of lizards declines, but that of endotherms increases. In terms of food-resource utilization, endotherms are very expensive organisms compared to ectotherms. If food supply limits population density in the field, one might predict that the biomass of a lizard population could be 50 times that of a mammal of the same size and diet living in the same habitat. This difference is much larger than previously anticipated, and could be relevant to the current debate about predator-prey ratios and endothermy in dinosaurs (Thomas and Olson, 1980) as well as considerations of competition and natural selection in terrestrial communities.

Population Energetics

Annual Energy Flow. Population energy flow was calculated using the rate values for individual *Uta* in Table 2.3, along with detailed demographic information for 1966–1967 from the same *Uta* population in Rock Valley, Nevada (Turner, Medica, and Kowalewsky, 1976). The calculated values, in megajoules (MJ) per hectare (ha) per year, are ingestion, 17.6; metabolizable, 13.8; respiration, 12.0; and production, 1.8.

Turner, Medica, and Kowalewsky (1976) estimated metabolizable energy flow to be 11.7 MJ $(ha\ yr)^{-1}$ in this population during the same year, a bit lower than my estimate. Their estimates of field respiration rates used laboratory measurements of minimum, resting oxygen-consumption rates (from Roberts, 1968) along with field measurements of temperatures in various places, changes in body mass of lizards, seasonal patterns of behavior, and estimated metabolic scopes. These estimates are also somewhat lower than the isotopic water measurements reported here. However, they included estimates for respiration of eggs, a parameter that was not included in this study. Alexander and Whitford (1968) estimated that population respiration in New Mexico *Uta* was about 10.4 MJ $(ha\ yr)^{-1}$, again in general agreement with this study. However, they estimated the population ingestion rate in New Mexico to be 31.0 MJ $(ha\ yr)^{-1}$, by assuming (probably incorrectly, see Table 2.2) that assimilation efficiency was only 30 percent.

A population of large (167-g) herbivorous lizards (*Sauromalus obesus*) living in the central Mojave Desert consumed about 37.7 MJ of food $(ha\ yr)^{-1}$ (Nagy and Shoemaker, 1975), which is about twice the ingestion rate of the *Uta* population in Rock Valley (eastern Mojave Desert). This difference is not surprising in view of the difference in trophic levels between these lizards, as well as the observation that *Sauromalus* populations occur only in rocky habitats but can be dense in these areas.

Production by *Uta* was estimated by Turner, Medica, and Kowalewsky (1976) to be 2.2 MJ $(\text{ha yr})^{-1}$, or 19 percent of metabolizable energy flow, a value that compares well with the value 1.8 MJ $(\text{ha yr})^{-1}$, or 13 percent of *M,* determined here—considering that fairly different methods were used to measure *P* in the two studies. These production ratios are in general agreement with those for other ectothermic vertebrates, and are about 10 times those of endothermic vertebrates (McNeill and Lawton, 1970; Humphreys, 1979; Turner and Chew, 1981). The absolute rate of production by the *Uta* population in Nevada is similar to or greater than that of many other vertebrates (including birds, mammals, and other lizards) in the deserts of the southwestern United States (Turner and Chew, 1981). The results of this study support and underscore the conclusion of Turner, Medica, and Kowalewsky (1976:2) that *U. stansburiana,* "and possibly other numerous and short-lived lizards, are of greater significance in the energy dynamics of natural communities than has heretofore been appreciated," as well as the more recent deduction by Pough (1980:104) that "their secondary production makes amphibians and reptiles as important as birds or mammals in terrestrial ecosystems."

Food Limitation. The relationship between food supply and food demand is of central importance in ecology and in the theory of evolution. The food demand of a *Uta* population was estimated herein, so it remains to assess food supply. The rate of food supply available to an animal is a very difficult variable to measure in the field, and it becomes more difficult as more is learned about the biology of the organisms involved.

Biomass or standing-stock measurements (of plants and arthropods, in the case of *Uta*) are inadequate because they do not represent *rates* of food supply. Even when rates of total plant and total arthropod production are measured, these cannot be equated with food supply for *Uta* lizards. Many plants or plant parts are not potential foods for *Uta,* because they may be unreachable, poisonous, too indigestible, too dry, too salty, too tough, or they may have other physical or chemical antiherbivore factors. Some plants that may be harmful if eaten exclusively could be consumed safely in small amounts along with other foods, which may contain chemicals that diminish or counteract the harmful effects of the former. Many of these aspects of plants change through the seasons and can change in response to damage inflicted by herbivores. Similarly, all arthropod production in Rock Valley is not available to *Uta* because some arthropods may be too big, too small, or inactive when *Uta* are active. Or, an arthropod could be active at the "right" time of day or season but unavailable because it flies, burrows, lives high in shrubs, or is toxic, distasteful, has a gut full of poisonous plant matter, and so on.

Again, certain mixtures of otherwise harmful arthropod species or arthropods and plants may be acceptable as food. Clearly, one must know a great deal about a consumer's behavior, morphology, nutritional requirements and limitations, and static and inducible capabilities for detoxifying harmful dietary substances in order to sort out potential food items from the total biomass produced in a habitat.

It does not appear that such a determination of food supply for *Uta* in Rock Valley will be forthcoming soon. However, many measurements of arthropod biomass have been made in Rock Valley, and I used these as the best indicators available to date of seasonal changes in the arthropod food supply for *Uta*.

The dry biomass of arthropods occurring in shrubs, collected by vacuuming (D-Vac Model 1), was measured monthly from March through September during 1971, 1972, 1973, 1974, and 1975 by Dr. E. Sleeper and his students (Department of Biology, California State University, Long Beach; F. B. Turner, personal communication). Biomass of ground-dwelling arthropods (collected in pitfall, can traps) was measured monthly in Rock Valley during most of 1976 (F. B. Turner, unpublished manuscript). Arthropod biomass presumably was relatively low during the cool months of November, December, and January, and no measurements were made then. Vacuuming and can-trapping results are underestimates of total arthropod biomass, but they are the best estimates presently available. I calculated mean biomass values (per hectare) on a monthly basis and converted from biomass to energy by multiplying by 22.8 kJ (gram dry matter)$^{-1}$.

The measured standing stocks of arthropods were two to six times the monthly rate of food consumption by *Uta* during most of the year (Fig. 2.4). In February, March, and April the biomass of trappable arthropods was relatively low, but *Uta* were consuming substantial amounts of plant material at that time. In autumn the biomass of arthropods declined precipitously, while food consumption by *Uta* was high due to the presence of many hatchlings. If food is an important limiting resource for *Uta* populations, these results suggest that the limitation may be most severe in October and November in Rock Valley.

Biological energy flow in the Mojave Desert is heavily dependent upon rainfall, and arthropod biomass can vary more than tenfold over a period of years (Turner, unpublished manuscript), so food supply to *Uta* may be at critical levels for a year or more during prolonged droughts. More sophisticated methods for measuring the availability of food to animals are needed before this question can be evaluated properly.

Summary

In this study all avenues of energy utilization were assessed in a wild population of small desert lizards (*U. stansburiana*) living in Nevada. Energy used in respiratory metabolism was measured using doubly labeled

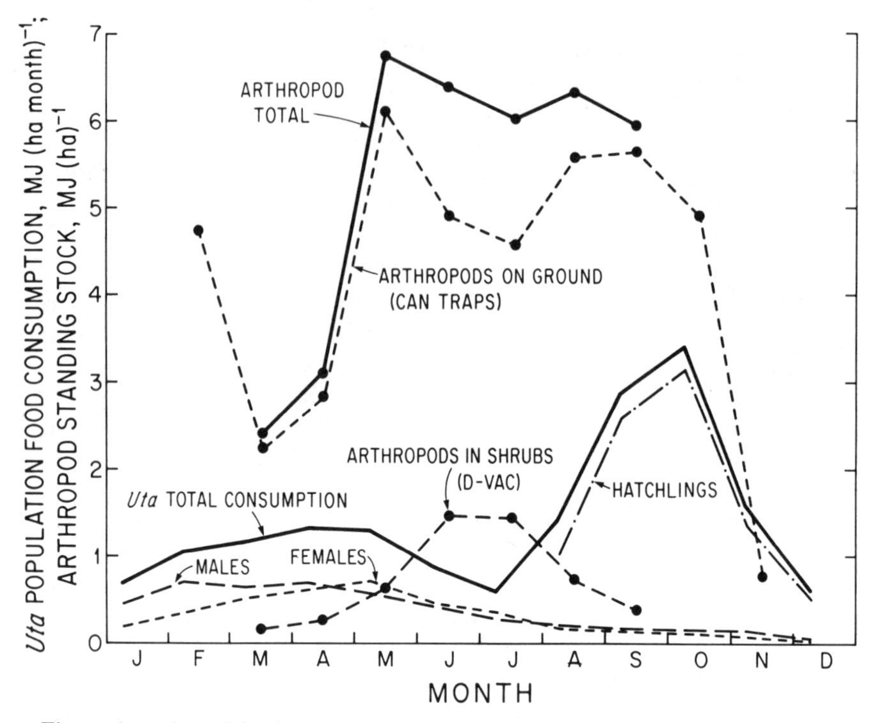

Figure 2.4 Monthly changes in rates of food consumption by *Uta stansburiana* (males, females, hatchlings, and total) and in standing stocks of arthropods (shrub-dwelling, ground-dwelling, and the sum of these) in Rock Valley, Nevada. Lizard consumption rates are based on energy flow rates per animal, measured in 1975–1976, and demography results for 1966–1967 (Turner, Medica, and Kowalewsky, 1976). Shrub arthropod values are 5-year means, but ground arthropod values are for 1976 only.

water. Energy allocated to growth of somatic tissue, somatic lipids, abdominal fat deposits, and reproductive tissue was measured by successive analyses of changes in body composition throughout a year. Metabolizable energy flow was calculated as the sum of these separate avenues of energy allocation. Laboratory measurements of assimilation and metabolic efficiencies, along with analyses of field diet composition and energy content, permitted estimation of rates of energy consumption and of energy loss via feces and urine. Adult males, one-year-old females, two-year+ females, and hatchlings were examined separately to reveal possible differences in energy allocation between these cohorts. Energy flow per animal (on a daily, seasonal, yearly, and lifetime basis), and energy flow per hectare of habitat were analyzed.

Rates of energy flow through *Uta* were highest in spring (April and May) and lowest in winter (December). During spring, energy utilization by males rose faster and peaked sooner than in females, coincident with the establishment and maintenance of territories by males. Energy flow in females peaked a month later, in conjunction with the time of egg laying.

These lizards fed during all months of the year. Energy stored in the body provided at most about 35 percent of the respiratory expenditure during winter, and food provided the remainder. In females, less than 25 percent of the energy in eggs could have come from stored energy in their bodies.

The portion of its usable energy flow that an animal diverts to reproduction (reproductive effort) is a central aspect of its biology. Reproductive effort was assessed as $(R_c + R_R)\ (M - S)^{-1}$, where R_c is the rate of chemical potential energy increase in eggs, R_R is the portion of total metabolic rate that was associated with reproduction, M is the rate of metabolizable energy input from the food, and S is the rate of chemical potential energy flow to storage depots in the animal. Annual reproductive effort of yearling females (0.357) was about the same as that of older females (0.349), and yearling females had a higher reproductive effort for their second clutch of eggs (0.779) than for their first clutch of the breeding season (0.463). This is in accord with theoretical predictions that are based on expectations of future reproductive success. However, in contrast to theory, reproductive effort by 2-year+ females declined from their first to their second clutch, despite increased feeding rates. This may be related to some physiological effect of age or previous reproductive effort.

Annual respiration expenditures associated with reproduction (R_R) in male *Uta* were about the same as those of females but annual total reproductive effort of males was lower than in females because the chemical potential energy investment by males in production of sperm (R_c) was very small compared to eggs. Annual production efficiencies (chemical energy flow to growth and reproduction divided by metabolizable energy flow) in individual female *Uta* in Nevada (0.144 to 0.203) were similar to those of female *Uta* in Texas, but a bit lower than those of a tropical lizard, *Anolis* (0.25).

Lizards require much less energy to live in the field than do endotherms. Daily metabolic expenditure (corrected for body-mass effects) in a free-living bird or rodent is about 20 to 40 times that of an iguanid lizard during its activity season, and mass-corrected energy metabolism over an entire year in a desert jackrabbit (*L. californicus*) is about 53 times that of a desert lizard (*Uta*).

Energy fluxes through a population of *Uta* in Rock Valley, Nevada, were estimated to be: ingestion, 17.6; metabolizable, 13.8; respiration, 12.0, and production, 1.8 megajoules per hectare per year. This population energy requirement and production rate is much higher than that of populations of birds, other lizards, and most mammals living in similar desert habitats, indicating that *Uta* lizards are an important component of energy flow in this desert ecosystem.

Comparison of monthly rates of food consumption of this *Uta* population with indicators of monthly food supply in Rock Valley suggests that food supply changes more radically than does food demand through a year. Food may be a scarce resource in October and November, when population food demand is high due to the presence of numerous hatchlings.

3 Biophysical Analyses of Energetics, Time-Space Utilization, and Distributional Limits

Warren P. Porter and
C. Richard Tracy

IN THIS CHAPTER we report and expand on selected recent biophysical developments that relate to the ecology and population biology of lizards. We emphasize the intimate relationships between (1) environmental effects on heat balance and (2) direct consequences for mass flows of food and water through an animal, and we discuss the resultant impact on potential growth and reproduction.

Norris (1967) presented the first biophysical analysis of the role of climate in the biology of lizards. He emphasized how background color matching and color change affect rates of bodily warming in still air and influence behavior and survivorship of lizards. Norris pointed out that Atsatt (1939), Klauber (1939), and Cole (1943) were aware that color change in reptiles could have the dual functions of concealment and thermoregulation. However, analytic tools and computers were not generally available to test these early ideas, and they remained relatively intractable hypotheses. Principles of heat and mass exchange (Bird, Stewart, and Lightfoot, 1960) were largely unused by the biological community with few exceptions (Winslow, Herrington, and Gagge, 1937; Church, 1960) until approximately 1962 (Gates, 1962), when Gates's book provided an explanation of heat-transfer engineering of use to biologists.

Norris (1967) quantified the relationship of color change to heat balance and thermoregulation after he observed what might be termed conflicting circumstances for background color matching. He noted that early in the morning some lizards were often very darkly colored (no matter what their background), presumably facilitating warming, but once these lizards reached the body temperatures that they maintained through the day, their color matched the background faithfully. And, under conditions which were too hot for predators to be present, the lizards became much lighter than their background, thereby increasing reflectance of incident solar radiation and reducing the heat load on them. There appeared to be a hierarchy of conflicting priorities wherein color was adjusted for particular body temperatures under some circumstances and for concealment under other circumstances.

Norris also observed a correlation between body size and color change. Parry (1951) had earlier observed that larger insects tended to have greater color change and higher reflectivities, and Norris showed that reptiles follow the same size-related capacity for color change (although some species of lizards cannot change color—for example, members of *Teiidae* and *Scincidae*—due to deposits of immobile melanin in the skin). The adaptation for color change is coupled with the presence of a black peritoneum that absorbs ultraviolet radiation transmitted through the skin when the animals lighten by aggregating mobile melanin granules (Porter, 1967; Porter and Norris, 1969).

Early observations and ideas of Cowles and Bogert (1944) and Cowles (1958) did much to stimulate experiments and field observations on lizard temperature relations and ecology (Hutchison and Larimer, 1960; Kleiber, 1961; Heath, 1962; Brattstrom, 1965; DeWitt, 1967; Regal, 1967; Tinkle, 1967; Turner, Medica, and Kowalewsky, 1976; Huey, Pianka, and Hoffman, 1977; and many others). The integration of some principles of physics and heat- and mass-transfer engineering, dubbed biophysical ecology (Gates, 1975), into the biological world by Norris (1967) and Bartlett and Gates (1967) promoted analyses of the physical mechanisms underlying temperature relations of lizards.

Healthy skepticism of energy-balance models and their limitations in trying to describe complex animals in complex environments has led from early steady-state models (Porter and Gates, 1969) to transient-state models of animals (Porter et al., 1973; Spotila et al., 1973; Tracy, 1976) and microenvironments (Beckman, Mitchell, and Porter, 1973; Porter et al., 1973; Mitchell et al., 1975). These models have been tested in a variety of habitats ranging from the deserts of the southwestern United States (Mitchell et al., 1975) to Africa (James and Porter, 1979; Porter and James, 1979) to a Michigan bog (Kingsolver, 1979) and the Galapagos Islands (Christian and Tracy, 1981). Simultaneously, sophistications and simplifications of models emerged that dealt with problems ranging from calculations of solar radiation (Gates, 1962; McCullough and Porter, 1971; Campbell, 1977), of turbulent convective heat transfer out of doors (Kreith, 1968; Wathen, Mitchell, and Porter, 1971, 1973; Kowalski and Mitchell, 1976; and Mitchell, 1976), to what constitutes an appropriate animal model (Bakken, 1976, 1981).

The focus of biophysical modeling seems to be broadening to include application to problems of population dynamics and to a hybridization of mechanistic and empirical models of growth and reproductive potential in complex, changing environments (Riechert and Tracy, 1975; Porter and Busch, 1978; Anderson, Tracy, and Abramsky, 1980; Muth, 1980; Porter and McClure, 1982; Roughgarden, Porter, and Heckel, 1981; Christian, Tracy, and Porter, 1982; Waldschmidt, 1982; Waldschmidt and Tracy, 1982; Nagy, Chapter 2; Schall, Chapter 4).

Interacting Models

The parameters and input variables needed to evaluate the energy balance of lizards (Fig. 3.1) are listed in Table 3.1. Solar radiation, air temperature and wind speed at reference height, and deep soil temperature (Fig. 3.1) are the four time-dependent boundary conditions that drive the microclimate heat-energy-balance model for the soil. Direct and scattered solar radiation can be measured or calculated (McCullough and Porter, 1971). Thermal radiation from the sky can be calculated from air temperature at 2 m using empirical equations developed by Brunt (1932), Swinbank (1963), or Idso and Jackson (1969). Deep soil temperatures can be estimated from the mean annual and monthly temperatures for any particular site (Van Wijk, 1963). Soil reflectivity can be calculated from measurements of reflected solar radiation or estimated from data in Sellers (1965). Soil roughness can influence profiles of air temperature and wind velocity. If data on velocity profiles are not available, profiles can be estimated from information on spacing, size, and density of bushes, trees, boulders, or other elements that can retard air movement (Mitchell et al., 1975).

Animal properties needed for biophysical models include the mass and solar reflectivity of the animal (Norris, 1967; Porter, 1967; Gates, 1980). Metabolic rates may be estimated from existing data (Bennett and Dawson, 1976). Rates of evaporative water loss have been measured for many lizards (Dawson and Bartholomew, 1958; Templeton, 1960). However, such measurements may not reliably predict rates of water loss in natural environments, since the measurements are typically done only in dry air and the functional relationships may change in moist air (Welch, 1980; Zucker, 1980).

To address questions of growth and reproduction one must consider the simultaneous interaction of at least four processes (Porter and McClure, in preparation; Fig. 3.2). The diagonal equation of Figure 3.2 is the heat balance between an animal and its environment. The energy inputs to the animal are above the equal sign to the left and the losses are below it to the right. Two mass-balance equations intersect the heat balance equation at the metabolism term. The horizontal equation is the mass balance of dry food. Food ingested minus that defecated is equal to that absorbed. Absorbed food is available as energy for metabolism, and the energy that is not used for maintenance or activity may be

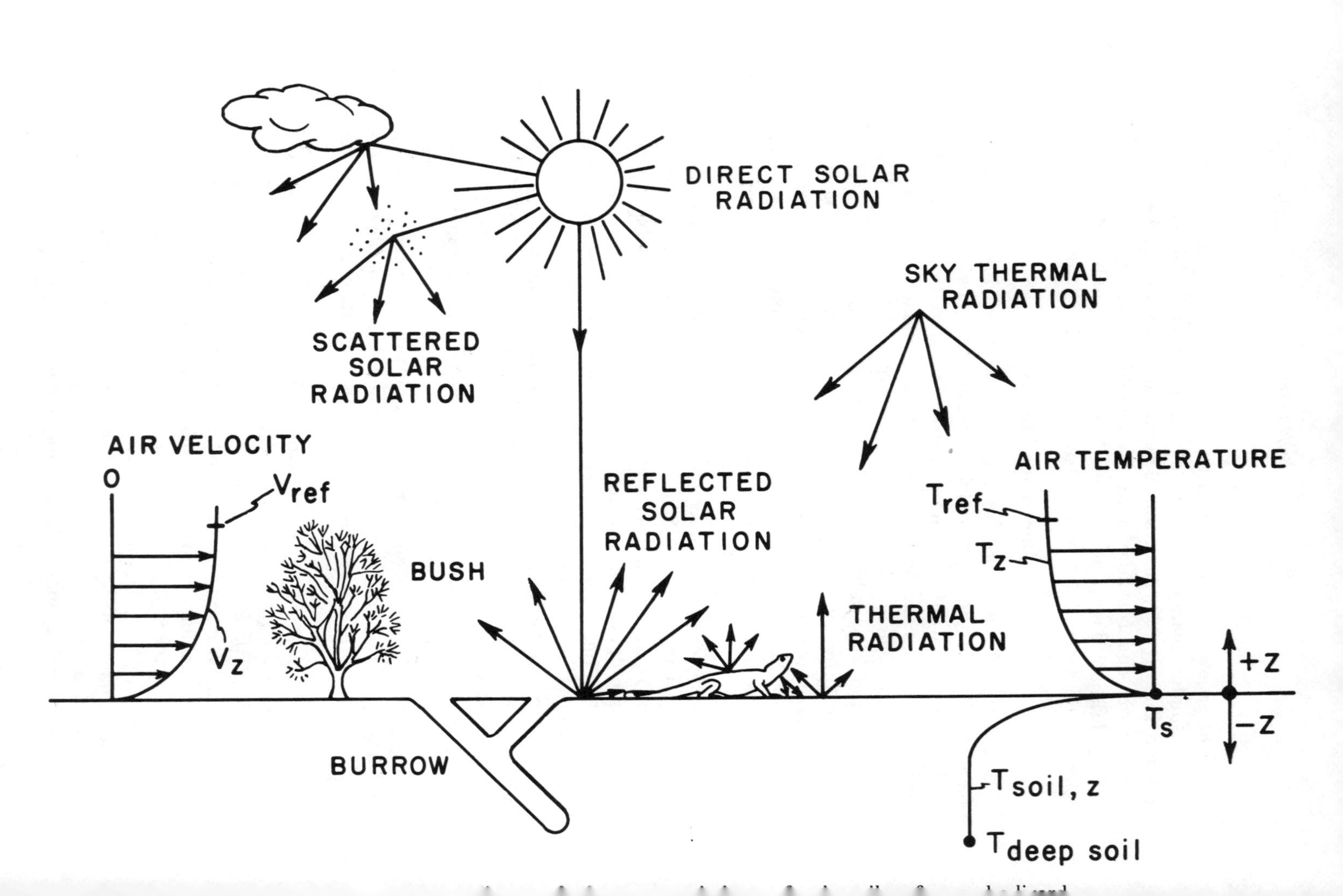

DIRECT SOLAR RADIATION
SCATTERED SOLAR RADIATION
SKY THERMAL RADIATION
AIR VELOCITY
0
V_{ref}
V_z
BUSH
REFLECTED SOLAR RADIATION
THERMAL RADIATION
AIR TEMPERATURE
T_{ref}
T_z
+Z
−Z
T_s
BURROW
$T_{soil, z}$
$T_{deep\ soil}$

Table 3.1 Parameters and variables needed to evaluate energy balance of lizards.

Symbol	Parameters	Values (sources and locations)				Units
		Indian Springs, Nevada	Smyrna, Washington	Palm Springs, California	Isla Santa Fe, Galapagos	
	Microclimate					
α_s	Soil solar absorptance	0.70	0.70	0.67	0.89	–
ϵ_s	Soil emissivity	1.0	1.0	1.0	0.92	–
ρc_p	Soil density times specific heat	2.093×10^6	2.093×10^6	2.093×10^6	1.988×10^6	$J\ m^{-3}K^{-1}$
σ	Stefan-Boltzmann constant	5.669×10^{-8}	(same)	(same)	(same)	$W\ m^{-2}K^{-4}$
k_s	Soil thermal conductivity	0.35	0.35	0.293	1.39	$W\ m^{-1}K^{-1}$
Z_o	Soil surface roughness					
	Top	3.67	11.16		16.8	cm
	Middle	3.29	10.57	0.05	6.42	
	Bottom	0.90	0.02		0.87	
Z_r	Reference height					
	Top	200	200	200	200	cm
	Middle	80	84	–	100	
	Bottom	60	13	–	30	
Z_a	Heights above the ground	1.5, 5, 15, 25, 35, 50, 100, 200				cm
Z_s	Soil depths	2.5, 5, 10, 15, 20, 30, 40, 50, 60				cm
	Time-dependent tables					
Q_s	Global solar radiation flux	SOLRAD (McCullough and Porter, 1971)				W/m^2
T_r	Air temperature at reference height	U.S. Climatological Data (U.S. Dept. Comm.; 1950–1960); Darwin Research Station Climate Data, Isla Santa Fe, Galapagos				°C
V_r	Wind speed at reference height	U.S. Climatological Data (U.S. Dept. Comm.; 1950–1960); Darwin Research Station Climate Data, Isla Santa Cruz, Galapagos				cm/min

Table 3.1 (continued)

Symbol	Parameters		Units
T_d	Deep soil temperature	Average annual temperature (U.S. Climatological Data, 1950–1960); Darwin Research Station Climate Data, Isla Santa Cruz, Galapagos	°C
Latitude	1°00′S	22.6°C Isla Santa Fe, Galapagos	
	33°59′N	22.5°C Palm Springs, California	
	36°34′N	16.2°C Indian Springs, Nevada	
	46°50′N	11.5°C Smyrna, Washington	
		J F M A M J J A S O N D	
Cloud cover for Isla Santa Fe, Galapagos		0.5 0 0 0 0 0 0.5 0.5 0.5 0.5 0.5 0.5	
Clear sky assumed for other sites			
	Lizard		
α_L	Solar absorptivity	95, 85, 75, 55 (91, 69—*Conolophus*)	%
ϵ_L	Infrared emissivity	1.0	–
A_L	Total surface area	$10.4713\ m^{0.688}$ (N = 17; $r^2 = 0.996$)	cm^2
A_{SIL},	Silhouette area normal to the sun	$3.798\ m^{0.683}$ (N = 30; $r^2 = 0.995$)	cm^2
A_{SIL},	Silhouette area pointing at the sun	$0.694\ m^{0.743}$ (N = 30; $r^2 = 0.98$)	cm^2
$F_{L\text{-}s}$	Shape factor for diffuse radiation between lizard and ground	0.4	–

$F_{L\text{-sky}}$	Shape factor for diffuse radiation between lizard and sky	0.4	–
m	Mass	0.32, 5.0, 40.0, 50.0, 100.0, 6500.0	g
Z	Height above ground	1.5, 5, 15, 25, 35, 50, 100, 200	cm
M	Weight and temperature specific metabolism (Bennett and Dawson, 1975)	$0.0056 \cdot 10^{(0.038T_L - 1.771)} m^{0.82}$	W
SVL	Snout-vent length	$2.998\ m^{0.333}$ (N = 42; $r^2 = 0.98$)	cm
T_L	Lizard activity temperature range	30–38.5 ° (*Uta*) 38–43 ° (*Dipsosaurus*) 18–40 ° (*Conolophus*)	°C

Convection

Nusselt-Reynolds correlations ($Nu = aRe^b$) for lizard convection coefficients determined from our lizard castings. The sphere correlation is from McAdams, in Kreith (1968), for comparison purposes. (L = snout-vent length for all species listed.)

Dipsosaurus dorsalis		
$Nu = 0.35\ Re^{0.6}$	Transverse	(Re = 2,000–20,000)
$Nu = 0.1\ Re^{0.7}$	Parallel	(Re = 2,000–20,000)
Agama agama		
$Nu = 1.03\ Re^{0.002}$	Transverse	(Re = 600–2,000)
$Nu = 0.35\ Re^{0.6}$	Transverse	(Re = 2,000–20,000)
Uta stansburiana		
$Nu = 0.088\ Re^{0.697}$	Elevated	(Re = 1,691–5,255)
$Nu = 0.185\ Re^{0.565}$	Prostrate	(Re = 1,731–5,380)
Callisaurus draconoides		
$Nu = 4.26\ Re^{0.320}$	Elevated	(Re = 400–3,500)
$Nu = 0.467\ Re^{0.589}$	Elevated	(Re = 3,500–10,100)
$Nu = 4.26\ Re^{0.273}$	Prostrate	(Re = 500–5,600)
$Nu = 0.072\ Re^{0.743}$	Prostrate	(Re = 5,600–10,250)
Sphere ($L = m^{1/3}$)		
$Nu = 0.34\ Re^{0.6}$	All orientations	(Re = 25–100,000)

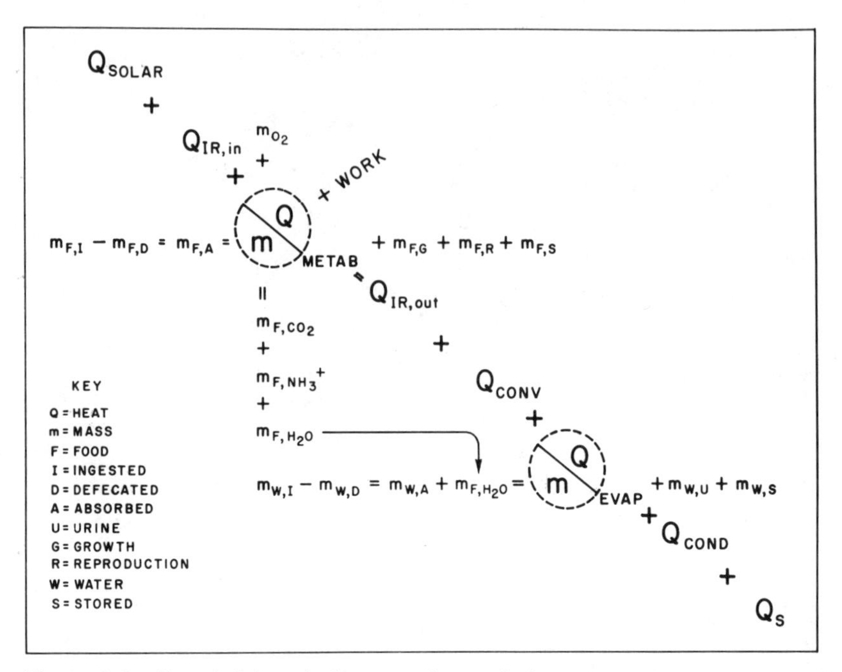

Figure 3.2 Coupled heat-balance and mass-balance equations for relating climatic variables to maintenance, growth, and reproductive potential (Porter and McClure, in preparation).

used for growth or reproduction, or it may be stored. The vertical equation intersecting the metabolism term is the mass-balance equation for respiratory gases. The mass of oxygen plus the mass of oxidizable substrate is equal to the mass of carbon dioxide produced, plus by-products of nitrogen containing waste compounds of protein metabolism, plus water. The mass of water produced by metabolic reactions is also an input to the water balance, which is the horizontal equation intersecting evaporation in the heat-balance equation. Ingested mass of water less that lost by defecation is equal to the mass of water absorbed plus metabolic water. Water may be evaporated, lost in urine, and/or allocated for growth and reproduction, or stored.

It is important to notice that a change in one parameter, such as solar radiation, results in readjustments in all the other terms in the equations. For example, an increase in solar radiation would lead to an increase in body temperature, which would increase evaporative water loss as well as convective and radiative heat losses. An increase in evaporation would increase consumption of water in the water-balance equation. Metabolism would be elevated, and additional mass of carbon would be used for metabolism. Less food would be available for growth, reproduction, or storage. Furthermore, an increase in oxygen required will often lead to higher respiration rates and more respiratory water loss. Because these equations are coupled, there are fewer degrees of freedom than if all terms or processes were independent. Therefore, even though the system is complex, fewer variables need to be measured than if each of the processes were treated as independent.

One method of depicting the interactions of these processes is seen in Figure 3.3 (Porter and McClure, in preparation). This three-dimensional figure serves as a heuristic representation of the complex interactions in Figure 3.2 and illustrates energy available for growth or reproduction. The radiation axis is based on a range of solar and thermal energies incident on a desert lizard. At low air temperatures and radiation, metabolic rate and the requirement for food are low. As the thermal environment becomes warmer, metabolic rate increases. The shape and absolute position of the curves of food processed that comprised the

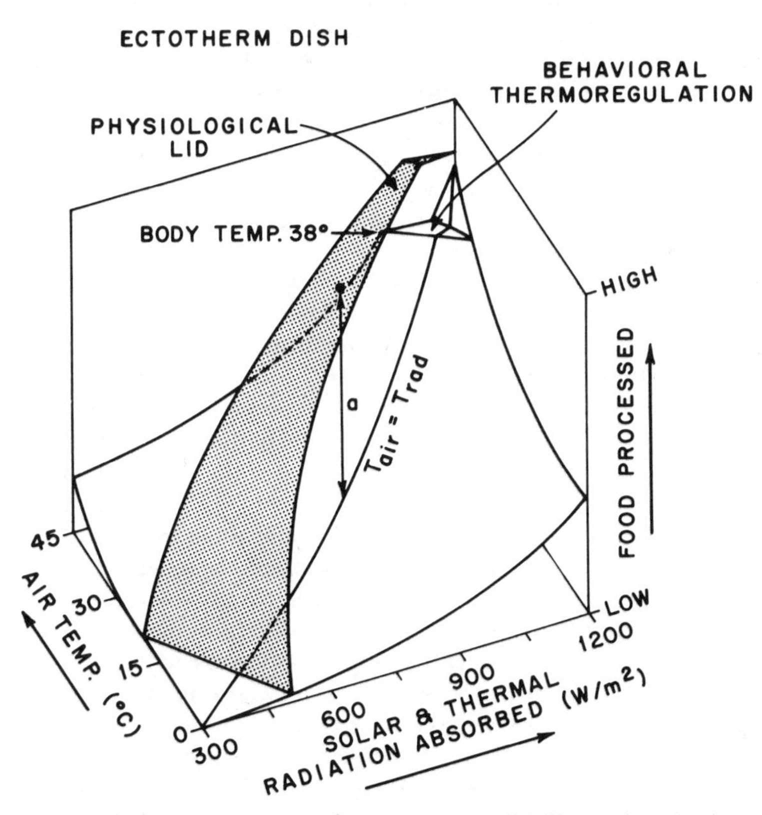

Figure 3.3 Three-dimensional floppy dish and lid illustrating the interactions of the equations in Figure 3.2. Potential for growth or reproduction or fat storage or activity for a lizard is represented by the space *a* (Porter and McClure, in preparation).

dish-shaped surface will vary among species depending upon such variables as activity level, health, and body temperature.

The flat shelf near the top of the surface in the back corner represents a constant body temperature maintained by combinations of different wind speeds and physiological and behavioral changes in heat-exchange properties such as absorptivity, posture, and orientation. Despite all that is unknown about the shape of the curves, we may define a lower and upper bound for the dish surface represented by resting and maximally aerobic animals, respectively (Bennett, Chapter 1). The position of the upper food-processing lid and lower maintenance surfaces relative to each other define the energy available for growth or reproduction in any environment for any size lizard.

The dish surface has been referred to as a "floppy dish" (Porter and McClure, in preparation) because it is a dynamic surface whose movements reflect the potential for growth or reproduction by affecting the space between dish and lid. For example, as wind speed increases, the whole lower surface rotates counterclockwise. This may be visualized by considering an animal on the dish shelf trying to maintain a 38 ° C body temperature at air temperatures of 30 ° versus 40 ° C. At an air temperature of 30 ° C, an increase in wind speed would require more solar radiation to offset the increased convective cooling. At an air temperature of 40 ° C, on the other hand, an increase in wind speed means more convective warming and a requirement for reduced radiation to maintain the same body temperature. The whole dish surface rotates clockwise as wind speed decreases, or animal size increases. The dish rotations are due to changes in the boundary-layer thickness of the animal (Porter and Gates, 1969; Porter and James, 1979). Since activity may raise the dish surface to higher levels at various temperatures, the dish not only rotates within limits but also rises and falls within limits. The line labeled $T_{air} = T_{rad}$ running along the diagonal of the dish from lower-left front to upper-right rear represents "blackbody" conditions, where the temperature of surrounding surfaces is the same as air temperature. This kind of environment is commonly found in metabolic chambers.

The stippled surface above the dish labeled physiological lid represents the digestive capacity of a lizard. Only a portion of the total lid surface is shown so that the underlying dish surface can be seen. The lid represents the maximum processing capacity of the animal, which is determined by the allometry of the gastrointestinal tract, the type of food consumed by the animal, and other physiological constraints on food processing. The lid surface goes to zero (mass of food processed) at some low body temperature. (The literature suggests digestion is not feasible for some lizards and snakes below about 15 ° C: Skoczylas, 1970; Harlow, Hillman, and Hoffman, 1976; Waldschmidt, unpublished paper.) The shape of the physiological lid is unknown for most animals, but the parabolic representation is based upon data from Harlow, Hillman, and Hoffman (1976) on the desert iguana.

The space between the two surfaces labeled *a* represents the food resources that can be used for growth and reproduction or retained in storage. This space is simply a three-dimensional representation of ideas proposed by Brody and coworkers (1945), Kendeigh (1949), and Wiens and Innis (1974). The physiological lid representing the maximum processing capacity of the animal may be suppressed by availability of food in time or space. That is why an ecological lid could be represented as lying below the physiological lid and decreasing the amount of food available under certain conditions, such as at very high temperatures. That suppression may be dynamic, but the physiological lid is intended to define the upper bound of available food. Competition and agonistic interactions may also suppress maximum available food. Thus, for some members of a population, a behavioral lid may lie below the ecological lid. Schall (Chapter 4) presents another way that the distance, *a,* may be compressed to reduce growth or reproduction. A disease like malaria may increase the height of the dish by adding to maintenance costs and may suppress the lid by interfering with obtaining or processing food.

The computation of food and water requirements for maintenance relies on integrated values of metabolism and water loss for each selected location in a microenvironment. For this computation one must know certain physiological parameters, such as digestive efficiency and various water conservation mechanisms that reptiles are known to possess (Templeton, 1960; Norris and Dawson, 1964; Schmidt-Nielson, 1964; Minnich, 1970; Minnich and Shoemaker, 1970; Murrish and Schmidt-Nielsen, 1970; Nagy, 1977). Although we now suspect that predicting water loss in reptiles in the field from laboratory data may not be nearly as clear cut as previously supposed (Welch, 1980; Zucker, 1980), estimates of metabolism and food consumption based upon Bennett and Dawson's equation (1976) show good agreement with observed food consumption in the field (Porter and James, 1979; Porter, Christian, and Tracy, in preparation).

Effects of Climate on Activity and Acquisition of Food

The Desert Iguana

Calculations of potential above-ground activity for an adult desert iguana (*Dipsosaurus dorsalis*) in Palm Springs, California, are given in Figure 3.4*a*. Microclimatic data driving the body temperatures of the lizards come from a revision of the model of Porter et al. (1973), modified to include the effects of free convection at high soil-surface temperatures (Garratt and Hicks, 1973), and from micrometeorological ex-

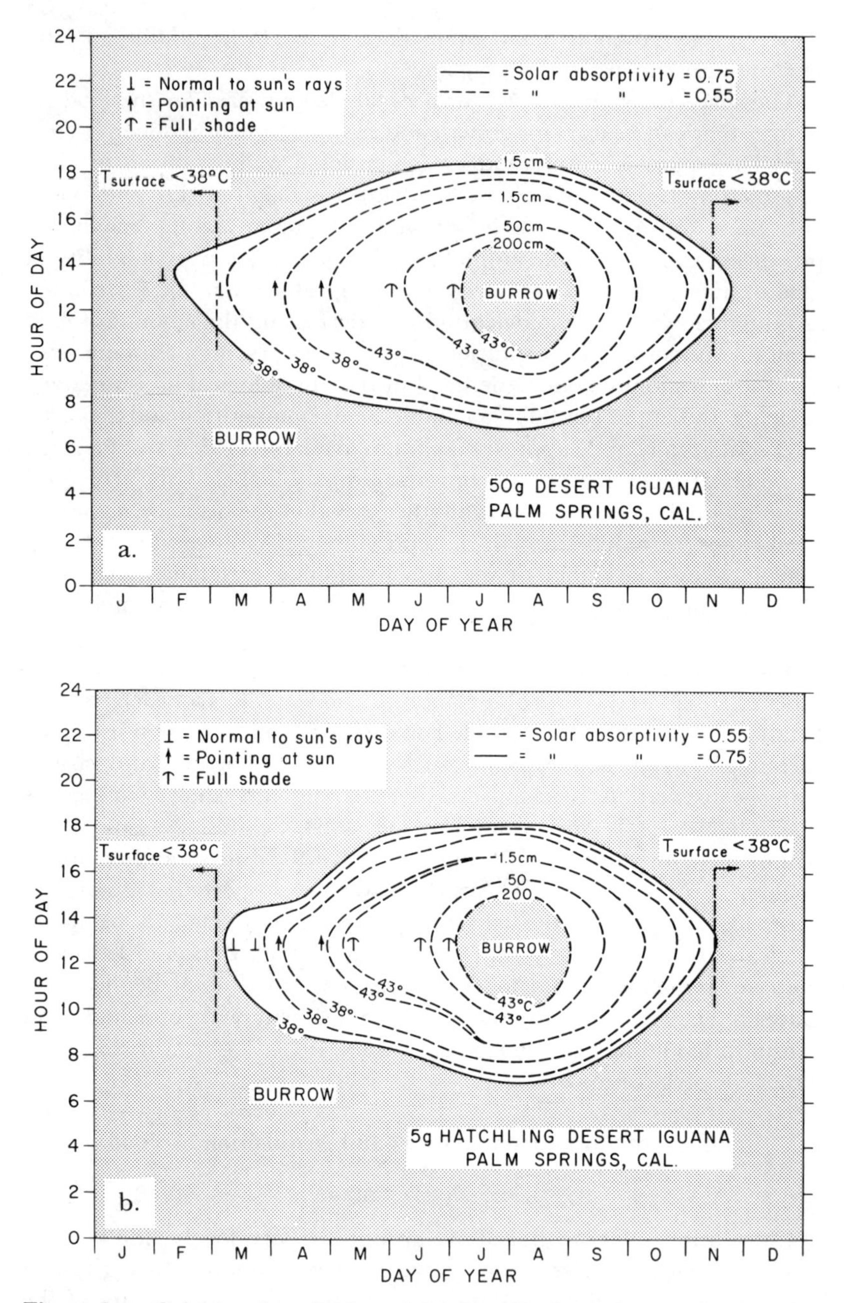

Figure 3.4 Calculated available activity time for desert iguanas in the Palm Springs, California, area for an average year. Effects of color change, posture changes, location in the microenvironment, and body size are illustrated: *a,* 50-g adult; *b,* hatchling.

periments in Deep Canyon, California (Porter and Mitchell, unpublished data). The posture of the animal relative to the sun is indicated next to each of the contour lines on the left edge of each ellipse. Potential increases in duration of activity due to change in color are shown by the space between the two outermost 38 ° C contour lines. The 38 ° C contour line third from the outer edge shows the potential increase in time for activity when the animal orients toward the sun. Note that the increase in time of activity by orienting toward the sun varies with season. A lizard may gain additional time for activity by retreating to the shade and climbing to a height of 50 cm as is represented by the middle 43 ° C contour or 200 cm (innermost contour), the greatest reasonable height to which the lizard can climb in this habitat. Each increase in height decreases air temperature and increases air speed (Fig. 3.1). Within the area bounded by the 200 cm, 43 ° C contour, it is too hot for the animal to be active under average summer conditions. This is consistent with observed activities (Mayhew, unpublished data in Porter et al., 1973).

Predicting patterns of activity for hatchlings provides an opportunity to explore the consequence of body size on available activity time. Figure 3.4*b* shows the calculations for a hatchling desert iguana. Data to calculate temperatures of hatchlings were obtained from newly hatched animals (Porter and Muth, unpublished data). Surprisingly, the calculated times of likely activity for hatchlings are similar to those of adults, but calculated times of emergence are later for hatchlings. Hatchlings have approximately a half-hour less time available for activity in any given day. This delay is due to greater convective heat transfer in the hatchlings due to their greater convective heat exchange. Time constants are so short even for adults at typical wind speeds that convective differences dominate the duration of potential activity time. It is also important to remember that this is an intraspecific comparison and the minimum activity temperatures are assumed equal in hatchlings and adults. Lower activity temperatures for hatchlings would allow them to be active earlier and later than adults, and we have observed this activity pattern in August and September (unpublished data).

The temperature contours may be translated to daily temperature-dependent metabolism—resting metabolism (Bennett and Dawson, 1972, 1976)—as shown in Figure 3.5. The cumulative daily metabolism is low in winter and increases to a peak in July, thus the greatest energetic demands are obviously in midsummer when subsurface temperatures are highest and above-ground conditions are hottest. On the order of 11 kJ per 100-g animal would be required simply for maintenance in this kind of environment (assuming the animals maintain a body temperature no greater than 43 ° C at any time). In March and April, energy requirements are only about 6 kJ per 100-g animal. If the animal proc-

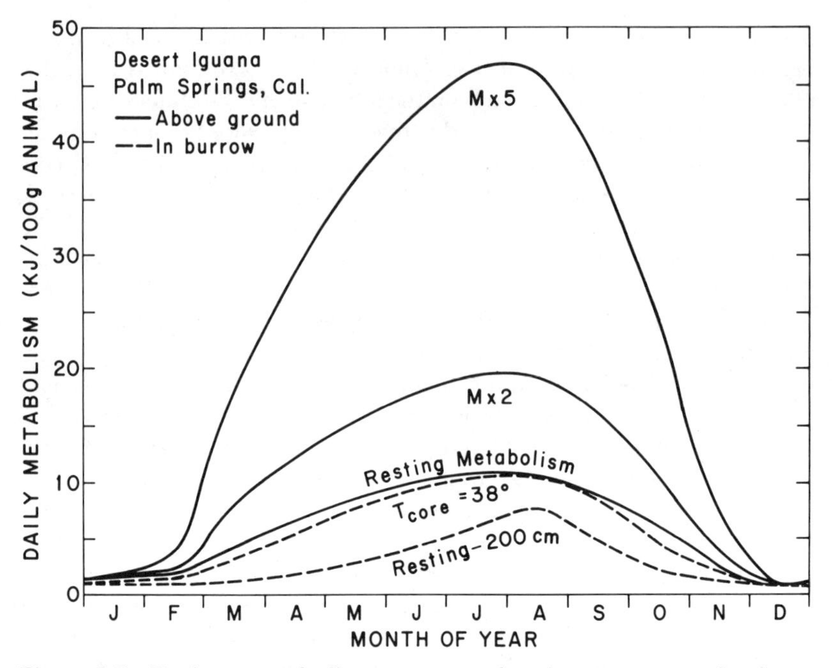

Figure 3.5 Environmental effects on seasonal maintenance costs for above-ground and below-ground locations and for different activity levels when activity is possible. The T_{core} = 38 ° line is for an animal thermoregulating at 38 ° C as much as possible while remaining in the burrow all day.

esses the same amount of food in spring and summer, then at least 45 percent of what the lizard consumes in late March and early April could be used for additional locomotion, growth, or reproduction. In late August and September, excess food could be converted to fat storage. Thus the mass of food absorbed but not used for maintenance is potentially much larger in a cool season than in midsummer when energy requirements are very high. These predictions are consistent with field observations of growth and reproduction (Norris, 1953) and fat storage (Moberly, 1962). However, the observed patterns of growth and reproduction could result from photoperiodic or circannual rhythms (Mayhew, 1965). Available microclimates might have little effect on growth and reproduction in an ectotherm like the desert iguana. Nevertheless, Muth (1980) has been able significantly to accelerate growth and reproduction of desert iguanas by maintaining favorable year-round microenvironments and by providing ample food. Desert iguanas reach reproductive maturity in about 7 months instead of the 4 to 7 years required in nature (Fig. 3.6; Muth, unpublished data). Under warm, spring-like laboratory conditions and with a fortified diet, adults breed all year, producing as many as four clutches annually rather than

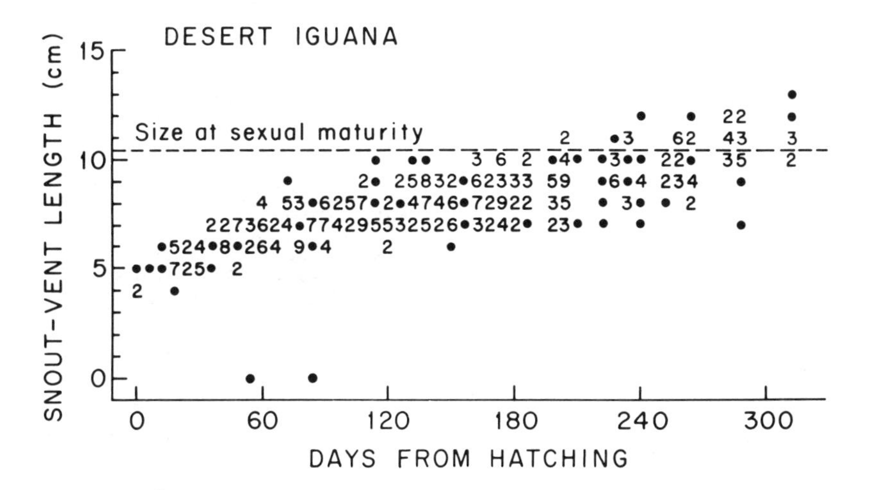

Figure 3.6 Laboratory growth rates of hatchling desert iguanas maintained on ad-lib food and year-round 35 °C air temperatures with photoflood and ultraviolet illumination (Muth, unpublished data).

the single clutch produced in nature. This suggests that these animals have a physiological capacity for growth and reproduction that is usually not realized in the field. We are thus led to question the traditional view that life-history traits are primarily or exclusively under genetic control (see Chapter 11). Life-history traits appear to be influenced significantly by the physical environment and the availability of food, as well as by genetics in the desert iguana, and we regard it as an important challenge to assess the extent to which this is the case in other species.

The Rainbow Lizard

Predictions for the times at which a 60-g adult rainbow lizard could attain the minimum and maximum temperatures in Cape Coast, Ghana (Porter and James, 1979; James and Porter, 1979) give a contour plot (Fig. 3.7) very different from the elliptical patterns seen for a temperate desert environment (Fig. 3.4). In this tropical climate, the predicted time for emergence from and retreat to overnight refuges are essentially the same throughout the year, though not synchronous with sunrise and sunset. A mild dry season centered in March and again in November raises temperatures slightly compared with the rainy season occurring in August. Observed duration of activity is indicated by the vertical lines.

The translation of these body-temperature predictions into cumulative metabolism (Fig. 3.8) shows a nearly constant requirement for maintenance throughout the year. If food-processing capacity and food availability remain constant through the year, the potential for locomotor activity, growth, and reproduction similarly remains nearly uniform

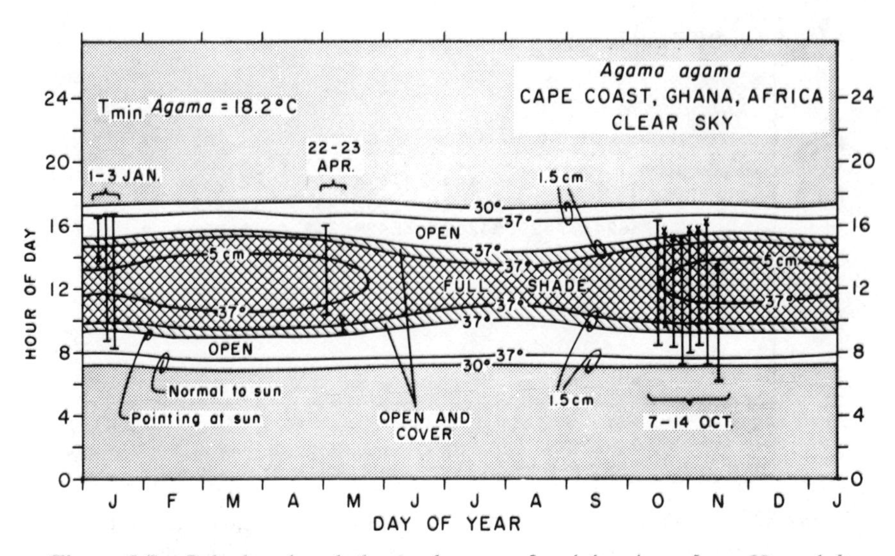

Figure 3.7 Calculated and observed range of activity times for a 60-g adult rainbow lizard in Cape Coast, Ghana (adapted from Porter and James, 1979).

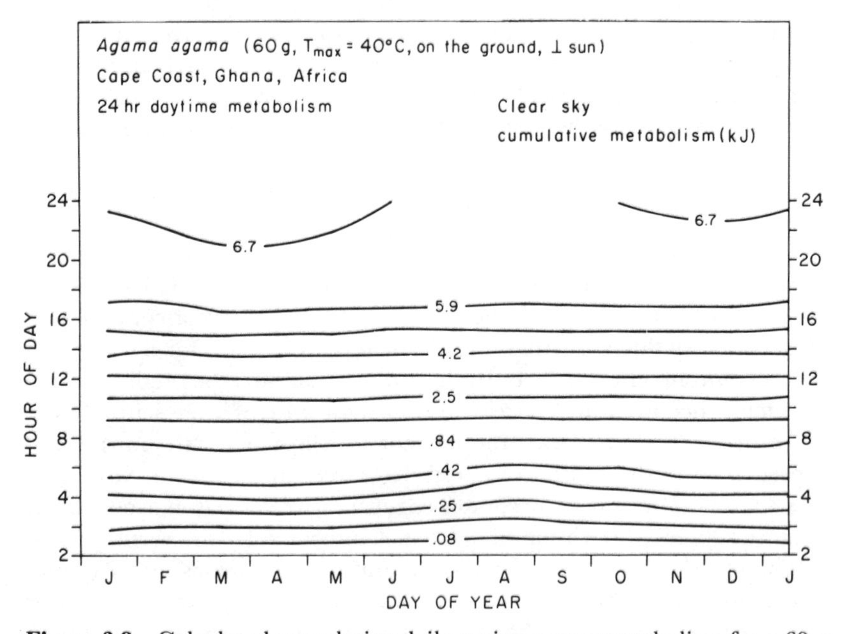

Figure 3.8 Calculated cumulative daily maintenance metabolism for a 60-g adult rainbow lizard active as long as possible in Cape Coast, Ghana. (Adapted from Porter and James, 1979.)

throughout the year. This is consistent with reproductive patterns for this animal reported in the literature (Daniel, 1960; Marshall and Hook, 1960; Chapman and Chapman, 1964); in areas of consistent rainfall, breeding may occur nearly all year long. Biophysical modeling and laboratory and field experiments suggest that the apparent marked contrast in patterns of activity and potential for growth and reproduction in desert iguanas versus rainbow lizards may be explained by a substantial, direct microenvironmental influence on maintenance costs and time available for foraging.

Time and Space Utilization

Computation of the energy beyond that necessary for maintenance which can be used for locomotor activity, growth, and reproduction requires a knowledge of (1) food availability as a function of season and (2) the environmental constraints on patterns of foraging. The herbivorous Galapagos land iguana, *Conolophus pallidus,* is an ideal subject for studying such questions; its behavior and natural history are well known (Carpenter, 1969; Christian and Tracy, 1981, 1982; Christian, Tracy, and Porter, 1982).

Food available to land iguanas declines to very low levels in the Galapagos dry season (June to February). The food and water available in the dry season are reduced to occasional pads of the cactus, *Opuntia echios* (Fig. 3.9, center and right), that drop from the plant. A predictable response to food shortage would be to extend the duration of activity in order to increase foraging time. Unfortunately, the dry season is also the cool season. However, these land iguanas extend their foraging time in the following ways.

First, these lizards are capable of substantial color change. The change in reflectivity of a 4-kg female observed on Isla Santa Fe one morning was from 9 to 31 percent reflectance over a 90-minute interval. This change was measured with a Kettering radiometer on a portable integrating sphere open on one side to the sun and on the other to the lizard or to a white barium sulfate reference plate. Such a large change in reflectance in a large lizard exposed to high air temperatures is entirely consistent with the pattern evident from the work of Norris (1967). He found that larger animals both within and between species have the greatest capacity to change reflectance. Greater change in reflectance for larger lizards is also entirely consistent with the expectations of heat-transfer theory. The larger the animal, the greater the convective insulation due to increased boundary-layer thickness. Decreased convective heat transfer with larger size means that radiative exchange and hence solar absorptivity has a greater impact on body temperature at larger sizes. Thus any color change in a large animal means disproportionately more to its body temperature than in a small animal with exactly the

Figure 3.9 Vegetation structure modifying air-velocity profiles on Isla Santa Fe, Galapagos. Opuntia trees, *Opuntia echios,* center and right, periodically drop a leaf pad, providing the only food for land iguanas in the dry season.

same color change in exactly the same environment (Porter and James, 1979). By being dark-skinned in the morning and late afternoon, a land iguana extends the time when it can be active. By lightening upon warming, the iguana can spend more of the time in the middle of the day in the sun without danger of overheating.

A second adaptation of land iguanas for lengthening periods of activity on Santa Fe is by behavioral exploitation of sunny north-facing slopes (Christian, Tracy, and Porter, 1982). Surprisingly, calculations for the sloping ground suggested that in the cool, dry season there should be less total activity on the slope than on level ground above (Fig. 3.10). Because of the steep angle of the north-facing slope, the ground absorbed less sunlight than did the level ground with the sun to the north. Calculations for both sloping and level ground were made using the same wind speed, but the prevailing wind is from the south, and the

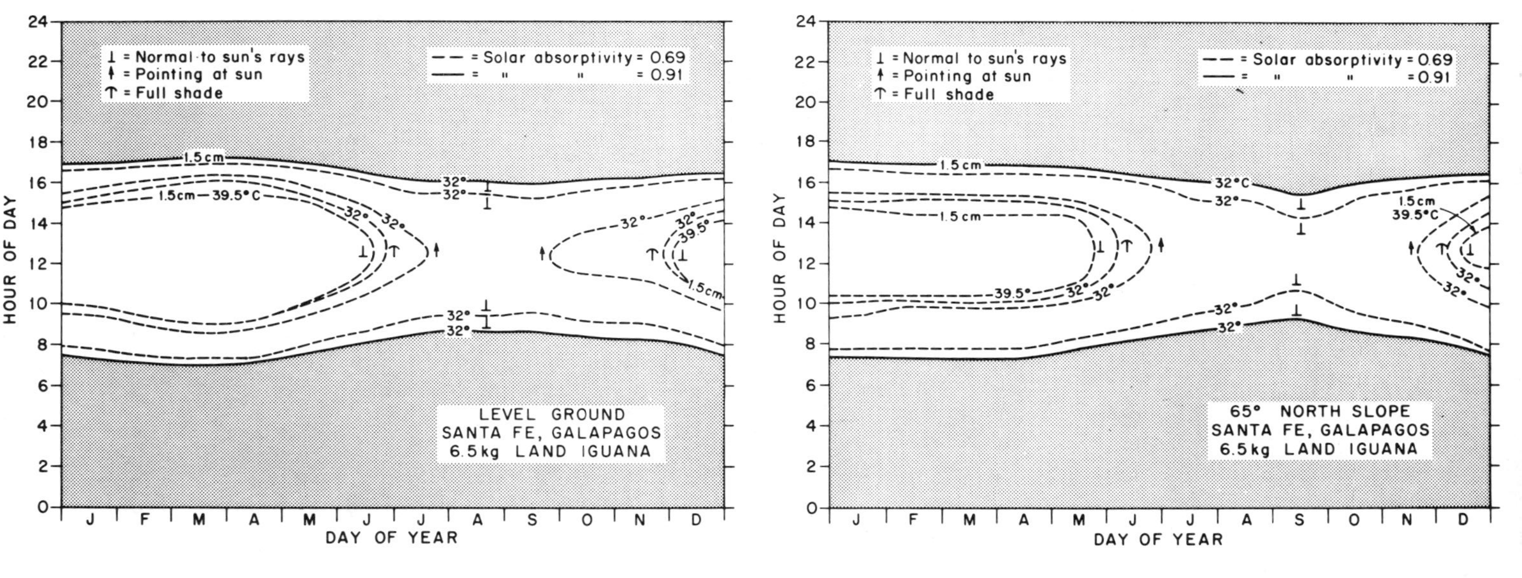

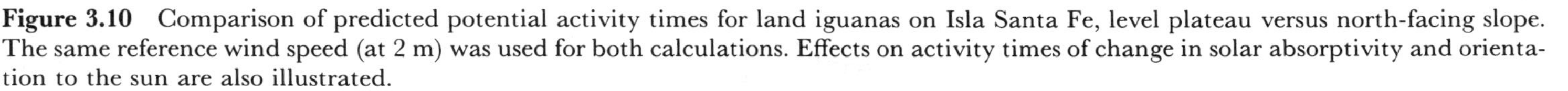

Figure 3.10 Comparison of predicted potential activity times for land iguanas on Isla Santa Fe, level plateau versus north-facing slope. The same reference wind speed (at 2 m) was used for both calculations. Effects on activity times of change in solar absorptivity and orientation to the sun are also illustrated.

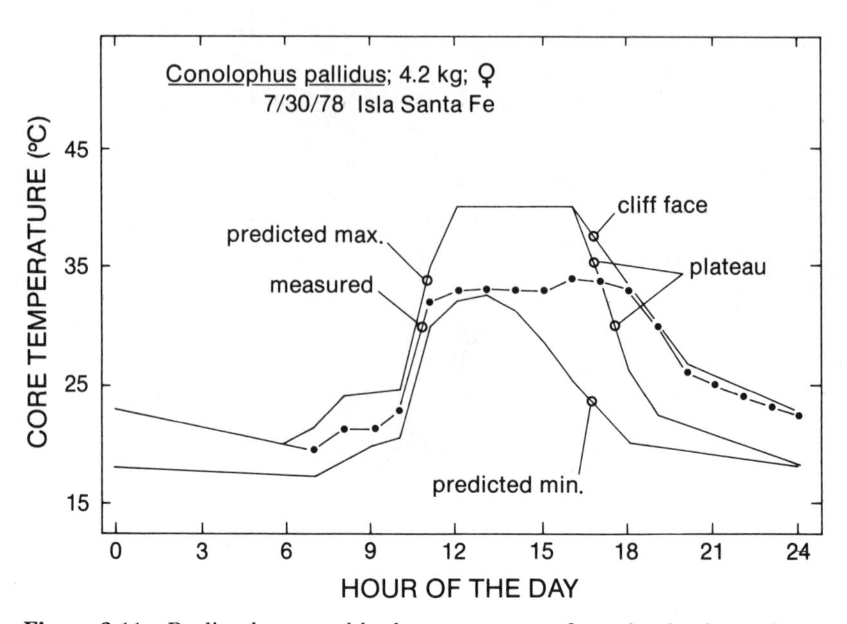

Figure 3.11 Radiotelemetered body temperatures for a day in the cool season for a 4.2-kg female land iguana, illustrating low body-temperature regulation and utilization of microenvironments to delay cooling. The predicted maximum curve has two cooling lines for late afternoon and early evening hours: one represents cooling if the iguana is over the edge of the cliff, the other represents cooling if the iguana is on the plateau (Christian, Tracy, and Porter, 1982).

wind speed on the north slope is much less than that on the flat ground (Christian, Tracy, and Porter, 1982). Thus, effects of wind and not sunlight appear responsible for a more favorable microenvironment on the north side of the cliff. Calculations of body temperature of an adult female on the flat ground and the north-facing slope using measured meteorological data show a different result (Christian, Tracy, and Porter, 1982). Specifically, the achievable body temperatures of the iguana are higher on the cliff face in late afternoon (Fig. 3.11). Indeed, the lizard modeled in Figure 3.11 was also the subject of an experiment in which the lizard's body temperature was continuously measured with temperature telemetry. The lizard, in fact, moved to the cliff face in late afternoon, and as a result, was able to maintain a constant body temperature for an additional hour. Thus, space utilization and duration of activity time seemed to be intimately coupled with the physical environment.

During the cool season the animals undergo another change. They reduce their selected body temperature to lower values, which probably serves to conserve energy, and likely water as well (Regal, 1967; Ballinger, Hawker, and Sexton, 1970; Nagy, 1977). Christian, Tracy, and Porter (1982) have been able to demonstrate that the lowered body temperatures selected by the lizards are not forced by the cooler weather but are due to deliberate thermoregulatory changes. The heat-balance model tested against radiotelemetry clearly shows that although land iguanas could maintain their highest body temperatures for a large portion of each day in the cool dry season, instead, they seek cooler microenvironments that result in lower body temperatures. This selection of lower body temperatures allows the lizards to remain homeothermic for the maximum length of time during the day.

The Galapagos land iguanas on Isla Santa Fe extend their period of activity by (1) lowering maintained body temperature thereby beginning activity earlier and extending it later, (2) exploiting warmer morning and late afternoon microenvironments, (3) changing skin surface reflectivity by darkening in morning and late afternoon, a feature which itself extends their foraging time approximately 2 additional hours by speeding warming and slowing cooling.

Distributional Limits

Climate might limit the distribution of animals by influencing activity and potential for acquisition of energy for growth and reproduction. Are there times when we can predict distributional limits of a species using biophysical models? The capacity to grow and reproduce is related to the habitats a species occupies (Schoener and Gorman, 1968; Roughgarden, 1979). Habitats in which the physical environment constrains the acquisition and processing of food should be outside the limits of distribution of a species. Nevertheless, factors other than consumable energy may limit geographic range. Consider the example of the desert iguana.

We have used biophysical models to compute the microclimates available to adult desert iguanas in Palm Springs, California, on the one hand, which is within the distributional boundaries of the species (Muth, 1980), and on the other hand, outside the present geographic limits in Rock Valley, Nevada, and Smyrna, Washington. We obtained the climatic data for selected localities in Nevada and Washington to match Parker and Pianka's study (1975). We obtained the meteorological data (Table 3.1) from the nearest weather stations, and the meteorological parameters for velocity and temperature profiles for those desert vegetation types as described in Mitchell et al. (1975). Soil properties for the Washington site were obtained from the Arid Lands Ecology Reserve at Hanford, Washington, and the same data for Nevada were obtained from Mitchell et al. (1975).

Model simulations suggest that desert iguanas could achieve sufficient body temperature for activity from about May 1 until the end of Sep-

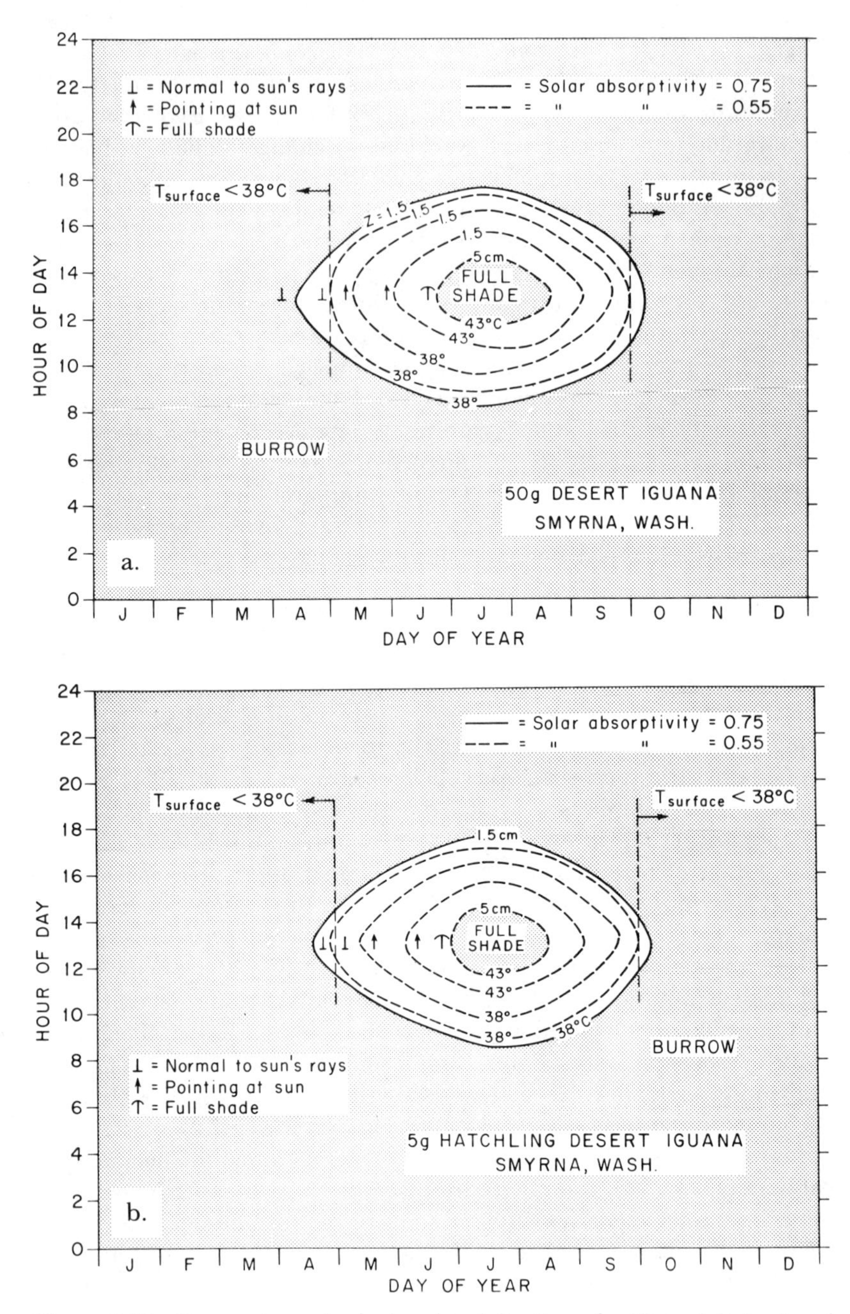

Figure 3.12 Comparison of calculated activity times for 50-g adult (*a, c*) and 5-g hatchling (*b, d*) desert iguanas outside their current geographic range. Effects of body size, color change, sun orientation, and location in the microenvironment are illustrated.

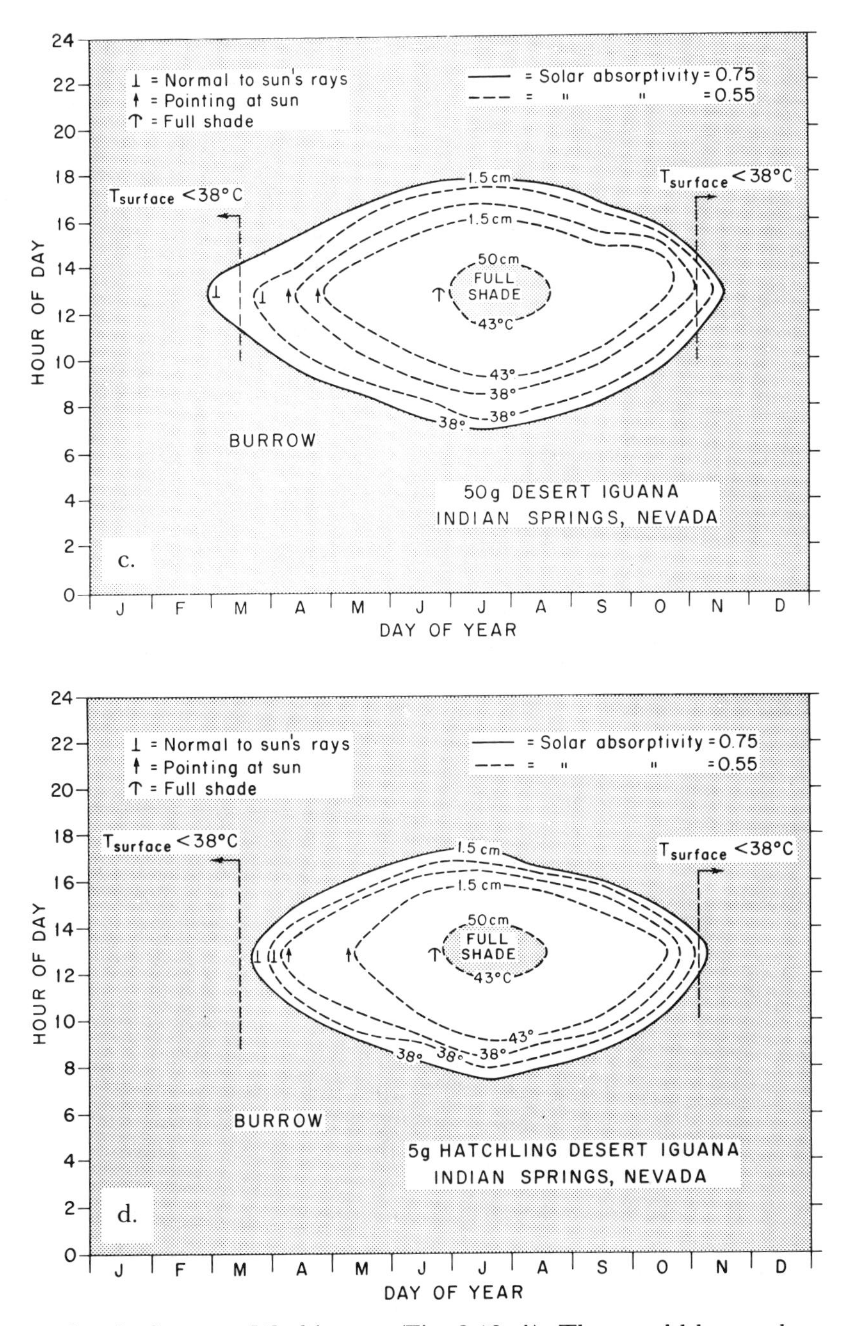

tember in Smyrna, Washington (Fig. 3.12*a,b*). They could have a long period of activity during a midsummer day that seemed more than sufficient to permit growth and the production of eggs during the course of the year for these primarily herbivorous animals. In midsummer the duration of potential activity is nearly as great in Nevada (Fig. 3.12*c,d*) as in California (Fig. 3.4). Calculations of activity for a lizard species not existing in an area cannot be tested directly. However, as an indirect test, we can compute activity patterns for *Uta stansburiana,* a lizard species that does survive and reproduce in these localities in Nevada and Washington (Fig. 3.13). The calculations are for sunlit level ground where there is no topographical shelter from normal ground surface winds. Some individuals with territories containing wind shelters and appropriate substrate orientations could extend the predicted activity times during the cold months of the year. However, we expect that the activity patterns of most *Uta* in these two areas are represented by our calculations. Although we are not aware of published data on *Uta* activity patterns in these areas, our computations are testable by observation and seem to be supported by the personal communications we have received from casual observers. Thus, our calculations for the desert iguana in Nevada and Washington suggest that the spring, summer, and fall climate is not especially limiting for the activity time of adults or hatchling desert iguanas.

However, when we consider physical environmental effects on survival rate of eggs of the desert iguana as defined by their temperature and moisture requirements, we reach an entirely different conclusion about habitat suitability. Figure 3.14 shows the effect of temperature on the length of successful incubation of desert iguana eggs (Muth, 1980). At 25 °C the eggs will incubate for 100 days, but will not hatch. Soil water potentials also affect embryogenesis. At water potentials less than −1500 kPa, survival chances of eggs are dramatically reduced. When this physiological information is combined with data on soil moisture and temperatures at varying depths (Muth, 1980), we discover that open washes in the Palm Springs area are suitable for successful incubation at depths below about 20 cm in midsummer (Fig. 3.15*a*). However, other soil microenvironments in the same area are unsuitable. For example, the soil beneath moisture-consuming shrubs is intolerably dry for the eggs of *Dipsosaurus* (Fig. 3.15*b*). Further north in Rock Valley, Nevada, (Fig. 3.15*c*), the soil is moist enough, but not warm enough to permit successful incubation of eggs in spring; yet in midsummer, the soil is warm enough but not wet enough. In the late fall the soil is wet enough but not warm enough. Further north the computed physical environment becomes even more intolerable for eggs (Fig. 3.16). Soil temperatures at a depth of 30 cm oscillate around 11.5 °C and never drop below freezing. Eggs of desert iguana laid at 30 cm would never develop because soil temperatures are always too cold. Eggs laid at 10 cm could start developing at about the third week of May; however, a substantial part of each day is too cold for development at this depth. Eggs laid at a 5-cm depth would only complete a maximum of 43 percent of their

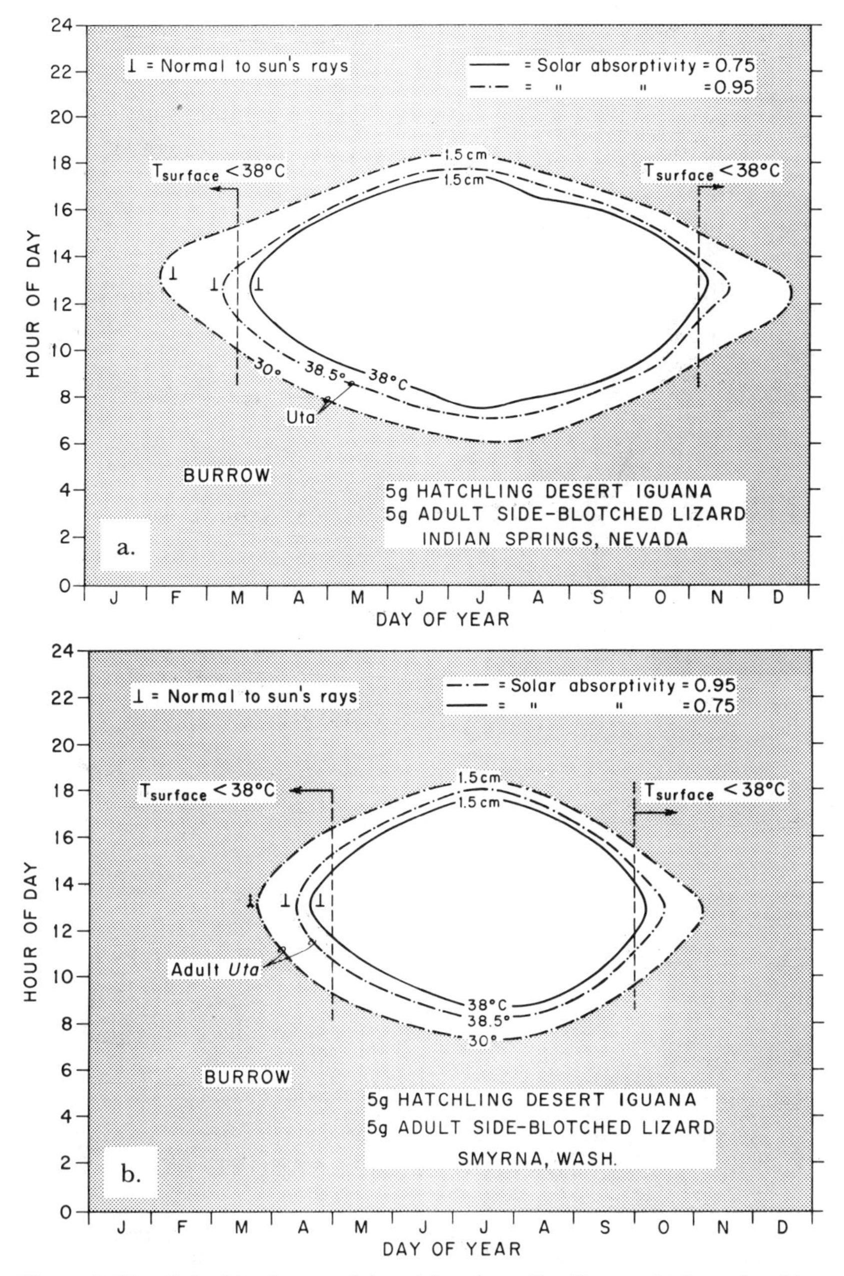

Figure 3.13 Calculated potential activity times for *Uta stansburiana,* the side-blotched lizard that lives in Nevada (*a*) and Washington (*b*) in the area of the sites used for desert iguana calculations. Higher solar absorptivities and lower body-temperature threshold for activity give the side-blotched lizard potential for approximately 2 hours earlier emergence and later retreat than a desert iguana of equal size in midsummer.

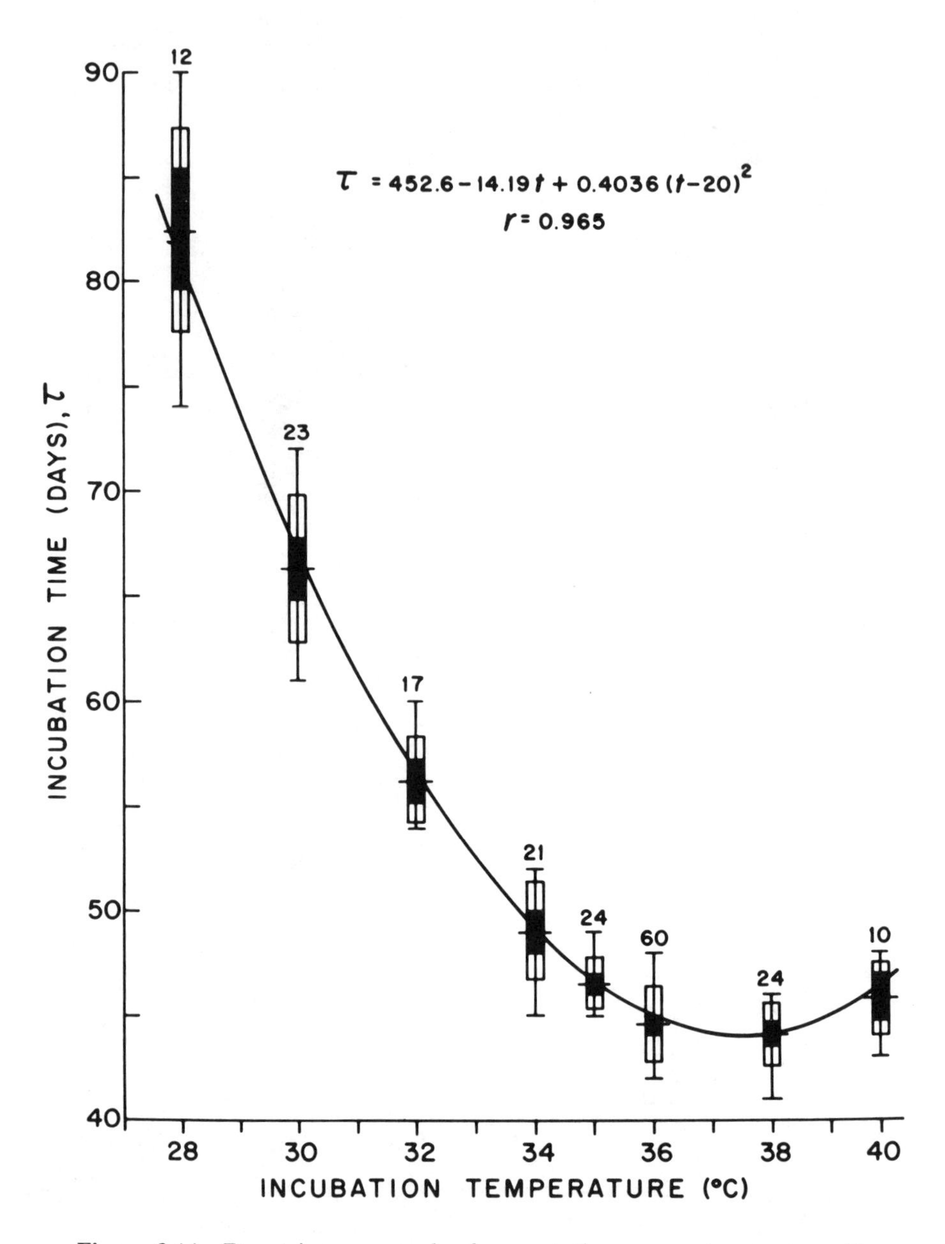

Figure 3.14 Desert iguana egg development, time versus temperature. No hormone injections or other artificial methods were used to induce egg laying. The eggs were laid in sand in laboratory cages and were placed in incubation chambers maintained at constant temperature and defined water potentials (Muth, 1980. Copyright © 1980, the Ecological Society of America).

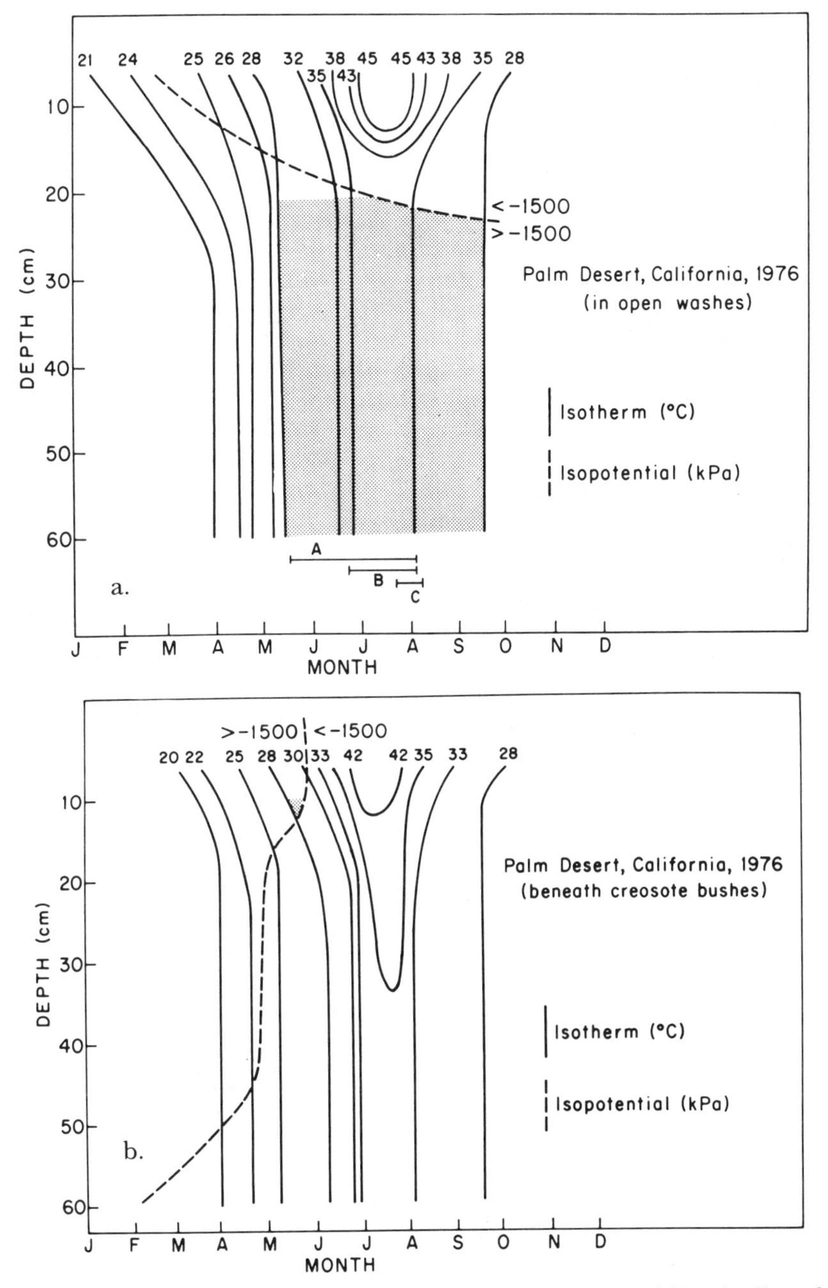

Figure 3.15 Measured soil moistures and temperatures within (*a, b*) and outside (*c*, overleaf) the geographic range of the desert iguana in 1976. Within their range eggs could hatch in open washes (stippled area) during times when sufficient moisture and temperature were available for successful hatching. (Muth, 1980. Copyright © 1980, the Ecological Society of America).

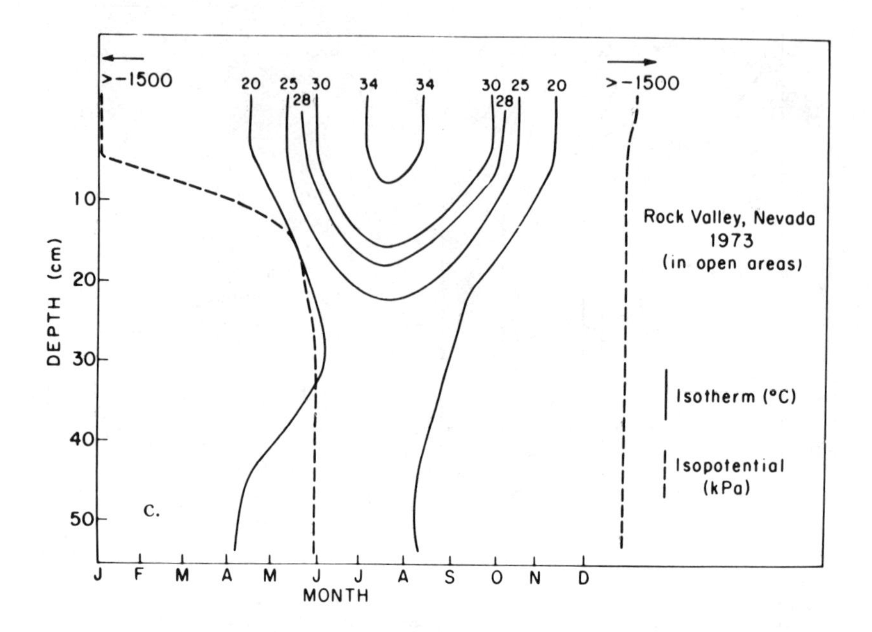

development before they would overheat and die. Hourly rates of development were computed from the data in Figure 3.14 (Keen, 1979; Muth, 1980). It is conceivable that eggs laid at exactly 7.5 cm might be able to hatch if laid as soon as possible and with a precision of depth measure-

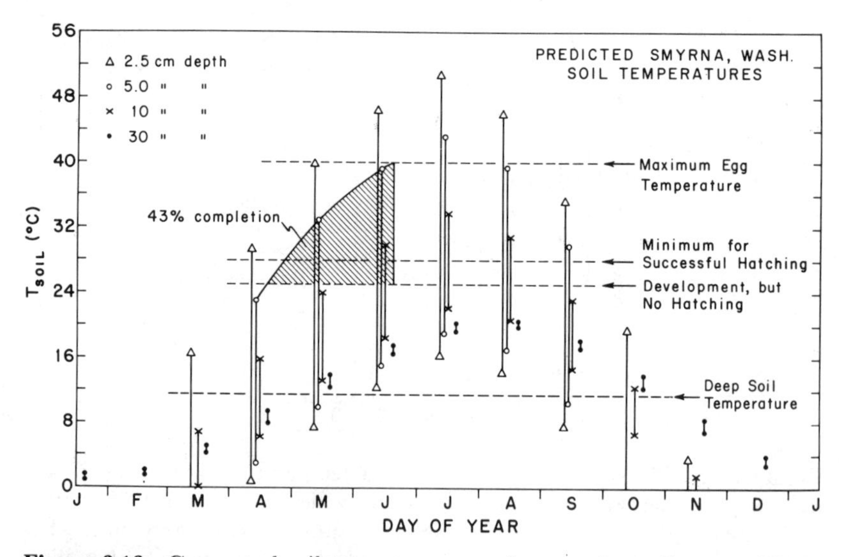

Figure 3.16 Computed soil temperatures at four depths in Smyrna, Washington, outside the range of the desert iguana. Vertical lines indicate computed daily temperature excursions at the specified depths.

ment by the female that was on the order of the dimension of the egg. In addition, there could be no unusual hot or cold spells during the summer. Otherwise, the eggs would either overheat or never complete development. In short, there is absolutely no room for error by the female desert iguana or for climatic inconsistencies.

These calculations suggest that large eggs that have long periods of incubation require deep deposition and high average annual temperatures. Shallow deposition of eggs must correspond with short incubation time and possibly be associated with retention of the eggs within the body of the female for part of the developmental period (Sexton and Claypool, 1978). The interaction between (1) potential nest environments and (2) adaptations such as viviparity, delayed oviposition, and nest-site selection are obviously more important than previously appreciated in population studies of reptiles.

Summary

We have shown that body temperatures can be calculated and that the influence of temperature on activity patterns can be deduced. Such predictions compare favorably with field observations for a variety of animals in a variety of environments. Climate appears to have an important effect on the potential for growth and reproduction through its effects on activity and foraging times and through associated costs of maintenance. Food shortages in the case of the land iguana result in adaptations to extend the length of activity, including expansion of home ranges to include favorable microenvironments, lowering the daily maintained body temperature, and undergoing a very large color change to increase morning and late afternoon solar absorptivity. While such information is useful in understanding the ecology of lizards, such knowledge is not sufficient to understand the distribution of many lizards. We also need to know temperature and moisture requirements of eggs and embryos. It appears that the distribution of the desert iguana is limited by the combinations of soil temperature and moisture that allow successful egg incubation. Although adults and juveniles may survive in a given area, they may not be able to reproduce successfully under the environmental conditions of that area.

4 Lizard Malaria: Parasite-Host Ecology

Jos. J. Schall

PARASITIC SPECIES COMPOSE a significant fraction of the world's fauna (Dogiel, 1966; Price, 1977). This remarkable richness of species is evident from the array of organisms parasitizing lizards. For example, Telford (1970) found an average of 11 endoparasitic species per host species in his survey of lizards from southern California. When bacteria and viruses are included, it is safe to assume that every one of the approximately 3,000 extant species of lizards is infected by several to many species of parasites. Thus, the potential number of parasite-lizard interactions must be vast, perhaps even eclipsing the number of competitive or predator-prey interactions.

A widespread and potentially very important group of lizard parasites are the saurian malarias (*Plasmodium*). Biologists generally are more familiar with malarias of birds and mammals, including the 4 species infecting humans. However, some 120 species attack nonhuman vertebrates, and about half of these are parasites of lizards (Ayala, 1977). Although new species of bird or mammal malaria are rarely discovered, surveys in previously uninvestigated regions frequently uncover new plasmodia of lizards. For example, Telford (1978) found 14 species of saurian malaria from a limited region of Venezuela, 5 of which were undescribed forms. *Plasmodium* appears to be primarily a parasite of lizards and perhaps originally evolved from a gut-dwelling lizard parasite (Manwell, 1955).

Although lizard malaria has been known to biologists for over 70 years (Aragão and Neiva, 1909; Wenyon, 1909), detailed knowledge of these organisms and their effects on hosts is scant. Recently Ayala (1978) catalogued the entire world literature on saurian malaria and found only 156 publications, mostly of a taxonomic nature or dealing with host and locality records. The literature of interest to ecologists consists of only a sparse handful of papers.

In contrast, the severe impact of human malaria on its host is well documented. These protozoans kill millions of humans worldwide each year, disable hundreds of millions of others, and have greatly influenced the evolution and history of our species (Livingstone, 1971; Harrison, 1978). Could lizard malarias have a similar effect on their hosts? Despite a lack of supporting data, the general view among malariologists has been that lizard malaria is a relatively benign parasitic disease (Russell et al., 1963; Telford, 1971). This view is based on the hypothesis that very old parasite-host relationships are likely to be benign compared to more recently established associations, because a fairly stable coevolutionary equilibrium has been reached (Burnet, 1962). Whatever the merits of this hypothesis, the data presented here provide the first reasonably complete inventory of effects of a malarial parasite on a lizard host (the first, in fact, for any lizard parasite) and demonstrate that these effects are substantial.

Since 1978 I have been studying a lizard malaria system in northern California. The host is the western fence lizard (*Sceloporus occidentalis*), and the parasite is *Plasmodium mexicanum.* When I initiated this research I asked the following questions: What are the costs (virulence) to individual lizards resulting from malarial infection? Can parasite-induced hematological pathology be related to physiological and behavioral effects and ultimately to reduction in reproductive success (fitness)? An ultimate goal of the research described here is to determine the variation in virulence among lizard malaria species and to discover the evolutionary origin of such variation.

Study System

Lizard malaria is widely distributed geographically and occurs in most lizard families (Ayala, 1977). Greatest species richness seems to occur in the neotropics where as many as 6 species of *Plasmodium* can be found in a very limited area (Telford, 1977, 1978), but this pattern may simply reflect unequal collecting efforts. Three species have been described in the United States: *P. mexicanum,* a parasite of *Sceloporus* in the western United States, *P. floridense* infecting *Anolis* and *Sceloporus* in the southeastern states, and *P. chiricahuae,* known from *Sceloporus* in Arizona.

Lizard malaria often has a locally patchy distribution; adjacent populations of lizards often differ greatly in parasite prevalance (Ayala, 1970; Jordan and Friend, 1971). For example, *P. mexicanum* has a patchy distribution in northern California. Infected lizards were found at 5 of the 13 sites I surveyed in Alameda, Sonoma, Contra Costa, Marin, Napa, Mendocino, and Sacramento counties. Even at sites where the parasite occurred, foci of infection were often very local, existing just a few hundred meters from similar but parasite-free habitat.

My primary study site was the University of California Hopland Field

Station, an agricultural research facility in southern Mendocino County, 90 miles north of the Golden Gate. The 5,300-acre Hopland Station is primarily foothill oak woodland with chaparral at higher elevations. Summers are warm and dry; precipitation falls primarily during the relatively mild winters. During the past decade approximately 10,000 trees have been felled to open the habitat for sheep grazing. These fallen logs are excellent lizard habitat, and *Sceloporus* are exceptionally abundant in such areas.

I sampled lizards during their seasonal activity period from April through September 1978–1980. Blood was drawn from a toe clip, and a smear made for staining and microscopic examination (Ayala and Spain, 1976). Approximately 2,500 lizards were sampled over the 3-year period.

Fence lizards at Hopland were infected by 4 species of blood parasites. Two were common: malaria and another protozoan, *Schellackia occidentalis.* Only sporozoites of *Schellackia* occur in the erythrocytes; other stages are in the intestinal epithelium (Bonorris and Ball, 1955). Lizards were rarely infected by a haemogregarine (only gametocytes are in blood cells) or a free-swimming trypanosome. As I wished to study the effects of malaria on lizard hosts, I have not included data for lizards infected with other blood parasites in the comparative analyses.

Plasmodium has a complex life cycle involving both a vertebrate and insect host. The invertebrate host of lizard malaria is unknown, although Ayala and Lee (1970) found sporogonic development of *P. mexicanum* in the psychodid fly, *Lutzomyia vexatrix.* This fly spends the day in burrows of ground squirrels at the Hopland Station, emerging at night to take blood meals from lizards and other ectothermal vertebrates (Ayala, 1973).

Whatever the insect host may be, *Plasmodium* sporozoites are passed onto a lizard and parasites eventually appear in the blood. There they follow a course of infection similar to that of bird or mammal malarias. In *P. mexicanum* a uninucleated small merozoite enters an erythrocyte and begins to feed. This feeding stage, or trophozoite, eventually undergoes nuclear division to form a multinucleated schizont. In mature schizonts (segmenters) of *P. mexicanum,* the 8 to 20 nuclei align themselves along the periphery of the cell mass just prior to cellular division. The red blood cell ruptures, freeing the daughter merozoites which reinfect other cells.

Rather than proceed through this schizogonous cycle, some merozoites mature into gametocytes ("females" are macrogametocytes, and "males" are microgametocytes). Gametocyte sex ratio in *P. mexicanum* is not constant but varies among infections (range: 20 to 70 percent males). Gametocytes are carried into the insect host when it takes a blood meal. Fertilization takes place in the insect's stomach and the resulting sexual cycle eventually produces new sporozoites.

Early in the spring at the Hopland Station, *Plasmodium* blood infections in lizards are of two kinds. Some consist only of trophozoites and schizonts (presumed "new" infections), and others consist of all stages, including many old gametocytes (large heavily vacuolated cells). These infections have probably persisted over the winter in the lizards. Active schizogony in blood cells occurs primarily during the early part of the season; by late August and September most infections consist mainly of gametocytes. The presumed vector, *Lutzomyia,* is not active at Hopland during April and May (personal observation of traps and communication from J. Anderson). Therefore, transmission probably takes place in late summer, and blood infections do not become patent until the next spring.

Parasite Prevalence

Parasite prevalence (percent of lizards infected) differs considerably between sexes (Table 4.1 and Fig. 4.1). A consistently greater proportion of males is infected at any time. Such higher prevalence of parasitic infection among male hosts is common in a wide variety of vertebrate parasites (Nawalinski et al., 1978). In this case, perhaps female lizards are killed by the parasite more often than are males. However, there appears to be no such differential mortality between sexes; sex ratio of adult lizards does not differ significantly from 1:1 (739 males versus 707

Table 4.1 Prevalence (percent of lizards infected) of *Plasmodium mexicanum* and *Schellackia occidentalis* in the western fence lizard at the Hopland Field Station, Mendocino County, California, 1978–1980.

SVL in mm	1978	1979	1980
		Plasmodium	
60–64	36 (87)	26 (50)	15 (26)
	13 (64)	14 (36)	5 (21)
65–70	37 (170)	27 (187)	26 (100)
	20 (143)	22 (159)	20 (46)
>70	40 (25)	33 (57)	35 (37)
	33 (63)	19 (111)	22 (64)
Total	37 (282)	28 (294)	26 (163)
	21 (270)	20 (306)	18 (131)
Grand total	29 (552)	24 (600)	23 (294)
		Schellackia	
Total	6 (552)	13 (600)	16 (294)
(Rainfall)	(126.2 cm)	(66.5 cm)	(108.1 cm)

NOTE: Data for males above those for females. Sample size in parentheses. Rainfall periods, October to May preceding each summer.

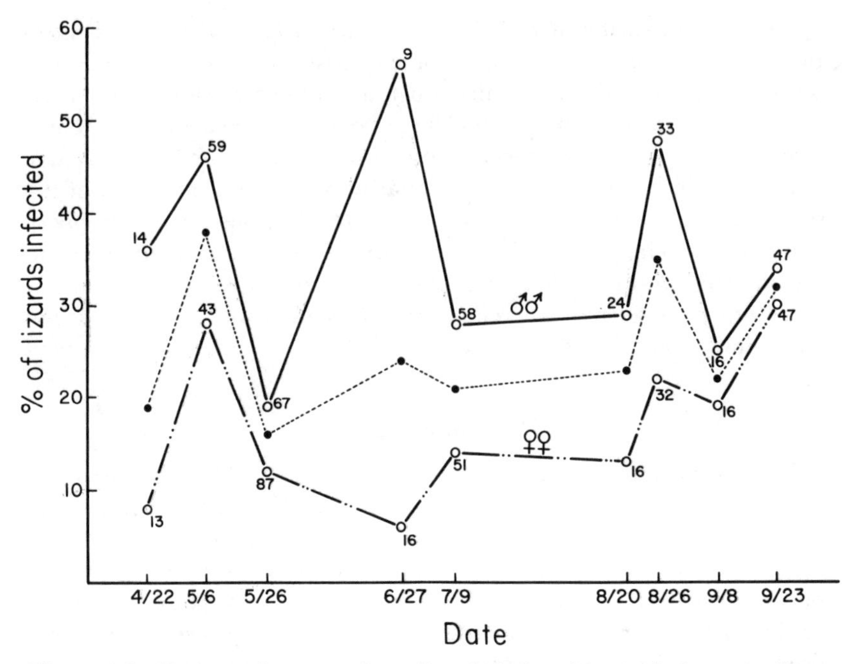

Figure 4.1 Percent of western fence lizards infected by malaria at the Hopland Field Station for 8 sampling periods (first day indicated) during 1978. Males indicated by upper line, females by lower, and overall percent by middle dashed line. Sample sizes indicated next to points.

females, $\chi^2 = 0.354$, $P > 0.05$), nor does the sex ratio of juveniles hatched from laboratory-maintained eggs (274 males versus 270 females, $\chi^2 = 0.015$, $P > 0.05$).

Although evidence is weak, the sexual trend in parasite prevalence seems to occur even in subadults (Fig. 4.2). Small juveniles (< 50 mm snout-vent length, SVL) are very rarely infected; however, larger (50–59 mm) subadult (prereproductive) males are more often infected than are females of this size. Sample size for the 50–59 mm group is small (only 89) and the difference between males and females is not statistically significant ($P = 0.26$). Other hypotheses explaining the sexual difference in parasite prevalence might be proposed, but any hypothesis should confront the possible difference in prevalence even among small, prereproductive animals. If the sexual difference in prevalence were apparent only in adult lizards then any one of many physiological, behavioral, or ecological differences between sexes could be implicated as potential causative factors. However, if the difference in parasite prevalence exists for juvenile lizards, there must be some important difference between sexes even at this early age.

The higher parasite prevalence in larger lizards suggests that an in-

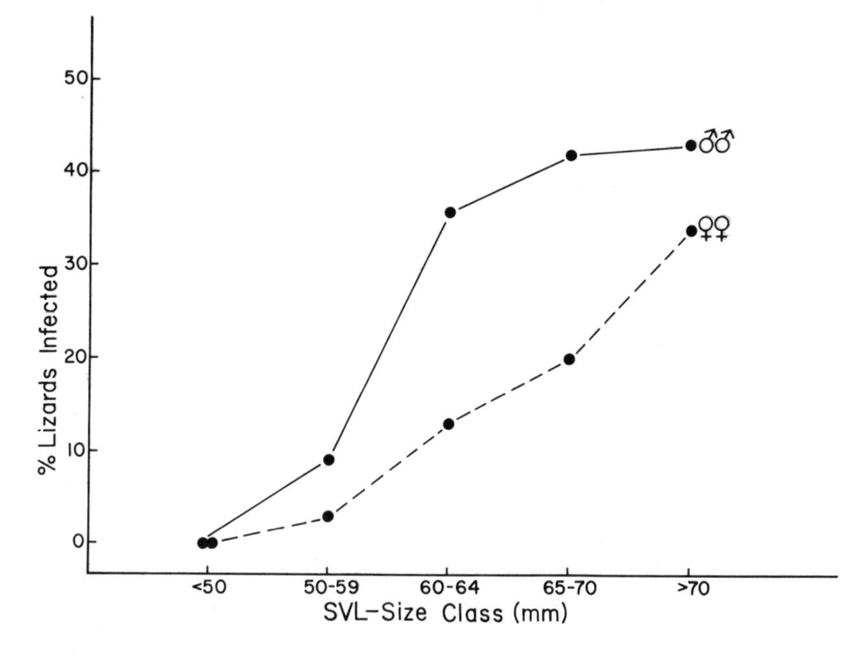

Figure 4.2 Percent of male and female western fence lizards infected with malaria, plotted against snout to vent length (SVL). Data are from 1978 sample.

fected lizard may maintain at least a low-level infection for years and perhaps for the duration of its life. Malarious animals kept in large laboratory pens never lost their infection, but parasitemias often dropped to very low levels. In Panama *Anolis limifrons,* a small, short-lived animal exhibits a similar pattern (S. Guerrero, personal communication).

Malaria prevalence did not vary significantly among years ($\chi^2 = 5.88$, $P > 0.05$) despite changes in environmental conditions resulting from variability in winter precipitation (Table 4.1). The winter prior to the 1978 season was wet and followed a long and severe drought. Consequently, vegetation growth in spring 1978 was exceptionally luxuriant compared to that of the next two seasons. This short-term constancy in parasite prevalence in the face of environmental perturbation suggests that lizard malaria may be a stable system. Ten years prior to my study, Ayala (1970) recorded a similar level of lizard malaria prevalence at Hopland. However, in Georgia *P. floridense* in *Sceloporus undulatus* remained at a fairly constant low prevalence over a 13-year period but exhibited long-term but dramatic changes (50 percent to 10 percent prevalence) in another host, *Anolis carolinensis* (Jordan and Friend, 1971). In comparison, *Schellackia* at my study site significantly increased in prevalence from 1978 to 1979 and 1980 (Table 4.1, $\chi^2 = 23.16$, $P < 0.001$).

The incidence of multiple infections of *Plasmodium* and *Schellackia* presents an equivocal pattern. In 1978 the two parasites occurred together in random frequency (8 multiple infections observed, 9.6 expected; χ^2 test, $P > 0.05$), but in 1979 and 1980 multiple infections were rare, suggesting some sort of interaction between parasite species (8 observed, 18.7 expected in 1979; 2 observed, 10.8 expected in 1980; χ^2 tests, P's < 0.05). The nature of such a possible interaction is intriguing, especially as *Schellackia* does not reproduce in blood cells and *Plasmodium* does not infect intestinal tissues.

Physiological Consequences of Infection

Larger lizards generally have a greater probability of being infected (Table 4.1). To eliminate bias I carefully matched body sizes of experimental groups in the following comparisons. Also, only animals with natural infections were used.

Lizards usually respond to malarial infection by producing copious numbers of polychromatic erythrocytes (Ayala, 1970; Scorza, 1971; Telford, 1972; Ayala and Spain, 1976; Guerrero, Rodriguez, and Ayala, 1977). These immature red blood cells (iRBC) are easily distinguished, as their cytoplasm has an affinity for the basic or blue portion of Giemsa stain, probably because they have less hemoglobin and more RNA in their cytoplasm (Diggs, Sturm, and Bell, 1978).

Infected *Sceloporus* have significantly more iRBC than do noninfected animals (U-test, $P \ll 0.001$). Typically, up to 2 percent of erythrocytes are immature in noninfected animals whereas the range is much greater (1 to 30 percent) for infected lizards (Fig. 4.3). Time lags may be responsible for the scatter; that is, infected lizards may have a high proportion of iRBC for some time after parasitemia has reached a "crisis" stage and declined. Increase in iRBC is probably a result of the lizard's hemopoietic response, which is mobilized to replace destroyed red blood cells (RBC). As parasitemia rises, the lizard may be forced to place immature erythrocytes into circulation to maintain RBC abundance. However, iRBC levels sometimes rise early in an infection, when parasitemia is low. *Plasmodium mexicanum* is primarily a parasite of mature erythrocytes and when *P. mexicanum* does infect iRBC, the parasites appear stunted. Thus, production of iRBC may function as an antiparasite tactic, weakly analogous to sickle cell anemia or the Duffy negative antigen defense (Friedman and Trager, 1981). This possibility deserves careful further investigation.

If iRBC contain less hemoglobin per cell than do mature erythrocytes, there should be measurable physiological consequences to infected lizards. The details of measurement of these effects are reported elsewhere (Schall, Bennett, and Putnam, 1982); here I will summarize the results. Compared to the noninfected lizards, blood of infected animals contains significantly less hemoglobin (5.5 versus 7.3 g/ 100 ml, U-test,

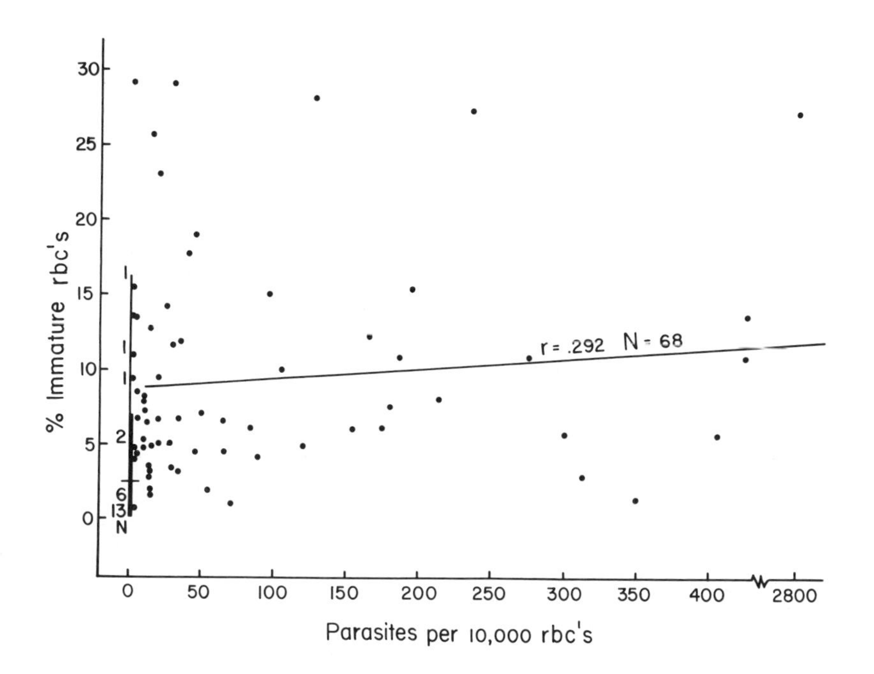

Figure 4.3 Malarial parasitemia plotted against percent of immature red blood cells (iRBC) for 68 infected *Sceloporus occidentalis;* also plotted is distribution for 24 noninfected lizards ($\bar{x}$ = horizontal line, S.D. = dark vertical bar, range = light vertical bar). Three lizards in "noninfected" class with high percent of iRBC (> 5) may be a result of false negatives.

$P < 0.001$). This does not appear to be a result of a decrease in numbers of RBC as neither hematocrits nor direct RBC counts differ significantly (U-test, $P > 0.05$). However, there is a negative correlation between percent iRBC and hemoglobin concentration ($r = -0.508$, $P < 0.01$, N = 49), confirming that decrease in hemoglobin concentration is due to the increased number of iRBC in circulation.

Except for lizards with overwhelming infections, malarious lizards in the laboratory typically appear healthy and behave "normally." This has led some observers to conclude that lizard malaria is a benign parasite. However, a reduction of blood hemoglobin concentration by about 25 percent should reduce the lizard's ability to deliver oxygen to body tissues and certainly should have important behavioral and ecological consequences.

Oxygen usage by resting lizards is very low, just a fraction of their maximal consumption (Bennett, 1978). As might be expected, resting O_2 consumption does not differ between infected and noninfected lizards. However, oxygen consumption during maximal activity is significantly lower for infected lizards (1.3 cc/g • h for infected versus 1.53 cc/g • h for noninfected, U-test, $P < 0.05$). The increment in O_2 consumption from resting to maximal activity, an indication of ability to support activity aerobically (aerobic scope), also differs between the two groups (0.71 cc/g•h for infected versus 1.00 cc/g•h for noninfected, U-test, $P < 0.01$). Both increments in oxygen consumption and maximal oxygen consumption are positively correlated with blood hemoglobin concentration ($r = 0.72$, $P < 0.01$, N = 28, and $r = 0.68$, $P < 0.01$, N = 28, respectively). This suggests that reduction in aerobic capacities of infected lizards is a result primarily of the deficit in hemoglobin levels, rather than of other pathological effects of infection.

A likely important consequence of reduced oxygen consumption during exertion would be a decrease in aerobically sustained locomotory ability. Very short bursts, or sprints, of activity are supported in lizards by anaerobic mechanisms (Bennett, 1978). Burst running speed, measured electronically in a 2-m track, does not differ between infected and noninfected lizards ($\bar{X}$ = 1.28 versus 1.44 m/s, U-test, $P \approx 0.10$). Running stamina, though, is supported in large part by aerobic means and is significantly reduced for infected lizards. For example, infected lizards, when forced to run continuously for 30 s, covered an average of about 17 m, whereas noninfected animals ran about 21.3 m (U-test, $P < 0.01$). A similar reduction in stamina was observed for lizards running a full 2 min (27 versus 32 m; U-test, $P < 0.05$).

Costs to Reproductive Success

Reproductive success, or Darwinian fitness, is notoriously difficult to measure, especially in males. Ideally, we should count lifetime number of offspring produced by an individual and determine survival of those offspring. In practice I used a number of measures, which when taken together provide a reasonable index of reproductive success. Results for males are equivocal but data for females clearly demonstrate that malaria infection reduces female fitness.

Testis Size. During 1978 I collected 8 samples of males (N = 266) and weighed the testes of each animal (Fig. 4.4). Testis mass declined during the reproductive season and began to increase in late August after reproduction ceased. The pattern for infected and noninfected males was very similar until late summer when noninfected males had significantly larger testes (U-test, $P < 0.001$). These results are confirmed by samples collected exactly 1 year later; again, infected animals had smaller testes (N = 22 infected, 41 noninfected, U-test, $P < 0.002$). In both years the average reduction in testis size was about 37 percent. I know of no evidence that male lizards with larger testes experience higher reproductive success, but larger testes may produce more gametes and hormones, and both are obviously important in male reproduction.

Stored Fat. Over 8 sampling periods in 1978, I weighed paired inguinal fat bodies of 530 *Sceloporus* (Fig. 4.5). Fat bodies remained small during the reproductive season and increased after reproduction ceased, a pat-

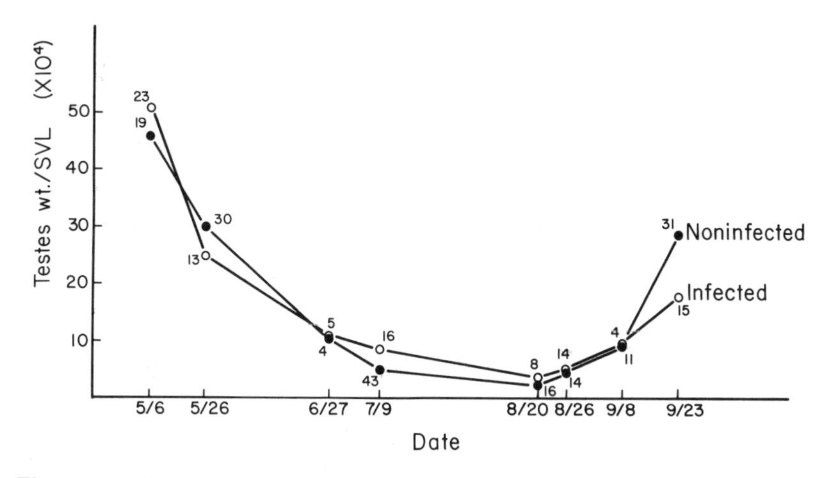

Figure 4.4 Mean testis size of western fence lizards that were infected (open points) or noninfected (closed points) by malaria for 8 sampling periods at the Hopland Field Station during 1978. Because body mass varies, depending on contents of the digestive tract, mass of paired testes is divided by snout-vent length for each animal to correct for body size. Sample sizes are indicated next to points.

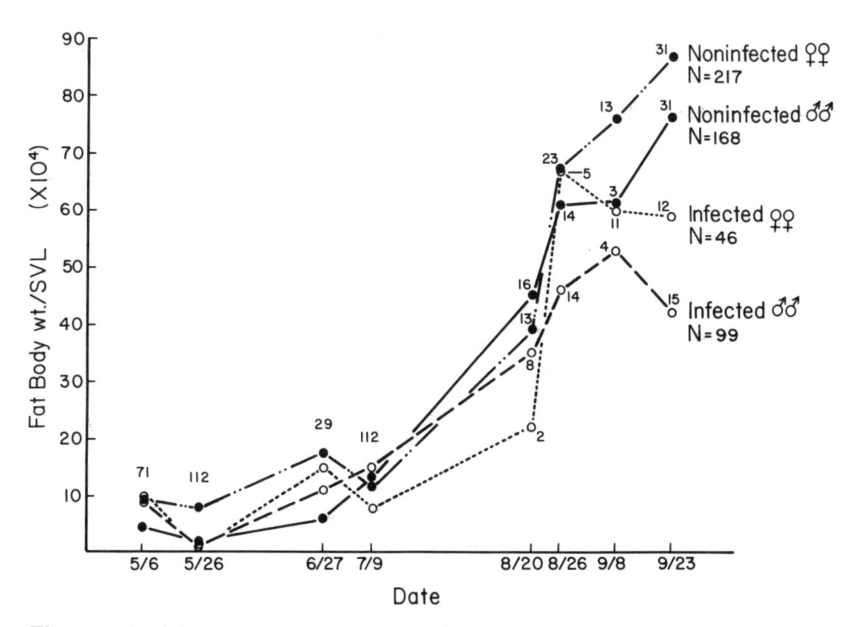

Figure 4.5 Mean mass of both fat bodies corrected for body size for western fence lizards for 8 sampling periods during 1978. Sample sizes indicated next to points.

tern typical of temperate zone lizards (Schall, 1978). By summer's end, females store more fat than males, and infected animals store less fat than do noninfected animals of the same sex. One year later I collected another sample. Once again, for both sexes, infected lizards store less fat (Table 4.2).

A comparison between years produces an unexpected result (Table 4.2). Both male and female noninfected lizards stored very similar amounts of fat by season's end in both years. However, in 1979, infected lizards of both sexes stored more fat than did infected animals the previous year. This suggests that resources may have been more abundant in 1979. Although I have no quantitative measures, insect density (especially grasshoppers) appeared much greater in 1979, a result perhaps of the lush vegetative growth the previous year. There may be an optimal amount of fat to be stored by these lizards, and they simply stop adding fat tissue once this level is reached. Therefore, in a very productive year, fat stored by noninfected lizards is about the same as during an average year. Infected lizards, though, have greater difficulty in storing fat, so the amount of fat they store varies to a greater degree on resource availability.

Does stored fat contribute directly to reproductive success? Fat bodies of males are assumed to be a source of energy during winter brumation (Gaffney and Fitzpatrick, 1973) and for establishing territories in the spring. Fat bodies of females are also used to produce the first clutch of eggs in spring (Hahn and Tinkle, 1964). Caloric content of fat bodies and of *S. occidentalis* eggs are known. Assuming that almost all calories in fat bodies are used in egg production, the deficit in fat stored by infected animals has been used to calculate a predicted decrease in clutch size (Schall, 1982). The decrement in fat stored by infected compared to noninfected lizards is equal to the calories in 1.42 eggs for 1978 and 1.00

Table 4.2 Mass of both inguinal fat bodies corrected for body size for western fence lizards infected and not infected by malaria, late September sampling periods.

	1978		1979	
Males				
Noninfected	0.0076 (0.0030, 31)	$P < 0.001$	0.0076 (0.0027, 42)	$P < 0.05$
Infected	0.0042 (0.0026, 15)		0.0059 (0.0038, 22)	
Females				
Noninfected	0.0087 (0.0031, 31)	$P < 0.01$	0.0092 (0.0029, 44)	$P < 0.05$
Infected	0.0059 (0.0032, 12)		0.0076 (0.0036, 21)	

NOTE: Mean fat-body mass in g/SVL; SD and sample size in parentheses. *P* for U-test given also.

eggs for 1979. Therefore, an infected lizard on the average should produce clutches with 1 to 2 fewer eggs than noninfected females.

Clutch Size. The reproductive season for *S. occidentalis* at Hopland extends from emergence in mid-April, when most adult females have enlarged yolked ovarian follicles, to about mid-July when at least some large adults produce their second clutch for the season (Schall, unpublished observations). The proportion of infected and noninfected females that were gravid at any time did not differ (χ^2 test, $P > 0.05$). Because fat bodies are utilized in only the first clutch, the following analysis is restricted to clutches from females collected in May and June.

Results for 1978 and 1979 are presented in Figures 4.6 and 4.7 respectively. Body size and clutch size are positively correlated so I compared samples by an analysis of covariance. For both years residual variance and regression slopes did not differ between infected and noninfected animals ($P > 0.05$), so I was able to compare regression elevations. Infected lizards produced significantly smaller clutches both years ($P < 0.01$). The difference was approximately 1 to 2 eggs as predicted above from the analysis of stored fat.

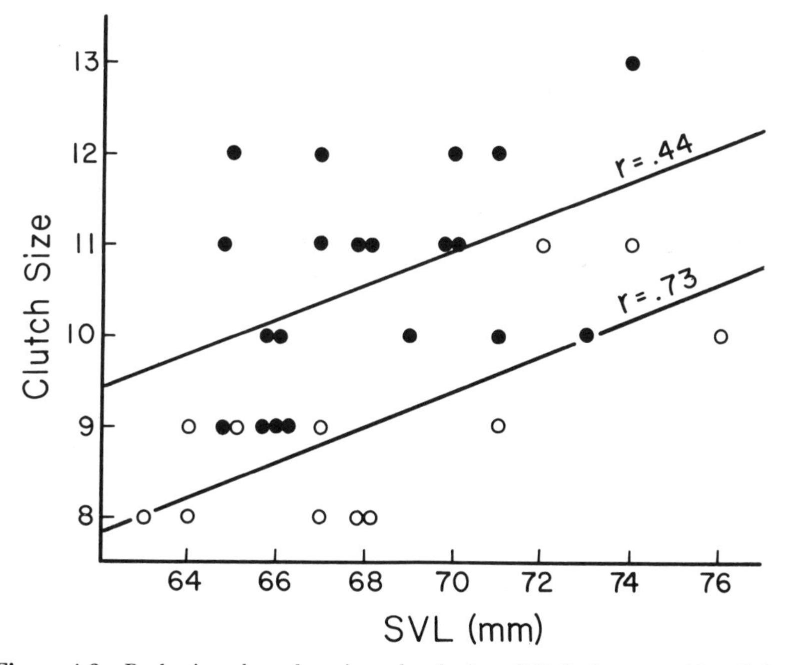

Figure 4.6 Body size plotted against clutch size of 12 *Sceloporus occidentalis* infected with malaria (open points, lower regression) and 20 noninfected (closed points, upper regression), collected at Hopland Field Station during spring 1978. Clutch size determined from counts of large yolked ovarian follicles, oviducal shelled eggs, or eggs laid in laboratory.

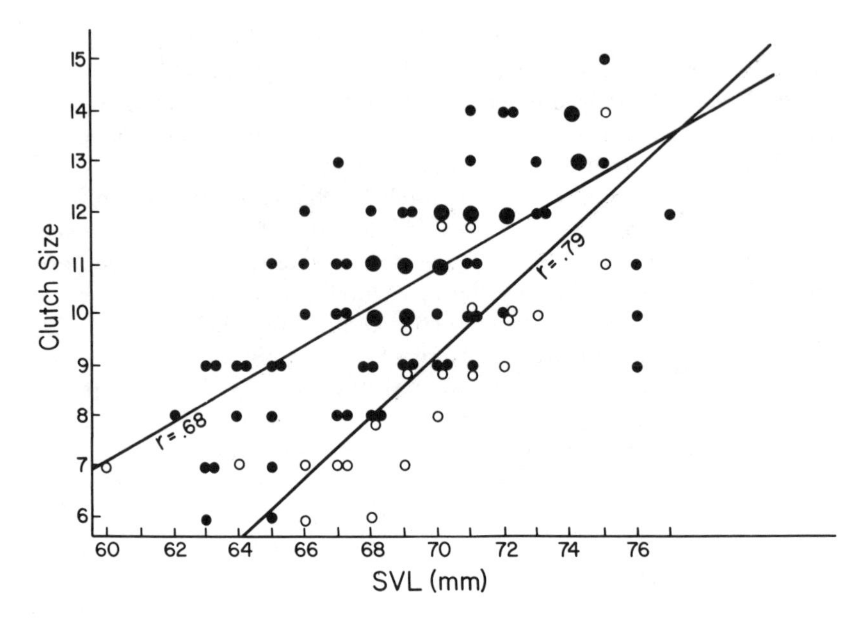

Figure 4.7 Data similar to those in Figure 4.6 except gathered during spring 1979. Large points indicate two or more overlapping data. Sample size was 23 infected and 99 noninfected lizards.

Hatching Success. Malarial infection could lower a female lizard's reproductive success by indirectly reducing survival of her offspring. For example, infected females may be unable sufficiently to provision eggs with a vital nutrient. I therefore measured a variety of indicators of hatching success and hatchling health. Hundreds of adult females were brought into the laboratory, and 74 clutches were gathered and incubated (Table 4.3).

Table 4.3 Various measures related to hatching success of western fence lizards.

	Infected	Noninfected	Test
Mean egg mass	0.346 g (0.063,14)	0.365 g (0.071,64)	U-test, $P > 0.05$
Clutch mass/female body mass	0.269 (0.052,15)	0.271 (0.040,62)	U-test, $P > 0.05$
Days to hatch	71.1 (3.76,8)	71.7 (4.84,50)	U-test, $P > 0.05$
Percent of clutches hatching	80 (10)	80 (64)	χ^2, $P > 0.05$
Percent of eggs hatching	93 (67)	85 (482)	χ^2, $P > 0.05$

NOTE: SD, when appropriate, and sample sizes in parentheses.

Hatching success was measured three ways: time to hatch when maintained at room temperature (21 to 28 ° C), percent of clutches producing hatchlings, and percent of eggs hatching from clutches that produced some hatchlings. Clutches produced by infected or noninfected lizards did not differ in any of these measures.

Mean egg mass and mean hatchling size in *S. occidentalis* are correlated ($r = 0.535$, $P < 0.01$, N = 51 clutches) suggesting that egg mass is a reasonable indicator of potential hatchling survival. To obtain a sufficient sample I weighed only oviducal eggs extracted from dissected lizards. Mean egg mass for eggs from infected and noninfected lizards did not differ and neither did overall investment by females in eggs (clutch mass/ female body mass).

These results demonstrate that, although infection with malaria results in significantly smaller clutches of eggs, eggs from malarious females are indistinguishable from those produced by noninfected lizards.

Growth Rate. Growth rate can affect a lizard's lifetime production of offspring in several ways. For example, clutch size and body size are positively correlated. Also, larger males may occupy larger or higher quality territories. I used a mark-recapture technique to study growth rate (Fig. 4.8). Growth rate declines as lizards mature, so each point in Figure 4.8 actually represents an average growth rate over the time between measurements (150 days). Also, some lizards scored on Figure 4.8 as "in-

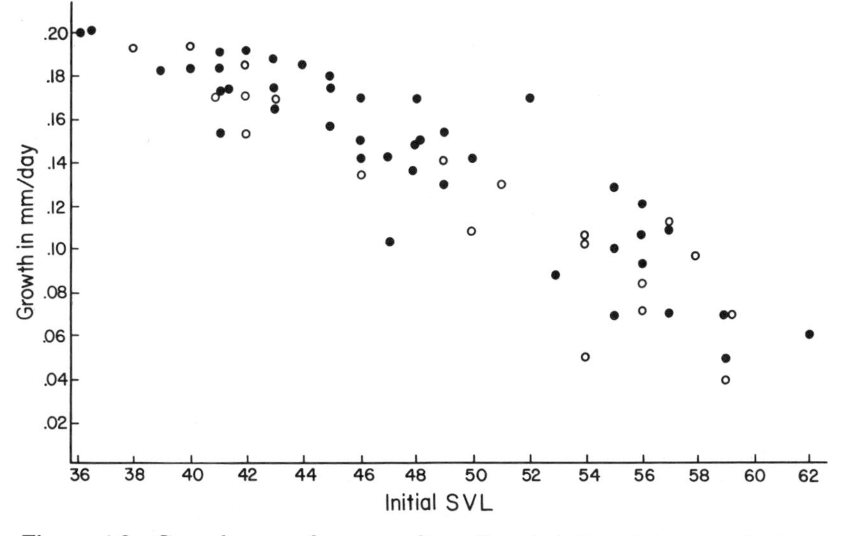

Figure 4.8 Growth rate of western fence lizards infected (open points) or noninfected (closed points) with malaria. Growth rate determined by marking lizards in spring 1979 (≈ 700 animals marked) and recapturing them about 150 days later (≈ 60 recaptured).

fected" were infected only at time of second capture (including most of lizards < 50-mm SVL at initial capture).

Infected lizards experienced only very slightly reduced growth rates, about 4 percent (analysis of covariance; residual variances and slopes, $P > 0.05$; elevations, $0.01 < P < 0.05$). Why should somatic investment (growth rate) be maintained at nearly normal levels by infected lizards when production of offspring is reduced? After all, offspring are the currency of natural selection. *Sceloporus occidentalis* may live as long as seven years (Ruth, 1977), so infected animals surviving peak parasitemia could maintain low-level infection for years. Perhaps infected lizards can maximize lifetime reproductive output by maintaining fairly rapid growth and reaching large size quickly rather than by maintaining a larger clutch size.

Parasite-induced Mortality. I have no estimates of mortality caused by the parasite in free-ranging lizards. However, infected animals maintained in laboratory cages were more likely to die than were noninfected lizards. Fence lizards were maintained for one week to several months in laboratory cages. There was no bias between infected and noninfected lizards in period of time animals were kept in captivity. Of 308 males, 15.1 percent of infected and 2.5 percent of noninfected animals died. Of 300 females, 28.6 percent of infected and 11.1 percent of noninfected animals died. For both sexes, χ^2 tests were significant (P's < 0.01). These results would suggest that, at least under stressful conditions, malarious lizards in the wild could suffer increased mortality.

Discussion

Despite their abundance, both in numbers of individuals and richness of species, parasites have elicited comparatively little interest among ecologists. Most general ecology texts devote only a few lines to several pages to the subject. A notable exception is a text used in the Soviet Union (Naumov, 1972) where parasite ecology is a very active discipline. A perusal of the literature on lizard ecology (including this volume) demonstrates that the parasite-lizard interaction is very rarely investigated. For example, although there are tantalizing hints that parasites can critically affect host populations (Warner, 1968; Barbehenn, 1969; Cornell, 1974; Anderson and May, 1979; Price, 1980), the effects of parasites on lizard populations has never been evaluated. An obvious starting point is to assess the impact of a parasite infection on individual lizards. Gross pathology of some lizard parasites is known (Telford, 1971), but overall impact on host individuals of any lizard parasite has not been reported prior to this study.

The data presented here demonstrate that lizard malaria can have substantial effects on its host. Malarial infection initiates a tractable sequence of hematological, physiological, behavioral, and reproductive consequences (summarized in Table 4.4). Infection results in release of elevated numbers of immature erythrocytes into peripheral circulation. These immature erythrocytes contain less hemoglobin than mature RBC, so blood hemoglobin concentration is reduced. This results in a decrease in oxygen transport capacity of the blood, measured as oxygen consumption during maximal activity. An important consequence of these physiological costs is a decline in aerobically supported locomotory ability (running stamina). Recovery from strenuous activity has not yet been compared for infected and noninfected lizards, but the trend described above suggests that infected lizards would require more time to recover from even relatively short bursts of activity. If so, malarious lizards would be less efficient in gathering food resources and defending territories.

Data on fat-body size support this possibility. Infected lizards store considerably less fat during late summer than do noninfected animals. The caloric content of the deficit in stored fat is approximately equal to the caloric content of 1 to 2 lizard eggs, which corresponds to the observed reduction in clutch size. Thus, reduction in reproductive success of infected female *Sceloporus* can be traced ultimately to the hematological effects of malarial infection.

These effects on individual lizards suggest that there might be populational-level consequences resulting from presence of the parasite. For example, could the malarial parasite limit the population size of *Scelo-*

Table 4.4 Summary of some costs to the vertebrate host (*Sceloporus occidentalis*) induced by a malarial parasite (*Plasmodium mexicanum*).

	Infected versus noninfected (percent decrease)
Hemoglobin concentration	24
Burst running speed	11
Active $\dot{V}O_2$	15
Increment $\dot{V}O_2$	29
Running stamina (30 s)	20
Running stamina (2 min)	17
Fat stored (females, 1978 season)	32
Fat stored (females, 1979 season)	17
Fat stored (males, 1978 season)	45
Fat stored (males, 1979 season)	22
Clutch size (1978 season)	16
Clutch size (1979 season)	13
Testis size (1978 season)	38
Testis size (1979 season)	37
Growth rate	4
Mortality	12–17 (increase)

porus at the study area? This might be possible as malarial infection increases mortality and decreases reproductive output of infected lizards. However, the density of fence lizards at Hopland is the greatest I have seen at any site. The effect of malaria on population size of lizards is probably very minor; other factors, such as availability of home sites, more likely play a more important role.

Nonetheless, ecologists comparing demographic or physiological characteristics of lizard populations should be aware that the patchy distribution of malaria (and perhaps other parasites) can add an unknown but important source of variation to their data. For example, clutch size might be observed to vary between two lizard populations, one infected with *Plasmodium,* and the other not. Without knowledge of the parasite, any explanation proposed to account for this life-history difference might be clever but spurious. Similarly, a physiologist might detect differences in blood hemoglobin concentration and oxygen consumption in *S. occidentalis* populations from high- and low-elevation habitats. These differences may appear to be a result of adaptation by the lizards to differing oxygen partial pressures or environmental temperatures. However, because lizard malaria does not seem to occur at higher elevations (presumably because of absence of the insect vector), observed physiological differences in the lizards may be simply a result of presence or absence of the parasite.

One last important populational effect of *P. mexicanum* on fence lizards must be considered: the coevolution of parasite and host. As demonstrated here, malarious fence lizards experience reduced fitness. Why has *S. occidentalis* not evolved appropriate mechanisms to reduce the impact of malarial infection? That is, what determines parasite virulence? A knowledge of variability in virulence of lizard malarias would cast light on this important problem. Preliminary evidence suggests virulence of lizard malaria does vary considerably, depending on which *Plasmodium* and host species are involved (Ayala, 1977). For example, although *P. mexicanum* has a substantial effect on its host, *P. tropiduri* and *P. balli,* which infect *Anolis limifrons* in Panama do not seem to affect the host's survival, weight, growth rate, or assimilation efficiencies (S. Guerrero, personal communication). Lizard plasmodia also vary greatly in number of merozoites produced by mature segmenters (Ayala, 1977). In some species, mother cells yield only 4 progeny, whereas in others more than 90 merozoites are produced. This hints that some species of lizard malaria may reproduce very rapidly, perhaps racing ahead of the host's immune response. These species might be exceptionally virulent, compared to more slowly reproducing forms. Lizard malaria appears to be an ideal model for studies on the evolution of parasite virulence.

PART II
BEHAVORIAL ECOLOGY

RESEARCH ON LIZARD behavioral ecology was initiated by G. K. Noble nearly half a century ago. His research was characterized by a great breadth of subject matter—taxonomy, anatomy, evolution, behavior, chemoreception, endocrinology, paleontology—and by an astoundingly modern and experimental approach. During the 1930s his behavioral work included an extensive comparative analysis of sexual behavior in *Ameiva, Cnemidophorus, Anolis, Leiocephalus,* and other lizards (Noble and Bradley, 1933); a study of the sensory mechanisms used by snakes (Noble and Clausen, 1936); research on the anatomy and function of heat receptors in snakes (Noble and Schmidt, 1937); articles on sexual behavior of frogs, migration of turtles, endocrinology of lizards, as well as studies on the behavior of birds and fishes. He experimented with hormonal implants in lizards and even described homosexual behavior in *Ameiva.* His accomplishments with *Anolis* were extensive, and his final study in lizard behavior was published posthumously (Greenberg and Noble, 1944).

Noble's research was outstanding in quality and represented a truly original pioneer effort in behavioral ecology. As curator of herpetology at the American Museum of Natural History, he produced a large and diverse body of work—including his famous *Biology of the Amphibia* (Noble, 1931)—and had the foresight to found the Department of Animal Behavior at the American Museum of Natural History.

The literature cited in the chapters of Part II provides some measure of the impact of various herpetologists on the behavioral ecology of lizards. Five chapters cite C. C. Carpenter and his extensive research on comparative social behavior of lizards. D. W. Tinkle's research in population and community ecology is well known, but his contribution to behavioral ecology is attested to by the fact that four behavioral chapters also cite his work. H. S. Fitch is the third most-cited author. Tinkle's and

Fitch's many publications on lizard ecology have built a substantive foundation for the current research in behavioral ecology.

As to the substance of the chapters, the work of Simon (Chapter 6) and of Crews, Gustafson, and Tokarz (Chapter 10) is a direct outgrowth of the research started by Noble. Simon reviews the importance of chemoreception. Previous research on sensory systems of lizards (particularly of iguanids) has focused almost exclusively on vision. Simon's arguments may well encourage a new emphasis on the complexities and diversity of sensory inputs in lizard behavior. Crews and his colleagues apply behavioral and endocrinological techniques to the study of a unisexual lizard and derive a new view of the psychobiology of sexual behavior. Although unisexuality in lizards is probably an ephemeral evolutionary phenomenon, Chapter 10 indicates that unisexual species may provide us with subtle opportunities to understand the role of sexuality in the absence of sex.

Regal (Chapter 5) demonstrates the strong physiological orientation that characterizes contemporary research in lizard behavioral ecology. Important ties between behavior and physiology have also been emphasized by the chapters in Part I, Physiological Ecology. Regal emphasizes the adaptive advantages and limitations of ectothermy (see also Pough, 1980, and Chapter 1 by Bennett in this volume). This point of view allows us to see that lizard behavior is not necessarily constrained by physiology, but rather that a parsimonious metabolism can sometimes be an ecological virtue.

Research on juvenile lizards reported in Chapters 7 and 8 by Ferguson, Hughes, and Brown and by Fox demonstrates the utility of lizards for field experiments on behavioral ecology. In both studies natural populations were manipulated to answer significant questions: how does food abundance affect territorial behavior of juveniles (Chapter 7), and what are the interactions among the social status, habitat quality, and survival of a lizard (Chapter 8)? Ferguson and his coauthors present the first experimental evidence that food abundance influences the development and extent of territorial behavior of juveniles. Fox exploits a battery of ecological, behavioral, and endocrinological techniques to demonstrate that survival of juvenile lizards is related to the quality of their home ranges and to their relative social dominance. Both chapters emphasize the importance of temporal variation in competitive and selective pressures.

Stamps's contribution (Chapter 9) is an important and multifaceted analysis of the social systems of lizards. She uses extensive comparative data on the behavioral ecology of lizards and theoretical models to evaluate the roles and interactions of sexual selection, intraspecific competition, and diet. Until recently the data base has been inadequate for such integrative analyses. Stamps's work sets the stage for an evaluation of how lizard social systems correlate with those of birds and mammals.

Population ecology and behavioral ecology overlap in many areas. One of the more significant is the determination of a lizard's home range or territory. The territory of a lizard is a behavioral measure. Field studies have involved a variety of techniques to determine the home range or territory of individual lizards. It is generally agreed that a defended territory can only be described by actual data on defense—the observation of aggressive interactions between lizards. Since such observation involves much time and effort, researchers do not often measure in this way. Instead, home range (the area utilized by a lizard) is the favored parameter. After obtaining observations of an individual, a researcher may choose to calculate a radius, a minimal polygon, or a convex polygon. Furthermore, various correction factors may be used to compensate for small sample size.

Rose (1982) and I have discussed some of the problems concerning home-range measurements. Two main deficiencies exist in the current literature:

(1) A lack of consistency in methodology, which prevents valid comparisons. Differing techniques can produce major variations in estimated sizes of calculated home ranges (see Waldschmidt, 1979).
(2) An absence of any consistent time factor in the estimates. What maximum time span should be used in determining home range—one day, a few days, a week, or a few months?

The absence of a time unit for our definition of home range is a serious deficiency. In determining a home range we are attempting to describe a space used by a lizard. Clearly the use of this space varies with time (Chapter 3). The measured home range must be an approximation to something in nature, not an arbitrary calculation. Furthermore, we must reach agreement on what constitutes an adequate sample size. As few as three sightings have been used to calculate a home range—obviously a perfect number, since both the minimal and the convex polygon are then identical. Under some circumstances the measurement of home range is actually a measurement of activity—the more active the lizard, the larger the home range. Which comes first, greater activity because of a larger home range or a larger home range because of greater activity? Activity and home range may not be separable.

Our knowledge of behavioral ecology should make us circumspect in determining a lizard's home range. Territory and home range are behavioral measurements, and consequently behavioral components should be incorporated in the definition, even when the measurement is for demographic purposes.

Rodolfo Ruibal

5 The Adaptive Zone and Behavior of Lizards

Philip J. Regal

THE APPROACH THAT I TAKE in this chapter begins in the spirit of comparative biology, though I shall stress the possible *basic* differences between lizards and other groups more than their similarities. I do this in the belief that hypotheses concerning the important functional differences between lizards and other groups will in the long run have the greatest heuristic value from an ecological point of view.

Extinction has been the rule in the history of life, and so it is of interest to wonder about any unique features that may have aided a Mesozoic group of reptiles to survive and indeed to thrive in modern times. My first step, therefore, is an attempt to suggest what the modern lizard adaptive zone might be. Adaptive zone is intended much in the sense of Simpson (1953); the concept is an attempt to deal with characteristic relationships between environment and organism in such a broad manner that it necessarily involves abstraction, analogy, and oversimplification, yet it can usefully orient our thinking and research efforts.

I shall not be able to discuss many aspects of behavior here. Such diverse phenomena as migration, orientation, habitat selection, antipredator behavior, foraging, prey recognition, prey capture, social organization, interspecific aggression, learning, thermoregulation, and locomotion are all parts of survival in nature. Fortunately there have been several recent and worthy reviews of those aspects of lizard and reptile behavior (Brattstrom, 1974; Burghardt, 1977; Carpenter and Ferguson, 1977; Heatwole, 1977; Stamps, 1977; Auffenberg, 1978; Carpenter, 1978, 1980; Ferguson and Bohlen, 1978; Greenberg, 1978; Jenssen, 1978; Marcellini, 1978; Rand and Rand, 1978; Regal, 1978, 1980; Huey and Stevenson, 1979; Crews and Garrick, 1980; Fitch, 1980; Schall and Pianka, 1980). Thus, here I shall be selective and concentrate on developing an integrated perspective for viewing the distinctions between lizards and other creatures. I shall also freely ask questions and mention some general items that may not have received attention previously.

My approach, by some definitions, is adaptationist. Adaptationism has become controversial in the last few years; however, I do not generally claim that creatures are, in fact, adapted in given ways, I only propose that it may be useful to look for certain sorts of adaptations and functional relationships. There is abundant evidence that adaptation is a major feature of evolution. That is, most of the basic features of organisms function to the advantage of the individual organism and are not mere ornamentation or for the benefit of other organisms, according to some metaphysically teleological plan. But differences between creatures are, of course, not necessarily adaptive, and systematists, for example, have long recognized classes of nonadaptive "character states" in their attempts to construct phylogenies of creatures. Yet the adaptationist's approach to structure, physiology, and behavior has been profoundly successful in its Darwinian version. Indeed, an adaptationist outlook has been so successful that there has been a growing tendency in recent decades to assume that all features of organisms must be precisely adaptive. To counterbalance this overenthusiasm, specialists have continually attempted to caution general biologists about the deceptive complexity of the subject and the pitfalls in assuming that features are always adaptive (Bock and Wahlert, 1965; Williams, 1966; Steenis, 1969; Waddington, 1969; Gans, 1974; Bock, 1977). Recently Gould and Lewontin (1979), have made a few of the same criticisms independently.

The general point is that not all features of an organism can a priori be assumed to be adaptive directly to prevailing environmental conditions; one must be cautious in interpretation. The loss of the middle ear in snakes, for example, might well have nothing to do with their adaptation to an environment without sound as has been usually assumed, but it may be lacking merely as the evolutionary result of mechanical constraints imposed in that part of the head by the peculiar feeding mechanism of snakes (Berman and Regal, 1967). The lack of elaborate insulating structures such as feathers on desert lizards may have little to do with the question of whether or not feathers would be useful as heat shields in the hot sun; rather the peculiar skin shedding mechanism of lizards probably precludes the evolution of such structures (Regal, 1975). Thus, the hypotheses below cannot be assumed to be true simply because they may be appealing. They call for further research and are offered in that spirit.

If we wish to distinguish lizards from other creatures, we may begin to define them as ecological entities by a process of exclusion. In the broadest sense a lizard is a terrestrial tetrapod vertebrate that is *not* endothermic as are birds and mammals, does *not* fly as do birds and bats, does *not* breathe through its skin as do amphibians, does *not* drag a large shell as do turtles, does *not* have the peculiar feeding and locomotory adaptations of snakes, and is *not* a large amphibious crocodilian or a relict tuatara.

Lizards Contrasted to Invertebrates

By virtue of having a backbone, a lizard is an animal that can potentially move fairly rapidly on land, as can other animals with skeletons such as the arthropods, in contrast to mollusks. Indeed, although some lizards are slow and sedentary, most can make speedy runs in pursuit of prey or escape from predators. Vertebrates differ from terrestrial arthropods, however, in having a complex "focusing" eye with high resolving power. So although some lizards are nearly blind, most may possibly have an advantage in prey and predator detection under some given conditions. Structural differences between the brains of vertebrates and invertebrates exist, but the functional significance of these is unclear. Only vertebrates are known to have low frequency brain waves, but the significance of this also remains obscure. Although lizards may in some ways have better eyes than do insects, insects have a very efficient oxygen transport system allowing them to engage in intense and sustained activity whereas lizards tire quickly.

Lizards Contrasted to Other Vertebrates

Habitats Poor in Food or Water. Lizards have low energy and water requirements compared to birds and mammals (Chapter 2). So they may starve and dehydrate more slowly than endotherms and may be expected in (though not restricted to) habitats where food or water is relatively scarce or uncertain in time (references in Regal, 1978; Schall and Pianka, 1978; Pough, 1980). Food and water requirements for growth, maintenance, and reproduction are many times lower for lizards than for endotherms, and starvation and dehydration may take many weeks or months rather than hours or days. Estivation in an endotherm should function best as an adaptive strategy to avoid harsh times when resources are predictably available upon periodic arousal. Otherwise the metabolic costs of repeated arousals to "test" for resource availability may outweigh the benefits gained in food and water by estivation. Any emergence tests by lizards are expected to be energetically inexpensive. If resources are uncertain then the advantage may shift in the direction of the ectotherm in any instances where there might be potential competition.

The uncertainty of rainfall goes hand in hand with increasing aridity, and this may have various ecological consequences (references in Regal, 1982). Aridity and uncertainty may constitute a relative set of problems for endotherms, a fact which in turn underscores the suitability, because of ectothermy and low metabolic rates, for lizards in such habitats. Small mammals in deserts are usually nocturnal, and the few diurnal species have special physiological and behavioral adaptations (Bartholomew and Hudson, 1961; Brown, 1974). Small mammals become rare during Australian droughts, and populations may die back to local refugia (Newsome and Corbett, 1975). The abundance and variety of lizards in deserts is of interest in these regards, although the usual explanation given for their flourishing in warm deserts is that abundant sunshine allows inexpensive thermoregulation. Studies that can assess the relative contributions of each factor are needed. The considerable abundance and diversity of reptiles in the tropics surely suggests that warm climate may be important in lizard biogeography (Huey and Slatkin, 1976). It is in the interests of precision to evaluate how particular components of climate interact with the particular physiological abilities and life histories of species.

It should not be assumed that competition is my focus or that I am taking the position that interspecific competition is necessarily the determining force in evolution or ecological diversity. Intraspecific competition and a variety of historical circumstances could also have led to the present differences between taxa. Yet in order to understand given taxa it remains instructive to ask questions about functional differences between them, whether these differences stem from a variety of traits accumulated over phylogenetic history or from constant long- or short-term interspecific competition and niche partitioning. Does the survival or flourishing of lizards in a given habitat sometimes result from the advantages of having low food and water requirements? Do lizards respond differently to drought than do local terrestrial endotherms? Do lizards use lower quality food than endotherms in some situations where both coexist? Are there situations where lizards use available food at lower rates than do endotherms? Finally, in situations where lizards might indeed be potential or actual competitors with other groups, such as Schall, Pianka, and Wright have discussed (Schall and Pianka, 1978; Wright, 1979; Pianka and Schall, 1981) could these last factors help us to understand any niche partitioning?

Small Size. Small objects have high surface-to-mass ratios, and this causes high heat-loss rates. Small body size and elongation are not energetic handicaps for ectotherms such as lizards, whereas mammals or birds the size of tiny skinks would be energetically disadvantaged. So very small, insectivorous lizards and other ectotherms may occupy (though need not be restricted to) habitats as very small tetrapod insectivores, whereas progressively smaller mammals and birds would become progressively disadvantaged energetically (Pough, 1980).

Varied and Abundant in Warm Climates. Many groups of organisms become species-rich and abundant in the tropics, and lizards are no exception. Many reasons for such trends have been proposed, and I shall not claim here to have solved the problem. But with regard to lizards, I do wish to make some points that are not usually discussed. Since Cowles and Bogert (1944) it usually has been claimed that reptiles flourish in warm climates because they can easily reach their activity temperatures by behavioral thermoregulation. I do not doubt that this is important. Lizards may have limited activity time to feed and grow and accumulate energy for reproduction in seasonally cool climates such as prevail in the northern continents (Avery, 1976). Cloudy days and unusually cool springs and summers may further limit the basking lizard. Birds and mammals, on the other hand, must also thermoregulate behaviorally, but because of endothermy and insulation they can be exposed to a wider range of thermal conditions than ectotherms before they begin to lose activity time to behavioral thermoregulation (Gates, 1975).

Yet how is it that amphibians, relative to lizards, remain abundant and relatively varied in moist regions at high latitudes even though they are ectotherms? They do have lower activity temperatures than do most reptiles (Brattstrom, 1979; Freed, 1980), and this may provide a partial answer, but some reptiles also have low activity temperatures. Why have such forms not proliferated in moist regions of northern latitudes? Many species of northern amphibians have aquatic larvae, and this might result in an energetic advantage, yet a variety of abundant plethodontid salamanders hatch as terrestrial life forms and so lack this advantage. Amphibians have evolved more simplified muscular and skeletal systems than lizards. Most amphibians have projectile tongues, whereas this is true among lizards only of the geographically restricted and highly specialized chameleons. Are the usable food resources of lizards thus more restricted than those of amphibians in some northern situations? Might competition between the two groups be involved? In my opinion this entire question and related issues merit detailed thought and study. Surely lizards, as ectotherms, have fewer constraints in tropical climates than in temperate climates but has the attractiveness of this generality distracted us from asking questions and making studies that could lead to a more exact understanding of lizard ecology and behavior?

Lizards also seem generally to become less common and varied at higher altitudes even in the tropics. However, insolation can be great in the thin air of high altitudes and so basking opportunities (Pearson and Bradford, 1976) can compensate for the cool air temperatures (though nocturnal life-styles might become restricted because of low night temperatures). Some of the decline in species-richness of lizards at high altitudes in low latitudes may be due to the fact that mountains may act as islands that decline in size with altitude (MacArthur and Wilson, 1967; Brown, 1978).

Thus, the physiological and behavioral strategies of lizards give partial insight into their ecogeographic distributions relative to those of other vertebrates, yet conspicuous questions remain.

Residency and Energy Storage in Ecosystems. Lizards are small and do not actively fly, and this limits their ability to migrate in and out of areas. Hence from a community point of view they can be a source of protein that remains a resident feature of a habitat. This could be important in providing a caloric "capacitance" in some ecosystems. For example, in Australian deserts lizards eat insects such as ants and may "store up" energy (since lizard starvation and dehydration are slow) that may support small marsupials, snakes, and monitor lizards through harsh times. If birds were instead to consume such insects and fly off during harsh times, then the marsupial omnivores, snakes, and small varanid lizards would have a very different and perhaps insufficient food base at critical times in the highly uncertain Australian climate. Do lizards, because of their ectothermy and size sometimes form distinctive links in food chains?

Xeric and Semixeric Habitats and Microhabitats. Since their skin is relatively unimportant in respiration and so can be relatively dry, lizards are not generally bound to moisture as are most amphibians. Hence, they flourish in dry habitats for this reason and perhaps in conjunction with others above. It is interesting that some microteiid lizards are said to occupy wet, salamander microhabitats in South America (for example, Fitch, 1968; Sherbrooke, 1975, and personal communication), where salamanders have only recently invaded. Some skinks may occupy similar microhabitats in Australia (Schall and Pianka, 1978). Why more lizards have not adopted amphibian life-styles remains an interesting question. Does the complexly glandular amphibian skin offer distinct advantages in terms of communication and protection in some habitats? Does amphibian cutaneous respiration, allowing gas exchange at low muscular effort, provide critical energetic savings for some life-styles? On the other hand, is amphibian cutaneous respiration, leaving little scope for increasing ventilation rate, disadvantageous for other more active life-styles? Here again, one wonders under what ecological circumstances particular lizard or amphibian specializations may be critical. Such questions pose challenges to physiological, population, and community ecologists alike.

A lizard, then, is a terrestrial tetrapod that can move quickly, that has good vision, that has low energy and water demands, that can accelerate physiological processes cheaply by basking, and that can invade drier habitats. Many are small insectivores and can prepare large food items using their heavy jaws and kinetic skulls.

This quick sketch does not convey the richness and diversity of lacertilian habits and morphology, but it does provide a point of departure for various sorts of functional analyses and it does give a general feeling for the ways in which, and the reasons for which, lizards may fit into given ecosystems. Despite the fact that some lizards are herbivores, top carnivores, or snail eaters, most lizards are relatively small insectivores, resident in the ecosystems of warm climates. Some can be very quick, whereas some are always very sluggish; some are arboreal, whereas some are terrestrial or fossorial. But most eat small invertebrates and are well suited to endure low-energy or low-water intake. Many species may bask and thereby accelerate their biochemical reactions using the heat of the sun. Yet, since many species live in deep shade where the thermal diversity can be minimal, the heliothermic lizard is exaggerated as a paradigm for all lizards, as is stressed elsewhere (Regal, 1968, 1980; Huey and Slatkin, 1976).

Physiological-Anatomical Basis for Lizard Life-Styles

Why is it that lizards can be generally characterized as above? Why are there no true endothermic lizards or no lizards that can fly under their own power? Moreover, why are most lizards such relatively sedentary creatures that vigorous, animated behavior seems uncharacteristic of them? I will attempt to address these questions not so much to advocate particular hypotheses as to develop a point of view about lizards that has helped me place their behavior in a mechanistic, testable framework.

Lizards have long been considered slow and sluggish because they are "cold-blooded and primitive," and this pedestrian view has even figured prominently in the argument over whether or not the dinosaurs were warm-blooded—that is, endothermic. The argument, in part, says that if the postures of dinosaurs reveal considerable and complex activity, then surely the dinosaurs would have to have been endothermic, since their modern analogues, the ectothermic lizards, are supposedly weak, slow, and simple. But this view of lizards is seriously flawed.

Neither the primitiveness nor the ectothermy of lizards explains their behavior, and recent work allows an interpretation of the underlying constraints on which their various behavioral strategies may balance. The seminal and elegant work of the late Walter Moberly (1968a, 1968b) on *Iguana iguana* revealed that a high fraction of strenuous exercise activity may involve anaerobic metabolism and the buildup of an oxygen debt that is paid back only slowly. Lizards may consequently engage in bursts of activity but they are sensitive to fatigue and unable to sustain high levels of activity. Importantly, the repayment of this oxygen debt is distinctly slower in lizards than in birds or mammals. A subsequent series of fine studies by Albert Bennett and his coworkers confirmed and conceptually extended this work for a variety of species (Chapter 1) and showed that even for the most active lizard species the aerobic capacity is inferior to that of mammals (extensive references in Regal, 1978; Bennett and Ruben, 1979; Coulson, 1979).

Why should the aerobic capacity of lizards be so limited and the anaerobic capacity be so high? Bennett and Ruben (1979) suggested that some unknown constraints exist that link basal metabolic rate, aerobic scope, and maximal aerobic output. Recent work of Fred White and others indicates that the peculiar, complicated hearts of reptiles seem to be incapable of delivering well-oxygenated blood at high rates and pressures, and this limitation may have been an important consideration in the evolution of characteristic low-sustained-activity reptilian behavioral strategies (Regal, 1978; Regal and Gans, 1980). In addition, most lizards have simple lungs, the intensively foraging varanids being a notable exception. So, in a manner of speaking, the conservation of behavioral activity and the high anaerobic capacity are behavioral and physiological strategies that have provided partial compensations for "weak" oxygen-support systems (Regal, 1978; Pough, 1980; Chapter 1).

Characteristic Lizard Behavior

Energy-Conserving Behaviors. Of course, the field biologist knows that many so-called inactive species of lizards can run with considerable speed and agility when necessary and that folk wisdom about "primitively sluggish" lizards is wrong. In the Dominican Republic I found that *Leiocephalus schreibersi* were immobile in the sun for over 99 percent of the day. They were certainly capable of short dashes after prey, and these were in fact so rapid that they were difficult to time (Regal, 1978). So their great inactivity is a misleading index to their capacity for activity. They are inactive, apparently, because there is some advantage to inactivity rather than because they simply cannot get up and move easily, as has been assumed. It has, indeed, long been known in the scientific literature that the burst speeds of some lizards are as great as for mammals (references in Bennett, 1980). So perhaps many lizards are sit-and-wait predators or normally have slow behaviors because their strategy is to "stay out of oxygen debt" as much as possible and to maintain a maximum reserve capacity for bursts of energy in escape, courtship, agonistic encounters, and pursuit of prey (Regal, 1978).

Restriction to Ectothermic, Behavioral Thermoregulation. Many lizards that live in open habitats are heliothermic, and much of their observed behavior is thermoregulatory. They shuttle in and out of the sun and shade and orient their bodies positively or negatively to the sun or to the hot or cold substrate or air (Cowles and Bogert, 1944, references in Heath, 1965; Huey and Slatkin, 1976; Greenberg, 1978; Regal, 1980). Because endothermy would impose high sustained metabolic costs, the lizard metabolic machinery and circulatory system seem ill suited for it. Thus, if it were endothermic a lizard might have little if any aerobic scope left for bursts of activity. So even if endothermy were physiologically possible, hypothetical endothermic lizards would be continually near the brink of oxygen debt or fatigue and would seem ecologically improbable. Perhaps this is why, in an evolutionary context, lizards have apparently been tied to behavioral thermoregulation (Regal, 1978). As stressed often above, ectothermy has some very positive advantages in producing an acceleration of biochemical reactions at low energetic expense by basking. So "restriction to ectothermy" is a phrase used here as one would say that porpoises are restricted to water, and it is not meant to imply that they are inferior in their element.

Foraging Styles and Their Effects on Diversity

The behavior of lizards is rather diverse and we can now begin to see patterns in some of the diversity and to discern some reasons for the patterns. Since the mid-1960s with the theoretical work of MacArthur, Pianka, Emlen, Levins, and Schoener it has become better understood that "sedentary" lizards are probably using a particular sit-and-wait foraging strategy in which very little energy is expended to obtain common food items that may be low in quality. Other species may use an active foraging strategy (see below) in which much energy is expended to obtain uncommon food of high quality. Recently Anderson and Karasov (1981) were able to confirm much of the theory using doubly labeled water (see Chapter 2) with *Callisaurus* and *Cnemidophorus* in the field. The actively foraging *Cnemidophorus tigris* had a higher rate of energy expenditure, but also a higher rate of foraging efficiency than the sit-and-wait *Callisaurus draconoides* did. Huey and Pianka (1981) report similar results for lacertid species using less direct techniques and outline several aspects of natural history associated with foraging strategy.

Although interspecific variations in foraging behaviors may be of considerable interest to ecologists if they are fundamental aspects of resource partitioning among lizards, no precise survey or taxonomy of these behaviors yet exists. I have modified the traditional division into sit-and-wait predators and active foragers (Schoener, 1971), and I have employed instead three categories for lizards—sit-and-wait, cruising foragers, and intensive foragers—a spectrum that is still probably inadequate (Regal, 1978). Moermond's recent (1979a,b) studies of several species of *Anolis* show nicely that sit-and-wait predation may grade into active foraging as species characteristically change position to varying extents and at varying intervals. For such reasons the term active forager generates confusion since some basically sit-and-wait and cruising species (in my terms) are literally active much of the time inasmuch as they switch positions frequently. I hope that the term intensive forager will best describe species that invest much energy in actually seeking out scarce or concealed prey having high food value. I have in mind as intensive foragers teiids such as *Ameiva* and *Cnemidophorus* that move continually and rapidly over a large area, constantly probing, tasting, digging, and exploring. A species that moves, stops, and merely scans the environment, then moves, stops and scans, and so on, would be a cruising predator, even if it were active most of the time. Such classifications are useful in making testable predictions about the behavior of species.

Mental Capacities Linked to Foraging. Despite much discussion among ecologists of fine- and coarse-grained responding organisms, there has been little attempt to analyze the behavioral capacities required of each. Lizards are relatively easy to work with and perhaps represent a system where we can begin studies and generate predictions. For example, we can predict that a prototypal sit-and-wait predator need be little more complex—in terms of information storage, processing, and behavioral output—than a mechanized mousetrap, with a neural organization for prey recognition and distance judgment and with neuromuscular patterns for pursuit and capture. On the other hand, the intensive forager would seem to require in addition mechanisms to produce sustained spontaneous locomotion, investigating and digging behavior, and an enlarged memory capacity for features of different prey and of the enlarged habitat (Regal, 1978).

Although the scheme will benefit from refinement and modification, it has already generated a few predictions. The foraging of *Ameiva chrysolaema* was studied and aspects quantified in the field. It is an intensive forager. In the laboratory, as predicted, it did have much higher rates of spontaneous locomotion and investigatory behavior than did a sit-and-wait predator, *Leiocephalus schreibersi* (Regal, 1978).

The significance of interspecific differences in brain size is not well understood (references in Regal, 1978), yet it is of more than passing interest that (macro) teiids proved to have the largest relative brain sizes of any lizards, with varanids also having very large brains, though no larger than some iguanids, agamids, and gekkonids (Platel, 1979). Likewise the intensive-foraging teiids and varanids have histologically nearly identical advanced brain structures. But they share several features with iguanids, agamids, and even chamaeleonids: this was termed the "iguanid pattern" by Northcutt (1978). The iguanid pattern is noted conspicuously by the enlarged dorsal ventricular ridge with a diffuse, rather than laminar, organization of its cells, as well as a number of other features. So brain size and organization agree only partially with feeding habits. Assuming that the possibility of relationship is valid and has any interest at all, various factors might account for the failure of complete correlation. Was advanced brain organization a prerequisite or preadaptation for the more advanced foraging behaviors? Have most iguanids and agamids secondarily and recently become sit-and-wait predators? Either factor might explain why many sit-and-wait lizards seem more or less to share relatively advanced brains with intensive foragers. Or perhaps we do not yet understand the subtleties of foraging in lizards very well. If we undertake more detailed studies of ecology and behavior, will our understanding of the correlation improve? Or, indeed, should we not have expected a relationship in the first place? Are the behavioral differences associated with foraging not, in lizards, of a degree that requires characteristic morphological reorganizations of the brain?

Obviously questions remain as to the behavioral and neuroanatomical substrates of foraging behaviors in different species. Yet, the issue of brain morphology aside, it should be useful to continue to attempt to define and identify the behavioral traits and abilities associated with the various foraging types. This should help us to understand the details of adaptive radiations within taxa, the details of patterns of resource utilization, and perhaps the basis for the coexistence of species that might appear to have similar diets.

Some Problems in the Study of Foraging Behavior. One of the most neglected and yet one of the most profoundly difficult problems in behavior is the fact that we are invariably selecting data when we describe the behavior of an organism (Hinde, 1970). Selected data are notoriously intractable if our goal is scientific objectivity. How can we fit any complex continnuum of activity and inactivity into man-made pigeonholes that have validity? For this reason and others, I suspect that it will be some years before we have accurate and useful ideas about foraging behavior in a variety of lizards. As one course I have suggested that we may approach the subject deductively and propose sets of hypotheses about behavioral capacities that we can expect lizards of various foraging types to have. We can observe the lizards and see if we are correct. If we are not, then we can go back and reexamine the hypothesis (Regal, 1978). The approach is somewhat like writing a computer program for a mechanical lizard that is able to find rare but rewarding resources and then comparing the actual behavior of a species in nature and in the laboratory to this model. *Ameiva* and *Cnemidophorus* behave very much as one would expect of a lizard making coarse-grained responses.

Such studies also provide a base of comparison for other species. For example, several herpetologists told me that Australasian skinks related to *Ctenotus* and *Emoia* are ecological counterparts of teiids and forage in the same manner. However, I was unable to confirm this by detailed questioning or by my own scant observations. Finally in December 1978 in Madang, Papua New Guinea, I was able to make extensive observations on *Lamprolepis* (tentatively *smaragdina*), a relative of *Emoia.* These quick green skinks forage mostly on the trunks of large vine-covered trees in a lowland tropical habitat. At first one may indeed think that they forage as do *Cnemidophorus* as they move along fairly constantly and frequently lick the substrate. But part of this activity is observer effect. If one stands back 15 or 20 m and uses binoculars to make observations, the lizards calm down and reveal a different level of behavior. They spend about 80 percent of the time sitting, and with occasional movements of the head apparently scan the massive epiphyte-covered limbs for prey. Every few minutes they will move from a few centimeters to several meters and may occasionally lick the substrate and may even poke their heads into cracks. But most cracks are ignored, and there is none of the intense investigative behavior that characterizes teiids. The *Lamprolepis* did not, then, behave as one would predict a coarse-grained responder might. Not only was the amount of time in activity very low, but the investigatory behavior was almost missing. In this case, I believe that a deductive categorization of foraging behaviors was helpful in prompting me to look at relevant activities that seem to distinguish between the skink and teiids.

Besides observer effect, a variety of other factors can give misleading results in foraging studies. Roger Anderson has pointed out to me that in studying *Cnemidophorus tigris* and *hyperythrus* it is possible to confuse foraging with mate seeking. By having animals individually marked and sexed and by following their behavior over the activity season, he is able to judge that what can appear to be exceptionally intense and extensive foraging is actually the result of males seeking females. He also warns that these lizards may not always move over large distances (perhaps if they find concentrated food supplies). These points again caution that what we see can be very much influenced by what we know or do not know, suspect or do not suspect. Ideally, foraging studies would be designed from a rather extensive knowledge of the species. Foraging may be influenced by the above factors and by weather conditions, seasonal cycle, local-habitat structure, and changes in food availability. Moermond (1979a,b) has discussed certain other problems in the study of lizard foraging behavior. Stamps (1976) describes female *Anolis aeneus* searching for egg-laying sites, snout poking, and digging. One wonders if one less familiar with the species could have incorrectly interpreted this behavior as foraging.

Foraging May Influence Social Systems. A relatively few species of lizards—notably teiids and varanids—seem to be somewhat like typical mammals in their foraging behavior. By this I mean that they move about widely and hunt selectively for hidden food, some species such as the Komodo dragon even utilizing ambush tactics (Auffenberg, 1978). *Ameiva chrysolaema* in the field move and hunt and probe for more than 70 percent of their activity period in the field, and their rates for spontaneous locomotion in a standard laboratory open-field test were in fact equal to those of a rodent (Regal, 1978). *Cnemidophorus tigris* are moving almost 90 percent of their activity period (Anderson and Karasov, 1981).

A widely moving ground animal can be expected to have a different social system than a sedentary species. Stamps (1977) has discussed the similarities of teiid and varanid social behavior to that of mammals—

with intensive foragers apt to show similarities to small mammals and sit-and-wait predators apt to approach classical territoriality. It is well to keep in mind that the conspicuous signal colors and stereotyped displays reported in so many studies of lizards are primarily common in species that perch much of the time while they are basking and practicing sit-and-wait predation. It is presumably also easier to monitor a territory, maintain territorial boundaries, and use such signals while perching than while moving about investigating holes and smells. Since the former species are easier to study than the latter, we may tend to bias our studies of social behavior toward the more territorial species with more conspicuous displays.

Thermoregulation Influenced by Foraging Strategy. Heliothermic lizards may use a variety of thermoregulatory behaviors (Cowles and Bogert, 1944; Heath, 1965). Such complex behaviors would seem to be compatible with sedentary basking and sit-and-wait predation. On the other hand, it would seem awkward for a lizard that is constantly moving about and investigating holes and cracks also to attempt elaborate thermoregulatory postures. So we might predict either a wider range of body temperatures in the intensive forager or else a restriction of intense activity to hours of the day when a satisfactory thermal equilibrium with the environment can be accomplished without frequent and elaborate postures (Regal, 1978). Garrick (1979) has postulated that for related reasons the thermostatic control in active foragers and sit-and-wait predators may be different. In any event, sedentary species seem to have more conspicuous and varied thermoregulatory behaviors. Moreover, reports indicate that the more active species may tend to remain exposed a relatively few hours and during hot hours of the day (references in Regal, 1978; Anderson and Karasov, 1981; Huey and Pianka, 1981). This may in part result from the likelihood that an active foraging mode involves greater risks of predation as Huey and Pianka (1981) have stressed and in part from the likelihood that complex behaviors might be necessary during cooler hours and these may be incompatible with active foraging as I have stressed.

We can ask many questions about the nature of interactions among foraging style and thermoregulatory behavior and ecology. Conflict situations involving necessary trade-offs may be of particular interest in analyses of such themes as discussed here. For example, in ecologically marginal areas, perhaps windy or cloudy, do active foragers often have to switch to more sedentary foraging modes? Does this result in longer exposure times and more varied basking behaviors? Is food quality and/or quantity sacrificed? Does this in turn result in reduced reproductive efforts or growth rates? Is social behavior altered and to what effect? Do typical sit-and-wait species under the same conditions have any advantage in a lower metabolic rate, better antipredator features, more efficient utilization of low-quality food, more efficient control of territories, or more effective basking behaviors and integumentary features resulting in more effective digestion, assimilation, growth, reproduction, and defense? The interrelationships among foraging, thermoregulatory, and social behaviors, and in turn their effects on energetics, growth, and reproduction, could in theory be complex for any species. How such factors may influence interspecific competition under varying conditions is also worthy of some thought.

6 A Review of Lizard Chemoreception

Carol A. Simon

LIZARDS ARE GENERALLY considered to be visually oriented animals. Nevertheless, many lizards routinely extrude their tongues, seemingly transporting chemoreceptive information from the environment to paired vomeronasal (Jacobson's) organs in the roofs of their mouths. Moreover, many species have specialized glands or similar organs that seem to be involved with intraspecific, chemical communication. This behavioral and anatomical evidence suggests that lizards may frequently use chemoreception. Yet the overall nature and significance of chemoreception to the behavior, communication, and ecology of lizards is unclear and understudied.

This review examines the types of chemoreceptors found in lizards, the types of chemoreceptive information potentially gathered, the use of chemical signals in intraspecific communication, and how patterns of variation in the importance of chemoreception correlate with anatomy, phylogeny, and ecology. Much of the information necessary to establish patterns is anecdotal or absent (Burghardt, 1970, 1980; Madison, 1977), and only recently has experimental and correlational evidence begun to elucidate the diverse uses of chemoreception in lizards. Although serious gaps prohibit a definitive synthesis at this time, enough is known to demonstrate that chemoreceptive systems in lizards deserve considerable attention.

Chemoreceptors of Lizards

Possible chemoreceptors of lizards include taste buds, the olfactory apparatus, and the vomeronasal organs. Limited anatomical information suggests that taste is unimportant in lizards. Taste buds are found primarily in the lining of the pharynx rather than on the tongue (Moncrieff, 1977). However, Guillette and Duvall (Duvall, personal communication) have tentatively identified taste buds on the tongues of *Sceloporus jarrovi* and *S. occidentalis.*

Experimental studies also suggest that taste is relatively unimportant (Rensch and Eisentraut, 1927; Gettkandt, 1931; Noble, 1937), but several of these studies may be inadequate (Burghardt, 1970). Since few anatomical and behavioral studies exist, the significance of taste to lizards is presently unclear.

The vomeronasal system and the olfactory system of lizards are well developed (Fig. 6.1) and are better studied. In all Squamata, the paired vomeronasal organs, which lie below the olfactory nasal passages and are anatomically separate from them, are connected to the mouth through ducts in the anterior palate. The vomeronasal organs and the olfactory mucosa are also separately innervated. The vomeronasal nerve is connected to the accessory olfactory bulb, whereas the olfactory nerve is connected to the main olfactory bulb (Parsons, 1959a,b, 1967, 1970).

The tongue and the vomeronasal organs are functionally related. The tongue collects various molecules from the environment and transports them to paired ducts leading to the vomeronasal organs (Broman, 1920; Kahmann, 1932; Wilde, 1938; Abel, 1951; Parsons, 1959a, b, 1967, 1970; Burghardt, 1970; Porter, 1972; Burghardt and Pruitt, 1975; Kubie, 1977). Cilia in grooves may help convey molecules to the interior of the vomeronasal organs (Bellairs and Boyd, 1950). Forked tongues may be

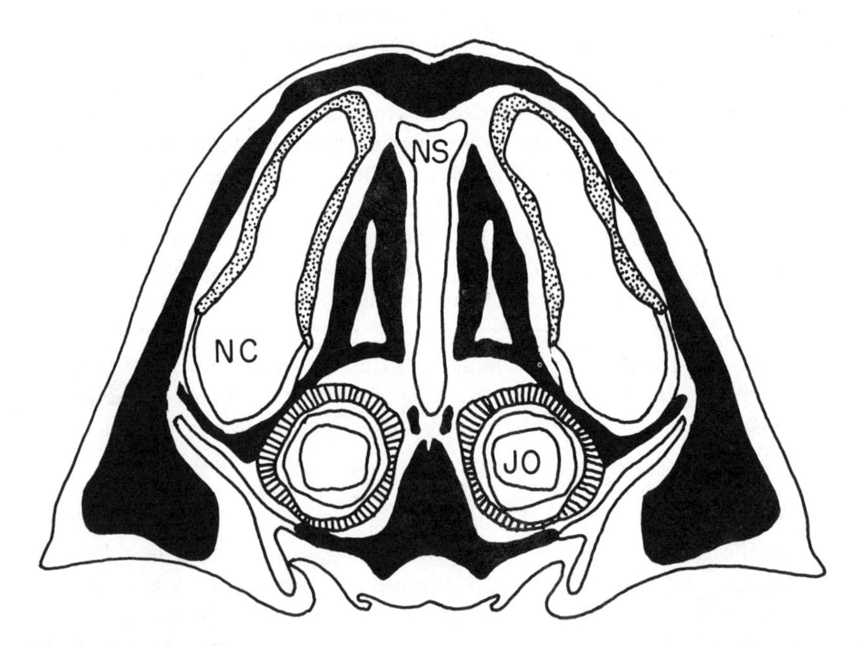

Figure 6.1 Diagrammatic cross section of snout of *Tiliqua scincoides scincoides.* Modified from Kratzing (1975). *NS,* nasal septum; *NC,* nasal cavity; *JO,* Jacobson's organ.

especially adept at delivering molecules since the tongue tips may be inserted into the paired ducts (but see Burghardt, 1980).

Parsons (1959a,b, 1967, 1970) reviews evidence that the vomeronasal organ of the Squamata is more highly developed than in any other vertebrate. This organ is highly differentiated in some lizards but poorly developed in others (Burghardt, 1970). This variation will be discussed later.

Almost all physiological work on the vomeronasal organ has been done with snakes rather than with lizards. In snakes materials collected via tongue extrusions do reach the vomeronasal organ (Kahmann, 1932; Kubie, 1977). Electrophysiological evidence demonstrates that the vomeronasal/accessory olfactory bulb system is activated following tongue flicks (Meredith and Burghardt, 1978). In one physiological study done with a lizard (*Iguana iguana*), electrical stimulation of the nucleus sphericus, the target of the fibers from the accessory olfactory bulb, as well as other structures, elicited tongue extrusions (Distel, 1978).

Chemoreceptive interactions among taste, nasal olfaction, and the vomeronasal organs have been suggested. Cowles and Phelan (1958) proposed that the nasal olfactory system initially detects the general presence of novel chemical stimuli and then "switches on" the vomeronasal system for acute discrimination (Duvall, 1980; Duvall, in preparation). For example, female *Sceloporus occidentalis* were placed in visually identical test situations containing either clean sandpaper or one of several chemically labeled types of sandpaper (Duvall, in preparation). In chambers containing clean sandpaper, tongue extrusions were initiated after approximately 11 minutes, whereas tongue extrusions in chambers containing sandpaper labeled with other odors were initiated at least 4 times more quickly (Fig. 6.2).

Possible Functions of the Vomeronasal System

Table 6.1 lists possible chemoreceptive functions for vomeronasal organs. Most of these functions, especially those proposed in early studies (review in Burghardt, 1970), are largely based on anecdotal rather than on experimental evidence. Nevertheless, available experimental analyses demonstrate that a variety of types of chemical information is brought to these organs via the tongue.

Several terms, some of them awkward and misleading, have been used to describe tongue extrusions in lizards (tongue tests, Parcher, 1974; licks, DeFazio et al., 1977, and others; lip smacking, Greenberg, 1977; tongue touches, Greenberg, 1977, and others; tongue flicks, Gove, 1978, and others; tongue extrusions, Bissinger and Simon, 1979, and others). In the following discussion I shall use "tongue extrusion" as a general term, "tongue flick" when the tongue samples air, and "tongue touch" when the tongue contacts a substrate or a conspecific.

Table 6.1 Suggested uses for the lizard vomeronasal system (often anecdotal).

Possible use	References	Species
Detection of conspecifics/ species identification	Berry, 1974	*Sauromalus obesus*
	Bissinger and Simon, 1981	*Sceloporus jarrovi*
	Carpenter, 1975	*Sauromalus obesus*
	DeFazio et al., 1977	*Sceloporus jarrovi*
	Duvall, 1979	*Sceloporus occidentalis*
	Duvall et al., 1980	*Eumeces fasciatus, Scincella lateralis*
	Fitch, 1954	*Eumeces fasciatus*
	Gravelle and Simon, 1980	*Sceloporus jarrovi, Anolis trinitatus*
	Hunsaker, 1962	*Sceloporus torquatus* group
Territoriality/spacing	Auffenberg, 1978	*Varanus komodoensis*
	Berry, 1974	*Sauromalus obesus*
	Bissinger and Simon, 1981	*Sceloporus jarrovi*
	Carpenter, 1962	*Cnemidophorus sexlineatus*
	Carpenter, 1975	*Sauromalus obesus*
	DeFazio et al., 1977	*Sceloporus jarrovi*
	Duvall, 1979	*Sceloporus occidentalis*
	Duvall et al., 1980	*Scincella lateralis, Eumeces fasciatus*
	Gravelle and Simon, 1980	*Sceloporus jarrovi, Anolis trinitatus*
Aggregations/hibernacula	Burghardt et al., 1977	*Iguana iguana*
	Duvall et al., 1980	*Eumeces fasciatus*
Undefined communication	Parcher, 1974	*Chamaeleo*
General exploration/responses to novel stimuli	Berry, 1974	*Sauromalus obesus*
	Bissinger and Simon, 1981	*Sceloporus jarrovi*
	Bogert and Martín del Campo, 1956	*Heloderma suspectum*
	DeFazio et al., 1977	*Sceloporus jarrovi*
	Gelbach, 1979	*Gerrhonotus coeruleus*

	Simon et al., 1981	*Sceloporus jarrovi*
	Wevers, 1910	*Heloderma suspectum, Trachysaurus rugosus*
Food seeking	Abel, 1951	*Lacerta*
	Bogert and Martín del Campo, 1956	*Heloderma suspectum*
	Burghardt, 1973	*Eumeces fasciatus*
	Burghardt, 1977	*Gerrhonotus liocephalus*
	Gove, 1978	*Ameiva corvina, Gerrhonotus multicarinatus*
	Kahmann, 1939	*Ameiva saurimanensis, Ophisaurus apus, Acanthodactylus scutulatus*
	Milstead, 1961	*Cnemidophorus tigris, C. tesselatus*
	Noble and Kumpf, 1936	*Ameiva exsul*
Predator detection	Berry, 1974	*Sauromalus obesus*
	Simon et al., 1981	*Sceloporus jarrovi*
Sex discrimination/courtship	Auffenberg, 1978	*Varanus komodoensis*
	Berry, 1974	*Sauromalus obesus*
	Duvall, 1979	*Sceloporus occidentalis*
	Fitch, 1954	*Eumeces fasciatus*
	Gravelle, 1981	*Sceloporus jarrovi*
	Greenberg, 1943	*Coleonyx variegatus*
	Hunsaker, 1962	*Sceloporus torquatus* group
	Noble and Teale, 1930	*Ameiva chrysolaema*
	Tinkle, 1967	*Uta stansburiana*
Maternal care	Duvall et al., 1979	*Sceloporus rufidorsum*
	Evans, 1959	*Eumeces obsoletus*
	Fitch, 1954	*Eumeces fasciatus*
	Noble and Mason, 1933	*Eumeces fasciatus, E. laticeps*
	Noble and Kumpf, 1936	*Eumeces laticeps*

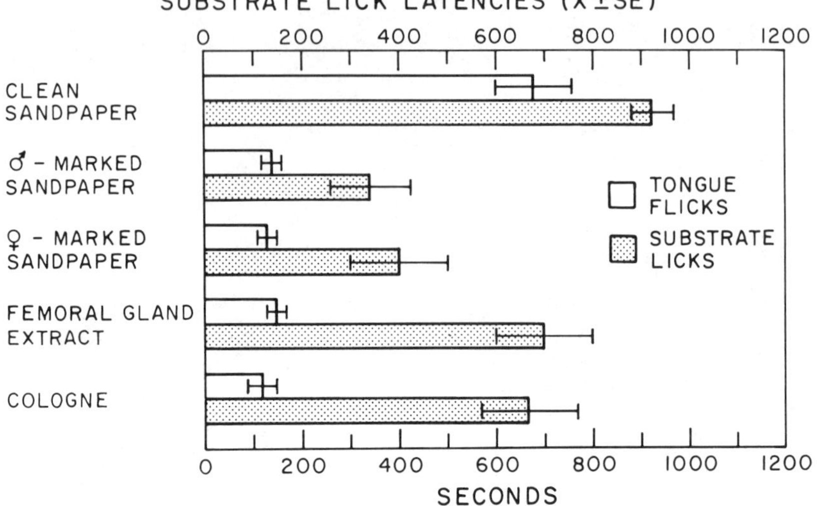

Figure 6.2 Latency to initial tongue flicking for 16 female *Sceloporus occidentalis*. From Duvall (in preparation).

The tongue may be extruded for purposes other than chemoreception. Virtually all lizards use their tongues to manipulate food, to drink, and to clean their faces and jaws (Gove, 1978). These tongue extrusions are, however, easily distinguished from chemoreceptive ones. In snakes the tongue may also provide tactile information (Klauber, 1956), detect airborne vibrations (Ditmars, 1931; Klauber, 1956), provide tactile stimulation during courtship (Gillingham, 1976), or even be used as a fishing lure in water snakes (Czaplicki and Porter, 1974). Behavioral evidence is available, however, only for the last hypothesis.

Tongues may also be used in signaling. The brightly colored tongues of some African snakes serve as visible threats (Mertens, 1955; Pitman, 1974), and the skink (*Tiliqua scincoides*) threatens conspecifics by presenting its cobalt blue tongue (Carpenter, 1978). Gove (1978, 1979) reviews the role of tongue extrusions in intra- or interspecific signaling. She documents characteristic patterns of tongue extrusion for given species and proposes a scheme for the evolution of tongue signaling.

The chemoreceptive functions of the lingual-vomeronasal system are numerous. For instance, it is used to detect food. *Ameiva exsul,* for example, could find hidden food, but this ability was diminished when their tongues were removed (Nobel and Kumpf, 1936). By touching its tongue to the substrate, a gila monster (*Heloderma suspectum*) was able to follow the trails of a pigeon egg and a mouse that had been dragged across the substrate (Bogert and Martín del Campo, 1956). Newly hatched five-lined skinks (*Eumeces fasciatus*) use chemical cues to identify species-characteristic prey (Burghardt, 1973). *Ameiva corvina* and *Gerrhonotus multicarinatus* increase rates of tongue extrusion when food is present (Gove, 1978). Not all lizards, however, appear to use the tongue to find food. Tongue extrusions almost never precede feeding in *Sceloporus jarrovi* (Simon et al., 1981).

Some lizards exhibit high rates of tongue extrusion just after emergence or when placed in unfamiliar areas and thereby may rapidly obtain information concerning conspecifics or predators (Simon et al., 1981). The frequency of tongue extrusion in *S. jarrovi* in nature is significantly higher during the first hour following emergence than at other times during the day. The frequency of tongue extrusion is also higher in displaced animals than in animals that have been active for at least an hour (Simon et al., 1981). The frequency of tongue extrusion in *Heloderma* also depends in part on the familiarity of the lizards with their surroundings (Bogert and Martín del Campo, 1956).

Chuckwallas (*Sauromalus obesus*), it has been suggested, may use the lingual-vomeronasal system to detect chemical deposits of predators in the area (Berry, 1974), but no clear evidence supports this premise. In a laboratory study, adult *S. jarrovi* were unable to detect a natural predator, the Arizona mountain kingsnake (*Lampropeltis pyromelana*) (Simon et al., 1981). Numbers of tongue extrusions were unaffected by whether or not a kingsnake had been in the cage, and the lizards did not avoid sections of the cage that had been used by the snake. The possibility of predator detection by lizard chemoreceptive systems needs further examination.

A few studies demonstrate that vomeronasal organs may detect information about conspecifics and thus be used for territorial recognition or in finding mates or both. *S. jarrovi* increase their rate of tongue extrusions when introduced to cages that have recently housed conspecifics (DeFazio et al., 1977; Bissinger and Simon, 1981). These studies were done at times other than the breeding season, suggesting that these lizards use their vomeronasal system to monitor territories.

Ground skinks (*Scincella laterale*) may use chemoreception to avoid conspecifics. These skinks are cannibalistic and engage in violent fights (Lewis, 1951; Brooks, 1967; Fitch, 1967; Duvall, Herskowitz, and Trupiano-Duvall, 1980). Other skinks can also use chemoreceptors to detect conspecifics. In a small animal olfactometer nonbreeding *E. fasciatus* significantly approached odors of conspecifics (Duvall, Herskowitz, and Trupiano-Duvall, 1980). However, *S. lateralis* (males and females) avoided male odors; and males, but not females, approached female conspecific odors. Skinks of both sexes responded randomly when presented with odors of other species. Tongue extrusion rates did not vary as a function of the sex or species of the odor stimulus. Perhaps the odors were strong, and the vomeronasal system was not necessary for crucial discriminations.

Hormone-treated and reproductively active male *S. occidentalis* also discriminate between unmarked surfaces and those labeled by conspecifics of the same sex (Duvall, 1979). Hormone-treated females do not appear to discriminate surfaces labeled by either sex. However, naturally breeding females respond with high rates of tongue flicks (Fig 6.3) and

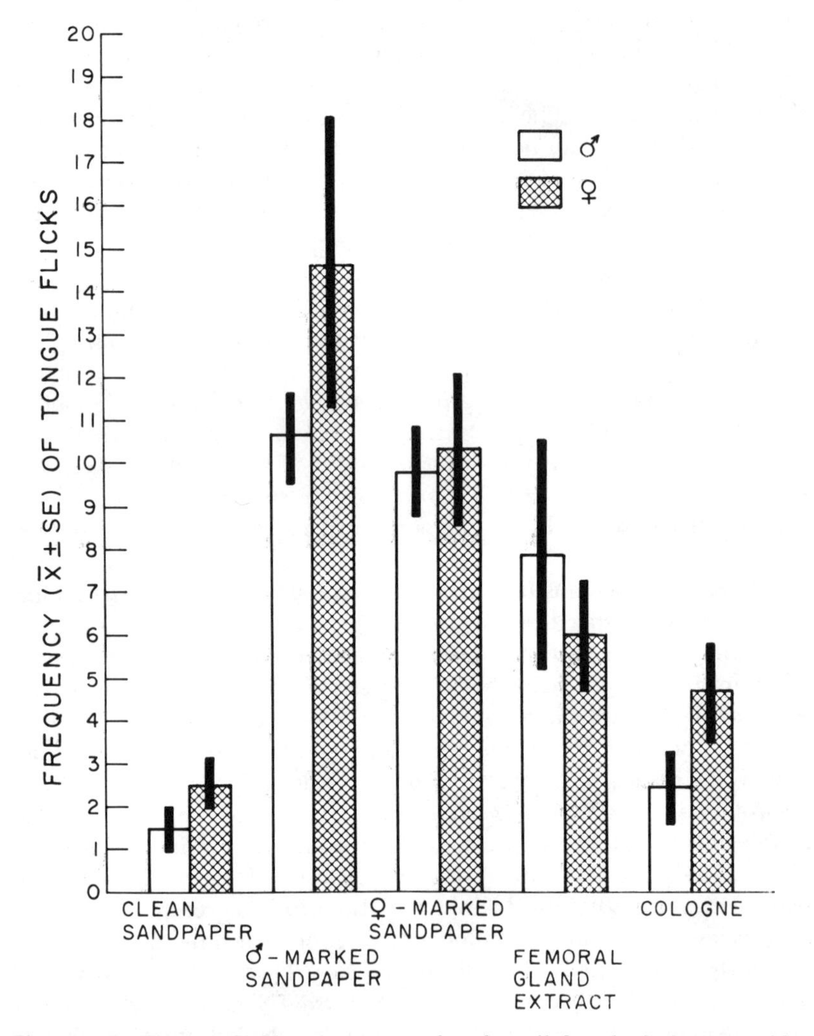

Figure 6.3 Tongue-flicking responses of male and female *Sceloporus occidentalis* to 5 test conditions. Tests lasted 18 minutes, and the experimental lizards were exposed to a successive discrimination procedure, on alternate days, to 5 types of sandpaper. From Duvall (1980, and unpublished data).

substrate touches to exudates collected from sexually receptive males and females (Duvall, 1980).

Tests with *S. jarrovi* in outdoor enclosures showed that naturally breeding males exhibited significantly more tongue extrusions in the part of the enclosure previously occupied by an adult female (Gravelle, 1981). Prior occupation of other parts of the enclosure by a male or a juvenile did not significantly influence the time spent in parts of the enclosure, or the defecation sites of the experimental males or females. However, pelvic wipes by males (see below) were more numerous in areas previously occupied by females.

Other studies have suggested the importance of chemical cues in lizard courtship and mating. Integumentary odors may influence sex discrimination in the gecko, *Coleonyx variegatus* (Greenberg, 1943). Males normally bite and hold tails of females during courtship. However, when presented with partially anesthetized male and female geckos that had had their tails surgically exchanged, courting males gripped and held "female" tails on male geckos but did not grip "male" tails on female geckos. Male Komodo dragons (*Varanus komodoensis*) touch their tongues to the body and head of a prospective mate, particularly around the sides of the head, between the eye and nostril and at the junction of the hind leg with the body (Auffenberg, 1978).

The vomeronasal system may also be important in maternal behavior. Female five-lined skinks (*E. fasciatus*) tongue flick frequently when finding and retrieving their eggs, but tongueless skinks are unable to find eggs (Noble and Mason, 1933; Evans, 1959). Intact skinks do not retrieve paraffin model eggs, shellacked eggs, and the eggs of other lizard genera (Noble and Mason, 1933). Female *Sceloporus rufidorsum* frequently touch their tongues to their eggs during egg burying (Duvall, Trupiano, and Smith, 1979).

Clearly, the lizard vomeronasal system has had many suggested uses. Some of these uses are backed by experimental evidence, but most need substantiation.

Chemical Signals and Associated Glands

Chemical secretions used in intraspecific communication in lizards may come from the body surface, the cloacal region, specific glands, or other organs such as the femoral pores. Behavioral evidence for the deposition of secretions from femoral pores or the cloaca comes from many sources. Hatchling green iguanas rub their thighs and bellies on the substrate immediately after emergence from nest openings (Burghardt, Greene, and Rand, 1977). Hatchlings often touched the substrate or each other with their tongues. Chuckwallas (Berry, 1974) and chameleons (Parcher, 1974) commonly rub their cloacas on the substrate following defecation, and chuckwallas touch their tongues to the cloacas of conspecifics (Berry, 1974). Defecations at specific locations by male chuckwallas may produce chemical signals which assert the male's presence (Carpenter, 1975). Substrate touching and rubbing movements of chuckwallas are typical of iguanids, and many iguanids use chemical signals to enhance or supplement visual signals (Carpenter, 1975). Conspecific exudates significantly increased the frequency of push-up visual displays (Duvall, 1979, 1980) in western fence lizards (*Sceloporus occidentalis*). In addition, cloacal exudates of both male and female *S. occidentalis* elicit tongue extrusions, and fecal boli on rocks and other surfaces serve as potent visual signals, suggesting some degree of coevolved chemical and visual signaling (Duvall, 1980).

Cloacal rubbing by male whiptails (*Cnemidophorus sexlineatus*) precedes sexual activity (Carpenter, 1962). Pelvic rubs by breeding *Sceloporus jarrovi* (Gravelle, 1981) are more common than in nonbreeding *S. jarrovi* (Simon et al., 1981). Moreover, males engage in pelvic rubbing more often in pens recently occupied by adult females than in pens recently occupied by males or left empty (Gravelle, 1981). Lizards may not have to make an effort to mark an area; resting the cloaca or femoral pores on the substrate may result in the passive deposition of chemicals.

Chuckwallas (Berry, 1974), *Iguana iguana* (Burghardt, Greene, and Rand, 1977), and *S. jarrovi* (Simon et al., 1981) rub their jaws on the substrate. *S. jarrovi* exhibited 227 chin rubs during 9,000 observation minutes during nonbreeding months. This behavior was not simply a way of cleaning the face after eating. Only 1.8 percent of these rubs occurred immediately after eating, whereas 30.4 percent occurred either immediately before or after movement to another location. However, the frequency of chin wipes did not increase during the breeding season (Gravelle, 1981). Simon et al. (1981) were unable to detect histologically any integumentary glands in the infralabial area of the chin. Large buccal glands did occur inside the mouth and opened medially into the mouth, and substances exuded from the mouth might be rubbed onto the substrate by the jaws. Alternatively, secretions from the Harderian gland, located near the eye (Bellairs and Boyd, 1950), may be deposited by face and jaw rubbing actions (Duvall, 1980).

Lizards have diverse types of integumentary glands (review in Madison, 1977). Gekkonid lizards have holocrine secretory structures (Taylor and Leonard, 1956; Maderson, 1967, 1968a, 1968b), called "generation glands" (Maderson and Chiu, 1970), whose secretions are directly associated with periodic skin shedding (Maderson, 1970).

Several macroscopically visible, discrete secretory structures in the preanal, posterior abdominal, and femoral regions are called "preanal organs" (Kluge, 1967). Such glands are widespread in lizard species and have been extensively studied in gekkonids by Maderson and coworkers (Maderson, 1972; Chiu and Maderson, 1975; Chiu et al., 1975; Menchel and Maderson, 1975). In geckos these organs are usually found only in males (Chiu and Maderson, 1975; Menchel and Maderson, 1975). Some females show macroscopically discernible depressions, whereas the males have preanal pores. These depressions of females may potentially differentiate into functional structures when androgens are provided (Chiu, Lofts, and Tsui, 1970; Chiu et al., 1975). Removal of trophic hormones in males decreased androgen production and produced glandular atrophy.

Many species of lizards have preanal pores, composed of branching tubes and tubules, in the femoral region along the midventral axis of each thigh. Femoral pores are generally present in equal numbers in both sexes but are usually larger or more complex in males (review in Cole, 1966b). In some species the pores are present only in males; each of the unisexual *Cnemidophorus,* however, has femoral pores. Hatchlings of all lizard species studied exhibited little sexual dimorphism in relation to these femoral organs. Femoral pores of males, but not of females, of the iguanid, *Crotaphytus collaris,* increase in size and complexity during ontogeny (Cole, 1966a). The relative length of these organs appeared to vary seasonally in adult males; the greatest size occurred during the breeding season (Cole, 1966b). In some species the pores secrete more actively in the breeding season, and their action seems to be testosterone dependent.

Cole (1966b) discusses five possible functions of femoral pores. Two concern chemical communication. The secretions may attract females to males during the breeding season or mark male territories or both. Unfortunately, these functions have only recently begun to be examined experimentally (Duvall, in preparation; see Fig. 6.3).

Many lizards have distinct cloacal glands (Gabe and Saint-Girons, 1965, 1967; Whiting, 1967, 1969). Some lizards possess a single, dorsal medial gland (Whiting, 1969; Kroll and Reno, 1971), whereas other species have dorsal and ventral cloacal glands (Gabe and Saint-Girons, 1965). *Sceloporus graciosus* males have a subcutaneous, sacular gland, just anterior to the vent (Burkholder and Tanner, 1974). The gland secretes material during and just after the breeding season. Similar glands occur in *Sceloporus undulatus, S. occidentalis,* and *S. magister.* Most postanal organs with integumentary ducts are found in scincid lizards (Kluge, 1967; Bellairs, 1970). Gekkonid lizards of both sexes possess bilateral cloacal sacs which open just posterior to the cloaca (Smith, 1935).

Glandular structures are clearly diverse in lizards. Nevertheless, an almost unexplored area in lizard chemoreception involves behavioral reactions to specific glandular secretions.

Variability in the Significance of the Vomeronasal System

The importance of the vomeronasal system varies somewhat with taxonomic affinities of lizards. Camp (1923) divided lizards into two major groups, the ascalabotans (Chamaeleonidae, Agamidae, Iguanidae, Gekkonidae, and Xantusiidae) and the autarchoglossans (Lacertidae, Scincidae, Teiidae, Cordylidae, Helodermatidae, Varanidae, and all other lizard families). An alternate scheme for grouping lizards (Northcutt, 1978) has also been related to lizard chemoreception (Duvall, 1980). The ascalabotans seem to rely most heavily upon vision, whereas most autarchoglossans use both vision and chemoreception (Camp, 1923; Evans, 1961, 1967; Underwood, 1970; Bissinger and Simon, 1979; Duvall, 1980; Duvall, Herskowitz, and Trupiano-Duvall, 1980).

Anatomical evidence supports this generalization. Jacobson's organs are usually better developed in autarchoglossans than in ascalabotans (Pratt, 1948). Nevertheless, some sceloporine iguanids have well-developed nasal and vomeronasal structures (Guillette and Duvall, personal communication).

Tongue morphology (McDowell, 1972) is also better developed in autarchoglossans. Teiids and varanids have long, thin, highly protrusible, and deeply notched tongues that can sample large areas (Burghardt, 1977; Gove, 1978) and deliver chemicals to the slits leading to the Jacobson's organs (Pratt, 1948). In contrast, cordylids and iguanids have short, broad tongues that are not especially protrusible and are scarcely or not at all notched at the tip.

Behavioral studies also suggest that chemoreceptive systems are most highly developed in autarchoglossans. Teiids, lacertids, and skinks rely more upon chemical stimuli. These lizards are often secretive, even if diurnal, and many are crepuscular or nocturnal; a few are subterranean (Evans, 1961, 1967). Chemoreception is probably their most important sense for finding food (Underwood, 1951). In contrast, chamaeleonids, iguanids, agamids, and gekkonids rely primarily on visual signals in courtship and territorial interactions (Underwood, 1951; Evans, 1961, 1967). These groups are mainly diurnal, conspicuous, and usually have color vision.

Baseline rates of tongue extrusions for 14 species representing 6 families (Table 6.2) also support the idea that the vomeronasal system is used more by autarchoglossans than ascalabotans (Bissinger and Simon, 1979). Teiids, which have well-developed Jacobson's organs and tongues, show high frequencies of tongue extrusion. The *Cnemidophorus* studied seem to have little visually oriented social behavior. The helodermatids, which have well-developed Jacobson's organs and tongues but reduced eyes (Underwood, 1951, 1970), also exhibit high frequencies of tongue extrusion. *Heloderma* exhume and eat reptile eggs that are buried in the soil and probably use chemical cues to find this food (Bogert and Martín del Campo, 1956). Skinks also extrude their tongues often. Skinks show various stages of eye reduction (Underwood, 1951, 1970) and have well-developed Jacobson's organs and slightly bifurcate tongues (Pratt, 1948).

Table 6.2 Freqency (mean ± S.E.) of tongue extrusions per 30 minutes in lizards.[a]

Taxon	Frequency
Autarchoglossa	
Cordylidae	
Cordylus giganteus	1.0 ± 1.00
C. warreni	5.8 ± 2.85
Gerrhosauridae	
Gerrhosaurus validus	28.8 ± 5.55
Zonosaurus madagascariensis	36.4 ± 9.33
Scincidae	
Egernia cunninghami	64.0 ± 15.88
Tiliqua nigrolutea	128.8 ± 45.01
T. scincoides	165.4 ± 28.86
Helodermatidae	
Heloderma horridum	201.0 ± 21.81
H. suspectum	235.4 ± 24.35
Teiidae	
Cnemidophorus tigris	228.2 ± 58.71
C. tesselatus	311.8 ± 58.05
C. exsanguis	358.4 ± 59.59
Ascalabota	
Iguanidae	
Sceloporus jarrovi	8.2 ± 3.67
Enyaliosaurus clarki	13.0 ± 5.41

a. Modified from data in Bissinger and Simon (1979).

Among the autarchoglossans only the cordylids had low frequencies of tongue extrusions (Table 6.2). Jacobson's organs of cordylids are similar to those of the skinks (Malan, 1946; Pratt, 1948), but their tongues are short and relatively nonprotrusible (McDowell, 1972). Many cordylids are seldom found far from their crevices and spend most of their time basking (FitzSimons, 1943). Consequently, they may monitor the small areas around their crevices visually, and chemoreception may be unimportant, except perhaps during the mating season.

The family Iguanidae was the only ascalabotan family studied. As expected, these lizards rarely extrude their tongues (Table 6.2), which are short, broad, and relatively nonprotrusible (McDowell, 1972). Jacobson's organs are reduced but functional (Pratt, 1948). Iguanids are highly visual animals and rely on vision in feeding and in territorial and courtship behaviors (Underwood, 1951, 1970; Bellairs, 1970). These patterns led some workers to suggest that chemical signals are unimportant for iguanids (Ferguson, 1966; Evans, 1967; Madison, 1977). Recent work, however, demonstrates that chemoreceptive systems of iguanids, though less well developed than in other lizards, do function in chemical communication (DeFazio et al., 1977; Duvall, 1979, 1980; Bissinger and Simon, 1981; Gravelle, 1981; Simon et al., 1981).

Although phylogeny seems to play a major role in the significance of the vomeronasal system of lizards, ecological factors—as with the cordylids—may also be important. Jacobson's organs are usually well developed in ground-living forms, intermediate in scansorial species, and reduced in arboreal lizards (Pratt, 1948). Arboreality may require massive orbital development, which in turn might limit development of the olfactory and vomeronasal organs (Pratt, 1948). Moreover, chemical senses may be more important to species living on the ground where vision is often blocked. The scansorial skink, *Egernia cunninghami,* extrudes its tongue less frequently than do terrestrial skinks of the genus *Tiliqua* (Table 6.2). Similarly, the arboreal iguanid, *Anolis trinitatus,* extrudes its tongue less frequently (Gravelle and Simon, 1980) than does a terrestrial iguanid, *S. jarrovi* (Bissinger and Simon, 1979).

Conclusions

Anecdotal, correlational, and experimental evidence suggests that chemoreception is important to lizards. Nevertheless, we have only just begun to understand how the vomeronasal system works, what its functions are, and when it is likely to be important for the survival and reproduction of an individual. We know far less about taste and olfaction. Many questions remain to be answered (Madison, 1977; Duvall, Herskowitz, and Trupiano-Duvall, 1980).

The use of taste and olfaction by lizards is poorly documented, especially with respect to the interaction of those sensory systems with the vomeronasal system. It seems logical to begin by systematically examining the histology of lizard tongues. If taste buds are present on the tongues of some lizards, we will need to examine carefully the interactions of taste and the vomeronasal system.

Lizards have well-developed olfactory systems, but the functional differences between the olfactory and vomeronasal systems are unclear. The hypothesis of Cowles and Phelan (1958), that olfaction "switches on" the vomeronasal system, is of interest here. Indeed, tongue flicking in female *S. occidentalis* is initiated more quickly when odors are present than when they are not (Duvall, 1980). Latencies of tongue extrusion, in the presence and absence of odor stimuli, could easily be measured in a variety of species to investigate the generality of Cowles and Phelan's idea.

The vomeronasal system clearly has a variety of functions. The preceding review suggests that some lizards, such as teiids and helodermatids, use this system often and for a variety of reasons, whereas other groups of lizards, such as iguanids, are more visually oriented and use the vomeronasal system less often. *Sceloporus jarrovi,* for example, uses this system to detect conspecifics (DeFazio et al., 1977; Bissinger and Simon, 1981; Gravelle, 1981; Simon et al., 1981) but not to detect predators or food items (Simon et al., 1981). Unfortunately, these latter studies are the only ones which focus on a variety of possible vomeronasal functions in a single species. We need to examine all possible functions of the vomeronasal system for individual species, and such detailed studies must then be repeated on many species to elucidate the influence of phylogeny and ecology. It would be interesting to determine, for example, all of the functions of the vomeronasal system for a group of closely related species that demonstrate variable life-styles.

Another major problem with the literature is that it is largely anecdotal. Nevertheless, anecdotes provide a reason to search for a particular behavior in a given species. Consequently, Table 6.1 provides a good starting point for potential experimental studies of vomeronasal functions such as maternal behavior. Of all the vomeronasal functions suggested in Table 6.1, only the idea that some lizards use the vomeronasal system to explore novel stimuli and detect conspecifics has been examined systematically. Even these studies are recent and few.

The knowledge that conspecifics can be detected via the vomeronasal system brings up another series of problems. Where do the deposited chemicals come from, where are they deposited, and what is their chemical nature? Are there seasonal differences in these depositions? The products of femoral pores as well as of cloacal and other glands must be isolated and tested for behavioral responses, such as rates and latencies of tongue extrusions and the amount of time spent in a marked area. Duvall's (1980) methodology seems promising here. Lizards can be presented with bricks covered with sandpaper that is clean or labeled with glandular exudates, and their behaviors noted. Seasonal and interspecific differences should be examined.

Behavioral ecologists cannot themselves answer all the important questions. Sensory physiologists, for example, must demonstrate that chemicals brought in by the tongues are delivered to the vomeronasal organs and stimulate the accessory olfactory bulb of the brain via the vomeronasal nerve. As mentioned earlier, these tasks have been accomplished for garter snakes (Kubie, 1977; Meredith and Burghardt, 1978), but not for lizards.

Obviously, many questions remain in this area of lizard biology. Studies have remained anecdotal and speculative too long. I hope that a clearer view of lizard chemoreception will emerge during the next decade.

7 Food Availability and Territorial Establishment of Juvenile *Sceloporus undulatus*

Gary W. Ferguson, John L. Hughes, and Kent L. Brown

TERRITORIAL BEHAVIOR, often defined as defense of an area or site, has been ascribed to various adaptive functions—for example, defense of mates, food, or nests (Nice, 1941; Hinde, 1956; Brown, 1964). In lizards all of these adaptive roles have been suggested. Territories defended by juvenile lizards are often thought to be related to defense of food supplies, refugia, or basking sites, and those of adult lizards may also be related to defense of mates or nest sites (Rand, 1967; Brown and Orians, 1970; Stamps, 1977).

Most decisions about the adaptive significance of territoriality are based on conjecture or indirect evidence, but sometimes such evidence is fairly convincing. For example, female territoriality that occurs near nesting areas where food for the female is scarce or absent and only at a time when other ovigerous females are seeking nest sites is hard to interpret as anything but nest defense (Carpenter, 1966; Rand, 1968; Yedlin and Ferguson, 1973). Territoriality by male elephant seals at sites offering little or no resources other than numerous females could be little other than harem defense (Le Boeuf and Peterson, 1969). Lekking territories of prairie chickens and other lekking species offer little else to the owner but mating rights (Robel, 1966).

In some instances, however, such indirect evidence is insufficient to explain the adaptive role of the territory. For example, territoriality may be confined to, or more intense during, the breeding season but may include a defended area where activities such as feeding, basking, and escape from predators occur at other seasons. Higher fitness may be due to defense of mates or nests. But breeding season is a time of increased energy demands and activity, and benefit due to defense of food supply or predator refugia cannot be ruled out.

Two field methods can help to elucidate the adaptive significance of territory for cases in which the roles are complex. (1) Territorial quality in terms of possible limited resources can be measured; then, the fitness of individuals on territories can be correlated with variations in territorial quality (Fox, 1978; Fox, Rose, and Myers, 1981). A positive correlation between fitness of the territorial individual and the territorial quality would indicate that the measured parameters act as an evolutionary stimulus for territorial behavior. (2) The presence or amount of a particular resource found within territories and suspected to be important in determining fitness can be manipulated. If the fitness of a territory holder is modified by this treatment, one can conclude that this resource may have served a primary role in the evolution of territorial tendencies.

Territory size has been altered by manipulating resources experimentally (Simon, 1975; Krekorian, 1976). This alteration would indicate that the manipulated resource served either directly or indirectly as a proximate modifier of territorial tendencies. While it seems unlikely that a resource that controls territorial tendencies would not also be necessary to maximize the fitness of the individual, this supposition may be incorrect if the effect of the resource on fitness is indirect. Thus, a juvenile male lizard may defend a log because there is always more food associated with logs in nature. His growth rate is enhanced; he is an adult sooner; he obtains more females; he leaves more surviving offspring. Addition of artificial logs may cause him to expand his territory to include more logs but reduce his fitness because the fake logs do not give him the nutrition necessary to sustain the cost of defending a larger territory. Conversely, the addition of food may increase his fitness without necessarily causing him to modify his territorial behavior because no direct behavioral association between food and fitness has occurred in the evolutionary past. When one uses the term "defended resources," one must clearly discriminate between the proximate role that resources have in stimulating or modifying territorial behavior of the individual and the ultimate role that these resources may have in modifying the fitness of an individual with inherited territorial tendencies (see also Rand, 1967; Brown, 1975). Direct measures of behavior are important in determining the proximate role; direct measures of fitness and behavior are necessary to determine the ultimate role. In 1978 an experimental field study was begun to assess the role of food shortage in the fitness of individual *Sceloporus undulatus.* The experimental design also allowed the assessment of the role of food availability in determining the future fitness of hatchlings and in modifying territorial tendencies. The effect of supplementary feeding on territorial establishment and fitness of hatchling northern prairie lizards, *S. undulatus garmani,* is reported here.

Materials and Methods

Sceloporus undulatus is perhaps the most widespread and abundant iguanid lizard in the United States. Various subspecies range from New Jersey to Florida, westward across the prairie, beyond the Rocky Mountains to the desert regions of Arizona and Utah. The prairie swift, *S. undulatus garmani,* contains the smallest sized individuals of the species (average adult snout to vent length 55 mm). They are more adapted to open sandy areas devoid of trees or canyon walls than the individuals of other subspecies. The demographic patterns of *S. u. garmani* have been well studied and compared with those of other subspecies (Ferguson, Bohlen, and Woolley, 1980). Prairie swifts tend to mature earlier and have a shorter life span than do members of other subspecies.

From June 1978 to October 1979 a study was conducted on the prairie swift in the sand prairie pastureland of Stafford County, Kansas. The study included 8 plots 900 m^2 and was designed to test experimentally a hypothesis generated by previous studies on the lizard conducted in Pottowattamie County, Kansas (Ferguson and Bohlen, 1978). The hypothesis attempted to explain the role of food competition as a selective agent causing larger hatchling size late in the season. Two experiments were conducted in each year; one was begun from July 5 to 21 in 1978, and from July 16 to August 1 in 1979; the other was begun from August 4 to 18 in 1978, and from August 9 to 25 in 1979. The experimental design included 3 study plots in the natural habitat of the lizard on which hatchling lizards of different sizes were provided supplementary food and 4 (1979) or 5 (1978) plots serving as controls. Plot treatment was rotated in 1979 so that the 1978 experimental plots served as controls and vice versa. Larger hatchlings did not exhibit a fitness advantage on any of the control or experimental plots in 1978 and 1979, but supplementary feeding had a dramatic effect on activity ranges and overall measures of fitness of the experimental versus the control lizards.

Study Area. The study region was selected because the prairie swift density was known to be very high in the sand prairie south of the Arkansas River (Smith, 1950). The soil is exposed, windblown sand with scattered patches of grasses and annual plants. Cottonwood trees (*Populus deltoides*), scattered or in groves, are the dominant shading factor. Groves of willow (*Salix humidis*), juniper (*Juniperus virginiana*), catalpa (*Catalpa speciosa*), black locust (*Robina pseudoacacia*), and chickasaw plum (*Prunus augustifolia*) are common in the pasture but are usually shunned by the lizards. The best single habitat indicator of high *Sceloporus undulatus* density is fallen logs (usually cottonwood or catalpa) that are exposed to sunlight most of the day.

The 8 plots required for the study were chosen from 15 preliminary plots selected in fall 1977. While it was impossible to select plots with identical habitat features and microclimate, it was possible to find plots that allowed duplication between experimental and control areas of such features as number of brush piles, sun exposure, and soil relief. The critical criteria for final selection were density of juvenile and adult lizards on the areas, presence of fallen logs, brush piles or other wooden structures, some degree of "insularity," defined here as the tendency of plots to be surrounded by perimeters of less favorable habitat (either plowed fields, heavily grazed, brush-free pasture, or dense tree groves). None of the plots was a true island of good habitat surrounded by bad, but each usually left one with the impression of semi-isolated concentrations of favorable habitat.

Collection Techniques. Because we were interested in accurately measuring hatchling size to correlate this variable with fitness, cohorts of hatchlings of known size were established on the plots as follows. Gravid females were collected from the plots in early May for the early experiments and in early June for the late experiments and returned to the laboratory until they laid their eggs in moist sand. Eggs were placed singly into plastic vials containing moist sand and incubated at a constant 27 °C in a Sherer constant temperature chamber (Ferguson and Brockman, 1980). All spent females were returned to their site of capture within three weeks. Eggs laid in May hatched in early July; those laid in June hatched in early August. Within three days of hatching, hatchlings were measured to the nearest millimeter, weighed with a Mettler top loading balance to the nearest 0.01 g, toe clipped, paint marked, and released on the study areas in cohorts of 15 to 23 individuals in July and 11 to 18 individuals in August. While hatchlings were not necessarily returned to the plot of their mother, the number returned to each plot approximated what the removed females would have produced if they had laid their eggs in the field. Thus, with one exception, densities were not appreciably manipulated to abnormal levels. The exception occurred in the July experiment in 1979. Unusually high laboratory temperatures in May caused several gravid females to die before depositing their eggs. Thus, there was a shortage of laboratory produced hatchlings to establish on all plots in July. Thirty-seven field-hatched juveniles of hatchling size were removed from area D. Fifteen lab-hatched lizards were released on area D, so July juvenile densities on that plot were lower than they might have been without interference. The release sites on all plots included habitat with cover sufficient to render the newly released lizards camouflaged. The release sites were the same for the July and August experiments (hereafter referred to as release I and release II periods, respectively).

Feeding Technique. The most effective means of feeding free-living hatchlings on the experimental plots proved to be the one-by-one delivery of one-quarter-grown commercial house crickets (*Achaeta domestica*) via a plastic soda straw "blow gun." Crickets carried in a vial could be sucked easily into the proximal end of the straw and shot accurately to within 2 to 10 cm of a resting or active lizard. Lizards nearly always readily devoured each projectile and would be fed to satiation. Indeed, some individuals became very docile and would run up to the investigators as if anticipating being fed. Some individuals eventually accepted crickets held between the fingers. There are several advantages of these techniques. First, one can enumerate quickly the actual supplementary food ingested by each marked individual. Second, location of the supplementary food depends on the location of the lizard immediately preceding each feeding. Lizards do not have to relocate to a "feeding station" with a location predetermined by the investigator. Third, introduction of insects into the ecosystem which could modify the insect community is minimized. Artificial feeding occurred 6 times per week per plot. Each lizard was fed to satiation an average of 1.7 times per week.

Collection of Data. Survivorship was assessed as persistence of marked individuals on the plots over time. An exhaustive census was conducted 3 weeks after the beginning of each experiment and again in the spring of 1979. However, survival of the marked hatchlings to the following spring was so poor (6.7 percent survival of experimentals; 5.6 percent survival of controls) that analyses of annual survival were meaningless. Clearly, some of the plot's "nonsurvivors" emigrated and possibly remained alive at other locations. The favorable habitat immediately adjacent to the areas was thoroughly searched, and any individuals emigrating considerably further across the less favorable habitat were assumed to suffer a high mortality.

Each day all lizards seen on the plots and not possessing a paint mark were captured, returned to the laboratory, processed in a fashion similar to that of the hatchlings, and returned to the field within 6 hours. This enabled the calculation of growth rates of previously toe-clipped lizards that had shed their paint mark and the determination of total numbers of lizards on the plots. This number included released lab-hatched lizards (both experimentals and controls) and those occurring naturally on the area. Each of the latter received a spot of tan paint on their tails and will be referred to hereafter as field-hatched juveniles. No field-hatched juveniles were fed on the experimental or control areas.

During walk-throughs, all paint-marked experimental and control lizards were tallied and their site of original detection plotted. From this, activity ranges could be determined by the minimum convex polygon technique (Tinkle, 1967). Percent overlap of the home range of lizard a with that of other lizards was determined by the formula $O = 100\ (b/\ B)$, where O is the percent overlap, b is the area in square feet co-occupied by lizard a and any others, B is total home-range area in square feet of lizard *a* measured by counting graph paper squares included within the polygon. Time did not permit territory mapping using nonresident introduction (see Yedlin and Ferguson, 1973; Simon, 1975) to demonstrate that exclusively occupied area was aggressively defended. However, vigorous aggression among unfed juveniles was observed on 6 occasions during the walk-throughs. Simon and Middendorf (1980) showed 80 percent of a juvenile *Sceloporus jarrovi*'s home range to be nonoverlapping and defended aggressively. Among juvenile *S. undulatus* on the control plots, nonoverlapping home-range area averaged 73.2 percent in 1978 and 71.6 percent in 1979. We assumed these areas to have been aggressively defended territories.

The estimate of home-range size by the minimum convex polygon technique has been shown to be positively correlated with number of recaptures (see Schoener, 1981). There were differences in the number of recaptures used to estimate home-range sizes in this study. In the 1978, release I period, the mean number of recaptures per home range was 12.5 on the experimental areas, 5.5 on the control areas, and 9.16 on the areas combined. In 1979 the means were: release I, 11.6 on the experimental areas, 6.9 on the control areas, 9.06 on the areas combined; release II, 8.2 on the experimental areas, 5.6 on the control areas, 7.0 on the areas combined. Thus, biases would be (1) to underestimate home-range size on control areas more than on experimental areas and (2) to underestimate home-range size during the release II period more than during the release I period. All significant differences discussed here were opposite to those potentially caused by these biases.

Insect abundance throughout the season was monitored in 1978 to detect seasonal changes in food availability. For purposes of this report, insects classified to size and order were trapped with Tree Tanglefoot applied to sheets of paper affixed to clipboards (Ferguson and Bohlen, 1978). Because ants avoided these traps, the visual census technique applied in the Ferguson and Bohlen study was used here also.

Standard parametric (analysis of variance, correlation and regression, Sokal and Rohlf, 1969) and nonparametric (Mann–Whitney U and χ^2, Siegel, 1956) tests were used to test for significant differences in home range, growth, survival, and densities.

Results

Home-Range Size and Overlap

Supplementary feeding, begun the day after release of the hatchlings on the experimental plots, caused them to establish a smaller mean home-range size than those of hatchlings on the control plots (Table 7.1). The effect was highly significant during the release I period, both in 1978 and when 1978 and 1979 data were combined. Whereas the mean

Table 7.1 Home-range size [$\bar{X}$ ± S.D. square feet (N)] of lab-hatched prairie swifts on experimental versus control plots.

a. Home-range size

Year	Release I (July)		Release II (August)	
	Experimental	Control	Experimental	Control
1978	26.8 ± 19.6 (35)	150.9 ± 452 (37)	–	–
1979	140.4 ± 128.5 (28)	167.5 ± 175.3 (30)	104.6 ± 106 (31)	134.8 ± 150 (26)

b. Statistical summary

Comparison	Test	Probability of H_o
Control versus experimental, home-range size, 1978 release I	Mann-Whitney U	<0.001
Control versus experimental, home-range size, 1979 release I	Mann-Whitney U	>0.05
Control versus experimental, home-range size, years combined release I	Mann-Whitney U	<0.002
Control versus experimental, home-range size, 1979 release II	Mann-Whitney U	>0.05
Control versus experimental, home-range s^2, 1978 release I	F-test	<0.001
Control versus experimental, home-range s^2, 1979 release I	F-test	<0.05
Control versus experimental, home-range s^2, 1979 release II	F-test	<0.05
Home-range size versus year, release I	Mann-Whitney U	<0.001

home ranges of the fed lizards were not significantly smaller in either the release I or II periods in 1979, home-range sizes were significantly more variable on the control plots during both periods. All hatchlings established significantly smaller home ranges in 1978 than in 1979 during the release I periods. Home ranges were not measured during the release II period in 1978.

Supplementary feeding begun the day after release of the hatchlings caused them to establish home ranges with a greater degree of overlap than those of the hatchlings on control plots (Table 7.2). Effects were highly significant during both release periods in 1979 and in the release I period in 1978. While aggression was observed among fed juveniles, only 2 of the 10 encounters included attacks or chases. Conversely, all 6 observed aggressive acts initiated by unfed lizards included a chase. We feel that supplementary feeding reduced the early expression of juvenile territorial behavior.

Fitness

Fitness is best defined as contribution to future generations. It is most completely measured as number of individuals surviving to reproduce and reproductive output of those survivors. In this study the fitness of all lab-produced hatchlings was very low. Very few released hatchlings survived to maturity (7 out of 103 experimentals, 9 out of 170 controls in 1979; spring survivorship could not be assessed in 1980 due to loss of access to the study areas). However, differences in early growth within the lab-produced hatchlings that in some years could translate into fitness differences were due to artificial feeding. Thus, faster growth should allow males to attain a larger size in spring, giving them a dominance advantage regarding establishment of a breeding territory. Faster growth among females should give them a size advantage in the spring allowing them to produce bigger clutches.

Supplementary feeding did not affect growth significantly in the release I periods of 1978 or 1979 (Table 7.3). However, growth was significantly enhanced in the release II periods of both years. In 1978, growth in length was enhanced and in 1979 growth in both length and weight was strongly enhanced. Furthermore, growth rates of controls were significantly lower during the release II period of 1979 than in 1978. Thus, food seemed to be in shorter supply late in the hatching season as suggested earlier (Ferguson and Bohlen, 1978). Also, food seemed to be in shorter supply in 1979 than in 1978.

Comparing growth in the field of lab-hatched juveniles with that of field-hatched juveniles revealed unexpected differences (Table 7.4). Only hatchling-size, field-hatched juveniles were used in comparisons. Field-hatched juveniles tended to grow faster than lab-hatched juveniles on plots where the latter were not fed. The trend was significant only in 1979. Supplemental feeding alleviated the growth differential. The rea-

Table 7.2 Overlap of home ranges of lab-hatched prairie swifts on experimental versus control plots (percent).

a. Home-range overlap

	Release I (July)			Release II (August)		
Year	Number with < 50% overlap	Number with > 50% overlap	Percent with > 50% overlap	Number with < 50% overlap	Number with > 50% overlap	Percent with > 50% overlap
1978						
Experimental	16	19	54	–	–	–
Control	29	8	22	–	–	–
1979						
Experimental	7	21	75	14	17	55
Control	24	7	23	17	5	23

b. Statistical summary

Comparison	Test	Probability of H_o
Control versus experimental, high versus low overlap, 1978 release I	χ^2	< 0.004
Control versus experimental, high versus low overlap, 1979 release I	χ^2	< 0.001
Control versus experimental, high versus low overlap, 1979 release II	χ^2	< 0.01

Table 7.3 Growth rates of lab-hatched prairie swifts on experimental versus control plots.

Year	Release I mm/day $\bar{X} \pm$ S.E. (N)	Release I g/day $\bar{X} \pm$ S.E. (N)	Release II mm/day $\bar{X} \pm$ S.E. (N)	Release II g/day $\bar{X} \pm$ S.E. (N)
1978				
Experimental	0.47 ± 0.018 (29)	0.034 ± 0.0013 (29)	0.49 ± 0.023 (16)[a]	0.027 ± 0.0019 (16)
Control	0.44 ± 0.018 (40)	0.031 ± 0.0016 (40)	0.42 ± 0.016 (28)	0.026 ± 0.0013 (28)
1979				
Experimental	0.44 ± 0.014 (22)	0.030 ± 0.0018 (22)	0.47 ± 0.022 [a] (19)[c]	0.030 ± 0.0009 [b] (19)[c]
Control	0.45 ± 0.017 (26)	0.028 ± 0.0015 (25)	0.37 ± 0.012 (31)	0.023 ± 0.0008 (31)

a. $P < 0.05$.
b. $P < 0.1$.
c. $P < 0.001$.

son for the advantage of the field-hatched juveniles is not immediately obvious, but it further suggests that food was in more limited supply in 1979 than in 1978.

There was little significant effect of feeding on survival through the experimental period (Table 7.5). The only effect was the negative effect that feeding release I hatchlings had on the survival of the release II hatchlings (also fed) on the same areas. Thus, combining data from both years, there was a significant difference between survival of lab-hatched juveniles in the early versus late release periods but only on the experimental plots where lizards were fed.

Survival of lab-hatched juveniles differed from that of field-hatched juveniles only on the nonfed areas during the release II period (Table 7.6). Survival of the lab-hatched juveniles was higher, whereas their growth rates were lower (Table 7.4).

Overall densities (Table 7.7) of lizards (lab hatched and field hatched) on plots were compared at the end of the release I experimental period (mid-August) and at the end of the release II experimental period (late September). Densities were estimated from the formula $y = 2.99 + 0.289x$, where y is the total number determined from exhaustive census, x is the mean number of lizards seen during 3 consecutive searches during the exhaustive census. The formula was then used to estimate y at other times when x but not y was determined empirically. There were no significant differences in densities between the fed and nonfed areas

Table 7.4 Growth rates of lab-hatched versus field-hatched prairie swifts on experimental and control plots.

Year and hatching location	Release I				Release II			
	Experimental		Control		Experimental		Control	
	mm/day $\bar{X} \pm$ S.E. (N)	g/day $\bar{X} \pm$ S.E. (N)	mm/day $\bar{X} \pm$ S.E. (N)	g/day $\bar{X} \pm$ S.E. (N)	mm/day $\bar{X} \pm$ S.E. (N)	g/day $\bar{X} \pm$ S.E. (N)	mm/day $\bar{X} \pm$ S.E. (N)	g/day $\bar{X} \pm$ S.E. (N)
1978								
Lab	0.47 ± 0.018 (29)	0.034 ± 0.0013 (29)	0.44 ± 0.018 (40)	0.031 ± 0.0016 (40)	0.49 ± 0.023 (16)	0.027 ± 0.0019 (16)	0.42 ± 0.016 (28)	0.026 ± 0.0013 (28)
Field	0.44 ± 0.025 (10)	0.034 ± 0.0033 (10)	0.46 ± 0.028 (9)	0.027 ± 0.0028 (9)	0.47 ± 0.057 (3)	0.029 ± 0.0045 (3)	0.48 ± 0.102 (4)	0.030 ± 0.0028 (4)
1979								
Lab	0.44 ± 0.014 (22)	0.030 ± 0.0018 (22)	0.45 ± 0.017 (26)[a]	0.028 ± 0.0015 (25)[a]	0.47 ± 0.022 (19)	0.030 ± 0.0009 (19)	0.37 ± 0.012 (31)[a]	0.023 ± 0.0008 (31)
Field	0.43 ± 0.002 (9)	0.026 ± 0.0017 (9)	0.54 ± 0.042 (7)	0.039 ± 0.0037 (7)	0.48 ± 0.027 (6)	0.028 ± 0.0022 (6)	0.52 ± 0.028 (2)	0.032 ± 0.0049 (2)

a. $P < 0.05$.

Table 7.5 Survival of lab-hatched prairie swifts on experimental versus control plots.[a]

Year	Release I		Release II		Periods combined	
	Number surviving	Percent surviving	Number surviving	Percent surviving	Number surviving	Percent surviving
1978						
Experimental	30	57	16	40	46	49
Control	47	50	28	43	75	47
Total	77	52	44	42	121	48
1979						
Experimental	22	46	19	35	41	41
Control	26	43	30	43	56	43
Total	48	45	49	40	97	42
Years combined						
Experimental	52	53	35	37	87	45
Control	73	47	58	43	131	45
Total	125	49	93	41	218	45

a. The only significant difference was between the proportion of survivors in release I versus release II for the experimentals with years combined ($P < 0.02$, χ^2 test).

Table 7.6 Survival of lab-hatched versus field-hatched prairie swifts on experimental and control plots.

	Survivors	Nonsurvivors	Percent surviving
Feed areas release I			
Lab hatched	52	47	53
Field hatched	17	14	54
Feed areas release II			
Lab hatched	35	59	37
Field hatched	7	10	41
Nonfeed areas release I			
Lab hatched	73	81	47
Field hatched	10	8	56
Nonfeed areas release II			
Lab hatched	58	76	56[a]
Field hatched	5	21	19

a. $P < 0.05$.

Table 7.7 Numbers of lizards per 900 m^2 plot: $\bar{X} \pm$ S.E. (N).

	Years combined		Areas combined		Years and areas combined
	Experimental	Control	1978	1979	
End of release I period	12.0 ± 1.4(6)[a]	8.9 ± 1.9(8)[a]	13.2 ± 1.3(7)[a, b]	7.1 ± 1.6(7)[a,b]	10.2 ± 1.3(14)[a]
End of release II period	80.5 ± 11.7(6)	76.0 ± 11.0(8)	79.4 ± 9.7(7)	76.4 ± 12.9(7)	77.9 ± 8.1(14)

a. $P < 0.001$, Mann-Whitney U. Release I vs. release II.
b. $P < 0.05$, H_o, ANOVA. 1978 vs. 1979.

although there was a tendency toward higher densities on fed areas. Densities were significantly higher after the release I period in 1978 than in 1979. Densities were very significantly greater ($P < 0.001$) after the release II period than after the release I period regardless of the treatment or year.

Discussion

Food availability clearly plays a proximate role in the modification of territorial behavioral tendencies in hatchling *S. u. garmani.* Greater dispersion from the release site and establishment of exclusive home ranges occurred on control plots but not on experimental plots (Tables 7.3 and 7.4). A similar but not identical effect probably occurs in red grouse (Watson and Miller, 1971; Moss, Watson, and Parr, 1975; Miller and Watson, 1978) in which adult territory size, a strong measure of aggression in this species, is negatively correlated with food availability during the maturation period. In grouse, however, nutrition affects the expression of aggression at a later time. Simon (1975) demonstrated a constriction of territory size in adult *S. jarrovi* in response to artificial feeding. Krekorian (1976) demonstrated an expansion of home-range size (but an *increase* in overlap) in *Dipsosaurus dorsalis* following a dust storm that reduced the available supply of food plants. To our knowledge this study is the first experimental demonstration of the importance of food in the development of territorial behavior of juvenile lizards.

Although total fitness of individuals was not measured, growth rates were, and food availability enhanced growth rates, which in turn can enhance fitness. During the release II period, fed, lab-hatched, free-living juveniles grew significantly faster than nonfed, lab-hatched juveniles (Table 7.3). In 1979 during both the release I and release II periods, feeding the lab-hatched juveniles reduced the growth advantage that field-hatched juveniles exhibited over lab-hatched juveniles on control plots (Table 7.4). The effect of food availability on territorial behavior and potential measures of fitness suggests that food availability may have been a prime factor in the evolution of territorial tendencies in juvenile *S. undulatus.* We know of no similar type of evidence for any other species. This conclusion does not preclude the existence of a different selective basis for the territory of adult *S. undulatus.* The nature of adult aggression changes, and more direct pressures associated with reproductive success and not with food availability may take over at that time.

Food shortage varies between years and may contribute to the evolution of a nutrition-dependent territorial response. Thus, two measures of resource competition were greater in 1979 than in 1978. Home ranges were larger (Table 7.1) and growth responses to artificial feeding were greater (Tables 7.3 and 7.4) in 1979. During a high-food year, it would be advantageous to curtail territorial activities. While we did not demonstrate that this will enhance survival, it could enhance growth of a passive individual relative to that of an aggressive individual who, under these conditions, would accrue only the energetic cost of territorial defense with no appreciable benefits. In contrast, during a low-food year the aggressive lizards would suffer the energetic costs of territorial defense but would probably obtain a richer share of limited food supplies than would passive lizards. The most fit lineage over many generations would be one whose individuals could modify their behavior to respond adaptively to both situations.

Resource competition increases toward the end of the hatching season. This has been suggested (Ferguson and Bohlen, 1978; Ferguson, Bohlen, and Woolley, 1980) and may form the basis for subtle seasonal shifts in reproductive adaptive strategies. Previous evidence on which this conclusion was based is indirect: lizard densities increased while food densities decreased. Our data reconfirm these facts and add to them. Thus, lizard densities were higher late in the season (Table 7.7). Between July 16 and August 2, 1978, an average of 8.8 insects and 138.5 ants were collected per trap per hour on the study plots (N = 12 traps); between August 13 and September 16, 1978, an average of 8.3 insects and 54.5 ants were collected per trap per hour (N = 12 traps). Among the lab-hatched juveniles growth responses to supplementary feeding occurred only during release II period (Table 7.3).

Mortality rates, while high (ca. 50 percent), are not related directly to food availability. There was no enhancement of survival among fed lizards (Tables 7.5 and 7.6). Survival between the low-food year (1979) and the high-food year (1978) did not differ significantly. There was no significant improvement of survival due to feeding during the period of most severe food competition (release II) (Table 7.5). Indeed, feeding seemed to inhibit survival of late recruits. Thus, the only significant difference in survival was between the relatively high survival of fed lizards (53 percent) during the release I period and the relatively low survival (37 percent) of the fed lizards during the release II period. While overall densities of the quarter-acre plots did not differ significantly (Table 7.7), obvious congestion of home ranges around the release sites (Table 7.2) did result in a higher concentration of lizards on experimental areas than on control areas. Mortality is due chiefly to predation. Some predators are known to concentrate their efforts in areas of high prey density (Croze, 1970). This may explain the effect and suggests a factor in addition to food competition that may have played a role in evolution of the territorial tendencies of juvenile *Sceloporus*. Thus, in high predation years dispersion may reduce the intensity with which predators search a given area and result in a higher probability of survival of a territorial individual.

8 Fitness, Home-Range Quality, and Aggression in *Uta stansburiana*

Stanley F. Fox

TERRITORIALITY IS MAINTAINED to secure advantages for individuals within populations (Rand, 1967a; Brown and Orians, 1970; Wilson, 1975). In lizards, as in other territorial taxa, these advantages include access to limiting environmental resources and increased mating potential (Rand, 1967a; Stamps, 1977), which in turn maximize survival (Boag, 1973; Fox, 1978) and reproductive success (Rand, 1967b; Tinkle, 1969; Trivers, 1976; Ruby, 1981). Little is known, however, about the patterns of territory acquisition and the influences that affect a juvenile's success during territorial procurement (but see Stamps, 1978; Simon and Middendorf, 1980; Chapter 7). My research on the ontogeny of territoriality among juvenile lizards was initiated to examine in detail these patterns and influences under a field experimental plan of study.

I have studied this development of territoriality in the lizard *Uta stansburiana* at the identical Texas field site of Tinkle (1967a,b). His long-term (1960–1966) field study of social, behavioral, and demographic aspects of the same species now stands as a landmark for lizard ecology. I have continued to study the behavioral ecology of this same population intermittently from 1971 through 1981. This population, then, has been extensively observed in the field, by investigators using much the same study methods, from 1960 to the present—a span of more than 20 years. As Tinkle noted, "Detailed [long-term] studies of natural populations, particularly of reptiles, are rare. Such studies are essential to disclose what ecological differences exist at the individual and population level in response to the environmental conditions prevailing in the area of study" (1967b:5). In this vein, I have continued to study this population, investigating home-range acquisition in juveniles and relating observed differences to the variation in environmental conditions (that is, variation in selection pressures) through time at this site.

This study of home-range acquisition was based on my prior study of natural selection (differential survival of juveniles) at the levels of behavioral ecology (Fox, 1978), morphology (Fox, 1975), and biochemistry (Fox, 1973). Natural selection was observed at all three levels, but it was mostly selection at the behavioral level that inspired my subsequent study of home-range acquisition. Lizards that survived their juvenile summer were behaviorally different than those that eventually perished. Survivors specialized their activity around habitat features that afforded maximum shade and shelter on warm, clear days, avoiding early-morning and late-evening exposure (Fox, 1978). Their daily period of activity was probably shorter than that of nonsurvivors, because they restricted their range of activity over these ecological variables, or they were not active every day.

These behavioral differences between survivorship classes are a product of the interaction between genotype and environment. For a juvenile *Uta,* the environment it experiences is its home range, and this suggested three questions: (1) Are home ranges of survivors different from nonsurvivors? (2) If so, what influences acquisition of a particular class of home range? (3) Does temporal environmental variation affect the nature of home-range acquisition? My research program, built upon Tinkle's extensive data base (1967b), has utilized field manipulation to help answer these questions.

In addition to the particular advantages of extending a study on such an already well-studied species, *U. stansburiana* in Texas has a generation time of only one year, a characteristic that maximizes its potential in an evolutionary study of the sort described here. This species, like many lizards, also offers other attributes that underscore its utility as a subject of ecological study. Many lizards are diurnal, easily observed, locally abundant, extremely sedentary, amenable to experimental manipulation, and exhibit a diversity of social systems (Brattstrom, 1974; Stamps, 1977), feeding strategies (Pough, 1973; Huey and Pianka, 1981), physiologies (Gans and Dawson, 1976), life-history traits (Tinkle, Wilbur, and Tilley, 1970), and community structures (Pianka, 1973, 1975).

Materials and Methods

Species, Study Site, and Plots. Aspects of the autecology of *U. stansburiana* in Texas are taken from Tinkle (1967b). It is a small lizard: adults range from 42 mm snout-vent length (SVL) to 55 mm, juveniles from 20 mm SVL. It is semelparous: females lay 3 or 4 clutches of 4 eggs the summer after they hatch; 90 percent of all adults fail to live into their second adult summer. Densities are high: adults average 42/ ha, juveniles (in July) average 210/ ha. Juveniles move little: average distance from the hatching site to the center of their eventual breeding territory is ≈ 15 m for females, ≈ 20 m for males. Adults are extremely territorial, males against males and females against females, resulting in a monogamous breeding system. Juveniles are aggressive from birth but show more home-range overlap than adults.

The study site, 8 km south of Kermit, Winkler County, Texas, is part of a larger study site originally described by Tinkle, McGregor, and Dana (1962). It is a sandy area at an elevation of 885 m along the ecotone of shortgrass prairie and the Chihuahuan Desert. The dominant plants of the area are mesquite (*Prosopis glandulosa*), yucca (*Yucca campestris*), broomweed (*Xanthocephalum microcephalum*), croton (*Croton dioicus*), and dropseed grass (*Sporobolus flexuosus*).

Originally, 4 plots measuring 0.2 ha enclosed by a 1 m galvanized sheet metal fence partially sunk into the ground contained the study subjects. In 1978 a common wall between two enclosures was removed, creating a 0.4 ha enclosed plot. Juveniles and adults were contained inside the enclosures. New juvenile stock was introduced from the surrounding area in 1971, 1978, and 1979. Natural breaks and holes in the fence preceding these years allowed movement of lizards into and out of the enclosures. Grid stakes 7.5 m apart were set onto each plot. An unenclosed, control plot of equal area was also established.

Home ranges of juvenile *Uta* are small. Tinkle (1967b) found home-range size of young juveniles to be about 12 m^2; the juveniles of my studies inhabit larger home ranges on the average—49 m^2 in 1978, 31 m^2 in 1979. The average home-range size of juveniles depends of course on the age of the lizards; both my measures are from late-summer juveniles, later in 1978 than 1979. Regardless of when home-range determinations of juveniles are made, the 0.2 ha plots (2,000 m^2) seem of sufficient size to enclose juveniles with minimal, artificial constraints. Density of juveniles inside the enclosures were usually slightly ($\approx$ 20 percent) higher than outside the enclosures, but well within the values of Tinkle (1967b).

Censuses. When first captured, each lizard was individually marked with a combination of clipped toes and paint spots that allowed continuous recognition throughout the season (Fox, 1978). Each day during the summer one or more plots were systematically censused for lizards. The time of day and direction of these walks were varied for each plot, although morning and late afternoon were generally emphasized. All vegetation was shaken and debris lifted on each walk to ensure as complete a census as possible, although many lizards were sighted ahead of the investigator and identified with minimal disturbance. Capture was by portable trap (Fox, 1973). At the end of each summer, 4–5 final censuses were made of the enclosed plots. A lizard was considered dead with 90 percent confidence if it was not seen in these censuses. Inspection of past observation records of lizards known to be alive verified that a minimum of 4 consecutive censuses was necessary to observe 90 percent of them at least once. This procedure identified the 2 fitness classes of lizards that were survivors and nonsurvivors over the summer.

Home-Range Quality. Home ranges were defined from subsequent relocations of marked individuals. This method describes home ranges and not territories, which are usually defined as exclusive areas that are defended against intruders (Brown and Orians, 1970). Although agonism between juveniles has been observed in the field, considerable home-range overlap occurs. Either exclusive territoriality gradually develops in juveniles as they age or is characteristic only among adults. Adults do defend exclusive territories (Tinkle, 1967b).

Home-range size and home-range quality (HRQ) were determined for juveniles seen at least 4 times. Tinkle (1967b) has shown that home-range size of juveniles increases very little when sightings beyond the first 4 are included. Lizards in my studies, however, were resighted in considerable excess of this minimum (overall mean of 9.4 sightings per lizard). Home ranges were described using the convex polygon technique (Tinkle, McGregor, and Dana, 1962) and measured with a planimeter. Counts within each home range were made of 3 plant taxa (bunchgrasses, mostly *Sporobolus flexuosus;* crotons, exclusively *Croton dioicus;* and broomweeds, exclusively *Xanthocephalum microcephalum*), surface debris (boards, stumps, and cow dung), insect holes, mammal burrows, nest openings of large ants, and nest openings of small ants. Mesquite and yucca were drawn to scale on each plotted home range and converted to a percent of cover. Counts were made within a 2- or 3-day interval of no rain, so elements such as insect holes and ant nest openings were comparable among different home ranges.

Dominance. Individuals of similar SVL were paired and allowed to interact for one hour or longer in a neutral arena in the laboratory to determine dominance (Fox, Rose, and Myers, 1981). In most cases they were also matched by sex. A sum of aggressive patterns (pushups, lateral displays, attacks, supplants, bites, and superimpositions) minus submissive patterns (flattening) was determined for each lizard; the lizard showing the higher positive sum was defined as the dominant member of the pair. At the close of each encounter, before individual tabulations, a subjective assignment of dominance was made. This was based on more subtle behavior such as avoidance of one lizard by another, general activity, hesitancy of approach versus direct approach, severity of encounters, and so on. In no case did the subjective assignment of dominance differ from the quantitative assignment.

Results

Fitness and Home-Range Quality

Home ranges of 17 lizards from one enclosure in 1972 that survived their juvenile summer were compared against 17 that died (Fox, 1978). Survivors' home ranges had significantly higher densities of yucca and insect holes than nonsurvivors' home ranges, whereas home ranges of nonsurvivors had a marginally greater density of croton (Table 8.1).

Table 8.1 Differences in density of features on home ranges of survivors and nonsurvivors in 1972.[a]

	Survivors (N = 17)		Nonsurvivors (N = 17)		
Home-range element	U	($\bar{X}$)	U	($\bar{X}$)	P[b]
Mesquite	164	(0.25)	125	(0.10)	NS
Broomweed	151	(1.00)	138	(0.47)	NS
Yucca	212	(0.46)	78	(0.14)	0.02
Croton	95	(2.27)[c]	195	(1.73)[c]	0.08
Grass	104	(7.44)[c]	185	(4.29)[c]	NS
Large ant nests	127	(0.01)	162	(0.05)	NS
Small ant nests	104	(7.39)[c]	186	(3.98)[c]	NS
Surface debris	147	(0.28)[c]	142	(1.41)[c]	NS
Mammal burrows	169	(0.57)	121	(0.28)	NS
Insect holes	220	(0.86)	70	(0.40)	0.01

a. Number of elements/m^2.
b. Two-tailed Mann-Whitney U-tests. Larger U indicates that most observations fall toward upper end of scale. Smaller U indicates that most observations fall toward lower end of scale. These more appropriate two-tailed tests yield significance levels (P) somewhat different from previously reported one-tailed tests (Fox, 1978).
c. Values for U show opposing trend to that suggested by means.

Shannon-Wiener diversity indices were calculated for each home range, using the 10 features counted. Survivors inhabited home ranges of greater diversity (H′median = 1.7) than nonsurvivors (H′median = 1.4; Mann-Whitney U-test: U = 214.5, $P < 0.05$).

Stepwise discriminant function analysis is a multivariate procedure that builds an additive function, appropriately weighting each variable so as to maximally separate known groups along a composite axis. It is especially appropriate when the variables may show mild intercorrelation. This technique also constructs a single, composite index of home-range quality that is employed below. Discriminant function analysis, applied to the home-range features of the 34 lizards of 1972, produced a function composed of 6 variables (densities of grass, croton, mesquite, yucca, and large ant nest openings; and home-range area) that significantly separated survivors from nonsurvivors (canonical correlation = 0.65, $\chi^2 = 16.2$, d.f. = 6, $P = 0.013$).

Dominance and Home-Range Quality

In 1976, 30 pairs of size-matched juveniles were resighted sufficiently often so that home ranges could be estimated and counts of home-range features made (Fox, Rose, and Myers, 1981). Dominance within pairs was then determined. Dominant lizards originated from home ranges with significantly lower densities of croton (Wilcoxon matched-pairs signed-ranks test: $T_s = 136$, $P < 0.05$) and grass ($T_s = 101$, $P < 0.01$). Home ranges of dominants were also significantly more diverse than home ranges of subordinates ($T_s = 138$, $P < 0.05$).

The several elements of home ranges were combined into one composite index of home-range quality (HRQ) using discriminant function analysis as above. A multivariate function of all 11 variables (10 element densities plus home-range area) was derived from the survivorship groups of 1972. Standardized coefficients of this function were then applied to the *z*-transformed variables from 1976. In this way home-range elements could be weighted according to their relative effectiveness in producing a separation between probable survivors and nonsurvivors in a different year when absolute densities of the various elements might differ. This process created a spectrum of multivariate indices of home-range quality based on features previously associated with survival. Dominant lizards originated from home ranges significantly more similar to home ranges of survivors (higher scores) than nonsurvivors (lower scores) (Wilcoxon matched-pairs signed-ranks test: $T_s = 134$, $P < 0.05$).

Removal Experiment

This experiment was conducted in the summer of 1978. Removal of territorial adults removes the resistance to colonization by other individuals; the vacant space is soon occupied by nonterritorial "floaters" or territorial individuals from other (presumably suboptimal) areas. This has been observed both in birds (reviewed by Wilson, 1975) and lizards (Boag, 1973; Ruibal and Philibosian, 1974; Philibosian, 1975; Stamps, 1978). Juvenile *Uta* were removed from their home ranges to see if (1) the vacated home range was usurped by another and (2) if the usurper upgraded its home-range quality by invading the vacant area. If both (1) and (2) proved true, then juvenile *Uta* on superior home ranges must be showing some resistance to the colonization of their home ranges by others, and *Uta* on inferior home ranges would prefer to invade better habitat.

Home ranges of 49 lizards in 2 enclosures were determined and their composite HRQ scores (based on features previously associated with survival) computed as before. Twenty-four of these from both superior (high discriminant scores) and inferior (low discriminant scores) home ranges were removed. Care was taken to remove lizards such that equal areas of superior and inferior habitat were laid vacant. This design furnished the remaining lizards with a choice of invading either extreme of habitat quality. After the remaining lizards were given 3 weeks to adjust their home ranges about the plots, home-range positions and HRQ measurements were again conducted.

Considerable reorganization of home ranges followed the removal; home-range overlap with both removed lizards' home ranges and currently occupied home ranges was extensive. The habitat following reorganization was divided into three categories: (1) vacated—home ranges of removed lizards, (2) currently occupied—home ranges of lizards still present, and (3) initially unoccupied—habitat previously uninhabited by any lizards. Invasion into these three habitat categories was not significantly different (Table 8.2). Pairwise tests, however, marginally showed that remaining lizards invaded vacated habitat over initially unoccupied (Wilcoxon matched-pairs signed-ranks test: $T_s = 91$, $N = 23$, $0.10 > P > 0.05$) but showed no differential invasion of vacated versus currently occupied habitats. So the remaining lizards could possibly distinguish between initially unoccupied (which was probably inferior since it was not initially utilized) and habitat that was once inhabited (vacated) or currently inhabited, but there is no evidence that the home ranges of the removed lizards were preferentially invaded over home ranges already occupied.

Did the remaining lizards show any tendency to upgrade their home-range quality? I compared the terminal HRQ and change in HRQ of lizards that invaded most into home ranges of removed lizards to those that invaded least. Both analyses showed that invaders did upgrade their HRQ over noninvaders, but not significantly so (Table 8.3). (But see discussion regarding temporally varying selection pressures.)

Testosterone Implant Experiment

Dominance and home-range characteristics are related just as survival and home-range characteristics are related among juvenile *Uta* (see above and Fox, 1978; Fox, Rose, and Myers, 1981). This experiment was a first step toward uncovering the causality of the relationship between dominance and home-range quality (HRQ). Does dominance help an individual attain a home range that ensures increased survival, or does an individual become dominant upon acquisition of such a home range?

This experiment was conducted over the summer of 1979, and is reported in more detail in Fox and Mays (in preparation). Home ranges of

Table 8.2 Percent invasion by each lizard of vacated, initially unoccupied, and currently occupied categories of habitat (standardized for individual home-range area).[a]

Habitat	N	Median	R^b
Vacated	26	0.060	1138.5
Initially unoccupied	26	0.032	942.0
Currently occupied	26	0.036	1000.0

a. $H = 1.53$, 2 df; $P > 0.05$ (Kruskal-Wallis test).
b. Sum of ranks for each category in ordered total sample.

Table 8.3 Home-range quality (HRQ) of invaders compared to noninvaders.[a]

	Invaders			Noninvaders			
HRQ comparison	N	U[b]	(Median)	N	U[b]	(Median)	*P*[c]
Change in HRQ over experiment	10	67.0	(−0.75)	10	33.0	(0.65)	≅ 0.10
Terminal HRQ at end of experiment	13	104.5	(1.29)	13	63.5	(−1.61)	≅ 0.16

a. An invader falls above, a noninvader falls below the medial percent overlap with vacated habitat.

b. Higher U denotes that group tends to rank higher than other group, regardless of trend suggested by medians.

c. One-tailed Mann-Whitney U-test, testing prediction that invaders would upgrade their HRQ.

67 lizards in 2 field enclosures were located from sightings of individuals as before. Home-range area and element densities were computed and transformed into standardized values as before. Using the original discriminant function generated from the 1972 survivors and nonsurvivors, HRQ composite scores for each individual were computed as before. All lizards were matched in pairs by sex and SVL. In one plot, individuals with the inferior (would-be nonsurvivors) HRQ scores were implanted subdermally just behind the left front leg with a pellet of testosterone propionate (1.14–2.17 mg testosterone/g live lizard) formed by repeatedly dipping a short section of surgical thread into molten testosterone (Crews, 1974). Its pairmate was implanted with a comparable length of undipped thread (sham) as an internal matched-pair control. All the lizards of the second plot were matched by pairs and implanted with undipped thread (sham) as an external control. Arena interactions of all pairs were performed before the implantations to determine relative dominance within pairs before the treatment. After two days of post-operative recovery, lizards of the first plot were transferred to the second and vice versa. In this way all lizards were exposed to novel habitat and were forced to establish new home ranges under the same impedimenta. After one month, home ranges and HRQ were measured for all surviving lizards as before. Lizards were captured, again allowed to interact by pairs in a neutral arena, and bled and sacrificed for endocrinological and histological inspection.

Over the same summer HRQ measurements of an enclosed, unmanipulated population of juvenile *Uta* were conducted. Survival over the summer was also monitored. These control data indicated that the discriminant function generated from my previous work was unsuitable for distinguishing home ranges of survivors from nonsurvivors in 1979 (see below). A new function was therefore generated from the home-range measurements and survival of the 1979 lizards. This function was retroactively applied to the home-range data of the experimental lizards of the same year. (Since survivorship cannot be ascertained until the very end of the study season, this current function could not be used to set up the implant experiment.) This current function showed that most of the testosterone-implanted lizards originated from high HRQ (as 1979 survivors) instead of low (7 out of 9 implant pairs), as intended.

Nevertheless, these data can still show if success at gaining a superior home range in novel habitat is independent of hormonal treatment. In fact, it is not: testosterone-implanted juveniles obtained significantly higher-quality home ranges than their sex- and size-matched sham pair-mates compared to the sham and sham pairs of the control plot. Home-range quality (HRQ) changes induced by the treatment of exogenous testosterone were analyzed by a nonparametric, paired design. For each plot, differences in final home-range discriminant scores of the sex- and size-matched pairs were computed. For the treatment plot, these differences were computed as testosterone implants (initially high HRQ) minus shams (initially low HRQ). For the control plot, the differences were computed as shams that originally inhabited high-quality home ranges, minus shams that originated from low-quality home ranges. The treatment paired differences were then compared to the control paired differences by a one-tailed Mann-Whitney U-test, testing the prediction that the testosterone implants would gain higher-quality home ranges. Individuals with testosterone treatment did reacquire higher-quality home ranges in novel habitat than shams with originally similar home-range experience ($U = 46$, $N_1 = 9$, $N_2 = 6$, $P < 0.05$). Testosterone implants also showed a significant increase in home-range area over the experiment (testosterone implants, $N = 9$, averaged an increase of 44.1 m^2, whereas all shams, $N = 21$, averaged an increase of 7.5 m^2; Mann-Whitney U-test: $U = 140$, $P < 0.05$). Body size of the 2 groups was not significantly different, and total lizard densities in the 2 plots were equivalent, so the increase in home-range size seems attributable to the hormone treatment.

Change in home-range quality of the same individual (pretreatment and posttreatment) also was compared for treatment versus sham lizards. In the treatment plot, both the testosterone implants with initially high HRQ scores and their pairmates with originally low scores improved the quality of their subsequent home ranges: testosterone implants showed a medial individual HRQ gain of 0.17, shams showed a medial increase of 1.60. Shams and implants improved their HRQ equivalently (Mann-Whitney U-test: $N_1 = N_2 = 7$, $U = 27$, $P > 0.05$). In the control plot, however, lizards with originally low HRQ subsequently gained high-quality home ranges (medial individual HRQ gain of 1.27), whereas those with initially high HRQ subsequently fell to low-quality home ranges (medial decrease of 2.13). These 2 groups differed significantly in their individual HRQ changes over the experiment (Mann-Whitney U-test: $N_1 = N_2 = 6$, $U = 34$, $P < 0.05$).

Before implantation, lizards selected for testosterone implantation were not socially different than their pairmates. At the end of the experiment, testosterone-implanted lizards had significantly increased their level of aggression (Table 8.4) and dominated their sham pairmates in arena encounters. Aggression of the shams was unchanged over the experiment. Implanted lizards also showed significantly larger gonadosomatic indices, hypertrophied androgenic target organs, and elevated blood titers of testosterone (Fox and Mays, in preparation).

Exogenous testosterone applied to intact juveniles enlarged their gonads; it increased their relative aggressiveness, so they rose socially over their unimplanted pairmates; and it allowed for successful acquisition of larger and superior home ranges in novel habitat compared to the performance of sham-implanted, intact controls.

The Field Experiments

Juveniles of most species encounter a competitive environment. In seasonal breeders that lay multiple clutches of eggs, competition increases as juvenile density increases and food supply diminishes (Ferguson and Bohlen, 1978). Individual survival through this competitive pressure may depend on characteristics of the parents or their home ranges (Tinkle, 1965, 1969), juvenile body size and/or aggression (Ferguson and Bohlen, 1978), juvenile patterns of behavior (Fox, 1978), or features of juvenile home ranges (above). The juvenile *Uta* I study show a diversity of behaviors across several environmental parameters and, specifically, a difference in the behavior of survivors and eventual nonsurvivors. Home ranges of survivors and nonsurvivors also differ and, in fact, may be the reason survivors and nonsurvivors show different patterns of activity (Fox, 1978). Those juveniles on suboptimal home ranges may have to generalize their behavior and resource-accruing activity

Table 8.4 Arena-determined aggressive scores before and after implantation.

	Before			After			
Treatment	N	U	(Median)	N	U	(Median)	P^a
Testosterone	8	12.5	(9.5)	7	43.5	(17.0)	< 0.05
Sham	12	84.0	(5.0)	11	48.0	(5.0)	NS

a. One-tailed Mann-Whitney U-test, testing prediction that aggression of testosterone implants would increase.

compared to juveniles on optimal home ranges. Even then, some lizards will be unable to find enough food and will die; those that are active more and under more variable weather conditions will further increase their risk of death through predation, especially in Texas where lizard predators are probably more common than further north (Pianka, 1970).

If a certain type of home range conveys increased survivorship, is there a struggle among juveniles for the better home ranges? Certainly the more dominant juveniles inhabit those home ranges that characterize survivors (Fox, Rose, and Myers, 1981; above). But is this dominance an indirect effect of inhabiting a superior home range, or does dominance help an individual acquire a superior home range—one which will allow time minimization (Schoener, 1971), ready resource harvest, and thus enhanced survival? On the one hand, the removal experiment did not demonstrate a competitive struggle for the better home ranges. Whereas juveniles slightly avoided areas where no lizards had previously resided, they did not prefer experimentally vacated habitat, some of putatively superior quality. Those that did invade vacated areas showed only a slight tendency to seek out and appropriate superior habitat patches so as to upgrade the quality of their home ranges. Yet both Tinkle (1967b) and I have observed agonistic encounters and supplants between juvenile *Uta* at this study site. Extensive overlap of home ranges of lizards remaining after the removal of others may indicate the failure of juveniles to defend the entire home range as found in some other lizards (Yedlin and Ferguson, 1973; Stamps, 1977). Or these juveniles may show spatial overlap but segregate activity temporally as in *Sceloporus jarrovi* (Simon and Middendorf, 1976). An important alternative explanation for the failure of my juvenile removal experiment to follow the expectations derived from previous removal experiments in other taxa may relate, though, to variations in selection pressure at this site (see below).

On the other hand, the testosterone implant experiment supports the notion that increased aggression aids the acquisition of superior home ranges. Testosterone implantation elevated the aggression and social status of those lizards, and when they and sham-implanted lizards were exposed to novel habitat and forced to establish new home ranges, the hormone-implanted juveniles acquired the superior habitat patches. This agrees with an analogous experiment by Fox, Rose, and Myers (1981). When dominant and subordinate juvenile *Uta* were introduced by pairs into small outdoor pens with disparate halves of habitat quality, the dominants frequented the superior end significantly more than the subordinates.

The sham and sham controls in the testosterone implant experiment present an apparent enigma, however. The initially high HRQ shams showed a significant fall in subsequent HRQ after relocation, while the initially low HRQ shams showed the inverse. If those juveniles with initially high HRQ had what it takes to acquire a superior home range, then why did they not retain that ability to reacquire superior home ranges in novel habitat? To some extent the differential response of the two sham classes is a statistical artifact. The initially high and initially low HRQ groups of the control plot were selected as external controls against the testosterone and sham implants of the treatment plot. Under the null hypothesis, the mean of the high-HRQ group would fall and the low-HRQ group would rise (since each was a nonrandom HRQ group). Both should converge toward an intermediate HRQ value. This was so: the final HRQ scores of the 2 sham groups in the control plot were not significantly different (Mann-Whitney U-test: $U = 28$, $N_1 = 6$, $N_2 = 6$, $P > 0.05$). Failure of the initially high HRQ lizards to reacquire superior home ranges again may relate to variations in selection pressure at this site (see below).

Variability of Selection Pressures

Behavior and Home-Range Quality

North American deserts are variable environments, particularly in rainfall patterns (Pianka, 1967, 1975). Plant production and insect densities are tied to precipitation and thus create temporally variable environments for the lizards of those deserts (Blair, 1960; Tinkle, 1967b; Hoddenbach and Turner, 1968; Turner et al., 1969a,b, 1970; Pianka, 1970; Nagy, 1973; Parker and Pianka, 1975; Dunham, 1978). Tinkle (1965) has shown variable recruitment of *Uta* over several years at the same site. He found that in years of high juvenile densities, offspring from certain parents were more likely to survive to maturity than those from other parents. In years of lower densities, there was not such a differential. I have found differential survival based on patterns of behavior one year and not the next among juvenile *Uta* (Fox, 1978). I have also found considerable variation in differences of home-range quality between survivors and nonsurvivors over the summer, comparing the same site over 3 different years. In 1972, 1978, and 1979, I measured HRQ of lizards on an unmanipulated, enclosed plot, using the same field techniques. At the end of each summer I determined individual survival as before. Complete, 11-variable, discriminant functions that maximally separated survivors from nonsurvivors on the basis of home-range features outlined above were derived. Only in 1972 was the 11-variable discriminant function relatively effective in separating survivors from nonsurvivors (Table 8.5). The 1978 function was the least discriminating, the 1979 function nearly as ineffective. All 3 showed considerable differences in weighting coefficents for the same variables. Using stepwise dis-

Table 8.5 Discriminant function analysis using all 11 home-range variables to separate survivors from nonsurvivors for 3 different years.

Year	Cumulative lizard density[a]	Canonical correlation[b]	Percent correctly classified	Wilks' λ^c	χ^{2d}	P^d
1972	520	0.69	82	0.53	16.8	0.11
1978	175	0.40	68	0.84	5.9	0.88
1979	290	0.51	77	0.74	5.7	0.89

a. Cumulative hatchlings/ha over entire summer (ca. June 20 to ca. September 20); considerably in excess of density at any one time, but still reflective of relative densities.

b. A direct measure of discriminating ability of function.

c. An inverse measure of discriminating ability of function.

d. Chi-squared value for the Wilks' λ and associated level of significance.

criminant function analysis, only in 1972 was a function of more than one variable created that showed statistical separation of the 2 fitness classes (the 6-variable function reported above). Recall that 1972 was the same year in which natural selection of behavioral phenotypes was observed; apparently that year was characterized by stronger selection than 1971, 1978, or 1979. No measure of selection for home-range features was attempted in 1976.

The weak selection pressure over the summer of 1978 may well explain the ambiguity of the removal experiment that summer. If conditions were such that no fitness differentials resulted from home ranges of different characteristics, I would not expect either (1) invasion of vacated areas preferentially over areas already occupied or (2) any tendency to appropriate one patch of habitat over another, since by definition, all home ranges conferred equal fitness.

On the other hand, social exclusion of some lizards from optimal habitat was seen in an experiment I conducted in 1972—the year of heavy selection pressure (Fox, 1978). The denser vegetation (mesquite and broomweed) was cleared from one half of an enclosed plot to produce a more open habitat. Counts were made of juveniles on both halves of the plot following habitat modification. Older lizards moved to the unmodified half; those retaining some activity on the modified half suffered significantly greater mortality than those confining their activity to the unmodified habitat. Survivors (frequenting the unmodified half) were significantly larger than nonsurvivors. Since dominance in lizards is tightly correlated to body size (Boag, 1973; Brattstrom, 1974; Ferner, 1974; Brackin, 1978; Ruby, 1978; Stamps, 1978), it seems likely that the larger, presumably dominant, juveniles of the unmodified side were socially excluding smaller juveniles from that side.

Since no measure of HRQ of survivors and nonsurvivors was attempted in 1976, I cannot know the relative selection pressures that summer. The dominant lizards inhabited home ranges very similar to the 1972 survivors' home ranges, though, so it appears that selection pressures for 1972 and 1976 were strong and similar. Also, strong selection in 1976 is suggested by morphological comparisons of survivors and nonsurvivors (Fox, in preparation, see also below).

Selection pressures apparently were weak in 1979—the year of the testosterone implant experiment. Yet the implanted lizards did become more aggressive and upgraded their HRQ in novel habitat and possibly excluded their sham controls from the superior habitat patches. But this response was neutrally adaptive if HRQ of survivors in that year was no different than HRQ of nonsurvivors. Why should the implanted lizards work to increase the quality of their home ranges in a year when HRQ was relatively unrelated to survival? My implantation of exogenous testosterone may have bypassed a step in the natural ontogeny of home-range acquisition in these *Uta*. Normally the lizards are responding to prevailing resource levels. If some resources one year are in short supply relative to demand, then competition over them will occur. More aggressive juveniles will be favored through interference competition, and ultimately these more aggressive juveniles will come to inhabit home ranges with more of these limiting resources than the less aggressive juveniles. Ready access to sufficient resources contributes a direct benefit and also allows these more aggressive juveniles to modify their behavior such that predation risk is lowered and survivorship thereby increased. Ferguson has indirect evidence that more aggressive *Sceloporus undulatus* enjoy increased survivorship, but only when resources are limiting (Tubbs and Ferguson, 1976; Ferguson and Bohlen, 1978; Chapter 7).

Differences in aggressiveness and relative dominance in natural dyadic encounters is provoked in *Uta* under natural conditions by limiting resources of some kind in the environment. Either the environment, or the animal's genome, or an interplay of both develops a certain level of aggression, but it is the condition of limiting resources that defines individual benefits to social struggle. Only in environments where some resources are limiting would interference competition (agonism over a resource) be adaptive. But in my implant experiment, certain lizards were artificially made more aggressive and their relative social rank in paired encounters raised by external means independent of the resource level. These more aggressive juveniles went ahead and usurped habitat features that were only minimally associated with survival that year; these lizards rather showed "irrelevant" aggression for the environmental conditions of that year. It would seem reasonable that increased aggressiveness would be adaptive only when the resources fought over are in short supply. In other, less-competitive years, no benefit accrues to the more aggressive individuals, so increased aggression is wasted time and energy (compare to "aggressive neglect" in Hutchinson and MacArthur, 1959, and Wilson, 1975). If the stimulus for increased aggression is in fact the very degree to which resources are limiting (proximately, the difficulty individuals encounter in gaining access to resources), then the adjusted individual level of aggression is a simple consequence and requires no higher-order decision making. Laboratory experiments on aggression in *Anolis aeneus* support this hypothesis (Stamps and Tanaka, 1981). At high densities, level of aggression was inversely adjusted to level of food provided. In the field, however, aggression was not altered by introduction of food in either the wet or dry season, implying that resources in addition to food might be influencing aggression of lizards in nature.

Morphology

I have shown above that selection pressure on behavior and home-range features is variable over years at the same study site. I also have evidence that selection pressure on morphological characters is variable over years at the same site (Fox, 1975; Fox, in preparation). The same scale characters of juveniles were measured in 1972 and 1976. In 1972 all juveniles on enclosed plots were measured at an early age and survivorship was then determined as before at the summer's end. Survivors' scale characters could then be compared with those of nonsurvivors, to infer selection. In 1976 a sample of hatchlings (before selection) from a nearby free-ranging population was compared with a sample of older, late-summer juveniles (after selection). Both these comparisons suggest natural selection only over the juvenile summer, but that is the same season over which my studies of home-range acquisition are conducted, and so relevant to them. If the survivors or the after-selection group show a mean shift or decreased variance compared to the nonsurvivors or before-selection group, directional or stabilizing selection, respectively, is suggested. In 1972, 4 characters (sexes separate) showed evidence of stabilizing selection (Table 8.6). One character among females showed disruptive selection (decreased variance of the nonsurvivors). Only 1 character showed evidence of directional selection. In 1976, 1 character showed stabilizing selection and 3 showed disruptive selection. On the other hand, 7 characters (sexes separate) showed directional selection. Of the total 17 instances of selection for the separate sexes and years, only 1 character (females, difference of supraoculars) showed the same type of selection both years. Mostly a character selected one year was not selected the other. One character (females, sum of upper labials) even showed a selection reversal: stabilizing selection was indicated in 1972, and disruptive selection in 1976.

Selection on specific scale characters over a whole year was also com-

Table 8.6 Summary of scale characters showing significantly different means or variances between survivors and nonsurvivors (1972) or early and late collected juveniles (1976) in at least one comparison, sexes separate.[a]

Sex	Character	1972 F test[b]	1972 t-test	1976 F test[b]	1976 t-test
Male					
	Gular fold	0	0	0	L > E
	Interfemorals	0	0	0	E > L
	Lamellae R4 digit	0	0	0	L > E
	Sum upper labials	+	0	0	0
	Sum lower labials	0	0	0	E > L
Female					
	Gular fold	+	0	0	L > E
	Ventrals	+	0	0	0
	Lamellae R4 digit	0	0	0	L > E
	Sum upper labials	+	0	−	0
	Difference supraoculars	−	0	−	0
	Difference upper labials	0	0	+	0
	Difference lower labials	0	S > NS	0	0
	Sum upper labials/sum lower labials	0	0	−	L > E

a. Plus (+) means s^2 of survivors or late juveniles $< s^2$ of nonsurvivors or late juveniles; minus (−) means s^2 of survivors or late juveniles $> s^2$ of nonsurvivors or early juveniles; 0 means no significant difference between variances at the 0.05 level, two-tailed. Inequalities give direction of significant t-tests (0 = not significant at 0.05 level, two-tailed): S is survivors' mean; NS is nonsurvivors' mean; E is early juveniles' mean; L is late juveniles' mean.

b. If means for a comparison were significantly different, F tests are based on ratio of squared coefficients of variation.

pared for 2 different years (Fox, 1975). In no character was selection the same both years. Directional selection occurred on at least 1 character every year inspected. Hecht (1952) and Pasteur (1977) have also shown directional selection for scale characters in 2 other lizards. Pasteur maintains that such mean differences between adults and juveniles of a vertical comparison through time is actually evidence for endocyclic selection, because the breeding adult character distributions do not change. Endocyclic selection requires early counterselection every generation and then consistent (directional) selection on the juvenile-to-adult stage. I have found, however, little consistency in the direction and intensity of selection on specific meristic characters over 3 different years in *U. stansburiana*.

Ferguson and Fox (in preparation), analyzing survivorship of different-sized hatchling *Uta* at the same Texas site during 6 different years from 1960 to 1976, have found evidence for varying selection pressures. Ferguson, Hughes, and Brown (Chapter 7) have found selection for larger hatchling *S. undulatus* only when competition is keen, removing such survivorship differentials when lizards are supplementally fed. Lack of concordance across years of selection on scale characters and hatchling size suggests that selection pressures are not uniform over different years. These lizards (especially Texas *U. stansburiana* with an almost annual population turnover) appear to show intimate tracking of environmental variation. Whereas character distributions shift through time and selection pressures on juveniles differ from year to year, long-term character distributions may be relatively unchanging. Environmental variations ultimately balance out, and lizard characters, adapted to prevailing environments, also balance out.

Genetics

I have analyzed the same *Uta* populations electrophoretically in two different years—1972 and 1979 (Fox, in preparation). Genotypic distributions for 7 variable loci of 1972 juveniles were compared to genotypic distributions of 1979 juveniles (Table 8.7). Three loci showed significantly different genotypic distributions between the years. These same three loci plus another also showed differences in allelic frequencies between years. Adults and juveniles were compared genetically over 8 polymorphic loci in 1972 (Fox, 1973). Whereas 3 loci showed significant heterogeneity between age classes, none of the 3 were those that differed among juveniles of 1972 and 1979. Allelic frequencies of the same loci reported for adult *Uta* from the same locality in 1969 (McKinney et al., 1972) also differed from adult allelic frequencies in 1972 (Fox, in preparation). Collectively, these genetic differences of the same *Uta* population through time indicate nonrandom changes that imply the same intimate tracking of environmental variation suggested by the variation of morphology, home-range quality, and behavior mentioned above.

Theoretical Considerations

This study, with data from three biological levels of organization (behavior, morphology, and genetics), indicates that selection pressures at this site are nonuniform among years. Only in certain years do particular phenotypes have relative survival advantages; fitness varies temporally. Maintenance of the observed variation at the three biological levels requires some sort of balancing selection. I have suggested that the temporal heterogeneity of the environment (fluctuating selection pressures among years) maintains the genetic, morphological, and behavioral variation. Differential fitnesses (which vary between years) have been observed between phenotypes differing in behavior and morphology. Changes in genotypic and allelic frequencies have been ob-

Table 8.7 Observed and expected (in parentheses) genotypic frequencies of juvenile *Uta* at the same site in two different years.

Locus	Genotype	1972	1979	χ^2	*P*
LDH-1	bb	128 (124.5)	76 (79.5)	5.95	0.01
	bd	2 (5.5)	7 (3.5)		
MPI-1	aa	35 (25.6)	7 (16.4)	11.00	0.004
	ab	57 (63.5)	47 (40.5)		
	bb	38 (40.9)	29 (26.1)		
LTPep-2	aa	25 (19.4)	7 (12.6)	8.26	0.02
	ab	59 (55.8)	33 (36.2)		
	bb	44 (52.8)	43 (34.2)		
NAD-MDH-2	aa	89 (90.9)	60 (58.4)	0.71[a]	> 0.05
	ab	39 (36.6)	21 (23.4)		
	bb	2 (2.4)	2 (1.6)		
GPI-1	aa	0 (0.6)	1 (0.4)	2.84[a]	> 0.05
	ab	2 (1.2)	0 (0.8)		
	bb	128 (128.2)	82 (81.8)		
PGM-2	aa	74 (72.0)	45 (47.0)	4.73	0.09
	ab	36 (41.7)	33 (27.3)		
	bb	17 (13.3)	5 (8.7)		
Es-3	cc	38 (36.0)	21 (23.0)	9.84	> 0.05
	ee	2 (3.7)	4 (2.3)		
	gg	23 (24.4)	17 (15.6)		
	bc	1 (1.2)	1 (0.8)		
	be	0 (0.6)	1 (0.4)		
	bg	1 (2.4)	3 (1.6)		
	ce	3 (2.4)	1 (1.6)		
	cg	49 (43.3)	22 (27.7)		
	eg	13 (15.9)	13 (10.1)		

a. Chi-squared test may lack validity due to excessive proportion of cells with expected value of less than 5.

served between years, although no annual survivorship differences have been directly observed by following a cohort through time. (But a significant survival advantage over the summer was seen for the heterozygotes at one locus; Fox, 1973.)

Although it seems intuitive that temporally fluctuating selection pressures could maintain genetic variation, there are some theoretical objections to that conclusion. A great amount of theoretical literature has appeared on the relationship between polymorphism and both spatial and temporal environmental heterogeneity (see reviews by Felsenstein, 1976; Hedrick, Ginevan, and Ewing, 1976). The theoretical results are difficult to draw together since both finite and infinite populations have been discussed, and in some cases theoretical predictions derived from finite populations are exactly opposite to those derived from infinite populations (Hedrick, 1976). Furthermore, it is difficult to apply these usually one- or two-locus models to polygenic traits of the sort I have studied. Nevertheless a few general principles of relevance to my study emerge.

Selection pressures varying in space or time can maintain genetic polymorphisms, within constraints. The constraints appear to be more severe for temporally varying pressures than spatially varying ones (Hedrick, Ginevan, and Ewing, 1976; but see also Bryant, 1976). The general instance where temporally varying pressures can maintain a polymorphism is when the geometric mean of the heterozygote's fitnesses in the different temporal environments exceeds the geometric mean of fitnesses of either homozygote (Haldane and Jayaker, 1963). The geometric mean of the fitnesses is the mean of the logarithms of the fitnesses, transformed back into a raw value. Such a mean tends to emphasize small values, so that the geometric mean is always less than the arithmetic mean for values that show variation. This is a kind of overdominance, called geometric mean overdominance, and it can maintain a polymorphism even if the arithmetic mean of fitnesses of either homozygote is greater than the average fitness of the heterozygote (Gillespie and Langley, 1974). In an additive model where the fitness of the heterozygote is intermediate between the homozygotes, the geometric mean will usually be greater over temporally varying selection pressures, since its average fitness varies less (less small values) than the fitnesses of either homozygote. Because of this dependence on a geometric measure of average fitness, some workers feel that temporally varying selection pressures can be quite important in the maintenance of some polymorphisms (Gillespie and Langley, 1974; Felsenstein, 1976).

I do not know the genotypes that correspond to the morphological and behavioral phenotypes of my study. These phenotypes do, however, have some genetic bases, and the idea of geometric mean overdominance may be crudely applied. Consider phenotypes for a given scale character or level of aggression. In some years medial types will be favored, in some years one or the other extreme. In years when one extreme is favored, the opposite extreme will be selected against, and selected against even more strongly than the medial phenotypes. Over many years, the variation in fitness of the medial types will be less than the variation in fitness of the extreme types. If the absolute differences in fitness are not large, the geometric mean of fitness of the medial phenotypes will exceed that of either extreme. This geometric mean overdominance will tend to maintain the medial types. If the medial phenotypes are largely heterozygous, then the extremes would also be maintained. I have seen such selection for extreme scale characters (directional selection) in the *Uta* I have studied (Fox, 1975). This maintenance of a polymorphism is easier in coarse-grained environments than fine-grained ones (Gillespie, 1974) and easier if migration from other temporally varying habitat patches occurs (Gillespie, 1975). *Uta* in Texas live but one year (Tinkle, 1967b), experience one year of selection pressure, and so inhabit coarse-grained environments with respect to selection differences between years. Immigration appears high (Tinkle, 1967b). It is likely, then, that temporally fluctuating selection pressures, different from year to year, are involved in the maintenance of the variation observed at the three biological levels in *Uta* at my study site.

It is not necessary that selection occur each year, only that it is not always the same when it does occur. The general condition may very well be that selection does not occur every year. Rather, the population may periodically experience strong selection during ecological crunches (Wiens, 1977), with years of relaxed selection in between. Indeed, I have observed years of strong and weak selection in Texas *Uta*, especially with regard to behavior, home-range features, and meristic scale characters.

In spite of this general theory of geometric mean overdominance, computer simulations of finite populations in temporally varying environments show that only under conditions of negative autocorrelation between environments will a one-locus, additive-fitness polymorphism be maintained (Hedrick, 1976). Any positive autocorrelation will speed the loss of genetic variation beyond that predicted by genetic drift. Even when a polymorphism is maintained by temporal changes in fitness, it is maintained at a rather low level of polymorphism (Hedrick, 1974). But there are field studies that link temporally varying environments with increased genetic variability of the organisms inhabiting them (Levinton, 1973; Bryant, 1974), and there are experimental studies that show temporally varying environments preserving genetic variance over many generations compared to constant environments (Beardmore and Levine, 1963; Powell, 1971; McDonald and Ayala, 1974). Theoretical studies and empirical studies of genetic variation in temporally varying environments are not always in agreement (Bryant, 1976). It will be intriguing to see how they are melded in the future and I hope studies on lizard ecology will help with that fusion.

9 Sexual Selection, Sexual Dimorphism, and Territoriality

Judy A. Stamps

IN THE PAST FEW YEARS many authors have presented models for the evolution of polygyny in terrestrial vertebrates (Verner, 1964; Orians, 1969; Goss-Custard, Dunbar, and Aldrich-Blake, 1972; Wittenberger, 1976; Bradbury and Vehrencamp, 1977; Emlen and Oring, 1977; Owen-Smith, 1977; Ralls, 1977; Wittenberger, 1980). These studies have attempted to identify the ecological and behavioral factors responsible for shifts from monogamous or promiscuous mating systems to polygynous mating systems in birds and mammals.

For simplicity, most of these models have divided mating systems into two groups, polygynous and nonpolygynous, and considered the ecological, behavioral, and morphological variables associated with each group.

This treatment of mating systems has been quite fruitful. In birds, polygyny has been attributed to uneven distributions of the resources necessary for females, so that males with territories in good habitats can achieve higher reproductive success than males in poor habitats (Verner, 1964; Orians, 1969; Wittenberger, 1976; Emlen and Oring, 1977; Orians, 1980). In mammals, most evidence suggests that female sociality is a key factor in polygyny; if females form groups, these groups can be defended by males (Bartholomew, 1970; Goss-Custard, Dunbar, and Aldrich-Blake, 1972; Ralls, 1976; Bradbury and Vehrencamp, 1977; Emlen and Oring, 1977; Owen-Smith, 1977; Clutton-Brock and Harvey, 1978; Wittenberger, 1980). In both birds and mammals, intrasexual selection appears to be more intense in polygynous than in nonpolygynous species, and sexual size dimorphism (males larger than females) tends to be more pronounced in polygynous species (Darwin, 1871; Selander, 1972; Trivers, 1972; Clutton-Brock, Harvey, and Rudder, 1977; Ralls, 1977).

Less information is available for terrestrial vertebrates other than birds and mammals. However Shine (1978, 1979) has shown that sexual dimorphism in snakes and amphibians is related to male combat and intrasexual selection in these groups.

Despite the utility of the polygyny-nonpolygyny dichotomy, it is clear that this dichotomy is really a reduction from a continuum. There are different degrees of polygyny in animals; a species with average harem sizes of 2 females may be considered less polygynous than one with harems containing 15 females. Thus far, few studies have considered the ecological and behavioral factors responsible for the degree of polygyny in terrestrial vertebrates (but see Emlen and Oring, 1977; Wittenberger, 1980).

Polygyny is widespread in lizards (Evans, 1951; Blair, 1960; Tinkle, McGregor, and Dana, 1962; Harris, 1964; Rand, 1967; Stebbins, Lowenstein, and Cohen, 1967; Blanc and Carpenter, 1969; Bustard, 1970; Andrews, 1971; Müller, 1971; Boag, 1973; Mitchell, 1973; Tinkle, 1973; Berry, 1974; Burrage, 1974; Fleming and Hooker, 1975; Ruby, 1976; Trivers, 1976; Bruton, 1977; Stamps, 1977b; Werner, 1978; Schoener and Schoener, 1980; Ruby, 1981; Dugan, 1982; Werner, 1982). More important, there is considerable variation in the degree of polygyny in lizards. Some species are, in fact, monogamous (Tinkle, 1967; Jenssen, 1970; Milstead, 1970); others have average harems of 2 to 3 females per male (Harris, 1964; Rand, 1967; Ruby, 1981); and still others average 4 to 6 females per male (Blair, 1960; Stamps, unpublished data).

The object of this chapter is to try to identify some of the factors responsible for the degree of polygyny and the intensity of sexual selection in lizards. This task was facilitated by several aspects of lizard natural history. First, most lizards are insectivorous, and nearly all insectivores are extremely catholic in their choice of prey. For exceptions see the ant specialists of the genus *Phrynosoma* and *Moloch* (Stamps, 1977b). As far as food type is concerned, insectivorous lizards are a fairly homogenous group. However, for comparative purposes, data are also available for a small group of herbivorous lizards. These species offer an opportunity to test the effects of diet on mating systems and sexual selection.

A second important point is that social behavior in lizards is closely related to phylogeny and foraging style. Extensive information on ecology and spacing behavior is available for only 5 lizard families. Of these, 2 families (Iguanidae and Agamidae) are composed of diurnal, sit-and-wait, visually oriented insectivores. In these families males usually defend most or all of their home ranges (home-range defense). In 3 other families (Scincidae, Lacertidae, and Teiidae) species are diurnal, actively foraging, visually and olfactorily oriented insectivores. In these species male home-range defense is extremely rare. Reasons for these differences in spacing behavior are explored in detail in a review of social behavior in lizards (Stamps, 1977b). The neat division into two groups, one territorial (Iguanidae, Agamidae) and the other nonterritorial

(Scincidae, Teiidae, Lacertidae), allows us to assess the impact of male territorial defense on the degree of polygyny and intensity of sexual selection. By also looking at the exceptional species within each group (for example, the territorial species of the Lacertidae and Teiidae), one can investigate the effects of phylogeny and social system on these factors.

Methods

Analyses in this chapter required information on size at maturity, sexual dimorphism, adult sex ratio, sexual differences in mortality, population density, and home-range sizes for both sexes. Since there is considerable spatial and temporal difference in these variables, data from different study areas or years were not combined in the analysis. As a result, there were few species for which all of this information was available (see Tables 9.1 and 9.2).

For convenience in analysis, species in the families Iguanidae and Agamidae are included in Table 9.1, and species in the Lacertidae, Teiidae, and Scincidae are included in Table 9.2. All of the species in Table 9.1 (with the possible exception of *Gambelia wizlizeni* in Nevada) exhibit home-range defense. On the other hand, home-range defense is highly unusual among skinks, teiids, and lacertids (Stamps, 1977b). Three species in which it does occur are indicated in Table 9.2; a fourth species (*Cnemidophorus arubensis*) is also suspected of exhibiting home-range defense.

For the purposes of this chapter lizards exhibiting home-range defense will be termed territorial. In the statistical analyses territorial species indicates the species on Table 9.1, nonterritorial indicates all species on Table 9.2 except *C. arubensis, C. lemniscatus, G. gallotia* and *L. muralis.*

Size at Sexual Maturity. Snout-vent lengths of the smallest sexually mature male and female during the breeding season were used to determine the ratio of size at maturity of males to size at maturity of females. The criterion for maturation was physiological (presence of eggs or sperm) not behavioral (participation in mating or defense of a territory).

Sexual Dimorphism. Average snout-vent lengths of mature males and females were used to determine the degree of sexual dimorphism in a species. Length was used instead of mass because the presence of eggs or young in females and fat bodies in either sex can affect estimates of dimorphism based on mass. Where data for different seasons were available, sexual dimorphism in the breeding season was used. In 3 studies, size at sexual maturity was given but average length of mature males and females was not. If sexual-size dimorphism at the time of maturity were highly correlated with overall sexual dimorphism, then the ratio of male to female size at maturity could be used to estimate sexual dimorphism. Sexual dimorphism in size at maturity (male size/ female size) was significantly correlated with sexual dimorphism based on average

Table 9.1 Territorial species (Iguanidae and Agamidae).

Species	Sex ratio (male/ female)	Sexual dimor-phism (male/ female)	Sexual maturity (male/ female)	Home-range ratio (male/ female)	Sur-vival (male/ female)	Female density (females/ hectare)
Agama agama	0.66	1.32	1.22	–	–	36
A. atra	–	–	–	1.65	–	–
	–	–	–	1.56	–	–
A. atra	0.70	–	–	–	–	109
Anolis aeneus	0.50	1.29	1.14	3.8	–	1163
	–	–	–	0.94	–	–
A. acutus	0.64	1.36	1.39	4.7	0.43	2683
A. angusticeps	0.96	1.19	1.21	3.2	0.78	–
A. cupreus	0.77	1.07	1.0	4.2	0.50	67.5
A. distichus	1.11	1.11	1.11	1.15	1.17	–
A. garmani	0.81	1.31	1.21	–	–	–
A. limifrons	0.79	1.09	0.97	–	–	50.3
A. lineatopus	–	1.4	1.35	4.2	–	–
A. nebulosus	0.98	1.16	–	3.2	–	454
A. oculatus	0.65	1.25	–	–	0.49	178
A. polylepis	–	–	1.23	4.2	–	29.2
A. sagrei	0.74	1.29	1.21	4.8	0.87	–
Conolophus subcristatus	1.23	1.26	–	–	–	9.4
Cyclura carinata	–	1.23	1.19	1.6	–	–
Dipsosaurus dorsalis	1.89	1.06	–	0.94	–	15.9
Gambelia silus	1.2	1.09	1.01	2.1	–	1.5
G. wizlizeni						
(Nevada)	–	0.93	0.97	2.1	–	–
(California)	1.9	0.87	0.97	–	–	0.48
Holbrookia maculata	1.06	–	–	1.7	–	20
Iguana iguana	1.2	1.17	1.0	0.32	–	20
Sauromalus obesus						
Great Basin (California)	1.47	0.99	1.0	2.4	–	5.7
Mojave (California)	0.86	1.1	1.2	3.4	–	3.8
Mojave (California)	–	–	–	1.2	–	–
Sceloporus graciosus						
(California)	1.0	0.99	–	–	–	4.0
(Utah)	0.71	–	–	–	–	74
S. jarrovi	–	–	–	3.3	–	–
	0.58	1.11	1.0	2.8	0.70	30.5
	–	–	–	1.8	–	–

Female home range (m^2)	Female home range/ $(length)^3$	Non-breeding season	Herbivores	Number of congeners	Reference
–	–	–	–	–	Harris, 1964
56	–	–	–	–	Burrage, 1974
77	–	–	–	–	
–	–	–	–	–	Bruton, 1977
2.8	1.61	–	–	1	Stamps, unpublished data
–	–	x	–	–	
0.38	.91	–	–	0	Ruibal, Philibosian, and Adkins, 1972; Ruibal and Philibosian, 1974
7.1	–	–	–	3	Schoener and Schoener, 1980, personal communication
8.3	2.25	–	–	–	Fleming and Hooker, 1975
13.5	2.48	–	–	3	Schoener and Schoener, 1980, personal communication
–	–	–	–	2	Trivers, 1976
–	–	–	–	–	Sexton et al., 1963
1.3	1.41	–	–	2	Rand, 1967
0.62	.99	–	–	–	Jenssen, 1970
–	–	–	–	0	Andrews, 1979
37	2.8	–	–	–	Andrews, 1971
3.7	1.86	–	–	3	Schoener and Schoener, 1980, personal communication
–	–	–	x	–	Werner, 1982
980	2.19	–	–	–	Iverson, 1977
1558	3.13	x	x	–	Krekorian, 1976
1000	3.2	–	–	–	Tollestrup, 1979
11050	4.11	–	–	–	Tanner and Krogh, 1974
–	–	–	–	–	Tollestrup, 1979
600	–	–	–	–	Gennaro, 1972
2500	2.10	x	x	–	Dugan, 1982
8100	3.38	–	x	–	Berry, 1974
1700	3.52	–	x	–	Johnson, 1965
–	–	x	x	–	Nagy, 1973
–	–	–	–	–	Stebbins, 1944
–	–	–	–	–	Tinkle, 1973
40	2.03	–	–	–	Simon, 1975
280	2.86	–	–	–	Ruby, 1978
–	–	x	–	–	Ruby, 1981

Table 9.1 (*Continued*)

Species	Sex ratio (male/ female)	Sexual dimor-phism (male/ female)	Sexual maturity (male/ female)	Home-range ratio (male/ female)	Sur-vival (male/ female)	Female density (females/ hectare)
S. magister	0.83	1.08	–	–	–	4.0
S. occidentalis	0.67	–	–	3.5	–	2.9
S. olivaceus	0.74	0.89	0.81	2.4	0.80	28.4
S. poinsetti	1.0	–	–	–	–	6.9
S. undulatus						
(Colorado)	1.2	0.93	0.92	2.3	–	13.6
(Utah)	0.71	1.03	–	–	–	73.8
S. virgatus	0.74	0.92	0.91	3.6	–	28.7
	–	–	–	1.2	–	–
Tropidurus albemarlensis	0.42	1.2	1.0	2.9	–	74.1
T. delanonis	0.92	1.33	1.37	6.3	–•	45.8
Uta stansburiana						
(Oregon)	0.90	1.04	0.98	1.2	–	37.9
(Colorado)	0.94	1.04	–	1.2	–	22.2
(Nevada)	1.07	–	–	3.1	–	16.0
(Arizona)	0.56	1.08	1.05	3.7	–	24.5
(Texas)	1.0	1.1	1.07	3.7	0.99	17.3
(New Mexico)	1.0	–	–	2.6	–	25.0

sizes (male size/ female size) in territorial lizards ($r = 0.88$, 24 df, $P < 0.001$), nonterritorial lizards ($r = 0.77$, 6 df, $P < 0.02$), and all lizards ($r = 0.87$, 32 df, $P < 0.001$). Sexual dimorphism was estimated for 3 nonterritorial species using the regression equation obtained from the other nonterritorial species:

$$\text{sexual dimorphism} = 0.75\ (\text{sexual maturity ratio}) + 0.26.$$

Sex Ratio. The number of mature males and females in a study area was used to compute the adult sex ratio for the species, where sex ratio is expressed as number of males/number of females. Where seasonal information was available, sex ratio in the breeding season was used.

Home-Range Sizes. Three types of techniques have been used to determine home-range and territory sizes in lizards. These are (1) the mini-

Female home range (m^2)	Female home range/ $(length)^3$	Non-breeding season	Herbivores	Number of congeners	Reference
–	–	–	–	–	Tinkle, 1976
1943	–	–	–	–	Tanner and Hopkin, 1972
291	2.75	–	–	–	Blair, 1960
–	–	–	–	–	Ballinger, 1973
363	3.25	–	–	–	Ferner, 1974, 1976
–	–	–	–	–	Tinkle, 1973
102	2.99	–	–	–	Rose, 1981
–	–	x	–	–	Rose, personal communication
129	2.8	–	–	–	Stebbins, Lowenstein, and Cohen, 1967
40	1.89	–	–	–	Werner, 1978
1750	3.4	–	–	–	Nussbaum and Diller, 1976
931	3.3	–	–	–	Tinkle and Woodard, 1967
490	–	–	–	–	Tinkle and Woodard, 1967
121	–	–	–	–	Parker, 1974; Parker and Pianka, 1975
121	31.8	–	–	–	Tinkle, 1967
393	–	–	–	–	Worthington and Arviso, 1973

mum convex polygon technique, (2) the minimum convex polygon technique with modifications for number of recaptures (Turner, Jennrich, and Weintraub, 1969), and (3) the probability-density function technique (Jorgensen and Tanner, 1963). The first method is preferred by workers who directly observe the movements of the lizards (Jenssen, 1970; Berry, 1974; Ruby, 1976; Stamps, 1978; Werner, 1978; Dugan, 1982). The second and third methods are typically employed when point sighting is used (Jorgensen and Tanner, 1963; Tinkle, 1967; Fleming and Hooker, 1975; Schoener and Schoener, 1980).

The 3 techniques can give different estimates of home-range size. In general, the same data can yield home-range estimates differing by a factor of 2 or 3 times, depending on the technique used (Fitch and von Achen, 1977; Schoener, 1981; Rose, 1982). Several authors have questioned whether any one method or any one correction factor is appropriate for all lizard species (Fitch and von Achen, 1977; Schoener, 1981;

Table 9.2 Nonterritorial species (Lacertidae, Teiidae, Scincidae).[a]

Species	Sex ratio (male/ female)	Sexual dimor- phism (male/ female)	Sexual maturity (male/ female)	Home- range ratio (male/ female)	Sur- vival (male/ female)
Ameiva quadrilineata	0.71	1.17[b]	1.21	2.5	–
Cnemidophorus arubensis	0.33	1.5	–	–	–
C. hyperythrus	1.0	1.03	–	0.47	–
C. lemniscatus	–	1.22	1.18	–	–
C. sexlineatus	1.0	1.0	1.0	0.80	–
C. tigris					
(Nevada)	1.5	1.0	–	0.55	–
(Colorado)	1.5	1.03	–	1.24	–
Emoia atrocostata	1.05	1.02	1.0	–	–
Eremias arguta	–	–	–	0.81	–
	–	–	–	0.94	–
E. persica	1.64	1.06	0.99	–	–
Eumeces egregius	0.64	0.92	0.94	–	–
E. fasciatus	–	1.04	–	1.8	–
E. obsoletus	1.09	–	–	2.7	–
	1.11	0.98[b]	0.97	1.1	1.26
E. schneideri	1.45	1.0	1.15	–	–
E. skiltonianus	–	0.95	0.95	–	–
Gallotia gallotia	–	1.22	–	–	–
Lacerta agilis	–	–	–	3.1	–
	–	–	–	1.5	–
	–	–	–	1.2	–
L. muralis	0.28	1.14	–	1.13	–
L. taurica	1.43	1.03	–	–	–
L. vivipara	1.03	0.95[b]	0.94	–	–
Mabuya buettneri	0.78	0.95	–	–	0.59
M. maculilabris	0.69	1.06	–	–	0.67
Panaspis nimbaensis	0.53	1.0	–	–	0.54
Scincella laterale					
(Florida)	1.0	0.91	–	3.1	0.88
(Texas)	0.88	0.90	0.93	–	–

a. Territorial members of these families (see x's in territorial column) were not included in statistical analyses. They are considered separately.

b. Sexual dimorphism is estimated from sexual maturity ratio for these species.

Female density (females/ hectare)	Female home range (m^2)	Female home range/ $(length)^3$	Territorial	Reference
14.6	415	3.57	–	Hirth, 1963
322	–	–	?	Schall, 1974
29.3	3968	–	–	Bostic, 1965
–	–	–	x	Müller, 1971
69.5	1012	3.67	–	Fitch, 1958
17.4	5178	–	–	Jorgensen and Tanner, 1963
6.9	6976	3.38	–	McCoy, 1965
20.3	–	–	–	Alcala and Brown, 1967
–	52	–	–	Tertyshnikov, 1970
–	16	–	–	
–	–	–	–	Bogdanov and Vashetko, 1972
37.5	–	–	–	Mount, 1963
–	–	–	–	Fitch and von Achen, 1977
–	410	–	–	Fitch, 1955
–	1170	3.0	–	Hall, 1971
–	–	–	–	Yadgarov, 1973
–	–	–	–	Tanner, 1957
–	–	–	x	Case, personal communication
–	31	–	–	Tertyshnikov, 1970
–	58	–	–	
–	108	–	–	
517	23	2.09	x	Boag, 1973
30.4	–	–	–	Cruce, 1977
–	–	–	–	Avery, 1975
6.5	–	–	–	Barbault, 1974
1.6	700	3.55	–	Barbault, 1974
0.42	250	3.56	–	Barbault, 1974
174	81	3.1	–	Brooks, 1967
251	–	–	–	Mather, 1970

Rose, 1982). Despite these cautionary remarks, I have not attempted to "correct" home-range estimates presented in the literature. In any event, any errors introduced into my analyses by this procedure should be minor, for several reasons. First, the home ranges of different species vary over a much greater scale (0.3 m^2 to 10,000 m^2) than the possible error factors of 2 to 3 times introduced by the use of different techniques. Second, there is no evidence of a systematic bias in the use of different techniques. For example, there is no indication of a relationship between the type of technique used and lizard home-range size (Schoener, 1981; Rose, 1982). Finally, most of the home-range data in this chapter are presented in the form of home-range ratios (male home range/ female home range). These intraspecific ratios should be free from error, since the same techniques would have been used for both sexes.

The home-range sizes of adult males and females during the breeding season are compiled in this chapter. In some species seasonal variation in home-range size occurs; in this case both breeding and nonbreeding home-range sizes are given.

Home-Range Ratio. The average home-range sizes of sexually mature males and females during the breeding season were used to calculate the relationship between male and female home-range sizes, where home-range ratio is male home-range size/ female home-range size. Occasionally workers presented data on male and female home-range sizes during the nonbreeding season; these are separately indicated in Tables 9.1 and 9.2.

Female Density. The density of sexually mature females per hectare was compiled. Wherever seasonal variation in density occurred, data from the breeding season were used.

Survival. Survival rates of adult males and adult females were unavailable for most studies. Only reports which determined the survival rates of sexually mature males and females over the same period in the same location were considered in this compilation.

Most of the parameters mentioned above can vary intraspecifically over rather small distances (Harris, 1964; Roughgarden and Fuentes, 1977; Schoener and Schoener, 1980). Because of behavioral interactions between males, discussed below, in territorial species, sex ratios may be more female biased and home-range ratios higher in areas with greater female density and smaller female home-range size (Schoener and Schoener, 1980). Unfortunately, nearly all of the available lizard studies have focused on a single study area, so that the range of intraspecific variation in population parameters is unknown for most lizards. If there is a systematic bias in the choice of study areas, it probably lies in the direction of greater lizard densities. Ecological or behavioral observers are probably more apt to choose study areas with greater than average densities, in order to facilitate data collection.

Male Superterritories and Polygyny in Lizards

Insectivorous Females

Male methods of female acquisition in terrestrial vertebrates depend on the temporal and spatial distribution of the females (Orians, 1961, 1969; Bartholomew, 1970; Goss-Custard, Dunbar, and Aldrich-Blake, 1972; Trivers, 1972; Bradbury and Vehrencamp, 1977; Emlen and Oring, 1977; Owen-Smith, 1977; Wittenberger, 1980). Since most lizards are insectivores, in this section the spacing and behavior of female insectivores will be emphasized. A small group of herbivorous lizards has been studied and found to exhibit highly variable social behavior and mating patterns. This group is treated separately below.

Insectivorous female lizards live in home ranges which they occupy for long periods of time (months to years). The only exceptions are the ant-eating *Phrynosoma,* which stay at an ant trail for several weeks, then move long distances to another trail (Baharav, 1975; Whitford and Bryant, 1979).

In general, female home ranges tend to overlap little with those of other females. Iguanid and agamid females are often territorial (Rand, 1967; Stebbins, Lowenstein, and Cohen, 1967; Tinkle, 1967; Jenssen, 1970; Milstead, 1970; Stamps, 1973; Burrage, 1974; Ferner, 1974; Ruibal and Philibosian, 1974; MacKay, 1975; Ruby, 1976; Bruton, 1977; Werner, 1978; Schoener and Schoener, 1980; Rose, 1981). In these species suitable habitat tends to be covered by a patchwork of nonoverlapping female home ranges. If this were generally true for iguanids and agamids, one would expect to find a negative correlation between female home-range size and female density among territorial species. In fact, there is a highly significant negative relationship between $\log_{10}$ female density and $\log_{10}$ female home-range size in territorial species ($r = -0.86$, 22 df, $P < 0.001$, Fig. 9.1).

Exclusive female home ranges also occur in some nonterritorial species (for example, *Scincella laterale,* Mather, 1970). Skinks are particularly apt to have small, widely spaced home ranges (Fitch, 1954, 1955; Barbault, 1974). In nonterritorial species there is no correlation between female density and female home range ($r = -0.003$, 6 df), but the sample size is too small for definite conclusions.

Overlap of home ranges leading to female dominance hierarchies has been reported for some species (review in Stamps, 1977b), but there have been no reports of an insectivorous species in which females always live in clumps with widely overlapping home ranges.

In contrast to the general lack of overlap among female home ranges, overlap between the sexes tends to be extensive. The only exception so far discovered is the chameleon *Chamaeleo namaquensis,* in which males

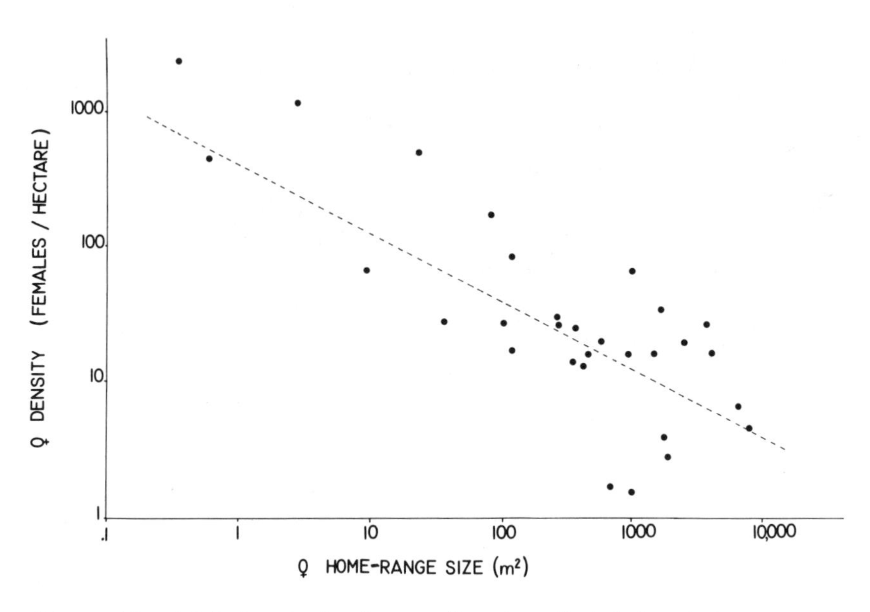

Figure 9.1 Adult female density as a function of adult female home-range size for both territorial and nonterritorial species.

and females occupy exclusive home ranges (Burrage, 1973). Home-range overlap between males and females is most extensive during the breeding season in many species (Ferner, 1974; Fleming and Hooker, 1975; Stamps and Crews, 1976; Ruby, 1978).

In order to copulate with a female, a male must find her within her home range and court her there. None of the copulations reported for insectivorous lizards has occurred outside of a female's home range. Courtship prior to mating appears to be required in all species, and many workers have commented on extensive courtship which rarely culminates in mating (Harris, 1964; Rand, 1967; Ruby, 1976; Werner, 1978). Crews (1975) has shown that male courtship can accelerate female receptivity in *Anolis carolinensis,* and evidence from other species suggests that females may require extensive courtship over a period of days or weeks before they will mate with a particular male. For example, in *A. aeneus,* the longer a male and female shared a home range, the less courtship was required prior to copulation (Stamps, 1977a). In *Sceloporus jarrovi* males courted each female in the home range daily, and females mated with the males which courted them most frequently (Ruby, 1981). After sufficient courtship female lizards usually signal their readiness to mate by remaining quiet and/or lifting the tail to facilitate the insertion of the male's hemipenes (Carpenter, 1977). Forced copulation in lizards can be defined as a case when a male physically subdues and mates with a female which attempts to retreat and which struggles throughout the copulation. Only one instance of this behavior has been described in any species of insectivorous lizard—*Tropidurus delanonis,* described by Werner (1978).

In most insectivores, female choice of mating partner is probably fairly limited. Since females do not leave their home ranges in order to mate, prospective male partners must have home ranges overlapping that of the female. A female with a home range on the border between 2 male home ranges might be able to choose between them, but this option is restricted in territorial species by the males' tendencies to arrange their territories to completely enclose female home ranges (Jenssen, 1970; Ruby, 1978; Stamps, personal observation). In some species males allow a small subordinate male to live within their territory (Fitch, 1956; Harris, 1964; Rand, 1967; Burrage, 1973; Stamps and Crews, 1976; Ferner, 1976; Trivers, 1976; Ruby, 1981). Females in these species might have a choice between the dominant and the subordinate. Subordinate males generally court much less frequently than dominants (Harris, 1964; Werner, 1978; Ruby, 1981; Stamps, personal observation), and on this basis alone they would be expected to have lower reproductive success.

Frequency of copulation has been measured in 4 species with two-step male-dominance hierarchies. In 2 cases dominant males did all of the mating (Stamps, 1977a; Ruby, 1981) and in 2 others subordinates achieved a few copulations—*Anolis garmani* (Trivers, 1976), *T. delanonis* (Werner, 1978). In *T. delanonis* the subordinate copulations appeared to be forced; the extent of female acquiescence in subordinate matings of *A. garmani* is not known (Trivers, 1976).

In many temperate species males set up territories or home ranges before females emerge (Crews, 1975). Females in these species might choose a home-range location based on the attractions of the available males. However, it is also likely that home-range location is determined by resource distributions, since the disappearance of a territorial male and his replacement by a subordinate neighbor or intruder rarely seems to lead to female desertion to join another male (but see Harris, 1964). Experiments similar to those developed for studying blackbird mating systems (for example, Searcy, 1979) could determine if female lizards' choices of home-range location were due to resource distributions or the courtship and characteristics of available males.

A Model for Polygyny

Given the preceding situation, male options for increasing their reproductive success are limited. There are no large groups of females that a male can sequester and defend. Nor do females travel to leks to choose a suitable mate. Instead, females must be courted (often repeatedly) within their own home ranges before they will mate with a particular male.

In insectivorous lizards, polygamous mating systems can only occur if the following conditions are met:

(1) Males must arrange their home ranges to overlap with female home ranges.
(2) Males must increase their home-range sizes so as to encompass more than one female home range.

In addition, a third mechanism must ensure that there is greater variance in mating success for males than for females. One obvious such mechanism is territoriality. If some males defend enlarged home ranges against other males, they would have exclusive breeding rights over the females within their territory. If male territories typically contained more than one female home range, then that species would have a polygamous mating system.

In the absence of male defense, other mechanisms might be evoked to produce differential male reproductive success. Consider a nonterritorial species in which males had enlarged home ranges overlapping those of females. If the adult sex ratio was 1:1, and males did not defend their home ranges, then the enlarged male home ranges would extensively overlap in the regions occupied by females, as for example, *Scincella laterale* (Brooks, 1967). One result of extensive male home-range overlap might be the formation of male-dominance hierarchies. If dominant males enjoyed higher reproductive success than subordinates, then a polygynous mating system could result. This scenario assumes that a dominant animal could prevent subordinates from courting and mating with the females in his home range, but this is unlikely for several reasons. Nonterritorial species typically live in habitats or microhabitats in which visibility is poor. Often they are unaware of the presence of a conspecific only a short distance away (review in Stamps, 1977b). In such species it would be extremely difficult for a dominant lizard to monitor activities of females and subordinates over his entire home range, especially since females themselves tend to be overdispersed across the available habitat. A dominant male could be assured of mating with a female in his presence but could probably not prevent subordinate males from courting and mating with females in other portions of his home range.

An alternate suggestion for nonterritorial polygyny is related to epigamic selection. Assume again that nonterritorial males have enlarged home ranges that extensively overlap in female home ranges. The result would be a wider choice of mates per female. Females might possibly prefer to mate with males possessing certain characteristics—large size, bright colors, enticing scents, flamboyant courtship techniques. If males were sufficiently variable, then more "attractive" males could achieve more matings than less attractive males. This could also result in a polygamous mating system.

But this female choice argument for lizard polygyny is unconvincing for several reasons. First, it would be difficult for a female to rank the characteristics of the available males unless they all courted her within a short period of time. Most evidence of female preference comes from species with male leks, in which females can see and make simultaneous comparisons of a number of males. A comparable situation (many males courting a single female) has never been reported in insectivorous lizards. In the absence of simultaneous courtship females would have to compare and remember the characteristics of males courting intermittently over a long period of time (weeks or months) in order to determine if a particular male were the most attractive of those available. It is doubtful that female lizards have the ability to do this.

A second problem with the female choice argument is that it allows for multiple female matings. Females could mate with several males if courted by two or more equally attractive males or if a newcomer was judged more attractive than a male with which she had previously mated. If multiple matings occur, paternity assurance would be reduced even for those males with characteristics judged attractive by females.

The difference between the territorial and nonterritorial systems can be formalized by considering the probability of paternity (P_p), where P_p is the probability that a male will sire the offspring of a female (p) living within his home range. Probability P_p is apt to be an important determinant of male reproductive success (male-RS).

In lizards, male-RS for a given breeding season can be approximated by

$$\text{male-RS} \simeq \sum_{p=0}^{n} P_p N_p, \tag{9.1}$$

where n is the number of females living within the male's home range, P_p is the probability of paternity for each female, and N is the number of offspring produced by each female.

In territorial species, P_p is apt to be high; it may approach unity in some species (Berry 1974; Trivers, 1976; Ruby, 1981; Stamps, unpublished data). In such species, the number of females within a male's territory multiplied by the number of offspring produced by those females should give a reasonable estimate of male-RS for that species (see also Wade, 1979). When P_p is high, as it appears to be in territorial species, conditions 1 and 2 above are likely to lead to polygynous mating systems.

In nonterritorial species, P_p would depend on the relative size of male and female home ranges. If males and females have home ranges of similar size, a male might be able to increase P_p by continuous attendance on a female during her receptive period. Such "tending" behavior has been reported in some teiids and skinks (Carpenter, 1977). On the other hand, if nonterritorial males expand their home ranges so as to encompass more than one female, then P_p would be reduced, because no male could be assured of exclusive mating rights over all of the females within his home range.

This model for polygyny suggests that the males of territorial species might have more to gain by home-range expansion than do nonterritorial males. From Eq. 9.1 we see that the reproductive benefit to a male of home-range expansion to include more females in his harem will be positively related to P_p. When P_p is close to 1.0, a male's expansion of his home range to overlap with one additional female will add that female's offspring to the reproductive total for that male. When P_p is low, the same amount of home-range expansion would only add a fraction of those offspring to the male's total. Hence, the benefits (in terms of male-RS) of home-range expansion are probably greater in territorial than in nonterritorial species. If the costs of home-range expansion were comparable in the two groups of lizards, one might expect to see larger home-range ratios in territorial than in nonterritorial species.

An interesting implication of this model is that male lizards may defend oversized territories larger than the size required for maintenance, growth, or survival in order to maximize their reproductive success (Andrews, 1971; Schoener and Schoener, 1980). This is in contrast to the trophic hypothesis, which suggests that male home-range size is determined by male trophic requirements (McNab, 1963; Turner, Jennrich, and Weintraub, 1969; Schoener, 1971, 1977). According to the trophic hypothesis, males should defend territories no larger than the optimal size for their energy and nutrient needs.

Testing the Model

To compare the trophic and the reproductive hypotheses for home-range size, we first need to consider the effect of sexual dimorphism on home-range ratios in lizards. If home-range size is related to body size among lizard species (Turner, Jennrich, and Weintraub, 1969), then sexual-size dimorphism should be positively related to home-range ratio for all lizards. That is, males of highly dimorphic species should require larger home ranges (with proportionally more food) than females of the same species.

The problem is to determine if sexual-size dimorphism necessarily leads to larger trophic home ranges in the males. One group of lizards provides data relevant to this question. These are the territorial species in the nonbreeding season. The home-range sizes of these lizards should be primarily determined by nonreproductive factors.

There is no indication of a positive correlation between sexual dimorphism and home-range ratio in the nonbreeding territorial lizards (Fig. 9.2). Home-range ratios are close to unity in this group ($\bar{X} \pm$ S.D. = 1.07

± 0.48, Table 9.1). Apparently, when male lizards are not defending breeding territories, male home-range sizes can be nearly the same size as those of females. This is true even for highly dimorphic species (for example, *A. aeneus,* Table 9.1).

The discussion of nonterritorial species suggests that those males would derive little reproductive benefit from home-range expansion. In this case, we would expect nonterritorial lizards to be similar to territorial species in the nonbreeding season. In fact, the home-range ratios of nonterritorial species map closely onto the values shown by nonbreeding territorial species (Fig. 9.2). Nonterritorial species do not show a positive correlation between home-range ratio and sexual dimorphism, and their home-range ratios are close to unity ($\bar{X}$ ± S.D. = 1.53 ± 0.90, N = 15). These results suggest that the home-range ratios of nonterritorial species, like those of nonbreeding territorial species, are primarily determined by nonreproductive factors.

The lack of a positive correlation between sexual dimorphism and home-range ratio in either the nonterritorial or the territorial lizards is surprising, since it suggests that sexual dimorphism per se does not lead to a requirement of more extensive home ranges in the larger sex. This implies that one or more of the assumptions of the trophic home-range hypothesis are incorrect. The best candidate for erroneous assumption is that food density is a simple function of home-range size. In many sexually dimorphic species, members of the larger sex can eat a greater range of prey sizes than can members of the smaller sex (Schoener, 1977). Hence available prey density (prey mass/unit area) may actually increase as a function of body size within a lizard species. A positive relationship between body size and home-range size only appears if lizard species of widely different sizes are compared (Turner, Jennrich, and Weintraub, 1969).

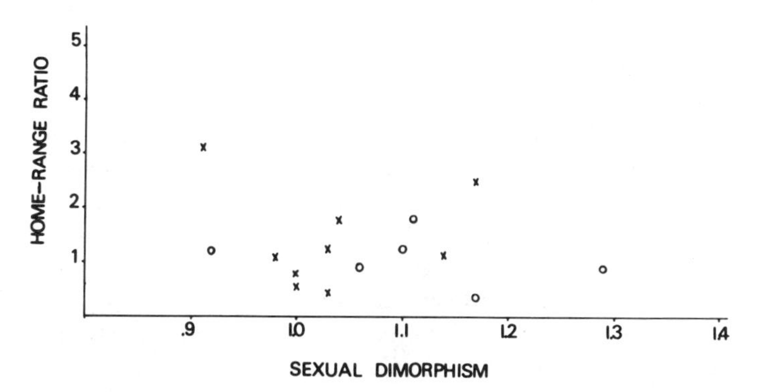

Figure 9.2 Relationship between home-range ratio (male home-range size / female home-range size) and adult sexual-size dimorphism. Nonterritorial species (*crosses*) and territorial species during the nonbreeding season (*circles*) have home-range ratios near 1.00.

The trophic and the reproductive hypotheses make different predictions about the relationship between territoriality and home-range ratio. As noted above, if male home-range sizes are primarily determined by reproductive considerations, then territorial species should have higher home-range ratios than nonterritorial species.

On the other hand, proponents of the trophic hypothesis usually assume that relationships between size, sex, and home-range size are the same in territorial and nonterritorial species. For example, Turner, Jennrich, and Weintraub (1969) pooled data from territorial and nonterritorial species when determining the relationships between mass, sex, and home-range size in lizards. This assumption of similarity is reasonable in view of the physiological literature, in which there are no apparent differences between the metabolic rates of territorial and nonterritorial species (Bartholomew and Tucker, 1964; Bennett and Dawson, 1976). Hence, the trophic hypothesis predicts that the home-range ratios of breeding territorial and nonterritorial species should be roughly similar.

Another test can be used to discriminate between the trophic and the reproductive hypotheses. This test involves territorial species with distinct breeding seasons, in which home-range ratios have been measured in the breeding and the nonbreeding seasons. If the reproductive hypothesis is correct, then territorial males should defend oversized territories in the breeding season, but not in the nonbreeding season. Hence, this hypothesis predicts higher home-range ratios in the breeding than in the nonbreeding season.

However, there is no indication from the physiological literature that territorial species have higher energy-requirement ratios (male/ female) in the breeding season than in the nonbreeding season. Indeed, available evidence suggests either that male and female energy requirements covary seasonally (Jameson, Heusner, and Lem, 1980; Heusner and Jameson, 1981; Nagy, Chapter 2), that there are no seasonal or sexual differences in metabolic rate (Bennett and Nagy, 1977), or that breeding females have higher energy needs than males or nonbreeding females (Dutton, Fitzpatrick, and Hughes, 1975). Hence the trophic hypothesis does not predict higher home-range ratios in the breeding season in territorial species.

The trophic hypothesis and the reproductive hypothesis make different predictions about home-range ratios. If males defend oversized home ranges in order to increase their reproductive success, then

(1) home-range ratios should be higher in breeding territorial lizards than in breeding nonterritorial lizards, and
(2) home-range ratios should be higher in breeding territorial lizards than in nonbreeding territorial lizards.

Alternately, if home-range sizes in lizards are determined by their trophic requirements, then

(1) home-range ratios should be similar in territorial and nonterritorial species, and
(2) home-range ratios should be similar in breeding and nonbreeding territorial lizards.

Data for breeding territorial lizards are presented in Figure 9.3. Note that territorial lizards tend to be more sexually dimorphic than nonterritorial lizards (Figs. 9.2, 9.3, and see below). Hence it was necessary to compare the home-range ratios of the two groups over comparable ranges of sexual dimorphism.

Home-range ratios of the breeding territorial lizards were significantly higher than those of nonbreeding territorial lizards ($P < 0.001$, $t = 4.2$, 25 df) when both were compared over the same range of sexual dimorphism (0.85 to 1.3). The home-range ratios of breeding territorial lizards were also significantly higher than the nonterritorial lizards—sexual dimorphism (0.85 to 1.3): $P < 0.001$, $t = 3.6$, 29 df; sexual dimorphism (0.95 to 1.2): $P < 0.01$, $t = 2.9$, 19 df.

In 4 cases information was available for both breeding and nonbreeding periods in the same species. In *A. aeneus, Sauromalus obesus,* and *Sceloporus jarrovi,* female home-range sizes were the same in both seasons but male territory sizes increased in the breeding season, resulting in a dramatic and significant increase in home-range ratio during breeding for each species (Fig. 9.3). In *Sceloporus virgatus* male territory sizes increased but female sizes decreased in the breeding season, again resulting in a higher home-range ratio at that time (Fig. 9.3). These results support the reproductive model for the evolution of polygyny presented in this chap-

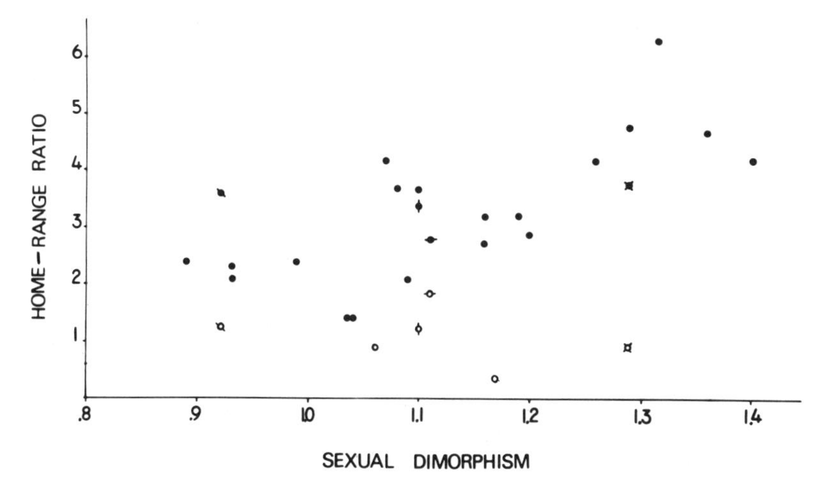

Figure 9.3 Relationship between home-range ratio and adult sexual-size dimorphism for territorial species during the breeding season (*dots*) and the nonbreeding season (*circles*). In four species (*special crossed or canceled symbols*) home-range ratios were available for both breeding and nonbreeding seasons.

ter. They do not support the predictions of the trophic hypothesis.

Other types of information from the literature also support the hypothesis that territorial males defend enlarged territories in order to increase their reproductive success. First, if territorial defense is related to reproduction, then one would expect territorial defense to be most vigorous in the breeding season. Territorial defense, male activity, and male display behavior are most vigorous in the breeding season in most territorial species (Fitch, 1940; Stebbins, 1944; Andrews, 1971; Mitchell, 1973; Berry, 1974; Ferner, 1974; Fleming and Hooker, 1975; Stamps and Crews, 1976; Ruby, 1978; Werner, 1978; Tollestrup, 1979; Rose, 1981; Dugan, 1982; Werner, 1982). Only two species have been reported to show vigorous territorial defense in the nonbreeding season—*Sceloporus merriami* (Milstead, 1970) and *S. graciosus* (Ferguson, 1971).

Several male *Anolis* do not defend territories at all during the nonbreeding season—*A. aeneus* (Stamps and Crews, 1976), *A. cupreus* (Fleming and Hooker, 1975), *A. polylepis* (Andrews, 1971). In the habitats occupied by these species (the island of Grenada and mainland Costa Rica), insect samples show that food abundance is lowest during the nonbreeding season, the dry season (review in Tanaka and Tanaka, 1982). Hence males abandon territoriality at the very time that food competition should be most intense.

If male territories are larger than the optimal size for their trophic needs, then males might show an energy deficit in the breeding season. An optimal-size trophic territory should, by definition, provide sufficient food to cover an animal's daily metabolic needs. Defense of an oversized territory could result in a daily energy deficit, if the costs of the added defense exceeded the extra food available in the larger area. In this case, one might expect male lizards to subsidize their territorial defense during the breeding season by metabolizing previously stored energy reserves.

Rose (1981) has compared the activity and energy allocations of male and female *S. virgatus* in the breeding and nonbreeding seasons. Males are much more active than females in the breeding season, and males use up more of their body fat than do females during the breeding season. However, males eat less than females while breeding, as measured by stomach and fecal weights. More important, males become leaner during the breeding season; the weight per unit length decreased 17 percent as compared to males of the same age in the nonbreeding season. Finally, if male territory sizes are set by mates in the breeding season and by food supply in the nonbreeding season, food provisioning should decrease territory size in the nonbreeding season but not in the breeding season. Provisioning of male *S. jarrovi* in the nonbreeding season resulted in decreases in male territory size (Simon, 1975). The same experiment in the breeding season had no effect on male territory size (Simon, 1973).

Sexual Dimorphism and Sexual Selection in Lizards

Two hypotheses have been advanced to account for sexual-size dimorphism in lizards. The first, called the competition-avoidance hypothesis, proposes that sexual-size dimorphism is favored because it reduces competition between males and females of the same species (Selander, 1966, 1972; Schoener, 1977). The second hypothesis assumes that sexual-size dimorphism among lizards is due to intrasexual selection for mates among males (Trivers, 1972, 1976; Schoener, 1977). These two hypotheses make rather different predictions about the relationship between sexual dimorphism and the other factors studied in this chapter.

Mating Systems in Territorial Species

The competition-avoidance hypothesis predicts that sexual dimorphism should be most pronounced when competition between the sexes is most intense. In general, food competition between the sexes will depend on the extent to which males and females forage in the same area. In lizards, competition between the sexes is apt to be greatest in monogamous species, in which one male and one female share a home range. In polygynous species, intersexual food competition should be less intense, since in these species one male spreads his foraging efforts across the home ranges of several females. Hence, the competition-avoidance hypothesis predicts the strongest sexual dimorphism in monogamous species (see also Selander, 1972; Clutton-Brock, Harvey, and Rudder, 1977; Searcy, 1979). In contrast, if sexual dimorphism is a product of sexual selection, then the most polygynous species should have the most pronounced sexual dimorphism.

Home-range ratios in lizards can be used to estimate the extent of polygyny in a given species. Since female lizards do not tend to show clumped distributions, home-range ratios give a rough estimate of the number of females expected per male harem. If females do not have contiguous home ranges, then the home-range ratio would tend to overestimate the average number of females per male. For example, to defend three noncontiguous female home ranges, a male would need to defend areas used by females plus the areas in between their home ranges. Hence, in general home-range ratios give an estimate of the upper limit for the average harem sizes in a given lizard species.

Using home-range ratios to estimate the degree of polygyny in lizards, we note that there is a positive correlation between the degree of polygyny and the extent of sexual dimorphism in breeding territorial lizards ($r = 0.57$, $P < 0.01$, 23 df, Fig. 9.3). This correlation is in the direction predicted by the sexual-selection hypothesis; species with high home-range ratios are more dimorphic than species with low home-range ratios.

Territoriality

According to my model for polygyny, intrasexual selection among males should be more intense in territorial species than in nonterritorial species. To the extent that the degree of sexual dimorphism in lizards reflects the intensity of sexual selection, this model predicts greater sexual dimorphism in territorial species than in nonterritorial species. In contrast, the competition-avoidance hypothesis makes no explicit predictions about the relationship between male territoriality and sexual-size dimorphism. It is difficult to see how the presence or absence of male defense would directly affect the intensity of resource overlap between males and females.

Among those species for which information on both social system and sexual dimorphism is known (Tables 9.1 and 9.2), territorial species have an average dimorphism of 1.13, nonterritorial species an average of 1.00 (t_{51} = 3.64, $P < 0.001$). There is also more variance in sexual dimorphism among the territorial species (territorial = 0.0203, nonterritorial = 0.00407; $F_{33,18} = 5.0$, $P < 0.001$).

Among teiids and lacertids, territoriality is highly unusual (Stamps, 1977b). Sexual dimorphism data are available for only 3 territorial species, *Cnemidophorus lemniscatus, Gallotia gallotia,* and *Lacerta muralis* (Table 9.2). In all 3 species males are markedly larger than females (Table 9.2). The extent of sexual dimorphism in each species is significantly greater than the average dimorphism for nonterritorial skinks, lacertids, and teiids (*C. lemniscatus* versus nonterritorial species, $t_{15\text{ df}} = 4.21$, $P < 0.001$; same for *G. gallotia; L. muralis* versus the nonterritorial species, $t_{15} = 2.72$, $P < 0.02$).

The territorial proclivities of a fourth species, *C. arubensis,* have not been investigated (Schall, 1974, Table 9.2). However, the extreme sexual dimorphism in this species suggests that it is also territorial.

Male Reproductive Success

If sexual dimorphism in territorial species is due to sexual selection, then within a territorial species male size and male reproductive success should be positively related. In contrast, the competition-avoidance hypothesis makes no explicit predictions about the relationship between male size and reproductive success.

Reproductive success of territorial males is often estimated by the number of females within their territories. All available studies of polygynous species have reported positive relationships between male size and number of females (Fitch, 1956; Blair, 1960; Harris, 1964; Stebbins, Lowenstein, and Cohen, 1967; Blanc and Carpenter, 1969; Berry, 1974; Fleming and Hooker, 1975; Trivers, 1976; Stamps, 1977a; Werner, 1978; Ruby, 1981, Dugan, 1982; Werner, 1982). Positive relationships between male size and number of females were not reported for two species, both of which appear to be monogamous (Tinkle, 1967; Jenssen, 1970).

A more accurate test involves counting the number of copulations achieved by males as a function of their size. In the species for which this has been done, there is a positive relationship between male size and number of copulations (Trivers, 1976; Ruby, 1981). In several other species, males were assigned to classes (territorial, subordinate, peripheral, and so on), and it was found that territorial males achieved the majority of the copulations (Werner, 1978, 1982; Dugan, 1982). Since territorial males were larger than the other classes of males, these observations also support the sexual-selection hypothesis.

Sexual Selection and Sex Ratios

Trivers (1972) considered in some detail the influence of sexual selection on adult sex ratios. In animals with little or no male parental investment, he suggested that intense male competition for mates should lead to higher mortality rates in males. According to this hypothesis, the intensity of intrasexual selection should be negatively related to the adult sex ratio (males/ females).

In lizards, sexual dimorphism can be used as a rough estimate of the intensity of sexual selection (see previous section). The relationship between sexual dimorphism and adult sex ratio was assessed for territorial and nonterritorial species. Among territorial species, there is a negative correlation between sex ratio and sexual dimorphism ($r = -0.46$, 28 df, $P < 0.01$).

Male-male competition for mates is expected to be less intense in nonterritorial species than in territorial species (see above). As expected, in nonterritorial species there is no significant relationship between sex ratio and sexual dimorphism ($r = 0.13$, 15 df).

If differential male mortality is responsible for skewed sex ratios in lizards, then sex ratios should be positively related to adult survival ratios. Data on adult survival are scant, but in those species for which data are available, adult survival ratios (males/ females) are highly correlated with adult sex ratios ($r = 0.83$, 12 df, $P < 0.001$, Tables 9.1 and 9.2).

There are two possible reasons for differential mortality in lizards. The first is that males must find and court females in the breeding season. Higher activity rates and excursions into unfamiliar areas by breeding males could increase their exposure to predators. This explanation does not necessarily involve sexual selection and applies to both territorial and nonterritorial species. In fact, breeding males are more active than breeding females in both groups (*territorial species:* Fitch, 1940; Stebbins, 1944; Tinkle, 1967; Andrews, 1971; Mitchell, 1973; Berry, 1974; Ferner, 1974; Fleming and Hooker, 1975; Stamps and Crews, 1976; Iverson, 1977; Ruby, 1978; Rose, 1981; *nonterritorial species:* Fitch,

1955, 1959; Tanner, 1965; Tertyshnikov, 1970; Danielyan, 1971; Barbault, 1974; Cruce, 1977; Fitch and von Achen, 1977).

A second cause of differential male mortality is directly related to territorial defense and sexual selection. In lizards, age and size tend to be correlated because of indeterminate growth. Territorial defense by old (large) males could lead to differential mortality in young (small) males. Higher male mortality is expected under two conditions:

(1) If females tend to occupy the more favorable parts of the habitat, and if the largest males have territories overlapping those females, then smaller males would be forced into less favorable or marginal areas. If the survival of males is lower in marginal areas, then intrasexual competition could lead directly to differential male mortality.

(2) If the aggression of large territorial males leads to injury, depressed feeding rates, exclusion from shelter sites, or stress effects in smaller males, then small males could suffer differential mortality even if not excluded from favorable habitats. These effects should be most pronounced when there are frequent aggressive contacts between males. High rates of contact are expected in areas of high density, as young males are chased from one territory to the next, and are also expected in species with strong male dominance hierarchies, in which subordinates live in the shadow of territorial, dominant males.

The mortality patterns of several insectivores suggest that territorial defense is related to differential male mortality. In *S. graciosus* and *S. jarrovi,* large (old) males survive better than small (young) males (Tinkle, 1973; Ruby, 1976). In *S. graciosus* large males actually survive better than females, while small males have much higher mortality than either females or old males (Tinkle, 1973). Ferguson, Bohlen, and Woolley (1980) studied mortality in *S. undulatus* and found that males had higher mortality than females, and that male mortality was greatest in early spring during the period of territorial establishment. In contrast, female mortality was highest later in the year, during the period of egg laying. More important, they found that male mortality during territorial establishment was density dependent.

Indirect evidence suggests that male exclusion to unfavorable habitats is partly responsible for differential male mortality. Many authors have reported seeing young males in marginal or unfavorable habitats (Rand, 1967; Blanc and Carpenter, 1969; Schall, 1974; Bruton, 1977; Ferguson, Bohlen, and Woolley, 1980; Schoener and Schoener, 1980; Ruby, 1981). Moreover, in a group of lizards in which young males are *not* excluded to unfavorable habitats, sex ratios tend to be male biased.

In herbivorous iguanids (the iguanines), young males are usually not forced into unfavorable habitats. Instead, they may be subordinates in the territory of a large, dominant male (Evans, 1951; Berry, 1974; Iverson, 1977; Wiewandt, 1977) or they may be allowed access to favorable habitat surrounding tiny, male breeding territories (Dugan, 1982; Werner, 1982; Boersma, in press). The average sex ratio for insectivorous territorial species is 0.96; the ratio for the herbivores is 1.33 ($P < 0.01$, $t_{36} = 3.41$). A male-biased sex ratio occurs even in species with strong male dominance hierarchies and frequent aggressive interactions between males, as, for example, *Sauromalus obesus* (Berry, 1974; Table 9.1). Hence, male aggression per se does not seem to lead to differential male mortality.

If territoriality in insectivorous species leads to increased male mortality, then territorial insectivores should have lower sex ratios than nonterritorial insectivores. The average sex ratio for insectivorous territorial species is 0.86; the average sex ratio for insectivorous nonterritorial species is 1.06 ($P < 0.05$, $t_{49} = 2.27$).

In summary, in lizards skewed adult sex ratios are related to sexual differences in adult mortality. In territorial species, male-male competition leads to female-biased adult sex ratios. These biased sex ratios may result from the exclusion of males, particularly small (young) males from areas of favorable habitat.

Female Home-Range Size and the Intensity of Sexual Selection

Territorial Species

According to the model proposed above, the intensity of intrasexual selection in territorial species depends on the ability of males to defend territories overlapping several female home ranges. Any factor affecting the expense or feasibility of defending a large territory should also affect the intensity of intrasexual selection in territorial lizards.

Although there are many ecological factors related to the costs of territorial defense, adequate data are available for only one: female home-range size. A simple graphical model illustrates the effects of female home-range size on degree of polygyny and on the intensity of sexual selection (Fig. 9.4). Assume that in a given species males and females each require a territory of size H for optimal growth, maintenance, and survival. Further assume that female territories are contiguous and nonoverlapping. The number of females within a male's harem will be a linear function of male territory size. If males have territories of size H, then they will overlap with one female, and the species will be monogamous. If males are capable of increasing their territory size above H, then they can overlap with more than one female, resulting in polygyny. Male reproductive success, B, resulting from defense of a territory of size T depends on female home-range size, H, and the probability of paternity P_p, so that

$$B = P_p(T/H).$$

The benefit curve may have a nonlinear shape if P_p changes as a function of T. For example, a male with a territory of 0.25 H might have no mating success with an overlapping female because of an insufficient opportunity to stimulate that female with courtship. Alternatively, as T became large, P_p could decline as males were unable to detect and expel courting male intruders from the edges of their territories. This model simply assumes that over some range of territory sizes P_p will be relatively constant, and the benefit curve will have a roughly linear shape.

Most theoretical models of territoriality assume that the costs of territorial defense increase monotonically as a function of territory size (Brown, 1964; Schoener, 1971, 1977; review in Davies, 1978). In this model we are concerned not with the total cost of male defense, but with the extra costs to the male which result from defense of an oversized territory. If H is the optimal territory size for growth, maintenance, and survival, then by definition, any increase beyond H would result in energy or nutrient deficits and/or an increase in the probability of mortality in males. Hence the costs of defending additional space increase as a function of the size of the extra area. Cost ultimately is measured in terms of the decrease in male reproductive success as a result of the male's defending an oversized territory. Cost, C, can be computed as

$$C = f(T - H),$$

where f is a positive monotonically increasing function. Optimal territory size for males of a particular species is that for which the value $B - C$ is maximized (Fig. 9.4).

Now compare two species: one has small female home ranges (H_1), whereas the other has large female home ranges (H_2) (Fig. 9.4). In each species the benefit of extra territorial defense increases as a function of male territory size, but the two benefit curves have very different slopes. The smaller the females' home range, the more females there will tend to be in the male's optimal territory.

The cost curves are identical in Figure 9.4 but the cost functions for territorial defense are not likely to be identical in different species. However, shapes and slopes of the cost curves can be varied over a wide range without changing the prediction that the optimal territory of species 1 should contain more females than the optimal territory of species 2 (Fig. 9.5). This is because benefit increases geometrically (as a function of T/H) while cost increases arithmetically (as a function of $T - H$). If two species have very different female home-range sizes, the disparity in slopes of benefit will tend to overwhelm interspecific differences in the costs of territoriality.

This model enables one to make specific predictions about the

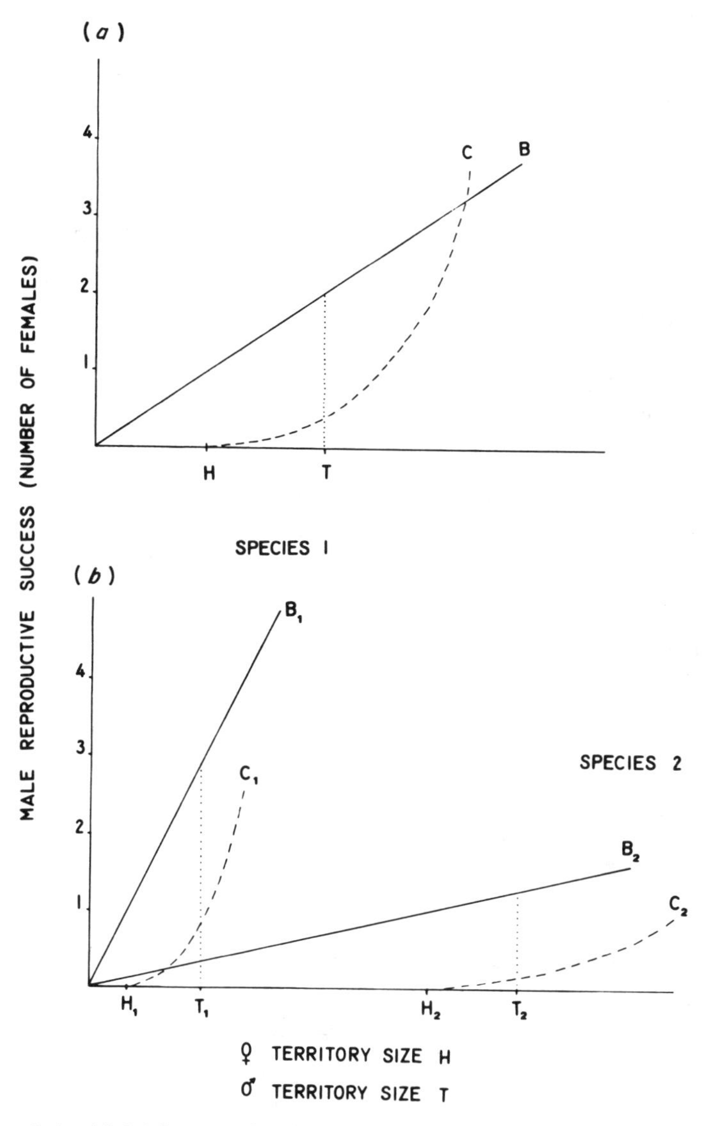

Figure 9.4 (*a*) Male reproductive success (*B*) depends on the benefits and costs of increasing individual territory size above the size optimal for energy balance and survival (*H*), where *B* (benefit) is proportional to the number of females defended; *C* is cost of defense for all territory sizes above *H*. The value $B - C$ indicates the male territory size (*T*) which will provide optimal reproductive success.
(*b*) The influence of female home-range size on the evolution of polygyny. A small female home-range size (species 1) permits a high home-range ratio (T_1/H_1) and the optimal male home range (T_1) includes 4 females. A large female home-range size (species 2) limits the home-range ratio and the optimal territory (T_2) includes only 1.2 females.

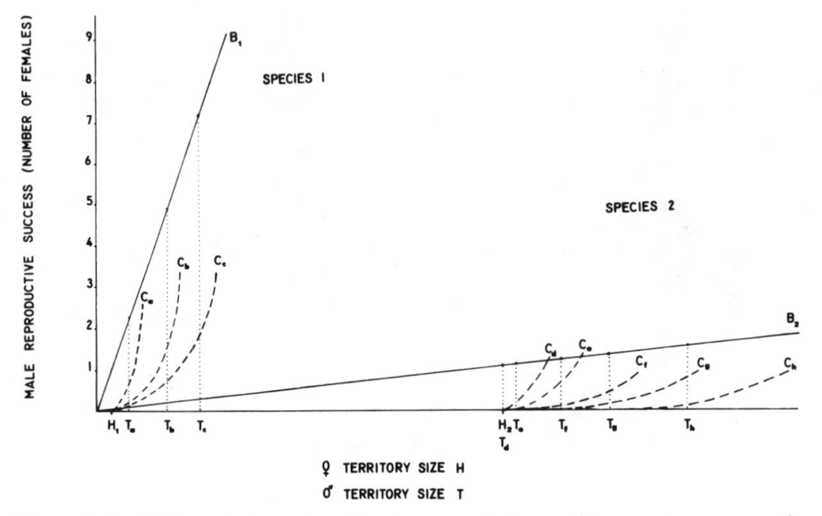

Figure 9.5 Effect of changing the shape and slope of the cost curve on the relationship between female home-range size, home-range ratio, and polygyny.

relationship between female home-range size, home-range ratio, sexual dimorphism, and sex ratio in territorial species.

(1) Home-range ratio is indicated in the model by (T/H). Home-range ratio is expected to decrease as female home-range size increases. Atchley, Gaskins, and Anderson (1976) point out that a ratio and the denominator of that ratio can be negatively correlated due to statistical artifact. However, they cite an equation which allows one to calculate the correlation due to artifact, if the relevant statistics are known. When X_1 represents variable 1, X_2 represents variable 2, p_{12} represents the parametric correlation between variables 1 and 2, and δ_X represents the coefficient of variation for variable X, and where $X_1/X_2 = Y$, r_{YX_2} can be approximated as follows (Chayes, 1949):

$$r_{YX_2} = \frac{p_{12}\delta_1 - \delta_2}{\sqrt{\delta_1{}^2 + \delta_2{}^2 - p_{12}\delta_1\delta_2}}. \tag{9.2}$$

There is a very strong positive correlation between male and female home-range size in territorial lizards ($r = 0.98$, 27 df, $P < 0.001$). The mean (± S.D.) for male home-range size is 2,443 ± 5,487; for female home-range size $\bar{X}$ ± S.D. = 1,136 ± 2,466. Substitution of these values into Eq. 9.2 yields a predicted r between female home-range size and home-range ratio of + 0.016. That is, if statistical bias were responsible for any correlation between female home-range size and home-range ratio, one would expect it to be a *positive* correlation. Instead, as is pre-

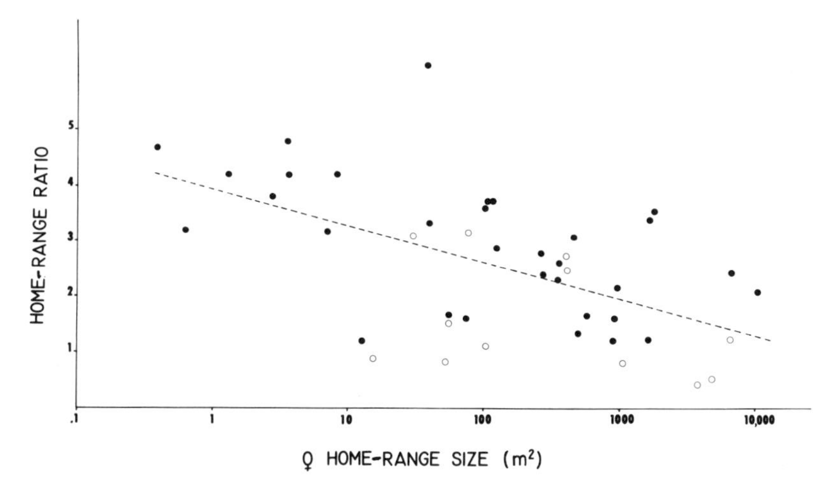

Figure 9.6 Home-range ratio as a function of female home-range size for territorial (*dots*) and nonterritorial (*circles*) lizards.

dicted by the model, home-range ratio is negatively correlated with $\log_{10}$ female home-range size in territorial lizards ($r = -0.52$, 29 df, $P < 0.01$, Fig. 9.6).

In lizards, heavier species tend to have larger home ranges (Turner, Jennrich, and Weintraub, 1969). Perhaps the correlation between home-range size and home-range ratio is due to a correlation of both these factors with body mass. To correct for this possibility, female mass was estimated for the animals in Tables 9.1 and 9.2 by assuming that (snout-vent length)^3 is proportional to mass (Stamps, 1977a). Female home-range sizes were divided by (female snout-vent length)^3 to give "mass-adjusted" female home-range sizes. Home-range ratios were significantly correlated with $\log_{10}$ of mass-adjusted female home-range sizes in territorial lizards ($r = -0.46$, 23 df, $P < 0.05$).

(2) From Figure 9.4 we can see that the intensity of sexual selection is expected to be greater for species with small female home ranges (which should be highly polygynous) than for species with large female home-range sizes (which should be monogamous). Since sexual dimorphism appears to be related to the intensity of sexual selection in lizards, the model predicts a negative relationship between sexual dimorphism and female home-range size in territorial species.

There is a significant negative correlation between $\log_{10}$ female home-range size and sexual dimorphism in territorial species ($r = -0.78$, 22 df, $P < 0.001$). There is also a significant negative relationship between $\log_{10}$ of mass-adjusted female home-range size and sexual dimorphism in territorial species ($r = -0.78$, 22 df, $P < 0.001$, Fig. 9.7).

In several animal groups, sexual dimorphism is positively correlated with average female body size (Rensch, 1959; Selander, 1966; Clutton-

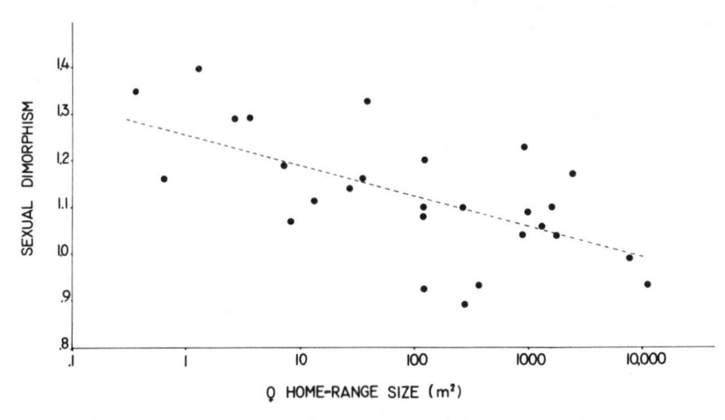

Figure 9.7 Sexual-size dimorphism as a function of female home-range size in territorial lizards.

Brock, Harvey, and Rudder, 1977; Ralls, 1977). However, there is no indication of a positive relationship between female snout-vent length and sexual dimorphism in territorial lizards ($r = 0.066$, 29 df). Substitution of the appropriate values into Eq. 9.2 indicates that one would expect a very slight negative correlation between female size and sexual dimorphism due to statistical artifact ($r = -0.041$).

(3) In territorial species, if male intrasexual selection leads to female-biased sex ratios and if female home-range sizes are related to male intrasexual selection, then female home-range size and sex ratio should be positively correlated. There is a weak positive correlation between $\log_{10}$ female home-range size and sex ratio in territorial species ($r = 0.43$, 24 df, $P < 0.05$).

Several authors have suggested that population density might be related to the intensity of sexual selection in lizards (Ghiselin, 1974; Dunham, Tinkle, and Gibbons, 1978). Since female density is highly correlated with female home-range size (Fig. 9.1), one might expect that female density would be correlated with many of the same factors as female home-range size.

In territorial species, $\log_{10}$ female density was positively correlated with sexual dimorphism ($r = 0.57$, 24 df, $P < 0.01$) and negatively related to sex ratio ($r = -0.55$, 31 df, $P < 0.001$). The relationship between home range ratio and $\log_{10}$ density was not significant ($r = 0.36$, 20 df, $P < 0.1$).

Nonterritorial Species

The model presented in Figure 9.4 was based on the assumption that males defend territories. To what extent can the model be extended to cover nonterritorial species?

Nonterritorial species might be able to increase their reproductive success by finding and courting more females in the breeding season

(Fitch, 1955, 1959; Barbault, 1974; Fitch and von Achen, 1977). One way to do this would be to increase their home-range sizes so as to overlap with more females. Of course, the benefit and cost functions would be quite different for territorial and nonterritorial species. Costs would tend to be lower, because males would not incur the expenses of territorial defense. But benefits would also be lower since P_p is apt to be smaller in nonterritorial species. However, in both groups one might expect a negative relationship between female home-range size and home-range ratio.

Substitution of appropriate values into Eq. 9.2 yields a slight negative relationship between female home-range size and home-range ratio due to statistical bias ($r = -0.038$). The actual correlation is stronger ($r = -0.40$), but it is not significant (11 df, $P < 0.2$, Fig. 9.6).

No data are available for nonterritorial species with very small female home-range sizes. This omission is not a result of sampling error. Several teiid and lacertid species with very high female densities and/or small female home ranges have been studied. In each case the males of these species have turned out to be territorial—*Cnemidophorus lemniscatus* (Müller, 1971), *Gallotia gallotia* (Case, personal communication), *Lacerta muralis* (Boag, 1973). That is, high female densities seem to trigger a switch to territoriality in groups which are ordinarily nonterritorial (see Stamps, 1977b). These territorial members of nonterritorial groups exhibit the amounts of sexual dimorphism and the skewed sex ratios predicted on the basis of their behavior rather than their taxonomic affinities (Table 9.2).

If moderately small female home-range sizes tended to lead to higher home-range ratios in nonterritorial species, then sex ratio should be positively related to female home-range size in this group. This is because one of the costs of male home-range expansion is presumed to be an increased risk of mortality (see above). As expected, there is a positive correlation between sex ratio and female home-range size in nonterritorial species ($r = 0.66$, 8 df, $P < 0.05$). However, in nonterritorial species there is little evidence of male-male contest competition for mates. As expected, there is no tendency for sexual dimorphism to be negatively related to female home-range size in this group ($r = 0.19$, 7 df).

Further Implications of the Model

1. Interspecific Competition and Sexual Selection. According to the model presented here, any factor which influences female home-range size or female density will have indirect effects on the intensity of sexual selection and the degree of polygyny. One factor that could affect female density is interspecific competition. Intense interspecific competition is expected to decrease female densities and/or to increase female home-range size. Since female home-range size is presumed to influence sexual selection, the intensity of interspecific competition should also be negatively related to sexual dimorphism and home-range ratio, and it should be positively related to adult sex ratios.

The relationship between competition and female density or home-range size has been analyzed in only a few groups of lizards. Dunham, Tinkle, and Gibbons (1978) studied the size patterns of island *Uta* populations. They measured a variety of biotic and abiotic factors and found that the only variable correlated with *Uta* population density was the number of potentially competing species. Schoener and Schoener (1980) found higher maximal densities of *Anolis sagrei* on Abaco where it occurs alone than on Bimini where it occurs with three congeners. Data on female home-range sizes for six West Indian *Anolis* species show a tendency for a positive relationship between the number of congeners and female home-range size ($r = 0.68$, 4 df, Table 9.1). These results all tend to support the hypothesis that interspecific competition tends to decrease female densities in lizards.

There are more data available on the relationship between competition and sexual dimorphism in lizards. Schoener (1966, 1968, 1969a, 1970, review in Schoener, 1977) has assembled a variety of evidence showing that the intensity of intrageneric competition is negatively related to sexual dimorphism in insular *Anolis* lizards. In general, species which occur alone exhibit greater degrees of sexual dimorphism than do species sympatric with many congeners. Fitch (1976) has reported a similar trend in mainland *Anolis.* Dunham, Tinkle, and Gibbons (1978) did not find a significant negative relationship between the number or density of competitors and sexual dimorphism among insular *Uta.* However, there was very little variation in sexual dimorphism among the *Uta* populations they studied ($\bar{X} = 1.08$; S.D. $= 0.02$, N $= 15$). Variance in sexual dimorphism is much greater among *Anolis* species (see Table 9.1).

2. Latitudinal Effects on Sexual Dimorphism and Home-Range Size. In several temperate New-World species there are latitudinal gradients in sexual dimorphism (Parker and Pianka, 1975; Pianka and Parker, 1975). In *Uta stansburiana* sexual dimorphism increases with decreasing latitude (Pianka and Parker, 1975). If intrasexual selection were responsible for this trend, and intrasexual selection is a function of female home-range size, then one would predict a positive correlation between latitude and female home-range size in *U. stansburiana.*

Available data on home-range size and latitude are assembled in Table 9.1. For these populations, latitude and female home-range size are positively correlated ($r = 0.85$, 4 df, $P < 0.05$). These results suggest that latitudinal changes in sexual dimorphism among temperate lizards could be due to the effects of female home-range size on intrasexual selection.

Diet and Social Behavior: The Herbivores

The effect of diet on social behavior in lizards is very striking. As we have seen, insectivorous iguanids and agamids are fairly uniform in their behavior. Males are territorial in the breeding season and often also in the nonbreeding season. Females tend to be aggressive and may defend territories or form dominance hierarchies with other females. Even juveniles are aggressive, and may form hierarchies or defend territories against other juveniles (Blair, 1960; Rand, 1967; Tinkle, 1967; Fox, 1978; Stamps, 1978; Simon and Middendorf, 1980; Fox, Rose, and Myers, 1981). In the insectivores, male reproductive success depends on the number of females within his territory, and intrasexual selection is probably more important than epigamic selection.

In contrast, the behavior of the herbivorous iguanids (the iguanines) is much more variable. Most male iguanines are territorial, but territory size and the intensity of defense depends on the species. In some species males defend small territories during the breeding season and females visit male territories to choose a mate: *Amblyrhynchus cristatus* (Eibl-Eibesfeldt, 1955; Carpenter, 1966; Boersma, in press), *Conolophus subcristatus* (Werner, 1982), *Iguana iguana* (Dugan, 1982). In other species males have larger territories overlapping female home ranges: *Dipsosaurus dorsalis* (Norris, 1953; Pianka, 1971), *Ctenosaura pectinata* (Evans, 1951), *Ctenosaura hemilopha* (Carothers, personal communication), *Cyclura carinata* (Iverson, 1977), *Cyclura stejnegeri* (Wiewandt, 1977), *Sauromalus obesus* (Berry, 1974). In some species territories are defended only in the breeding season: *A. cristatus, C. subcristatus, I. iguana, D. dorsalis;* in others territories are defended in both seasons but more strongly while breeding: *C. carinata, S. obesus.* In *C. stejnegeri* some males defend territories all year; others defend only in the breeding season (Wiewandt, 1977). In *Sauromalus varius* there is no sign of male aggression or territoriality at any time (Case, 1982).

Even though male iguanines are territorial, they are much more tolerant of subordinate males than are insectivorous males. In species with small mating territories, subordinate males are kept outside of the territorial borders, but the territories are so small that there is plenty of favorable habitat in which the subordinates can live (Dugan, 1982; Werner, 1982; Boersma, in press). In species with large territories, subordinates are often found within the territory of the dominant territory holder (Evans, 1951; Berry, 1974; Iverson, 1977; Wiewandt, 1977; Carothers, personal communication). Subordinate males can form complex, multistep hierarchies within the territory of the "tyrant" (Evans, 1951; Berry, 1974; Carothers, personal communication); in contrast, the most elaborate dominance hierarchy seen in breeding insectivorous males is a simple two-step hierarchy (review in Stamps, 1977b; Ruby, 1976). We have already noted that the tolerance of subordinate males by dominant males may be responsible for the high adult sex ratios in herbivorous species.

Female herbivores are remarkably passive. Aggressive interactions among females have been noted in the context of nest-site defense

(Rand, 1968; Rand and Rand, 1976; Iverson, 1977; Werner, 1982) but at other times herbivorous females usually show little or no antagonism. Female dominance interactions have been observed only in *C. hemilopha* (Carothers, 1981).

Juveniles have been watched in only two herbivores. In both, juveniles have a clumped distribution, with several juveniles in the same plant or retreat (Burghardt, Greene, and Rand, 1977; Iverson, 1977). In *I. iguana,* juveniles actually are socially attracted to one another (Burghardt, Greene, and Rand, 1977).

The intensity of intrasexual selection in iguanines is probably as variable as their mating systems. The largest males are the territory holders in all iguanines studied so far. Small males may be subordinates within a male's territory (Evans, 1951; Iverson, 1977; Wiewandt, 1977; Berry, 1974) or may wait around the periphery of a small mating territory for a chance to force copulations with females or take over the territory (Dugan, 1982; Werner, 1982).

Epigamic selection is clearly stronger in many herbivores than it is in most insectivorous iguanids or agamids. In species with small mating territories the food supply within the territory is not a factor in female choice; number of females per territory is unrelated to the food supply of the territory (Dugan, 1982; Werner, 1982; Boersma, in press). Female choice of males in these species is quite clear; females may visit several before staying and mating with one (Dugan, 1982; Werner, 1982). Even in herbivores with more conventional mating systems there is some evidence of female choice. When a tyrant *S. obesus* male disappeared, his territory was inherited by his two subordinates; whereupon one female moved out and became the mate of a neighboring tyrant (Berry, 1974).

Social patterns in the nonbreeding season are also highly variable in herbivores. In some species males and females live in stable, clumped distributions. This pattern is characteristic of species living in rocky habitats—*C. pectinata* (Evans, 1951), *S. obesus* (Berry, 1974). In other species, groups of mixed sex and age wander through the habitat together—*I. iguana* (Dugan, 1982). Marine iguanas form dense sexually segregated aggregations along the shore next to the feeding grounds (Boersma, in press), whereas in *C. subcristatus* adults scatter thinly over a wide area during the nonbreeding season (Werner, 1982). There is even a suggestion of monogamous pairs in *C. carinata* (Iverson, 1977).

The differences between the herbivores and insectivores can be summarized as follows:

(1) Herbivores in general are less aggressive than insectivores. Male herbivores defend no territory, defend a small territory, or allow subordinates within their territory. Females usually confine their aggression to defense of nesting sites, and the juveniles of at least two species have clumped distributions.

(2) Herbivorous males have a wider variety of mating systems than insectivores. These range from leks (*A. cristatus*) to exploded leks (*I. iguana*) to large territories with subordinates (for example, *S. obesus*) to no territory at all (*S. varius*).

(3) The social behavior of herbivores in the nonbreeding season is more variable than that of insectivores. Herbivores can be highly clumped in dense aggregations (*A. cristatus*) or thinly dispersed over wide areas (*C. subcristatus*). Clumps of lizards can move through the habitat (*I. iguana*) or have stable distributions (*C. pectinata, S. obesus*).

How do these behavioral differences relate to the difference between a herbivorous and insectivorous diet? It is axiomatic in behavioral ecology that resource distribution affects the feasibility and form of resource defense (Brown, 1964; Brown and Orians, 1970; Wilson, 1975; Emlen and Oring, 1977). This principle is particularly evident in the present case.

The arthropod prey of insectivores seems to be relatively predictable in time and space. Three long-term studies have examined seasonal changes in arthropod prey abundance within lizard habitats—*Sceloporus jarrovi* (Simon, 1973), *Sceloporus merriami* (Dunham, 1978), *Anolis aeneus* (Stamps, Tanaka, and Krishnan, 1981). In each case there was relatively little variation in prey abundance through the lizard's activity season (coefficients of variation are 39 percent for *S. jarrovi,* 56 percent for *S. merriami,* 63 percent for *A. aeneus*). Two studies have examined small-scale spatial differences in prey abundance in lizards (Simon, 1973; Stamps and Tanaka, unpublished data). In both cases spatial variation in prey abundance within a study area was low (coefficient of variation is 23 percent for *S. jarrovi;* for *A. aeneus* it is 19 percent).

Comparable figures on plant distribution and abundance are not yet available for herbivorous lizards. However, the descriptive accounts leave no doubt that fluctuations in food abundance and distribution are much more dramatic for many herbivores. Many authors cite examples of favorite fruits and flowers which are available in great abundance but only in widely scattered locations for short periods (Norris, 1953; Berry, 1974; Iverson, 1977; Dugan, 1982). This situation typically leads to non-aggressive aggregations of lizards using the food source (Norris, 1953; Berry, 1974; Iverson, 1977; Dugan, 1982). Seasonal and annual changes in plant productivity can be extreme, particularly in xeric habitats (Johnson, 1965; Berry, 1974; Iverson, 1977; Werner, 1982; Case, 1982). A uniform, abundant food resource has been reported for only one species, the marine iguana *Amblyrhynchus cristatus* (Eibl-Eibesfeldt, 1955; Carpenter, 1966; Boersma, in press).

It is reasonable to assume that in lizards, as in other animals, a uniform predictable food source (arthropods) is more conducive to the evolution of territoriality than food resources which fluctuate in distribution and abundance (flowers, fruits, new leaves). The general passivity of herbivores as compared to insectivores may be directly attributable to the differences in their diets.

On the other hand, herbivory does not seem to have led to a single type of social system. Instead, it seems to have opened the door to a wide variety of social systems. Variations in resources (especially food and shelter sites) are probably responsible for much of this behavioral variability (see Dugan, 1982). However, most authors have not carefully measured resource distributions, so that any model accounting for herbivores' social systems would have to be descriptive in nature. Model building is also hampered by the fact that every species studied so far seems to have a different social system. More studies of these unusual iguanids are needed to determine the evolutionary origins of their social behavior.

If nothing else, the herbivores serve to dispel the notion that lizards necessarily have simpler social behavior than birds and mammals (Wilson, 1975). It is easy to see how this idea could have arisen, given the focus on insectivorous species. As we have seen, these species tend to be remarkably uniform in their social behavior and mating systems. That this uniformity is due to the constraints of a certain type of food supply and not to a lack of neural sophistication, is illustrated by the variety of social behavior and mating systems seen in a group of iguanids that switched to a more variable type of food resource. The iguanines are a numerically small and insignificant offshoot of the insectivorous family Iguanidae (Avery and Tanner, 1971), but they will be very important in our effort to understand the relationship between ecology, evolution, and social behavior among lizards.

10 Psychobiology of Parthenogenesis

David Crews, Jill E. Gustafson, and Richard R. Tokarz

ALL-FEMALE PARTHENOGENETIC SPECIES present a unique opportunity to test hypotheses regarding the nature and evolution of sexual behavior. Whereas the majority of vertebrates are gonochoristic (fertilizing female ova with male sperm), at least 27 species of squamate reptiles representing 7 families consist mostly or entirely of females and produce clones of all-female offspring (Cole, 1975). Behavior patterns have been observed in 5 unisexual species (Teiidae and Gekkonidae) that are remarkably similar to the courtship and copulatory behavior of closely related sexual species.

In this chapter we have two aims. First, we briefly review parthenogenesis in vertebrates and the mechanisms of sex determination and sexual differentiation in reptiles. Second, we provide the first progress report of our research program on the biological bases of pseudocopulatory behavior in an all-female, isogenic lizard. We have chosen for this research *Cnemidophorus uniparens* because of its availability and adaptability to laboratory conditions, and because of the growing literature on the species' ecology and physiology. In addition to performing ethological studies describing and quantifying species-typical behavior patterns, we have examined the physiological underpinnings of this male-like behavior. We show also that the presence and behavior of cagemates have a significant influence on a female's fecundity. Finally, we present results of experiments on the effect of steroid hormones on the determination and differentiation of primary and secondary sexual structures.

Parthenogenesis in Vertebrates

Parthenogenesis refers to the development of a new individual from an unfertilized egg and can be accomplished in several ways (see Fig. 10.1). *Accidental* parthenogenesis can occur in frogs by physical or chemical stimulation of the eggs (White, 1973). In domestic turkeys, eggs

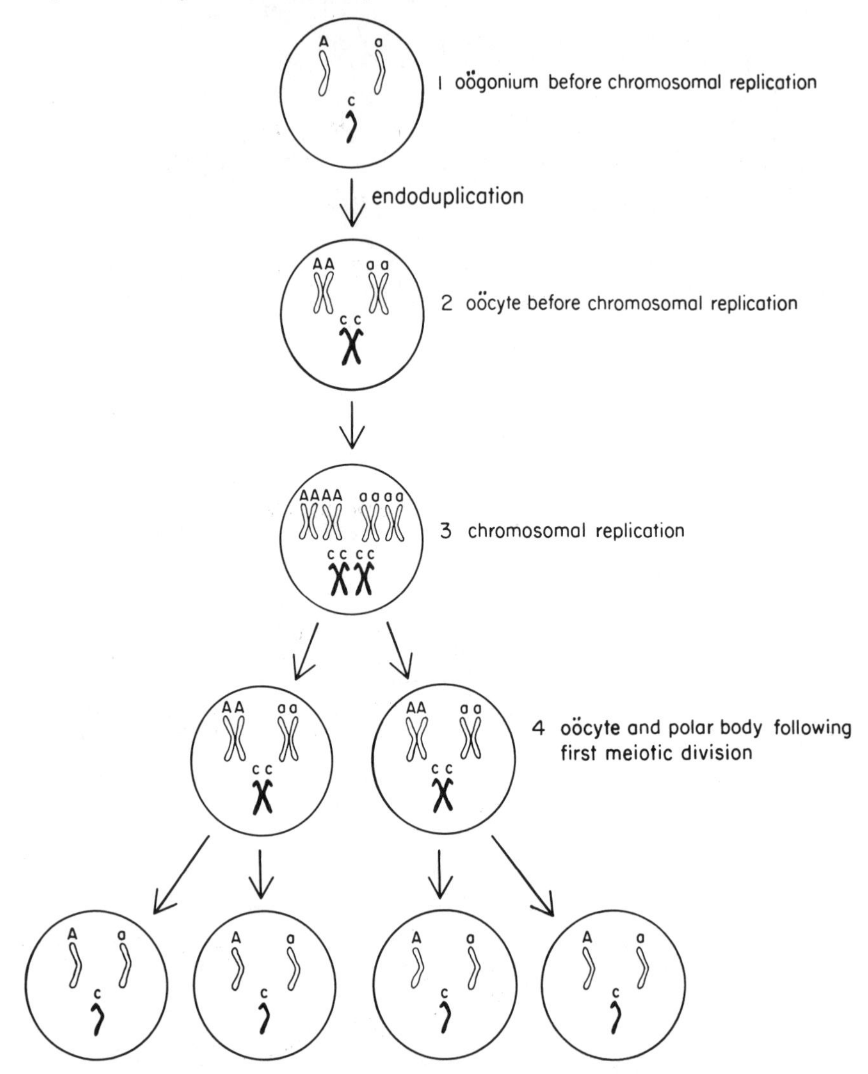

Figure 10.1 Oögenesis in the triploid parthenogenetic lizard, *Cnemidophorus uniparens.* In contrast to sexual species, no oögonial cell division (cytokinesis) occurs during the mitoses preceding meiosis. Consequently a primary oöcyte is hexaploid which after completion of meiosis yields a triploid ovum genetically identical to the mother's oögonium. For clarity, only 3 chromosomes are illustrated, and the relative sizes of germ cells are not indicated.

will occasionally develop without fertilization. Selective breeding for frequency of hatchings from unfertilized eggs has produced a line of turkeys that reproduces parthenogenetically 65 times as often as the average (Olsen, 1960), indicating that the propensity toward parthenogenetic reproduction has a genetic basis in these animals. These and other studies also indicate that the triggering mechanism for cleavage of the egg must be contained, to some extent, within the egg itself.

In some species, parthenogenesis is *facultative,* occurring only when this mode of reproduction is necessary as a last resort to produce offspring. For instance, in the phasmatids *Clitumnus extradentatus,* two-thirds of a female's eggs will develop parthenogenetically if not fertilized (White, 1973). It is important to note, however, that in those species where facultative parthenogenesis occurs, it does not replace sexual reproduction, but rather ensures that, if all else fails, some of the mother's genes will be transmitted to the next generation where successful fertilization may occur.

In a small fraction of animals, however, reproduction is *exclusively* parthenogenetic. Our research program has as its object the development of a model for the evolution of sexual behavior in vertebrates, and we shall restrict the following review of parthenogenesis to those species of fish, amphibians, and reptiles that exhibit this unusual form of reproduction.

Pathways leading to parthenogenesis with a complete complement of chromosomes can be ameiotic, in which case no reduction in chromosome number takes place, or meiotic, in which somatic ploidy level is maintained by omission of the first or second meiotic division or by pre- or postmeiotic restitution involving karyokinesis without cytokinesis (see Cuellar, 1971). Ameiotic parthenogenesis leads to an oöcyte genetically identical to the somatic cells of the mother. Premeiotic doubling of chromosome number results from suppression of the cytokinesis of the last mitosis before the oögonium enters meiosis (premeiotic endoduplication). Sister chromosomes undergo synapsis in the subsequent meiosis. Thus, ova are genetically identical to each other as well as to the mother, thereby producing isogenic individuals (clones) from the first generation forward. Postmeiotic restitution occurs when cytokinesis does not accompany the first zygotic karyokinesis. The first parthenogenetic generation differs genetically within itself and from the mother, but each line is homozygous at every chromosomal allele. Thus, second-generation individuals constitute a clonal line beginning with their mother. Uzzell (1970) points out that, since recessive lethal genes are common in vertebrates, such a parthenogenetic pathway would result in death for most ova. A second mechanism of postmeiotic restitution is the fusion of two daughter nuclei to form a gamete with the typical somatic ploidy level.

Exclusive Parthenogenesis

Fishes. At least two modes of normal parthenogenesis occur in fishes, the more common of which is termed gynogenesis. Gynogenesis differs from true parthenogenesis in that ovum development requires an external stimulus such as sperm penetration (Uzzell, 1970). Gynogenesis is included in our discussion of parthenogenesis since the cytogenetic problems of producing an ovum with a complete set of chromosomes as well as the ecological and evolutionary considerations are similar to those in true parthenogenesis.

The Amazon molly, *Poecilia formosa,* is a diploid ($2n = 46$) unisexual occurring in freshwater streams and coastal lagoons of northwestern Mexico and southern Texas (Schultz, 1971, 1979). The species lives in sympatry with either *P. mexicana* or *P. latipinna,* and the males of these species court and mate with *P. formosa.* Although the sperm activate ovum development in *P. formosa,* they make no genetic contribution to the offspring (as determined by skin and internal organ transplants between members of the same population).

Poecilia formosa can be viewed as a sexual parasite on *P. mexicana* and *P. latipinna,* for mating with a unisexual is a waste of reproductive energy for the male of the sexual species. The host species has apparently evolved some degree of defense against this exploitation. By tabulating pregnancy ratios of wild-caught sympatric females of *P. formosa* and *P. latipinna* from Brownsville, Texas, Hubbs (1964) concluded that females of *P. latipinna* have a reproductive advantage. When wild-caught females were tested in the laboratory with *P. latipinna* males, males from areas of sympatry with the unisexuals began courtship with *P. formosa* only 24.4 percent of the time; allopatric males, however, directed 39.7 percent of their courtship advances toward *P. formosa.* These results are understandable if there is selective pressure against a male's propensity to mate with a unisexual. In allopatric sexual species, the selective pressure to avoid useless matings with unisexuals has not been exerted, and the preference for conspecific females is not as pronounced.

The genus *Poeciliopsis* (family, Poeciliidae) also contains at least two species from northwest Mexico known to be gynogenetic. Both species consist of triploid ($3n = 72$) females and probably arose as hybrids between *P. lucida* and *P. monacha.* This led Schultz (1969) to designate the groups as *P. 2 monacha-lucida* and *P. monacha-2 lucida.* The sperm needed to initiate ovum development come from males of either of the two parental species. Isogenity and lack of paternal genetic influence was determined by Schultz (1969) by repeated matings of the unisexuals with *P. monacha* and *P. lucida* as well as with *P. latidens,* a species with particularly distinctive coloration. Lack of evidence of divergence among any of the lines led him to conclude that the sperm did not contribute any genetic material in mating.

Schultz (1971) also cites evidence (from Cimino, unpublished data) that parthenogenetic zygote production in the *Poeciliopsis* triploids consists of endomitotic doubling to produce 6*n* oöcytes which then undergo normal meiosis. Since the identical sister chromosomes replace true bivalents in meiosis, there is no genetic variability introduced by crossing over, and each offspring from the parthenogenetic generation forward is a clone of the mother.

Another triploid fish ($3n = 141$) also known to be gynogenetic is the goldfish *Carassius auratus gibelis* that commonly mates with wild carp and goldfish of bisexual groups (Cherfas, 1966).

Amphibians. Parthenogenetic amphibian species have been described in the *Ambystoma jeffersonianum* complex (reviewed in Bogart, 1979). *A. tremblayi* and *A. platineum* are triploid ($3n = 42$) gynogenetic salamanders that mate with sympatric males of *A. laterale* and *A. jeffersonianum,* respectively. According to electrophoretic studies, the triploid species, generally intermediate in morphology, have complete sets of chromosomes from each host species. *A. platineum,* for example, has two sets from *A. jeffersonianum* and one from *A. laterale.* This evidence points to the conclusion that the triploid unisexuals arose as hybrids of the two species now serving as host species in this gynogenetic system (Uzzell and Goldblatt, 1967).

The classical problems of parasite-host relationships apply to the *A. jeffersonianum* complex. If the parasite is too prolific, it will drive the host to extinction, thus destroying itself. Uzzell (1964) suggests that decreased fecundity in the unisexuals, a longer transformation time for the larvae, and perhaps a courtship preference on the part of the males, keep the unisexuals in check. However, ratios of females (parthenogenetic and sexual) to males taken in some areas where triploid and diploid species live together indicate drastic sex-ratio changes over time. For example, in Scio Township, Michigan, the percentage of the salamander population that was female went from 67 percent in 1928 to 89 percent in 1959. Such figures suggest that the *A. jeffersonianum* complex is an unstable system, with parthenogenetic females replacing sexual females.

Careful chromosomal studies have indicated that the oöcytes in these triploid animals are hexaploid before entering meiotic division. The mechanism for parthenogenesis appears to be premeiotic restitution. Since sister chromosomes replace bivalents, genetic variability introduced by crossing over does not occur, and each individual in a population is a clone of each other and of the mother (MacGregor and Uzzell, 1964).

Reptiles. Probably the most extensively studied parthenogenetic vertebrates are the unisexual reptiles. Species in which parthenogenesis is the norm include 3 Pacific island geckos (*Hemidactylus garnotii, Lepidodactylus lugubris,* and *Gehyra variegata agasawarisimae*), some populations of the Panamanian xantusiid lizard *Lepidophyma flavimaculatum,* several species of Lacertidae, and many species of closely related New-World teiid lizards.

The 4 parthenogenetic species of Lacertidae are *Lacerta armenica, L. dahli, L. rostombekovi,* and *L. unisexualis* (Darevsky, 1966). These animals are found in the Caucasus from the Black Sea to the Caspian Sea. The individuals of these species are diploid ($2n = 38$) and resemble females of closely related sexual species in morphology and behavior. Interspecific matings between unisexuals and males of sexual species do take place and result in both parthenogenetic diploid females and sterile triploid females.

Abortive embryos are male in *L. dahli* and *L. rostombekovi* (Uzzell, 1970). This suggests that heterogametic sex-determining characters are segregating during oögenesis, and that the altered step resulting in parthenogenesis must be after Meiosis I. Darevsky (1966) suggested initially that Meiosis II is the crucial step in maintaining somatic ploidy level in the zygote, although later studies do not support this. Variation in the genotypes of unisexual *Lacerta* probably occurs through recombination prior to independent assortment during the first meiotic division. Parthenogenetic *Lacerta* exhibit less phenotypic variation than do their sexual relatives and accept skin grafts as do unisexual *Cnemidophorus.*

Maslin (1962) first suggested that parthenogenesis was the normal reproductive pattern in some species of North American whiptail lizards, *Cnemidophorus.* In extensive collections of these groups, only females were found. Parthenogenesis was later confirmed by laboratory hatchings of unfertilized eggs (Maslin, 1971). Subsequently, 13 parthenogenetic *Cnemidophorus* species have been described. These lizards occur in the southwestern United States, Mexico, and South America. One group, *C. lemniscatus,* is of interest because it includes both parthenogenetic and sexual populations (Cole, 1975; Serena, 1980). Some species, for instance *C. neomexicanus,* are diploid ($2n = 46$) like their sexual relatives; however, most groups, like *C. uniparens,* are triploid ($3n = 69$) (Cole, Lowe, and Wright, 1969).

Although parthenogenesis may evolve spontaneously (Cuellar, 1977b), most parthenogenetic groups of *Cnemidophorus* probably arose as hybrids between sexual species. For example, *C. neomexicanus* appears based on morphology and karyotypes to be a hybrid of *C. tigris* and *C. inornatus* (Lowe and Wright, 1966). This was confirmed recently by the demonstration of equivalence of restriction sites in mitochondrial DNA (Brown and Wright, 1979). Triploid species are thought to have arisen by the backcross of a diploid unisexual species with one of its parental species or another species. Cuellar and McKinney (1976) have reinforced this theory in a study using naturally occurring hybrids between *C. neomexicanus* and *C. inornatus,* the sexual species considered to be one of the parental species of *C. neomexicanus.* Because xenografts from *C.*

neomexicanus to the offspring were not rejected, the full complement of chromosomes must have been passed from the unisexual to the offspring. However, a set of paternal, as well as maternal, chromosomes must also have been contributed to the offspring, since reciprocal grafts among the offspring were rejected as well as were grafts from the offspring to *C. neomexicanus*. These results establish that the offspring of a cross between the unisexual *C. neomexicanus* and the sexual *C. inornatus* had chromosomes contributed from both parents. Thus, the offspring were confirmed to be hybrids.

In studies of *C. neomexicanus* and *C. tesselatus*, Neaves (1969) determined that heterozygosity existed in both groups at the lactate dehydrogenase-b locus on the chromosomes and that the alleles from each of the 2 (in *C. neomexicanus*) and 2 or 3 (in *C. tesselatus*) genomes were equally active. This heterozygosity suggests that in meiosis, identical sister chromosomes must pair rather than homologous chromosomes; hence, premeiotic doubling of chromosomes may be the mechanism for ensuring full somatic ploidy level in the zygote. If oögenesis had involved pairing of homologous chromosomes and a postmeiotic restitution of ploidy level, the offspring would be homozygous at every chomosomal allele. Or, if incomplete meiosis were responsible for maintaining ploidy level, homologous chromosomes would pair and independent assortment would produce nonclonal lines, as in *Lacerta*. By using skin grafts Maslin (1967) and Cuellar (1976) have, indeed, shown isogenity among individuals of the same or different populations of *C. uniparens*. Cuellar (1977a) had also reported that skin grafts among individuals of separate populations of *C. neomexicanus* from New Mexico (San Antonio, La Jolla, Albuquerque, and Espanola) were accepted more than 99 percent of the time. This suggests that *C. neomexicanus* may be genetically uniform over its entire range.

In a detailed cytogenetic study Cuellar (1971) found evidence for premeiotic restitution of chromosomal number as the parthenogenetic pathway in *C. uniparens*. The number of bivalent chromosomes of the first polar body equals the somatic triploid number ($n = 69$). A second polar body was also observed, indicating a normal second meiotic division.

Sex Determination and Differentiation in Reptiles

Gonads and secondary reproductive structures in vertebrates are bipotential and their ultimate adult form depends in part upon intrinsically programmed developmental processes and in part upon substances secreted by the embryonic gonads. Thus, in both sexes of all vertebrates (except bony fishes), embryos have a Wolffian (mesonephric) and Müllerian (paramesonephric) duct system. In female lizards, as in mammals, subsequent development of the Müllerian ducts into the oviducts (uteri) and the cloaca (upper vagina) and the concomitant regression of the Wolffian ducts is an intrinsic process (Adkins, 1980). In contrast, differentiation and development of the Wolffian ducts into the epididymides and vas deferens in males requires androgen secreted by the fetal testis (Raynaud and Pieau, 1971). It is not known if regression of the Müllerian ducts in reptiles is dependent upon a second substance secreted by the testicular Sertoli cells as appears to be the case in mammals (Ohno, 1979). The development of the male external genitalia in mammals and reptiles, on the other hand, is mainly dependent upon the presence or absence of testicular androgens (Forbes, 1961; Ohno, 1979).

Mesonephric kidneys are present in reptilian embryos but partially or totally degenerate soon before or after hatching in all species studied (Fox, 1977). In genetic males, derivatives of the mesonephric (Wolffian) duct system are retained in adults as the epididymis and vas deferens. In genetic females of a few reptilian species, the Wolffian duct remains as a rudimentary structure in adult females; in other species, it completely disappears. In no case is the mesonephric kidney maintained in the adult female.

Steroid hormones control the development and activity of the reproductive structures in reptiles (reviewed in Forbes, 1961; Crews, 1979; Adkins, 1980). For example, in all lizards the distal portions of the mesonephric kidneys are modified. This structure, the renal sex segment, is homologous to the seminal vesicles of mammals and produces the seminal fluid. In many lizards, including *Cnemidophorus,* glands located on the medial surface of the thigh secrete a waxy substance. These femoral pore exudates communicate species identity and reproductive status (Cole, 1966; Duvall, 1979). Both the renal sex segment and femoral pores are sexually dimorphic in size and are most active in breeding males. Androgen administration stimulates hypertrophy of both structures in sexually inactive intact, or castrated, male lizards and in female lizards (Crews, 1979).

Temperature appears to be the sex-determining mechanism in some oviparous reptiles lacking identifiable sex chromosomes (Bull, 1980). The sex ratio of hatchlings in species of turtles from 5 different families has been shown to be influenced by egg incubation temperatures (Bull, 1980). For example, in the snapping turtle (*Chelydra serpentina*), constant incubation temperatures of 31 ° C and above or less than 24 ° C produce female hatchlings while intermediate temperatures (24–27 ° C) produce male hatchlings (Yntema, 1976, 1979). Recent studies with 4 turtle species (*Chrysemys* and *Graptemys*) also reveal temperature-dependent sex determination under natural conditions (Bull and Voigt, 1979). In the oviparous lizard *Agama agama,* raising embryos at 29 ° C results in all-male broods; if the clutch is raised at 26 to 27 ° C, all-female clutches result (Charnier, 1966). Wagner (1980) has obtained similar results with the oviparous gecko, *Eublepharis macularis.* It is important to note that in these latter studies, the animals are functionally rather than genetically sex reversed.

Sex chromosomes are difficult to identify in reptiles (Gorman, 1973). Lizards exhibit a number of sex chromosome mechanisms, ranging from no apparent heteromorphic pairs of chromosomes to extreme heteromorphism. Only one sexually reproducing cnemidophorine species, *C. tigris,* has been found to have X and Y chromosomes that differ in the position of the centromere and heterochromatin (Cole, Lowe, and Wright, 1969; Lowe et al., 1970; Bull, 1978); sex chromosomes can be identified in the karyotypes of those parthenogenetic species that resulted from hybridization between *C. tigris* and other species. Sex chromosomes have not been identified in other *Cnemidophorus* species, including *C. uniparens* or its sexual ancestral species, *C. inornatus* and *C. gularis* (Cole, Lowe, and Wright, 1969; Lowe et al., 1970; Bull, 1978).

Parthenogenesis in *Cnemidophorus uniparens*

Housing and maintenance of lizards as well as the reproductive measures and surgical and hormone treatments used in our studies have been described in detail elsewhere (Gustafson and Crews, 1981).

Behavioral Repertoire

An accurate and quantitative description of the behavioral repertoire of a species is an essential prerequisite for psychobiological research. Observations in both the field and the laboratory can identify natural categories of behavior (McBride, 1976), making it possible to conduct more detailed analyses of sociosexual behavior and to provide a context in which to interpret physiology-behavior interrelationships. Just as the field is the logical and necessary testing ground for hypotheses that have arisen from experimental work with laboratory populations, so might the behavior of animals housed in the laboratory under seminatural conditions call attention to behaviors that may ordinarily be overlooked or dismissed. This interdisciplinary approach has proven very effective in investigations of the biological bases of reproduction in the green anole lizard, *Anolis carolinensis* (Crews, 1975, 1980).

Although the reproductive behavior of some sexual *Cnemidophorus* species has been described (Noble and Bradley, 1933; Fitch, 1958; Carpenter, 1962), little is known of the behavior of unisexual *Cnemidophorus* species. Currently 13 parthenogenetic *Cnemidophorus* species are known to exist (Cuellar, 1977a,b; Cole, 1978) and have been the subject of various ecological, evolutionary, and cytogenetic studies (Lowe and Wright, 1964, 1966; Maslin, 1967, 1971; Wright and Lowe, 1968; Cuellar, 1974, 1977a,b; Cole, 1975, 1978; Serena, 1980). Information on the behavioral repertoire of these species is critical in light of our recent demonstrations that social interactions play a major role in the reproductive biology of one species, *C. uniparens* (Gustafson and Crews, 1981). Captive *C. uniparens* have been observed in our laboratory under a variety of housing arrangements and social situations for 2 years. The major behavioral categories observed in this species are as follows:

Maintenance Behavior

General postures and stance

1) full body up, limbs extended supporting body
2) full body down, limbs fully flexed, held at sides
3) head on plane with body, parallel with horizontal
4) head tilted down from plane of body
5) head tilted up from plane of body
6) head cocked to side
7) alert stance—anterior torso up, anterior limbs extended, head parallel with horizontal
8) freeze

Investigative behaviors

1) tongue touch
2) tongue flick
3) scratch at substrate

Locomotor patterns

1) walking, body up
2) walking, body down
3) running
4) backing up
5) climbing
6) jumping
7) dropping to lower level
8) plowing
9) squirming

Basking postures

1) resting on venter, limbs out, moderate
2) resting on venter, limbs sprawling
3) eyes closed
4) lying on side
5) feet held away from substrate
6) limbs raised, body resting on venter
7) single limb raised
8) tail up off substrate

Feeding and drinking

1) food grab
2) food toss
3) food shake
4) expulsion of noxious substance from mouth

5) mouth wipe
6) chewing
7) swallowing
8) undulating body after swallowing
9) licking lips
10) vomiting
11) gape following vomiting
12) licking moisture
13) sucking water
14) eating substrate

Defecation
1) tail arch
2) posterior body raised, hind limbs extended
3) pelvic scrapes

Shedding
1) head scrape
2) body rub
3) scratch

Burrowing and digging
1) digging with front limbs, entire venter down or posterior venter up
2) digging, whole body at once
3) digging with hind limbs
4) pushing
5) lifting object with anterior body or head
6) turning in a confined area
7) body curled
8) resting with body under object or in burrow, head out on surface

Oviposition

Social Behavior

Agonistic behaviors
1) armwaving
2) head nodding
3) bobbing
4) chasing
5) biting
6) aggressive gape
7) tail flick
8) face-off
9) tug-of-war over food
10) chain biting
11) egg eating

Sexual behaviors
1) cloacal rubbing

2) preliminary bites to side or tail
3) shoulder bite
4) straddle
5) cloacal apposition
6) pelvic thrust
7) release

The occurrence of many of these behavioral categories has been verified in field observations; they are not abnormal behaviors induced by captivity.

Field observations in the summers of 1979 and 1980 revealed that in those areas where they exist, *C. uniparens* are very dense, often with four or more females in the same mesquite bush (see also Mitchell, 1979). Whereas intense social interactions were observed repeatedly in nature (Crews, personal observations), their functions in determining social dominance, population density, resource utilization, or fecundity are as yet unknown. Another very important question is the extent to which behavioral differences between parthenogenetic and sexual *Cnemidophorus* can be accounted for by ecological differences and vice versa (Cuellar, 1977b, 1979; Leuck, 1980; Serena, 1980).

Behavioral observations were conducted on animals living in 20-gallon aquaria in groups of 4; aquaria were also partitioned to accommodate 2 pairs of lizards or 3 isolated lizards (Gustafson and Crews, 1981). Observation periods lasted 25 minutes with 6 cages being observed daily over the course of a month, beginning the second week of captivity, so that each cage was observed for 10 to 13 periods at different times of the day. In addition to the taking of descriptive notes, the frequency of the following behavior patterns for each individual was recorded: armwaving, biting, chasing, and bobbing. To facilitate observation, individuals were marked on the back with white nail enamel.

Maintenance Behavior

When *C. uniparens* are approached in the field or in the laboratory, they will abruptly stop and assume a resting posture. Continued alertness is evidenced by open eyes and a raised head. In the *alert stance* the anterior torso is held away from the substrate on extended front limbs while the posterior torso rests on the surface. In this posture the head is held horizontal, sometimes swiveling to survey the vicinity.

Investigative Behavior. Cnemidophorus uniparens are highly active in the field and in the laboratory, continually sticking their snouts into and around corners and into crevices. Tongue flicking is also frequent during this time. Much scratching at the substrate occurs with the lizard shifting its weight to 3 legs and using the front limb to brush at the surface sand, displacing only the topmost grains. This behavior is easily distinguished from actual digging in which 2 or all 4 limbs are utilized and enough sand is displaced to leave a noticeable depression (see below).

Locomotor Patterns. Movement can take a number of forms. The two extremes are represented by a sliding movement across the substrate with the limbs moving paddle-like at the animal's sides, versus walking with limbs extended and the body raised above the substrate. *Cnemidophorus uniparens* are extremely quick and capable of running at high speeds. *Cnemidophorus uniparens* are agile climbers but do not appear to rest on vertical surfaces. Jumping can occur as either a hopping, up-and-down movement or as leaps across gaps. In both instances, all 4 legs are tensed and quickly extended, springing the animal into motion.

A common behavior observed in both the field and laboratory is *plowing.* When in this posture the animal presses its venter into the substrate, pushing a quantity of sand ahead by the snout and shoulders between the two front limbs. This behavior is seen most often during the excavation of burrows.

Basking Postures. Much of an animal's time on the surface is spent basking. Lizards emerge shortly after lights on in the morning and take up basking positions under the heat lamp, a pattern corresponding to the early morning emergence of animals in the field. During basking the venter is flat against the substrate and the limbs are usually abducted from the side and the eyes closed. When directly under the heat lamp the lizards often lift their feet off the substrate or raise the entire limb (see fig. 1 in Gustafson and Crews, 1981).

Burrowing and Digging. Cnemidophorus uniparens are swift diggers. While digging, the forelimbs or both sets of limbs may be used sequentially or even simultaneously. Sand is moved under and around the body and out of the developing burrow. Once inside the burrow the animal turns around and pushes sand forward with its front limbs; in this way a lizard can excavate from within the burrow. Lizards may also use this method to close off the burrow entrance.

C. uniparens maneuver easily within the confines of the burrow by curling themselves into tight circles and turning until they attain the orientation desired. Lizards always sleep in a burrow or under a piece of bark in a curled posture. At night 2 or more lizards may share the same sleeping site, each curled into circles and lying side by side.

Animals were often seen to rest with their bodies inside the burrow and their snouts or whole heads protruding from the burrow entrance. When prevented from emerging by an aggressive cagemate in the basking area, individuals may stay in this position for an hour or more. This behavior is nearly always observed during periods of the day when other animals were basking. Thus it may represent a thermoregulatory strategy in which interaction with an aggressive lizard is avoided.

Oviposition. The oviposition behavior includes many of the digging patterns described previously. Oviposition typically occurs in late afternoon or early evening. The gravid lizard first digs a wide-mouthed hole at the cool, moist end of the aquarium, usually in a corner. These holes are deep, most often reaching the floor of the aquarium. The animal then backs into the depression and begins laying eggs. An individual typically lays 3 eggs over the course of an hour. Between eggs the animal lies with eyes closed and is lethargic in its response to cagemates or human intruders. When the last egg is laid the lizard covers the clutch with sand and crawls into a burrow for the remainder of the day.

Social Behavior

Agonistic Behavior. Cnemidophorus uniparens exhibit agonistic behaviors both in captivity and in the field (Mitchell, 1979; Leuck, 1980; Gustafson and Crews, 1981). The most frequently observed agonistic behavior is the *armwave* (see fig. 1 in Gustafson and Crews, 1981) in which the lizard raises the arm closest to the cagemate to which the activity is directed, waving it in a forward, overhand motion. Armwaving is a submissive display (Carpenter, 1960; Brattstrom, 1974) and is most frequently observed when an animal attempts to enter a basking area dominated by an aggressive animal. The intruding animal moves haltingly into the area and, when first noticed by the resident lizard, will raise its head and wave the arm closest to the resident. The intruding lizard will persist in this behavior until the aggressive lizard looks away or gives chase. The armwave thus appears to be typical of subordinate animals, reducing the likelihood that a dominant animal will act aggressively (see also Brattstrom, 1974).

The dominance rank of individuals can be assessed according to the frequency of armwaving; lizards that wave most being the lowest-ranked individual, whereas animals that wave least being the highest-ranking individual. In all of the groups, and 6 of the 8 pairs, dominance was clearly established and reflected in the frequency of armwaving (Table 10.1). Isolated animals were never observed to armwave. We have also observed behaviors similar to those described by Cuellar (1971), who

Table 10.1 Frequency of armwaving in captive *Cnemidophorus uniparens* housed in groups of 4 or in pairs (mean and standard error of mean).

	Dominance rank			
Housing	1	2	3	4
Pair	8.50 (3.23)	25.50 (9.95)		
Group	4.88 (2.70)	11.38 (4.82)	25.12 (7.47)	38.25 (7.62)

mentioned that *C. uniparens* initially act aggressively toward one another in captivity (see also Leuck, 1980). Hardy (1962) also observed an "approach order" in sexual *C. sexlineatus* housed in outdoor pens. Animals low on the approach order retreated at the approach of an animal higher in the order, thereby relinquishing basking or foraging sites.

Our observations raise the question of what differentiates a dominant animal from a subordinate animal with which it is genetically identical (see also Leuck, 1980). Dominance clearly is not determined by body size. The size of animals (SVL in millimeters) does not correspond significantly to the number of arm waves (group: $r = -0.05$; pair: $r = -0.30$). Since size cannot be used as a predictor of dominance, other factors must be considered.

Dominance seems related to reproductive activity. The average time between the beginnings of vitellogenesis and oviposition in *C. uniparens* is 18 days and the maximum reported is 29 days (Cuellar, 1971). Thus in one experiment we assumed that an animal that did not lay eggs within 30 days from the beginning of the experiment must be reproductively inactive at its start (Gustafson and Crews, 1981). Analysis of the group animals revealed that 25 percent of the dominant animals fell into this category, while 62 percent of the lowest-ranking animals were reproductively inactive at capture. All of the pair animals ($n = 12$) laid a clutch of eggs within 30 days of the start of the experiment. Research with other vertebrates (Leshner, 1979) suggests that a female's endocrine state and her level of aggressiveness are causally related. A second possibility is that an animal that has begun its reproductive cycle has a more urgent requirement to secure a basking site. If the female is unable to thermoregulate properly, the follicular cycle will be protracted or, worse, the reproductive effort already expended may be wasted. Further investigation of the relationship between initial reproductive state and subsequent dominance rank is necessary before these and other questions can be answered.

Head-nodding behavior is also observed frequently. In its more vigorous form this behavior takes the form of a bobbing movement involving the anterior torso as well. Bobbing and head nodding become most frequent during an aggressive exchange.

Chasing and biting are far less frequent than armwaving or head nodding. When these behaviors do occur, fighting animals chase and bite each other vigorously. The attacking lizard may occasionally gape its mouth before lunging at its opponent (see fig. 49 in Crews, 1980). Biting involves a quick, directed lunge and quick release of the jaws. Animals involved in a fight are easily distracted.

On 3 occasions *tail flicking* was observed during fights. In each instance the lizards were engaged in a moderately aggressive bout consisting mainly of darting chases and hopping retreats without biting. Suddenly one lizard quickly flicked its tail sharply over its head in the direction of the opponent. Since activity continued as before, tail flicking did not seem to play a decisive role in the aggressive bout.

A *face-off* occurs when two animals are either facing one another or are closely parallel. Both lizards momentarily stop, tense and then, upon any movement of the opposite animal or the sudden passing of a third animal, both jerk into another orientation. This results either in the lizards coming face to face again (in which case another face-off may occur) or in placing the animals sufficiently far apart that they commence other activities.

Fighting over food is frequent but usually it is quickly resolved by the prey item breaking in two. Even when prolonged these conflicts never result in any additional aggression.

Chain biting, as described by Cuellar (1971) was only observed when lizards from a group cage were temporarily housed in a small bucket for transport or during maintenance of their cage. The occurrence of this behavior in this context is similar to Carpenter's observation (1960) of aggression among sexual *C. sexlineatus* housed temporarily in a small trap. This behavior appears to be a function of extreme crowding. In its simplest form, chain biting consists of animals biting and holding onto each other. Occasionally the bitten animal bites a third animal and the third a fourth. The fourth lizard usually closes the chain by taking a jaw hold on one of the other lizards. Chain biting is distinct from the biting attack (see above) in that the grip is maintained for several minutes during which the animals are unresponsive to interference from the observer. A closed chain can be picked up and handled without any animal loosing its hold. Chain biting was observed in a cage only once, in this instance between the two animals of a pair.

On 3 occasions lizards were seen eating the recently deposited eggs of a cagemate. In 2 cases the eggs were being eaten by one animal alone, but in one group cage 2 animals were rapidly devouring a cagemate's clutch, at times tugging together on the same egg. At no time was an animal observed to eat its own eggs.

Sexual Behavior. We have discovered recently, from observations of captive populations of 3 parthenogenetic *Cnemidophorus* species (*C. uniparens, C. velox,* and *C. tesselatus*), behavior patterns remarkably similar to the courtship and copulatory behaviors of closely related sexual species (Crews and Fitzgerald, 1980); similar behaviors have been observed in *C. exsanguis* (Shine, personal communication) and in the parthenogenetic gecko, *Lepidodactylus lugubris* (Werner, 1980; Falanruw, personal communication). A sexual encounter in all of these species, perhaps more accurately termed a pseudocopulation, follows precisely the behavioral sequence observed during mating in captive *C. tigris* (Crews, personal observation) and that reported for *C. sexlineatus* (Noble and Bradley, 1933; Carpenter, 1960), both sexual species.

Briefly, the sequence of sexual behavior in *C. uniparens* begins with pursuit and mounting attempts by one animal (see Crews and Fitzgerald, 1980, for a complete description). The male-like female grips the tail or side of a preovulatory female. This appears to pacify the pursued preovulatory female which then holds still while the male-like lizard grips a fold of skin between its jaws and straddles her. The female-like animal sometimes carries the mounting female around the cage in this position. When the mounted female stops the mounting animal will attempt to curl its tail beneath the tail of its partner so as to oppose the cloacal regions (Fig. 10.2). During this time, the mounting lizard also intermittently rubs her cloaca against the dorsal pelvic region of the female beneath it and strokes her back and neck with the jaws and forelimbs. Immediately upon cloacal apposition, the mounting female shifts her grip to the iliac region, thereby forming the "doughnut" copulation

Figure 10.2 Pseudocopulation in the captive parthenogenetic lizard, *Cnemidophorus uniparens*. Following lunging attacks directed at the smaller female, the larger female approaches the now passive small female, first gripping in her jaws the female's foreleg. This is accompanied by mounting and riding behavior (*a,b*), during which the active female scratches the side of the mounted female with her forelegs and hindlegs and strokes the back of her neck with her jaw. Shortly afterward, the active female twists her tail beneath the other's tail (*c*), apposing the cloacae, and assumes the copulatory posture characteristic of sexual cnemidophorine lizards (*d*). (From Crews and Fitzgerald, 1980.)

posture characteristic of sexual cnemidophorine lizards (Noble and Bradley, 1933; Carpenter and Ferguson, 1977). Indeed, in 2 sequences the mounting female was seen to evert partially her cloacal lining; in sexual lizards the male everts 1 of 2 hemipenes through the cloacal opening during copulation (Crews, 1980).

In 4 instances in which pseudocopulation was timed, the copulatory posture was maintained for an average of 2 minutes, 19 seconds. Pseudocopulation is terminated when the mounted female shakes the mounting female off or the latter releases her grip and dismounts of her own accord. Complete mating sequences were observed on 9 occasions in 6 different cages. Intermittent observations throughout the day further indicate that the occurrence of pseudocopulation is not rare.

Carpenter (1960) reported cloacal rubbing by the male and suggested that it served to stimulate the male sexually and was a sign of sexual arousal. This behavior, in which the cloaca is rubbed on the substrate while the tail moves sinusoidally, was observed on 4 occasions in the present study. On 2 of these occasions the behavior did indeed precede a pseudocopulation, but in the other 2 instances no further sexual activity was apparent.

Physiological Bases of Pseudocopulatory Behavior

We can assume that the regular pattern of follicular development, ovulation, and oviposition through 3 or more successive clutches during the breeding season in *C. uniparens* reflects corresponding cyclical changes in the secretion of ovarian hormones. Although the identification and pattern of secretion of hormones during different stages of the ovarian cycle have yet to be determined for *C. uniparens,* studies of other oviparous lizards and turtles indicate that there is a sharp preovulatory peak in circulating estrogen followed by a rise in progesterone (Callard et al., 1978; Licht et al., 1979; Crews, 1980; Tokarz and Crews, unpublished data). In *Uromastix hardwicki,* the only oviparous lizard studied to date, Arslan and coworkers (1978) report an elevation of plasma androgens following ovulation.

One female in a pair of *C. uniparens* will commonly be in an early reproductive stage (ovaries containing small follicles), while her cagemate will be in the final stages of a follicular cycle (ovaries containing large, preovulatory follicles). Further, the former female will typically behave as a male, courting and mounting the latter (Crews and Fitzgerald, 1980). During the breeding season, then, females will undergo physiological and behavioral cycles that complement one another. This raises the intriguing question of what hormones are secreted during the "masculine" and "feminine" behavioral phases of the follicular cycle.

That these behaviors are hormonally mediated is clear. In addition to the relationship between a female's behavioral role during a pseudocopulation and her reproductive state (Crews and Fitzgerald, 1980), we have found that intact, reproductively inactive *C. uniparens* frequently exhibit mounting behavior upon implantation of silastic capsules containing androgen, either testosterone or dihydrotestosterone (in 8 of 9 cases, the courting individual had been implanted with androgen). Females receiving estrogen (estradiol benzoate) implants also court, but not as frequently (in only 3 of 13 cases, the courting individual was implanted with estrogen). Dihydrotestosterone was administered in subcutaneous 1 cm implants prepared by mixing 0.5 ml of Silastic Medical Adhesive with 14.5 mg DHT and extruded through a 0.5 ml Glaspak syringe. Estrogen was administered in daily injections of 0.2 μg EB for 3 days followed by a single injection of 0.8 μg EB. Behavioral testing began 5 days after the beginning of hormone treatment and continued through day 8.

Functional Significance of Male-like "Sexual" Behavior

It is important to distinguish between the essential and facilitatory functions of males. In gonochoristic species, gonadal sex is essential for reproduction while behavioral sex is facilitatory; the sperm produced by the testes are necessary for successful reproduction, whereas male courtship and copulatory behavior have a facultative role in stimulating and coordinating female reproductive activity. Studies with representatives of all vertebrate classes show that male sexual behavior synergizes with other environmental stimuli to induce ovarian recrudescence in conspecific females (reviewed by Adler, 1974; Crews, 1975, 1980; Vandenbergh, 1975). This separation of the fertilization and facilitation functions of the male is best illustrated by Adler's research on the social and behavioral cues governing pregnancy in the rat. In a series of elegant experiments, Adler has demonstrated that the copulatory behavior of the male has at least two distinct functions in addition to effecting transfer of sperm. First, the male's intromissions terminate further sexual receptivity of the female, thereby assuring the male's paternity. Second, the vaginocervical stimulation the female receives initiates a neuroendocrine reflex that results in an increase in progestational hormones which prepare the uterus for implantation of the fertilized eggs.

Male facilitation of female reproduction is well documented in other vertebrates as well. In the green anole lizard (*Anolis carolinensis*), ovarian functioning can be altered differentially by specific male displays. For instance, the stimulatory effects of the environment either can be facilitated or inhibited depending upon the male's behavior (Crews, 1974). The consequences of this behavioral regulation of female pituitary gonadotropin secretion by specific male displays are that ovulation and oviposition occur at the most opportune time for survival of the young.

Although male-like "sexual" behavior is not essential for reproduction in *C. uniparens,* it will facilitate reproduction, as in sexual vertebrates. For example, laboratory observations indicate that mounting serves to synchronize ovulation and oviposition. In our colony it is highly unlikely that any 2 isolated females will lay eggs on the same day. However, if females housed in pairs or groups are mounted on the same day, they go on to lay eggs simultaneously 2 weeks later (Crews, unpublished data).

"Sexual" or pseudocopulatory behavior also influences the rate of egg laying as well as the total number of clutches laid in a reproductive season. Although isolated females will lay eggs that develop normally and hatch, females will lay clutches more frequently if housed in social groups. We have noted that isolated female *C. uniparens* average 32.1 days ($N = 9$) between clutches, whereas paired females have a significantly shorter interclutch interval of 26.6 days ($N = 9$) (Crews and Gustafson, unpublished data). Fecundity is also dramatically influenced by pseudocopulatory stimuli. A female can be categorized as a "poor" reproducer if she lays 1 or no clutches and a "good" reproducer if she lays 2 or more clutches. Isolated females as well as females housed with an ovariectomized cagemate are all "poor" reproducers (Gustafson and Crews, 1981) (Table 10.2). However, females housed with ovariectomized females that are receiving exogenous androgen treatment lay significantly more clutches. These data indicate clearly that this male-like or pseudocopulatory behavior is serving as a neuroendocrine primer much as male sexual behavior does in sexual species.

Hormonal Organization of Secondary Sex Structures

Gonads of parthenogenetic *Cnemidophorus* lizards develop as ovaries. The isogenic nature of these lizards thus provides a potentially useful preparation for examining the mechanisms controlling sex determination and differentiation. Of primary interest is whether these mechanisms can be altered by environmental factors such as temperature and hormones. The ability of environmental factors to alter primary gonadal sex is common in fish and amphibians and has also been described re-

Table 10.2 Clutches laid by *Cnemidophorus uniparens* females housed with intact and ovariectomized conspecifics, treated with cholesterol (CHOL) and dihydrotestosterone (DHT).

Housing/treatment	Good reproducers	Poor reproducers
Isolate	2	10
Intact, untreated	1	7
CHOL treated	3	9
DHT treated	8	3

NOTE: Isolate versus intact, untreated, $P > 0.05$; DHT versus CHOL, $P < 0.05$ (both Fisher's exact probability test).

cently in reptiles (reviewed by Bull, 1980). Any environmentally induced changes in primary and secondary sex structures in the all-female *Cnemidophorus* (for example, the formation of testes and retention and development of the mesonephric kidney) would be easy to identify and study in comparison to similar changes in sexual species.

Adult, unisexual, parthenogenetic lizards appear to retain the mesonephric kidney and a male-like appearance of secretory renal sex segments and femoral pores. Adult females of the parthenogenetic gecko, *Lepidodactylus lugubris,* possess fully developed mesonephric kidneys (R. E. Jones, personal communication); the same is true for juveniles at least of the unisexual lizard *C. velox* (K. T. Fitzgerald, personal communication). Also, unisexual female *C. lemniscatus* exhibit skin coloration patterns that are similar to those of male sexual forms (Serena, 1980). These kinds of evidence suggest that unisexual female lizards may have higher androgen levels during at least some stages of their life cycle than do females of sexual lizards.

Considering the role hormones and temperature play in morphological sex differentiation in sexual reptile species and taking into account the anomalies already noted in unisexual *Cnemidophorus,* we have been investigating what role, if any, steroid hormones or environmental conditions such as temperature play in the sexual differentiation of parthenogens. That is, is it possible to create males experimentally?

Eggs produced by captive *C. uniparens* have been dipped in solutions containing either estradiol-17 β (1 $\mu g/\mu l$), testosterone, dihydrotestosterone, or cholesterol (5$\mu g/\mu l$), or an equal amount of vehicle (steroid suspending vehicle); all eggs were treated once at 12, 27, 34, 41, 47, 52, or 61 days following oviposition. Of the 35 eggs that hatched, 14 had been treated with testosterone, 3 with dihydrotestosterone, 7 with estradiol, 7 with cholesterol, and 4 with vehicle; with the exception of dihydrotestosterone-treated and vehicle-treated eggs, all treatment ages were represented in each hormonal condition.

Both androgens stimulated Wolffian duct development without any noticeable effect on the Müllerian ducts (Fig. 10.3). Not only were the kidneys larger, but they were darker and similar in coloration to the kidneys of males of sexual cnemidophorine species. Estradiol, on the other hand, stimulated Müllerian duct development but left the Wolffian ducts unaffected. Cholesterol and steroid-suspending vehicle had no discernible effect on either tissue. Histological examination of the urogenital system revealed a similar pattern (Fig. 10.4). Estradiol stimulated marked oviductal hypertrophy. All of the gonads appeared to be undifferentiated and in the bisexual stage of development (Forbes, 1961). There was insufficient material in each hormone treatment group for statistical comparison of measures of the mesonephric kidney or the gonad. Thus, it is unclear at this time what developmental effects can be

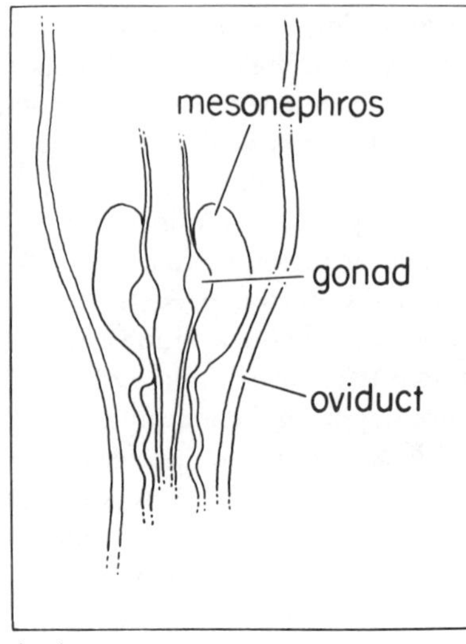

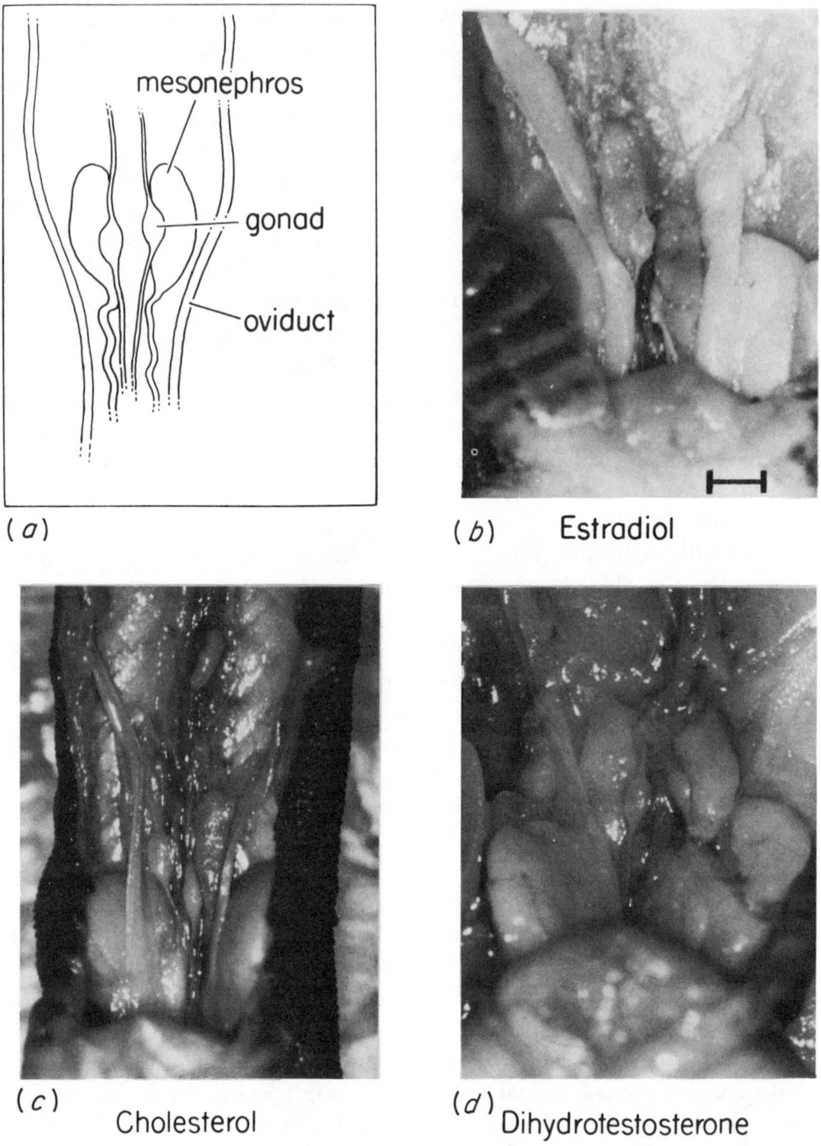

(*a*)

(*b*) Estradiol

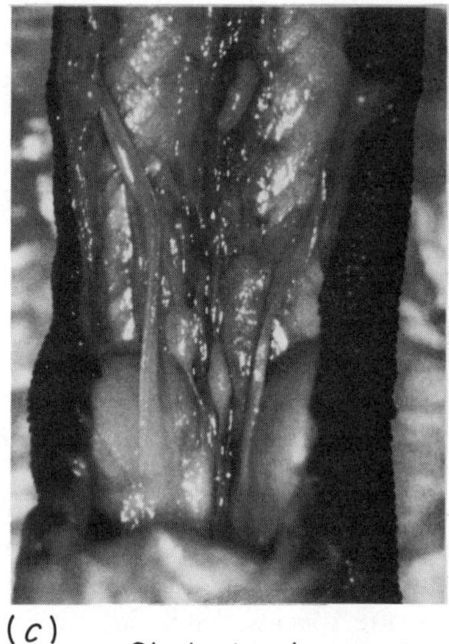

(*c*) Cholesterol

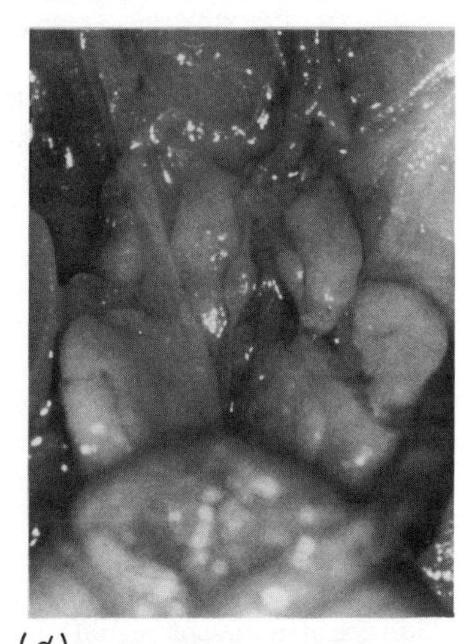

(*d*) Dihydrotestosterone

Figure 10.3 Effect of embryonic steroid treatment on the urogenital system of recently hatched *Cnemidophorus uniparens.* The line drawing (*a*) indicates the relative location of the mesonephros, gonad, and oviduct. Each photomicrograph illustrates the effect of a different steroid treatment. Oviducts are conspicuously enlarged in the estradiol-treated embryo (*b*), whereas no such development is evident in either the dihydrotestosterone-treated (*d*) or cholesterol-treated (*c*) embryos. Note the hypertrophy of the mesonephros in the dihydrotestosterone-treated embryo (*d*) and the lack of a similar effect in the other treatment groups. Scale: bar on photomicrograph (*b*) equals 66 mm.

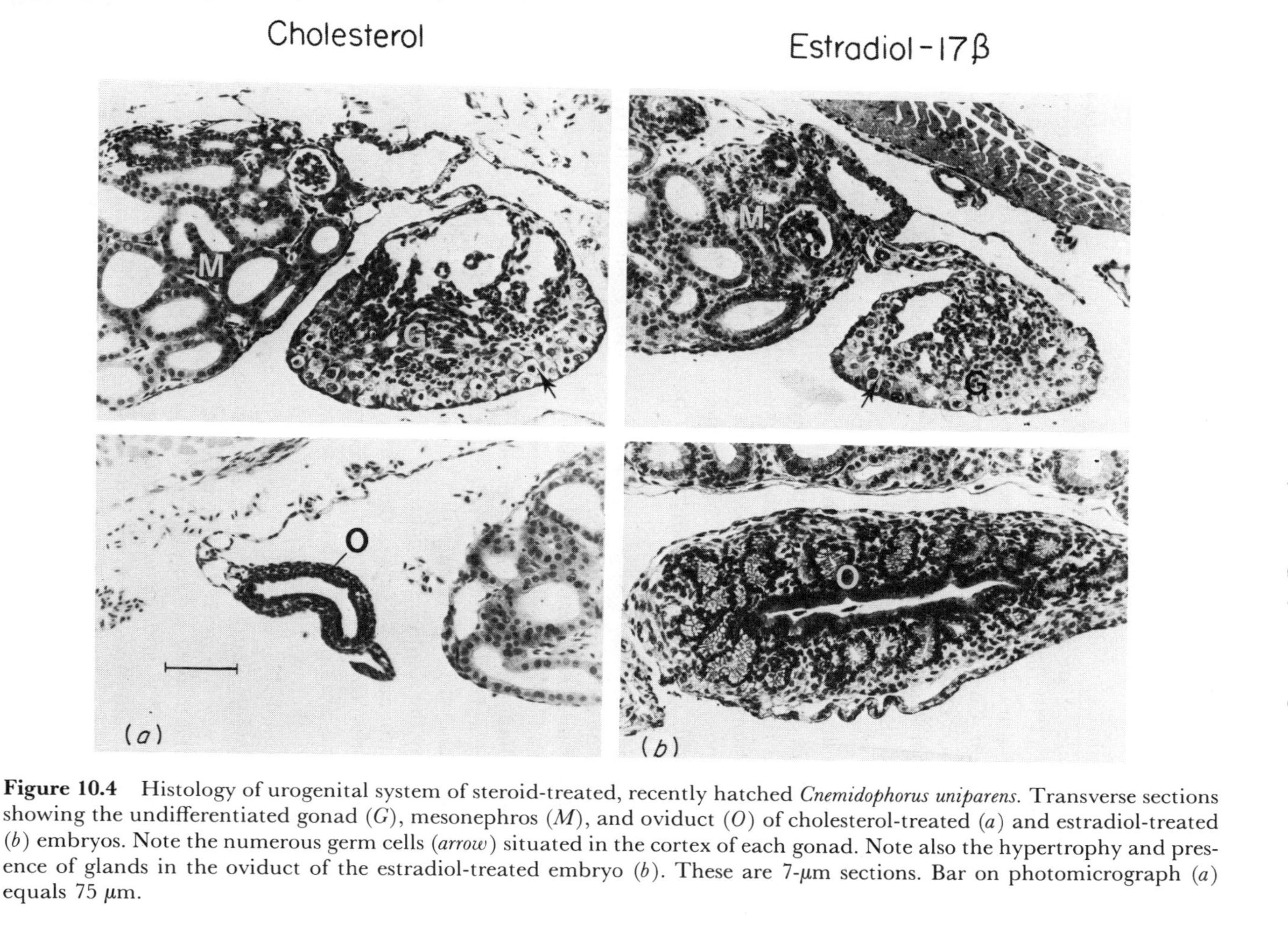

Figure 10.4 Histology of urogenital system of steroid-treated, recently hatched *Cnemidophorus uniparens.* Transverse sections showing the undifferentiated gonad (*G*), mesonephros (*M*), and oviduct (*O*) of cholesterol-treated (*a*) and estradiol-treated (*b*) embryos. Note the numerous germ cells (*arrow*) situated in the cortex of each gonad. Note also the hypertrophy and presence of glands in the oviduct of the estradiol-treated embryo (*b*). These are 7-μm sections. Bar on photomicrograph (*a*) equals 75 μm.

induced by longer hormone treatment or if the effects of hormone treatment are revealed later in ontogeny.

Is Pseudocopulatory Behavior in an All-Female Species "Normal"?

Is male-like sexual behavior normally a component of the behavioral repertoire of an all-female species or merely an artifact of captivity? Cuellar (1981) has argued that such female-female matings result from the unnatural constraints imposed by captivity. In support of this interpretation, Cuellar states that he has "observed such behavior in *C. uniparens* and other [unisexual] species in the laboratory for 15 years, but only sporadically." But it is significant that Cuellar, as well as other workers, has observed male-like sexual behaviors in parthenogenetic *Cnemidophorus.* That these observations have gone unreported in previous studies should not be too surprising. Since the function of these courtship and copulatory behaviors is not obvious, these workers most likely felt that this behavior was an abnormal manifestation of captivity. Preconceptions, however, guide perception, and one does not very often see what one is not looking for. Although perceptions, being subjective, are not readily changed by argument and riposte, we will address the issues raised by Cuellar.

Cuellar's argument is open to question on other grounds as well. First, many wild animals stop reproducing in captivity. That unisexual *Cnemidophorus* in our laboratory, as well as in others (for example, Cole and Townsend, 1977; Cuellar, 1981), reproduce regularly is compelling evidence that our housing conditions duplicate the essential features of nature. Second, observation of a behavior in captivity, but not in nature, is not in and of itself evidence that the behavior is abnormal or brought about by the conditions of captivity. In many cases, greater control over the environment and ease of behavioral observation allows discovery of the species' complete behavioral repertoire for the first time; there are numerous examples of secretive or wary species exhibiting behaviors in captivity that were later verified in field studies.

That male-like sexual behavior has not been observed in unisexual *Cnemidophorus* lizards in nature does not mean that it does not occur. Anyone who has worked with cnemidophorine lizards in the field knows how difficult they are to observe. *Cnemidophorus uniparens* are extremely active foragers and spend much of their time above ground in thick mesquite and creosote bushes. They are very wary of humans and, if approached too closely, will retreat quickly into extensive burrow systems. Furthermore, the literature indicates that matings even in sexual *Cnemidophorus* are observed in nature only infrequently.

As described above, the courtship and copulatory sequence in sexual *Cnemidophorus* is identical in form to the behavior we have observed in parthenogenetic *Cnemidophorus.* In both, the mounting lizard will swing its tail beneath the female's tail to appose the cloacal regions. Once this is achieved, the mounting lizard will shift its jawgrip from the neck or foreleg to the iliac region of the mounted female, forming the doughnut posture characteristic of all cnemidophorine lizards. Following mating, a temporary V-shaped abrasion commonly forms where the jaws grip the skin but soon disappears. Male *Anolis* also maintain a firm jawgrip on the female's skin throughout mating, leaving a temporary bite mark.

Cuellar (1981) cites the rarity of these "copulation bites among nearly 2,000 individuals collected in the wild" as evidence for the idea that male-like sexual behavior in captive parthenogenetic *Cnemidophorus* is an aberration. However, this conclusion is based on the presence or absence of a temporary mark at the time of capture. Cuellar's observations themselves raise doubt about the validity of his criticism. Furthermore, he provides no information about the frequency of such marks in sexual versus unisexual *Cnemidophorus.* Examination of 1,100 adult female *C. tigris,* a sexual species, collected during the breeding season and deposited in the Museum of Vertebrate Zoology, University of California, Berkeley, revealed that only 3 percent had marks on the back and side; further, the same frequency of males (N = 1,100) possessed such marks (Crews, unpublished data). On the basis of our behavioral observations of both sexual and unisexual cnemidophorine lizards, we would suggest that it is equally likely that these marks reflect intraspecific aggression, predation attempts, or accidents.

It is important to keep in mind also that Cuellar's observations were incidental to the main purpose of his studies. Unfortunately, he makes no mention of observation schedules, the time and length of the observations, or the number of times the behavior was observed. Without such information, no accurate statement can be made about the frequency of male-like sexual behavior in unisexual lizards.

The often reported observation of same-sex mating in sexually reproducing species has also been cited as evidence that this behavior in parthenogenetic lizards is biologically unimportant. In reviewing this literature, Beach (1968) concluded that in some species such behavior serves a social function (for example, dominance) while in other species it has physiological consequences. Thus, observation of females mounting other females or males allowing themselves to be mounted does not say anything meaningful about the "normalcy" of this behavior (Beach, 1979). As pointed out in the following section, these observations are of fundamental importance to our understanding of the organization and activation of the physiological substrates underlying sexually dimorphic behaviors.

Finally, Cuellar (1981) has stated that Crews and Fitzgerald (1980) "proposed that such pseudocopulations may be necessary for successful reproduction," and others have echoed this statement. But this is a misinterpretation of that paper. The purpose of that initial report was to document the alternation of male-like and female-like sexual behaviors during specific stages of the follicular cycle in 3 unisexual *Cnemidophorous* species. It concluded that "it is likely that social interactions play an important role in the reproductive biology of parthenogenetic *Cnemidophorus*" and raised the question of whether "this behavior may be necessary for successful reproduction in the species (for instance, by priming reproductive neuroendocrine mechanisms) as has been demonstrated in sexual species." Obviously, Cuellar and others have chosen to interpret this to mean that Crews and Fitzgerald (and the present investigators) believe pseudocopulatory behavior to be essential for reproduction. We would like to emphasize that this is not our intention. Rather, we are suggesting that the presence and behavior of conspecifics may act as a neuroendocrine primer and facilitate reproduction in parthenogenetic lizards as does male courtship in sexual lizards. It has long been known that eggs laid by isolated unisexual female lizards will hatch (Maslin, 1971), a finding confirmed in our laboratory.

The question then is not whether parthenogenetic lizards lay eggs in isolation, but whether isolated females lay at a "normal" rate. But what is "normal"? Our laboratory experiments indicate that the parthenogenetic *C. uniparens* are capable of laying more clutches more frequently when housed together than when housed alone. The actual reproductive rate, the reproductive fitness, of the species will vary with fluctuations in the environment. In this context it is important to keep in mind that *C. uniparens* are found in high densities and are known to interact socially. Therefore it is possible that the effects we observe in the laboratory may very well occur in nature and that the level of reproduction in isolated females may represent the lower limit of reproductive effort.

We cannot yet say whether the *presence or absence* of male-like sexual behavior in captive populations of unisexual lizards is abnormal. However, the following observations are not subject to controversy and still require explanation. (1) Five species of parthenogenetic *Cnemidophorus* perform complex behavioral displays that are identical in form to male courtship and copulatory behavior in sexually reproducing species. (2) This male-like sexual behavior is directed exclusively toward preovulatory females. (3) This courtship and copulatory behavior shares similar physiological mechanisms and functional outcomes as male courtship and copulatory behavior in sexually reproducing vertebrates. (4) Pseudocopulatory behavior has been documented in wild populations of the parthenogenetic mourning gecko, *L. lugubris* (Falanruw, personal communication; Werner, 1980). Obviously, laboratory studies as well as field studies are necessary before these and other observations can be fully accounted for.

Directions for Future Research

Parthenogenetic lizard clones promise to be valuable subjects for investigations of the biological bases of reproduction. Study of the psychobiology of parthenogenesis of *C. uniparens* has only just begun and the number of questions is far greater than answers. For example, information on the social organization and behavioral ecology of both parthenogenetic and sexual *Cnemidophorus* is much needed. With the identification of the natural behavioral categories of *C. uniparens,* we can ask how the form and frequency of these behaviors differ between individuals. How does the behavior of *C. uniparens* differ from that of its sexual ancestral species, and is it more similar to the behavior of one than the other? What are the circulating levels of sex steroids during the follicular cycle and how do these fluctuations correspond to the exhibition of male-like and female-like sexual behavior? Are the brain mechanisms that mediate male and female sexual behavior differentially activated in the course of the follicular cycle? What specific behaviors are acting as neuroendocrine primers and what is their mechanism of action? Does the loss of males (sperm) but the retention of male-like behavior (courtship and copulatory behavior) mean that the gametogenic function of the male and the facilitatory function of the male are under different selective pressures?

The greatest potential contribution of this research, however, lies in its implications for our understanding of the biological basis of sex determination and sexual differentiation. It is clear that male reproductive morphology and physiology developed to support their facilitating and, thus, gene-controlling function. The behavioral traits associated with these functions depend on sensory perception of effective stimuli. We can assume that the chain of events underlying specific behaviors (that is, the need for selective perception, the consequent integration of external stimuli with internal state, and the subsequent activation of a response) also affect the structures underlying that response. That specific stimuli will selectively activate or even determine the development of specific neural structures is well known (Greenough, 1976; Hubel et al., 1977; Gorski, 1979). This fundamental question of the organizational influences of the internal and external environments is addressed by this kind of research.

PART III
POPULATION AND COMMUNITY ECOLOGY

LIZARD POPULATION ECOLOGY began in the 1960s with two monographs, one by Blair (1960) on *Sceloporus olivaceus,* the other by Tinkle (1967) on *Uta stansburiana.* The general objectives of the two studies were similar. Blair (p. 4) wrote, "We have attempted to gather every possible bit of information pertinent to the biology of the local population." Tinkle (p. 5) wrote that his primary aim was "to quantify as many aspects of the life history . . . as possible." One can well believe them: Blair marked about 3,000 lizards, Tinkle 3,729! It seems that both investigators shared the opinion that if anything worthwhile was to be learned, it would only be through a detailed study of every knowable aspect of a population's ecology.

The specific conceptual issues of the studies, however, were not so similar. Blair (p. 5) wished to determine "how one population exists as a continuing system in time"—specifically, as it happened, how it recovers from drought. His hope was that the knowledge so gathered would be relevant to natural populations in general. Population regulation was a central theme in Blair's study, as it was in most nonlizard studies during the 1950s and 1960s. Tinkle's main conceptual interests were features of the lizard's life history; he cited Cole's (1954) paper, one of the first explicit theoretical treatments. Tinkle called for a comparative life-historical approach to population ecology, similar to the approach of comparative psychology to behavior. But in his 1967 study he did not test specific predictions. Rather, Tinkle expressed the inductionist's hope that the data would suggest predictions and thereby be of value to theoretical ecology. Subsequently, it was largely Tinkle who answered his own call, testing specific predictions in a series of papers on comparative life-history phenomena.

Of the many developments in lizard population ecology since these first works, three areas are prominent.

Population Density and Regulation. Lizards vary enormously in density: anoles and a few other species range up to $1/m^2$ (Schoener and Schoener, 1980), whereas most other species are orders of magnitude lower (Turner, 1977). Exceptional tropical productivity, speculated some time ago to be the factor accounting for the high densities of anoles (Heatwole, 1967), cannot be solely responsible. Insular tropical populations exceed those of mainlands by about a hundredfold—at the least, the differential importance of competition and predation seems to be involved (Andrews, 1979). But, as Turner (1977) lamented, we still cannot explain in any convincing way why the densities have their particular values. Here it would seem that a combination of population modeling and experimental manipulation is necessary. While the former is inchoate for lizards (compare fisheries' modeling for example), the latter has already begun to prove itself and holds much additional promise. More on this topic is given below in the section on communities.

Sex Ratio. Blair's (1960) monograph reported that the sex ratio in *Sceloporus olivaceus* at birth is unity but subsequently increases with differential male mortality. He considered this phenomenon an adaptation for reproductive efficiency, presumably selected for at the population level. Turner, in his 1977 review, stated that lizard sex ratios generally approximated 1. However, recent data on *Anolis* (Schoener and Schoener, 1980) show that sex ratios can be extremely variable, even (and especially) within the same species. As in certain territorial polygynous birds such as redwing blackbirds (for example, Orians, 1979), sex ratios of *Anolis sagrei* vary with apparent habitat quality. Good habitats have relatively more females than poor ones. The significance of this variation is probably not to be interpreted in terms of group selection. Rather, differential selective pressures on individuals of the 2 sexes seem likely: males are affected primarily by numbers of females, whereas females are affected by habitat quality, for example, food availability. Certain males are forced out of good habitats by other males and accumulate in poor habitats, with concomitant differential mortality; females, although territorial, must have a ceiling on the food they can process and so defend smaller territories in good habitats, where they tend to congregate. The differential selective pressures on males and females and their variation between kinds of lizards were discussed by Stamps in Chapter 9.

Life-History Traits. By far the most active area of population ecology, and the one most represented in this book, deals with life histories or reproductive strategies. In a series of synthetic papers (Tinkle, 1969; Tinkle, Wilbur, and Tilley, 1970), Tinkle and his colleagues searched for patterns in the existing information on lizard life histories. The 1969 study showed that 14 species populations fell onto a continuum from large per-unit-time reproduction but short survival to the reverse. The

1970 study included more populations and excluded survival data. Two distinct clusters were found, one of early-maturing and multiple-brooded populations, the other the reverse. This supports the theoretically intriguing idea (for example, Christiansen and Fenchel, 1979) of discrete adaptive strategies.

The present status of our knowledge of certain species has so improved during the last 15 years that we now can compare populations occurring over a wide geographic and environmental range. This is Ballinger's approach in Chapter 11. He attempts a preliminary partitioning of the variance in clutch size for *Urosaurus ornatus.* The most important explanatory variable is a composite of proximate environmental factors, accounting for 43 percent of the variance. Design constraints account for 19 percent, and ecotypic variation is apparently responsible for much of the rest. Because only the last directly reflects fine-tuned genetically controlled adaptation, Ballinger argues that his results illustrate the incompleteness of an entirely adaptationist approach.

Let us now consider community ecology. In the early 1960s this area was flourishing for a limited set of taxa, most notably birds. Indeed, virtually the only published work on the community ecology of lizards was that of Milstead (1957, 1961) on *Cnemidophorus.* Although Gause and others had laid the groundwork in their studies of species interactions long before, and Lack and Elton had advanced community ecology significantly in the 1940s and 1950s, it was primarily Robert MacArthur's approach that stimulated the rush of bird studies during the 1960s. Many of the key questions of the species-diversity approach to community ecology (as opposed to the energy-flow approach) were asked and partially answered first by researchers on birds. However, it has become evident since 1965 that the same questions could often be investigated in a more satisfactory way using lizards, as we shall see here.

Species Diversity. An enormously popular area for research in the 1960s was the study of species diversity, particularly stimulated by MacArthur (1965) and Whittaker (1965). Diversity studies by Pianka and coworkers on lizards (reviews in Pianka 1977; Pianka, Huey, and Lawlor, 1979) have outstripped work on most if not all other kinds of organisms. First, Pianka (1967) attempted to elucidate the factors responsible for diversity gradients among North American desert lizards. He concluded that variation in annual rainfall was most important and that this factor largely controlled lizard species diversity indirectly, via its effect on vegetational heterogeneity. Not satisfied with this study of a relatively simple desert saurofauna, Pianka and coworkers explored most of the major warm desert systems of the world, gathering data on community characteristics such as mean lizard niche overlap, mean lizard niche breadth, and bird diversity, as well as on individual species that might be ecological equivalents (for example, *Phrynosoma* versus *Moloch*, Pianka, 1971). The resulting massive body of data has been related to models of limiting similarity. As one example, evidence for the importance of diffuse competition was obtained by comparing niche-overlap values from 3 desert saurofaunas (Pianka, 1974).

Resource-Partitioning Studies. In contrast to species-diversity studies, resource-partitioning studies deal with a more ecologically restricted group of animals, frequently one that is congeneric. Such studies attempt to determine how mostly similar species differ ecologically and, by searching for patterns, attempt to implicate particular causal factors. Motivating most such studies is the hypothesis that food competition is important.

Numerous resource-partitioning studies have been performed on lizards since 1965, many on *Anolis* (review in Schoener, 1977). The earlier studies of Ruibal (1961) and Collette (1961) on particular niche dimensions prepared the way for Rand's (1964) prototype study of Puerto Rican anoles. As data accumulated, Rand and Williams (1969) were able to recognize ecomorphs, which as Williams defines them (Chapter 15), are species similar in color, size, proportion, and behavior but frequently unrelated phylogenetically. The 3 largest islands of the West Indies contain as striking examples of convergence as are known for any kind of species. Entire local faunas are not always totally parallel, that is, different combinations of ecomorphs can exist in different localities, but similarities here too are rather great. Many of the Lesser Antilles, a group of much smaller islands, are inhabited by a single species of *Anolis*. While varying in color and pattern, such species are, with one striking exception, very similar in size (Schoener, 1969; Williams, 1972, Chapter 15). Two-species islands are more variable, but repeated patterns of size difference, and to a lesser extent habitat segregation, exist (Schoener, 1970: Williams, 1972). Roughgarden and colleagues (1982) have now studied this situation in detail, and Moermond has also worked in this area. In Chapter 16, Roughgarden, Heckel, and Fuentes document the size and habitat differences of most Lesser Antillean anoles and construct a coevolutionary theory that accounts for a large part of the picture—even the sole exception to the solitary-size rule. An intriguing aspect of the theory is the proposition that some of the insular communities in this system are not at evolutionary equilibrium.

Chapter 13 by Huey and Pianka deals with the rationale of certain data collected for resource-partitioning studies. In particular, these authors question whether temporal separation implies resource separation. While their own extensive lizard data show that synchronously feeding species are somewhat more likely to be similar in diet than nonsynchronously feeding species, published data on raptors and snakes do not support this contention. Huey and Pianka interpret these patterns as cautioning against the use of data on activity time in resource-partitioning studies. I should like to suggest two interpretations, one of which is an alternative to theirs.

The Huey-Pianka rationale can be argued as follows. Optimally, we should like to know what the true resources are in terms of, say, more or less independently renewing populations (as in Schoener, 1974a). In theory (Haigh and Maynard Smith, 1972) resources could be prey species, or classes of individuals within species, or even parts of individual organisms. If the prey types Huey and Pianka distinguish are the true resources, then their study implies that activity time is not likely to be a useful dimension where the true resources are not known, because in their data differences in activity time result in minimal differences in prey type. Moreover, certain theoretical arguments (Schoener, 1974b) support their thesis that activity times are often not likely to be involved in resource partitioning.

An alternative interpretation of the Huey-Pianka data exists, one that does support the use of activity time. Suppose, as must be always at least partially true, we do not know what the true resources are. For example, in the Huey-Pianka study suppose age classes of mammals or species of birds were closer to the true resources for raptors than the categories Huey and Pianka use. Then we attempt an indirect description, indexing resource units by dimensions such as habitat, time, and type (prey size or prey taxon, broadly defined). For this multidimensional description a particular dimension is useful if it is not dependent on another dimension; if two dimensions are strongly correlated, then there is redundancy and one dimension should be discarded. According to this rationale, the temporal dimension would be useful in addition to the prey-type dimension if overlap in the former were *not* strongly related to overlap in the second. This is indeed exactly what Huey and Pianka found.

A major characteristic of resource-partitioning studies, as I have said, is the documentation of niche differences between sympatric species. Such differences in themselves are not proof that some selective pressure causing divergence between species, such as interspecific competition, was operating (Schoener, 1974a). Recently, various resource-partitioning studies have been attacked for failing to demonstrate statistically that ecological and/or morphological differences between species are regular rather than random (for example, Strong, Szyska, and Simberloff, 1979). Critics employ a variety of null models (Connor and Simberloff, 1979; Strong, Szyska, and Simberloff, 1979; Simberloff and Boecklen, 1981), frequently showing that purported patterns cannot in fact be distinguished from the output of a null model. The whole issue was debated in detail at a conference held in 1981 (Strong, Simberloff, and Abele, 1982). In Chapter 14 Case points out some inadequacies of previous null models and uses new ones to evaluate, among others, patterns in *Cnemidophorus*. These lizards allow rejection of certain null models but not all; size displacement can be accepted for the only region examined and size assortment for 1 region out of 3.

Experimental Studies of Competition. In 1965 very few experimental studies directed toward competition existed for any kind of animal, and none existed for lizards. Even now, only 4 studies on competition in lizards are published (reviewed by Dunham, Chapter 12). Given the definitive nature of some of the studies, their relative rarity is perhaps due to the enormous amount of hard work they involve.

Dunham's (1980) own study on *Sceloporus merriami* and *Urosaurus ornatus* is of considerable importance not only for lizard ecology but for ecology as a whole. He was able to detect competition during only 2 years of a 4-year study, suggesting that in some systems competition may be intermittent rather than regular, occurring only during times of resource shortage (see also Wiens, 1977). A second result is that overlap in diet and microhabitat is greatest during the time of least competition (as determined experimentally) and vice versa. The result can be related to the data of Smith and associates (1978) on Galápagos finches and to their review of other systems: species overlap most during times of resource abundance and least during times of resource scarcity. Lack (1946) and Svärdson (1949) have argued that this should be so, but their rationale is not obvious. My own view is that periods of overall food abundance frequently result from increases in the abundance of especially desirable food types—if food were nonrandomly distributed, simple clumping of a given food type could reduce search time and thereby make profitable inclusion of this type in an optimal-feeding strategy (for example, Royama, 1970). During times of food scarcity each species shifts to that portion of the resource spectrum to which its morphology and behavior are evolutionarily most adapted. This seems to imply, and here my view differs from Wiens (1977), that strong directional selection reducing interspecific competition takes place during such times.

Dunham's result on overlap may at first glance seem disconcerting to those who advocate limiting-similarity models, but it need not be. I interpret the result to say that we cannot assume resource-overlap values determined for a single or a few seasons are those that should be plugged into a limiting-similarity model to test its validity. Morphological measures such as ratios of trophic appendages might indeed be superior for indexing intensity of the important competitive pressures; ironically so, as such measures traditionally have been considered inferior to data on resource use itself. Moreover, for a large number of species, relative overlap measures taken during a single season may still indicate relative intensity of competition.

Competition experiments with lizards have also been explicitly tied to observational data on resource partitioning. Pacala and Roughgarden (1982) have recently shown experimentally that certain dissimilarly sized *Anolis* in the Lesser Antilles compete less than do similarly sized species. This result was clearly evident only a short time after experimental initiation, suggesting either that the investigators luckily selected the right year or that tropical insular lizards compete more chronically than do temperate continental ones—indeed densities of the latter are several orders of magnitude lower and their climatic environment appears more variable.

We have seen in the Introduction that community-ecological studies have begun to level off, in contrast to the exponential growth of population-ecological studies. Perhaps the former's differential shift toward experimental work provides an explanation, as field manipulation is extremely time intensive. Ultimately, however, the new combination of experiment and observation should provide a level of understanding of lizard communities not available and indeed unattainable for many other ecological systems.

Thomas W. Schoener

11 | Life-History Variations

Royce E. Ballinger

INTEREST IN LIFE HISTORIES has increased rapidly since Lack's extensive studies (1947, 1950, 1968), and since age at maturity was implicated as a key component of fitness (Cole, 1954). The essence of these issues was raised much earlier by R. A. Fisher, whose insightful curiosity caused him to wonder "what circumstances in the life history and environment would render profitable the diversion of a greater or lesser share of the available resources towards reproduction" (1930: 47). Williams' elegant essay (1966) on adaptation and natural selection served to crystallize discussions of the evolutionary ecology of life-history traits around an adaptationist paradigm.

An extensive literature on the theory of life-history evolution has developed in the past decade (see Stearns, 1976, 1977, for reviews). This theory has attempted to explain patterns of variation observed in nature and to predict what types of life histories are to be expected in different environments and why. Researchers have employed various organisms (see Stearns, 1977), including lizards, as model systems. Lizards are well suited to life-history studies because of the interesting variations they exhibit and because of the ease with which their life-history characteristics can be quantified. In particular, it is reasonably easy to measure age-specific survivorship probabilities and reproductive characteristics for many lizards in their natural environment.

The purpose of this chapter is to evaluate the nature of life-history variations in lizards by (1) examining the sources of biological variations which lead to causal hypotheses relevant to life-history traits, (2) assessing these causal hypotheses in view of the literature on lizard life histories, (3) combining them in a single model to assess their relative importance, and (4) providing an initial test of the model.

The Question

Lizards vary considerably in their life-history features (Tinkle, 1967; Fitch, 1970; Tinkle, Wilbur, and Tilley, 1970; Ballinger, 1978). In general, lizards have a reasonably short prereproductive period, often characterized by high mortality, followed by a reproductive period of variable length with little or no postreproductive period. Lizards exhibit no major postembryonic morphological changes and grow continuously but at a decreasing rate throughout life. Within groups, particularly within species, fecundity increases with body size. Otherwise, few generalizations apply to lizards as a group. Lizards may be oviparous or viviparous. Fecundity varies from 1 to over 70 eggs per clutch, and both single- and multiple-brooded species are common. Continuous reproduction is common in tropical species, whereas well-defined seasonal reproductive patterns are the rule in temperate species. Age at maturity is as early as 2 months in *Anolis poecilopus* and *A. lionotus* (Andrews, 1976) or as late as 7 years in large species such as *Cyclura carinata* (Iverson, 1979). Lizards may have short (less than 1 year) or relatively long life expectancies.

Attempts to categorize species within general theoretical constructs such as r-K selection (MacArthur and Wilson, 1967; Pianka, 1970a) or into generalized patterns such as long- versus short-lived groups (Tinkle, 1969; Tinkle, Wilbur, and Tilley, 1970) have served to increase the size and delineate the components of the data base necessary for understanding life-history variation in lizards. However, such theoretical frameworks are generally incomplete, and this led Wilbur, Tinkle, and Collins (1974) to recommend detailed studies and analyses of specific populations "to identify causal mechanisms in the evolution of their life histories." But therein lies one of the assumptions that restricts our understanding of life-history variations. We must know to what extent observed life-history variations are a result of evolutionary processes and to what extent these variations reflect nonevolutionary sources. Because we do not ask this fundamental question, it is no wonder that both our theory and data are inadequate (Stearns, 1976, 1977).

Pure adaptationists (in the sense of Gould and Lewontin, 1979) argue that all phenotypic characteristics, including the limits on the scope (plasticity) of responses to proximate environmental factors by physiological and behavioral mechanisms, have an evolved, genetic basis. However, phenotypic plasticity and phenotypes per se have both an adaptive component determined by genetic mechanisms and a proximate component determined by mechanisms of direct chemical and structural interactions (Stearns, 1980, 1982). Thus, it is insufficient simply to argue that because phenotypic plasticity can evolve, all observed life-history responses have a direct evolutionary causal basis. Some responses may in fact be environmentally induced epigenetic responses. It is more instructive initially to define causal sources of variation and to ask to which of these possible sources is a specific phenotype more sensitive. This fundamental question must be answered before the question of the evolution of phenotypic plasticity can be explored precisely.

Sources of Variation

Biological characteristics, including life-history traits, result from response to genetic and nongenetic factors ("environmental variance"). In the language of genetics, phenotypic variation (V_P) is the sum of the variation due to genetic and environmental factors:

$$V_P = V_G + V_E. \tag{11.1}$$

Geneticists have further divided V_G into V_A (additive), V_D (dominance), and V_I (interaction or epistatic) components (Falconer, 1960).

Alternatively, ecologists and particularly evolutionary biologists have considered biological variation to result from ultimate and proximate factors (Mayr, 1961; Lack, 1965). The former represents a cause-effect pathway mediated through the genetic system (V_G), and the latter results in a direct response to immediate environmental stimuli (V_E). Genetic sources of variation belong to adaptive ecology, and environmental sources of variation belong to functional ecology (Orians, 1962).

Evolutionary ecologists have not been as inclined to subdivide the V_G sources as have geneticists, whose additive, dominance, and epistatic components are of limited use in the context of evolutionary questions. These components (V_A, V_D, V_I) describe the genetic properties of a population at a given time and permit precise formulations under Hardy-Weinberg equilibrium conditions. Such formulations are useful in predicting the effects of selection, migration, and drift on gene frequencies. However, population genetics does not explain the origin of genetic differences between populations or whether a specific genotype exists because of historical or design constraints or recent natural selection. Such sources of genetic variance are of particular interest in explanations of differences between populations, species, and higher categories.

I suggest that genetic variance be divided into evolutionary components of V_Y and V_C such that

$$V_P = V_Y + V_C + V_E + V_R. \tag{11.2}$$

This model is intended to be descriptive of the sources of variation in life-history traits rather than predictive (in the sense of quantitative genetics). Component V_E represents the proximal environmental source of variation as before, and V_R is the residual variance left unexplained by the other variance components (omitted before as is often customary). Component V_Y is the phylogenetic variance which represents genetic variation resulting from phylogenetic and design constraints—constraints of phenotypic variation which result from limited genetic information contained in a particular phylogenetic lineage or the constraints of design which result from what is developmentally possible (Stearns, 1982). In terms of life-history traits, V_Y would represent the "phylogenetic design or developmental constraints" (Stearns, 1977, 1980; Gould and Lewontin, 1979), such as placental versus marsupial reproductive modes in eutherian and metatherian mammals or monoallochronic ovulation in *Anolis* (Smith et al., 1973). Component V_Y would tend to be manifested by characteristics that are extremely conservative through evolutionary time. Component V_C is the "ecotypic variance" which represents the genetic variation resulting from adaptations to specific environmental conditions. Characteristics exhibiting high V_C should include those which are easily and rapidly changed by natural selection such as color and anatomical proportions. Life-history traits that fall into this category include clutch size and aggression.

I do not suggest that V_Y and V_C arise from different evolutionary mechanisms or that one is more genetic or more adaptive than the other. Rather the separation of genetic variance into evolutionary components is intended to facilitate the interpretation of phenotypic variation. This approach allows evolutionary and functional genetic questions to be addressed with a perspective comparable to that which Orians (1962) delineated for evolutionary and functional ecology. Whether or not the evolutionary components of V_G can or need be quantified in the same manner as the classical components remains to be seen. They at least provide a conceptual perspective which permits a broad-based assessment of total phenotypic variation and leads directly to causal hypotheses for this variation.

Causal Hypotheses of Life-History Variations

The major sources of variation in life-history traits in lizards include (1) those proximal variations of the environment that alter allocations in the time and energy budget and (2) those genetic factors that, through evolution, have resulted in different adaptive patterns to maximize fitness or have affected morphological or developmental characteristics that canalize or constrain life-history patterns. Each of these major sources of variation represents a fundamental hypothesis as to the basis for observed variation in life-history traits. Examples from lizard life histories that support each hypothesis are outlined below.

Variation Due to Design Constraints. HYPOTHESIS: Differences in life-history traits among species result from morphological or developmental constraints that establish limits beyond which an organism cannot operate.

These constraints limit both adaptive and functional aspects of the ecology of a species. Certain adaptive characteristics predicted to be evolutionary optima will not be found regardless of natural selection because those characteristics do not exist in, nor can they be produced by, the genetic variation of the species. Similarly, certain functional characteristics, for example those dependent upon responses to proximal factors such as increased reproductive potential during periods of high resource availability, may never be formed because the individual organism is not equipped to process superabundant food. Design constraints result from two very different phenomena. These are constraints associated with phyletic relationships (Stearns, 1977) and general nonadaptive "architectural" designs (Gould and Lewontin, 1979). Although one could divide V_Y into two components (V_{Y_p} and V_{Y_a}), I have chosen not to do so here.

Phyletic constraints limit the characteristics a lizard may possess. Variation in life-history characteristics of this kind is considered part of V_Y of the phenotypic variance. Examples of these constraints in lizards include those that arise from polyploidy and from characteristics that are unique to specific taxonomic groups.

Approximately 25 species of lizards are polyploids. All 25 are triploid and are known or presumed to be all-female, parthenogenetic species (Darevsky, 1958; Maslin, 1971; Cole, 1975; Cuellar, 1977; Bogart, 1980). Most of these species apparently arose through hybridization between divergent species (Cole, 1979). Regardless of their origin, parthenoform descendants will be constrained in their life-history characteristics.

Except for mutations that result from errors in gene copying or occur during developmental processes, descendants of parthenoform lizards are genetically identical to their parents. Cole (1979) has observed chromosome aberrations as well as life-history characteristics such as continuous rather than seasonal egg production (personal communication) which are retained in individual lines of parthenogenetic *Cnemidophorus* for at least three generations. Clearly, the polyploidy-parthenogenesis phenomenon can act as a constraint on life-history variations in lizards.

Numerous examples of variation in life-history characteristics of lizards represent design constraints due to developmental, morphological, or physiological limits associated with phylogenetic position. About 20 percent, over 700 species, of the living lizards in the world belong to the family Gekkonidae. As far as is known, all species produce an invariant number of eggs per clutch, either 1 or 2 (Fitch, 1970). Tinkle, Wilbur, and Tilley (1970) excluded geckos from their evaluation of lizard reproductive strategies because of this nearly constant clutch size. Geckos are pantropical in distribution and evolved in tropical environments. Although reproduction is generally cyclic, species occurring in nonseasonal tropics reproduce continuously (Inger and Greenberg, 1966; Kluge, 1967) so that the small clutch size does not greatly decrease reproductive potential. The small clutch size may have evolved as one of many adaptations to the tropical environment (1) to facilitate decreased reproductive risk through greater mobility of females in their frequently arboreal habitat, thereby decreasing the risk of mortality, (2) to permit production of larger eggs which would result in larger young better suited to the relatively competitive conditions in tropical systems, and (3) to reduce predation on nests because small clutches would be less conspicuous than large one. Regardless of the evolutionary origin of the small clutch size in the Gekkonidae, it is reasonable to assume that the invariant number of eggs indicates a phylogenetic constraint on clutch size. If this were not so, geckos that have ventured into other environments (for example, seasonal tropic or temperate zones) would have evolved clutch sizes typical of lizards in these areas. The inability of geckos to produce larger clutches represents a phylogenetic constraint which helps explain why geckos are relatively less abundant in temperate zones. Geckos may not be able to compete favorably with lizards which do produce large clutches in seasonal environments.

A similar constraint regarding clutch size is found in species of the genus *Anolis* which ovulate a single egg alternately from one ovary then the other. Smith and colleagues (1973) termed this ovulatory pattern "monoallochronic." Their data suggest that all species of *Anolis* exhibit monoallochrony, although reproductive cycles are known for less than 20 percent of this group of approximately 251 species. Furthermore, this unique pattern of ovulation is found only in 4 other genera in the world (*Chamaeleolis, Chamaelinorops, Phenacosaurus* and *Tropidodactylus*), all of which are closely related to *Anolis.* These lizards occur in tropics of the New World. This reproductive pattern is likely to be well suited to their arboreal habits in nonseasonal tropical environments. Although some species exhibit cyclic reproductive patterns (Sexton et al., 1971; Gorman and Licht, 1974) and occur in seasonal environments, the monoallochronic reproductive pattern probably developed early in the evolution of this group in environments favoring continuous reproduction. Thus, there would be no disadvantage to having a small clutch size in the nonseasonal tropics. I suggest that the phenomenon of monoallochronic ovulation was adaptive to the arboreal life-style of the ancestor of these tropical anoline lizards and now represents a phylogenetic constraint on the ovulatory pattern. The inability of lizards of this group to produce larger clutches may now prevent them from competing favorably with lizards which do produce large clutches in seasonal environments.

Another component of phylogenetic constraint is parity condition. Bellairs (1970) pointed out that there is an almost continuous gradation between typical oviparous species and true viviparous species and suggested that ovoviviparity not be used for lizards. Numerous authors

(Weekes, 1935; Sergeev, 1940; Neill, 1964; Packard, 1966; Fitch, 1970; Packard, Tracy, and Roth, 1977; Tinkle and Gibbons, 1977; Guillette et al., 1980) have considered the evolution of viviparity in reptiles, and there are conflicting views regarding certain details. Apparently viviparity has evolved several times in different phylogenetic groups presumably from oviparous species through various intermediate stages of egg retention and placental development. Nevertheless, subsequent speciation within viviparous or oviparous groups has probably resulted in a retention of the ancestral parity condition. Identifiable taxonomic groups are uniformly of one parity condition (for example, Xantusiidae, viviparous; Iguaninae, oviparous; *torquatus* species group of sceloporine iguanids, viviparous), although Tinkle and Gibbons (1977) pointed out many cases of both parity types within certain taxonomic groups. Guillette and associates (1980) recently considered the evolution of viviparity in *Sceloporus.* Of 15 distinct species groups (based on characters other than parity type), 11 groups contain only oviparous species whereas 3 groups contain only viviparous species. Only 1 group (*scalaris* species group) has mixed parity, including 1 viviparous species (*Sceloporus goldmani*), 1 oviparous species (*S. scalaris*) that retains its eggs almost until hatching (Newlin, 1976), a condition considered ancestral to true viviparity, and 1 species (*S. aeneus*) that exhibits both parity types (Guillette, 1981).

Viviparity requires or permits considerably different life-history patterns than oviparity. Thus parity condition does constrain life-history variations regardless of whether the constraint is phylogenetically based or not. Table 11.1 depicts the clutch size for several species of *Sceloporus* in 2 species groups of moderate to large-size species. The *spinosus* group contains only oviparous species, whereas the *torquatus* group contains only viviparous species. Viviparous species generally produce fewer young per reproductive episode and furthermore produce only 1 litter (clutch) per year. Oviparous species produce larger clutches for a given body size and all are probably multiple brooded. Viviparous species produce about 1 young per 10 mm of snout-vent length each year, whereas oviparous species produce 3 to 4 times this number. There are many other life-history differences between viviparous and oviparous reproductive modes such as mating season (fall versus spring), appearance of young (spring versus summer), and prehatchling mortality; Blair (1960) reported 69 to 80 percent in *S. olivaceus,* and Ballinger (1973) indicated 4.5 percent in *S. jarrovi.* Although parity undoubtedly has an adaptive basis, there can be little doubt that parity condition constrains life-history variations in lizards.

Another type of design constraint that can limit life-history variation results from general nonadaptive architectural characteristics. Gould and Lewontin (1979) argue that the adaptationist paradigm is often ex-

Table 11.1 A comparison of clutch size, body size, and clutch frequency in 2 species groups of *Sceloporus* having different parity types.

Group species	Adult body size (mm SVL)	Clutch size, mean (range) sample size	Clutches per year[a]	References
Spinosus (oviparous)				
S. clarki	88	14 (4–26)	M	Fitch, 1978; Ballinger, unpublished data
S. horridus	90	12 (5–15)	M	Davis and Dixon, 1961
S. magister	89	8 (2–19)	M	Fitch, 1970; Parker and Pianka, 1973; Tinkle, 1976
S. melanorhinus	91	12 (5–20)	?	Davis and Dixon, 1961; Fitch, 1970
S. olivaceus	93	18 (8–30)	M	Blair, 1960; Fitch, 1978
S. orcutti	92	11 (6–15)	S	Mayhew, 1963; Fitch, 1978
Average	90.5	12.5		
Torquatus (viviparous)				
S. torquatus	103	10 (6–12)	S	Fitch, 1978
S. cyanogenys	93	10.8 (4–18)	S	Hunsaker, 1959; Ballinger, unpublished data
S. jarrovi	74	7.6 (2–16)	S	Ballinger, 1973, 1979
S. macronatus	91	9 (5–14)	S	Werler, 1949; Fitch, 1978
S. poinsetti	104	10.4 (6–26)	S	Ballinger, 1973
Average	93.0	9.6		

a. M = multiple brooded; S = single brooded.

tended to provide unnecessary explanations of observations that need no adaptive explanation. In their charming nonbiological example of the spandrels of San Marco they point out that these elaborately ornamented spaces between the arches, which often seem the focus of artistic design, "are necessary architectural by-products of mounting a dome on rounded arches." Furthermore, they say, "evolutionary biologists, in their tendency to focus exclusively on immediate adaptation to local conditions" often ignore similar architectural constraints in biological systems. Their argument is elaborated for emphasis, but their point is well taken.

The relationship between clutch mass and body shape (Vitt and Congdon, 1978) exemplifies the influence of architectural design constraints on life-history variations in lizards. General body shape in lizards varies for many reasons, including phylogenetic heritage as well as habitat occupied or escape and foraging techniques employed. Arboreal species tend to be thin with long and fragile appendages, whereas grassland forms are typically elongate with short or no appendages. Species that utilize wide-foraging techniques to obtain food or speed to escape predators tend to have elongate, torpedo-shaped bodies (for example, *Cnemidophorus*) or extraordinarily well-developed rear legs (for example, *Crotaphytus* and *Basiliscus*). Cryptic species or sit-and-wait foragers are generally more robust. Studies on the ecotypic basis of adaptive morphology in lizards (Hellmich, 1951; Lundelius, 1957; Hurtubia and di Castri, 1973; Sage, 1973; Jaksić and Núñez, 1979; Jaksić, Núñez, and Ojeda, 1980) suggest that these differences in basic body plan are adaptive.

Body shapes also affect reproductive potential in nonadaptive ways (see Gould and Lewontin, 1979) because different architectural shapes make available different volumes for reproductive materials, in spite of the fact that the shapes may have adaptive bases. Andrews and Rand (1974) suggested that the arboreal habits of *Anolis* placed constraints both on lizard size and weight of reproductive materials compared to the more terrestrial *Sceloporus*. Pianka and Parker (1975) argued that the spinily armored tank-like body form and cryptic behavior of horned lizards (genus *Phrynosoma*) facilitated the evolution of large clutches. Vitt and Congdon (1978) extended these ideas and demonstrated that body shape was related to escape and foraging strategies and placed constraints on relative clutch mass. They developed a model to predict reproductive volume with increasing body size given a knowledge of escape and foraging strategies.

The patterns of body shape are striking. The arboreally adapted body plan of anolines constrains increases in clutch mass considerably as compared to the torpedo body shape of sceloporines. Shine (1980) showed that escape potential (running speed) decreased and vulnerability to predation increased so much in gravid female skinks that a body plan designed for predator escape was selectively favored over one for large clutch size. Vitt (1981) demonstrated that clutch size in *Platynotus semitaeniatus* was significantly reduced and eggs were elongated because of morphological adaptations to its crevice-dwelling habits. Thus, some variation in clutch size (volume) is nonadaptively related to general body architecture which imposes a design constraint on life-history variations. Because body plans generally correspond to broad adaptive designs in morphology within larger taxonomic groups, variations associated with architectural design constraints can be considered a part of phylogenetic variance (V_Y) of the total phenotypic variance.

Variation Due to Proximal Environmental Factors. HYPOTHESIS: Differences in life-history traits among populations result either from physiological and behavioral responses of individuals to immediate environmental factors or from simple numeric responses in populations.

Such variations in individuals constitute the V_E component of phenotypic variation in life-history traits including reproductive characteristics, maturation, growth rates, and behavior. Examples of proximate environmental factors affecting life-history traits potentially include much of functional ecology. Heatwole (1976) has provided a general review, and the discussion here is intended to be illustrative (rather than exhaustive) of the types of proximate variations occurring in life-history characteristics.

Among all proximate factors, food availability is perhaps the single most important because it directly affects the energy allocation process. Reproductive cycles are often timed to correspond to the period when food is most available for the young. This is particularly evident in the tropics where maximal reproduction corresponds to maximal food levels during the rainy season (Marshall and Hook, 1960; Sexton and Turner, 1971; Sexton, Bauman, and Ortleb, 1972; Barbault, 1974, 1976; Stamps and Crews, 1976) as well as the desert where increased reproduction has been correlated with increased food following high rainfall periods (Mayhew, 1967; Hoddenbach and Turner, 1968; Kay et al., 1970; Pianka, 1970b; Martin, 1977).

Unfortunately, most of these studies only infer food abundance from precipitation records. Ballinger (1977) measured relative food abundance and demonstrated reduced reproduction (smaller and fewer clutches) in *Urosaurus ornatus* in a low food year. Dunham (1978) found reduced growth rates and foraging success in *Sceloporus merriami* in low food years. In *S. jarrovi,* reduced availability of food resulted in increased territory size (Simon, 1975), altered food composition (Ballinger and Ballinger, 1979) decreased growth in yearling males, and reduced adult activity (Ballinger, 1980). Reduced food abundance also decreased the general nutritive state (lower fat reserves and smaller weight/ SVL ratios) in *Urosaurus* and *Sceloporus* (Ballinger, 1977, 1980). By supplemental feeding, Licht (1974) significantly increased body, fat-body, and liver weights in natural populations of *Anolis cristatellus.*

The importance of fat-body size to reproduction has long been recognized (Hahn and Tinkle, 1965; Smith, 1968). Derickson (1974) indicated that lipids stored in the corpora adiposa are the first reserves to be utilized for reproduction and maintenance. In his excellent review of lipid cycles, Derickson (1976a) indicated 4 patterns in lizards: (1) species that lack lipid cycles such as tropical species with continuous reproduction (Licht and Gorman, 1970; Ruibal, Philibosian, and Adkins, 1972), (2) species whose lipid cycles are associated with winter dormancy (Dessauer, 1955), (3) species whose lipids cycle with reproduction (many examples including Ballinger, 1977), and (4) species in which lipids cycle with both reproduction and brumation (Gaffney and Fitzpatrick, 1973; Derickson, 1976b; Ballinger and Congdon, 1981). Derickson (1976a,b) suggested that the ability of *Sceloporus undulatus* to store greater quantities of lipid allowed it to extend its reproductive season to 3 clutches as compared to 2 in *S. graciosus*. There is little doubt that variation in food causes proximate variation in life-history characteristics. Unfortunately most of the literature focuses on qualitative rather than on quantitative effects on life-history traits. This is even more evident for other proximate factors outlined below.

Humidity, moisture, and particularly rainfall are important to life-history functions. Variations in rainfall cause variations in food abundance that affect lipid cycles and reproduction as reviewed above. Because moisture also directly affects other ecological aspects of reproduction, variations in precipitation and humidity will likely result in some proximate variation in reproductive or life-history patterns. Andrews and Rand (1974) suggested that short-term fluctuations in climatic factors, particularly rainfall, were of prime importance in the evolution of opportunistic reproduction in *Anolis*. Many species are known to wait for rainfall before laying their eggs (Ayala and Spain, 1975; Vinegar, 1975; Newlin, 1976; Stamps, 1976; Ballinger, 1977), or otherwise center their reproductive cycle during the wettest part of the year, presumably because conditions then are most favorable for oviposition or hatchling success (Heatwole and Sexton, 1966; Bustard, 1968; Licht and Gorman, 1970; Sexton et al., 1971). Stamps (1976) demonstrated that the initiation of nest digging was directly stimulated by rain in *Anolis aeneus*. High humidity stimulated ovarian follicle development in *A. sagrei* (Brown and Sexton, 1973).

The length of the growing season can greatly affect life-history characteristics. Tinkle (1969) suggested that shorter breeding seasons at high latitudes and altitudes restricted clutches and had a positive effect on adult life expectancy, and Tinkle and Ballinger (1972) indicated that a longer growing season in Texas (as compared to Ohio) increased mortality in *Sceloporus undulatus* because of higher predation. Pianka (1970b) provided a similar explanation for geographic differences in *Cnemidophorus tigris*. The effect of length of season on number of clutches is documented for several North American species. For example, 2 clutches are produced by *C. tigris* in Texas, whereas only 1 clutch is possible in the shorter season in Colorado (McCoy and Hoddenbach, 1966). Similarly, 2 clutches occur in *Crotaphytus collaris* in Texas (Hipp, 1977) but only 1 in Kansas (Ferguson, 1976). Also, 1 or 2 clutches occur in *Uta stansburiana* in Washington (Parker and Pianka, 1975), 3 clutches in the San Gabriel Mountains of California (Goldberg, 1977), and as many as 5 clutches in southern Nevada (Turner et al., 1970). A longer growing season provides increased oppportunity for within-year variation in life-history traits. A number of researchers (for example, Martin, 1973) have noted that clutch size tends to be larger early in the reproductive season than later. Ferguson and Bohlen (1978) reported that size of eggs was greater in later clutches in *Sceloporus undulatus* in Kansas which they suggest to have an adaptive basis rather than direct proximate result. In *Anolis carolinensis,* young that hatched in July grow faster than those that hatched in August or September (Michael, 1972).

The impact of the length of the growing season may be mediated by direct proximate effects through changing photoperiod and thermal regimes. Seasonal variation in photoperiod in temperate climates serves as an environmental cue by which certain life-history events are timed. This explains some variation between species or populations occurring in different locations. However, the regular patterns of photoperiodicity explain variations at the same locality between years or individuals. Increased photoperiod affects testis development in several species (Mellish, 1936; Miller, 1948; Bartholomew, 1950, 1953; Mayhew, 1964; Licht, 1967, 1971), increases appetite (Fox and Dessauer, 1957), stimulates female reproduction (Mayhew, 1961), alters thermoregulatory patterns (Ballinger, Hawker, and Sexton, 1969), and promotes growth (Fox and Dessauer, 1957; Mayhew, 1965). Licht (1969) showed that light intensity for photoperiodic responses could be low (0.05 to 12 foot candles) and that white light was more effective than red, which was more effective than green, which was more effective than blue. Mayhew (1965) found incandescent lights more effective than heat lamps. Other investigators failed to find stimulation of reproduction by photoperiod in other species; *Uta* (Tinkle and Irwin, 1965), *Lacerta* (Licht, Hoyer, and van Dordt, 1969), *Sceloporus* (Marion, 1970). Bartholomew (1959) pointed out that the effects of photoperiod are not always clear because of the confounding effects of temperature. Both photoperiod and temperature have effects on male and female gonadal development in *Anolis carolinensis* (Licht, 1967, 1973a). Gorman and Licht (1974) suggested that reproductive cycles of Puerto Rican anoles were more responsive to photothermal events than to rainfall patterns, unlike many other tropical anoles.

The literature on temperature relations and thermoregulation in lizards is extensive (Cowles and Bogert, 1944; Brattstrom, 1965; Templeton, 1970; Dawson, 1975; Huey and Slatkin, 1976). By altering rates of physiological processes, temperature affects many if not most ecological characteristics including reproductive cycles (Tinkle and Irwin, 1965; Marion, 1970; Licht, 1971, 1973b), growth (Blair, 1960; Bustard, 1968; Michael, 1972), development (Fitch, 1964; Licht and Moberly, 1965; Sexton and Marion, 1974; Huey, 1977), and age to maturity (Ballinger, 1979). This is only a modest sample of the literature, but in spite of the extensive documentation of temperature effects, little attention has been given to the importance of temperature as an explanation of the variance observed in life-history features. Huey and Stevenson (1979) drew attention to this void and suggested it resulted in part because thermal biology of lizards has been approached primarily from a physiologist's point of view rather than from that of an ecologist (that is to say, studies have centered on physiological rather than ecological performances and goals). Huey (1974) noted that precision of thermoregulation in *Anolis cristatellus* depended upon costs and benefits associated with habitat type. Thus, it seems clear that temperature can influence variations in life-history characteristics in proximate ways although its relative importance to other factors has been little studied in an ecological context.

Behavior patterns in lizards are complex and varied (see Carpenter and Ferguson, 1977, for review). Basic functions of the behavioral repertoire of species (Ruibal, 1967; Ferguson, 1971) and of particular behavioral acts (Ferguson, 1966; 1977a; Rand and Williams, 1970; Ruby, 1977) have begun to be understood. There is significant behavioral variation within species (Jenssen, 1971; McKinney, 1971a) and among species (Hunsaker, 1962; Carpenter, 1963) as well as between seasons and ecological contexts (Evans, 1951; Carpenter, 1967; Stamps and Crews, 1976; Stamps, 1977). Crews (1975a, 1978) has shown that ovarian recrudescence in *Anolis carolinensis* is greatly stimulated by normal male behavior and is inhibited by aggressive behavior in dense aggregations. Furthermore, the simple extension of the dewlap (Crews, 1975b) is an important component in stimulating female development and mate selection. The close integration between behavioral and neuroendocrinal functions (Crews, 1979) will undoubtedly prove to be the mechanism of operation and control between internal and external proximate factors. Thus, behavior functions as a proximate factor not only in eliciting behavioral responses but also in affecting reproductive development.

Other factors such as altitude, age, size, density, and stress may operate in a proximate manner to alter life-history phenotypes. Effects of these are relatively less common or significant than effects of those factors discussed above except for responses resulting from aspects of population dynamics such as density, age structure, sex ratios, and inter- and intraspecific interactions. Turner (1977) recently provided an excellent review of the dynamics of lizard populations, so no attempt will be made to detail effects which these parameters have on life-history traits. Suffice it to recognize that characteristics of group behavior, dynamics, and interactions can cause proximate variance (V_E) in life-history phenotypes in much the same way as can the specific, largely external, environmental factors outlined above.

Variation Due to Ecotypic Adaptations. HYPOTHESIS: Variation in life-history traits of individuals between populations results from adaptations to specific ecological conditions at a specific time and place.

Life-history characteristics, in other words, are a direct expression of genetic factors which were preserved by natural selection because of their contribution to fitness. These are explained by the V_C component of phenotypic variance. The adaptive basis of life-history traits poses one of the most interesting questions in ecology and has received considerable attention in lizards (Tinkle, Wilbur, and Tilley, 1970). Unfortunately, the genetic basis of life-history characteristics has been little studied in lizards and remains one of the major assumptions to be evaluated in the consideration of the evolution of life histories in general (Stearns, 1977).

Comparative studies within species (Jenssen, 1971; Tinkle and Ballinger, 1972), between closely related species (Ballinger, 1973), or between species within a limited geographic area such as southeastern Arizona (Ballinger, 1973, 1979; Simon, 1975; Vinegar, 1975; Newlin, 1976; Smith, 1977; Congdon, Ballinger, and Nagy, 1979; Ballinger and Congdon, 1981) have revealed differences in life histories that appear, at least in part, to be genetic and thus adaptive. However, studies are needed that consider the inheritance patterns of specific traits or the responses of different populations (or species) in identically controlled environments. Some such work with lizards has been done, but it has been mostly anecdotal.

Ferguson (1977b) has reviewed the studies that suggest a genetic basis for behavioral characteristics. These traits include both behaviors observed in newborn young, of which there are many examples, and intermediate behavior patterns observed in hybrid individuals (Gorman, 1969; Ferguson, 1971; McKinney, 1971b). Some physiological studies suggest that certain behavioral (or related) characteristics have a significant genetic component. For example, lizards with seasonal reproductive patterns typically breed at particular times of the year. These seasons can sometimes be extended under laboratory conditions but some period usually remains during which reproductive activity cannot be induced. This is known as the refractory period (Tinkle and Irwin, 1965; Licht, 1967; Marion, 1970; Crews, 1977). This period may be adaptive, "setting the clock" to provide an intrinsic control to time reproductive events. Refractory periods have been observed under a variety of experimental laboratory conditions and thus they probably have a genetic basis. Bowker and Johnson (1980) recently reported on the thermoregulatory behavior in 3 species of *Cnemidophorus* kept under identical environmental conditions. A genetic basis is suggested by the distinct differences in the precision of thermoregulation they found among species.

In other experiments bearing on the genetics of life-history traits, Ballinger (1979) transferred neonate *Sceloporus jarrovi* from a high-altitude site to a low-altitude site. In nature *S. jarrovi* mature in 4 months at low altitude but require 16 months at high altitude. In the transfer experiments, the neonates from high-altitude parents did not mature as did the control neonates from low-altitude parents. Therefore, they conclude that the difference is probably genetic. Ferguson and Brockman (1980) grew hatchlings of *Sceloporus graciosus* and 3 subspecies of *S. undulatus* under identical laboratory conditions and observed significant differences in body growth rates that probably represent genetic differences. Genetic differences in life-history traits which explain ecotypic variance (V_C) must exist, but much more work needs to be done to quantify this component of phenotypic variance of life-history characters.

A Model of Life-History Variations

Equation 11.2 can be used as a base for a model to explain variations observed in life-history traits. The goal of such an approach is to understand phenotypic variance within the ecological setting of an observed life history and thus to reduce V_R to the smallest possible component. The goal is also to understand total phenotypic variance rather than only one or two of its components. Through experimentation it should be possible to determine values for each of the variance components and therefore quantify both the causal bases of any life-history trait as well as the relative importance of the different causes in determining the nature of that trait. However, such studies will require considerable effort and experimental technique and even then may produce less profound results than simple discovery of the possible factors responsible for the variation.

The model that can be used is a typical nested analysis of variance,

$$P_{ijkl} = \mu + Y_i + E_{j(i)} + C_{k(ij)} + R_{l(ijk)}, \qquad (11.3)$$

where P is the life-history character of interest; μ is the overall mean value of that character; Y_i is the phylogenetic and design background within which P is constrained: E_j is the proximal environmental component; C_k is the genetic component that reflects the adaptation of P to specific ecological conditions (location); and R_l is a random variable. Usually the Y's would be chosen because of their biological interest and thus would represent fixed effects. However, these Y_i's can be assumed to be representative of the possible Y's available, thus the variance component can be visualized within a random-effects model. In this model both E and C are random with the E's nested within Y and the C's nested within E. Thus, the model requires that two or more experimental units with uniform genetic content are each observed within each E and Y. The model requires that C_k remains constant and thus is not suited for organisms with short generation times capable of adapting rapidly enough to track short-term environmental changes. Likewise the model could not be used if polymorphisms of C_k exist such as those found by Bradshaw (1971) in growth rates of *Amphibolurus ornatus* in western Australia. Techniques for analyzing and estimating variance components using this model are well known (Gunther, 1964).

This experimental design requires (1) that Y_i's be chosen because of their interest, such as particular taxa (if more than one taxon is used, then ideally from a specific location) or populations (from different locations within a species)* and thus represent fixed effects† in the ANOVA design, (2) that E_j's also be fixed, although this can be approximated by observing random environmental differences in nature over a sufficient period for all E's to be represented at each Y_i, and (3) that C_k's be randomly chosen individuals from Y_i and be subjected to each E_j. Thus the model has some practical limitations. Such an experiment has not been done with lizards. It could most easily be done with parthenogenetic lizards so that absolute control of the C component would be assured. For example, it should be possible to produce strains in parthenogenetic lizards in which individuals have the same genetic constitution. These lizards could then be used for experimentation. Although data for lizards that could be used in this analysis of variance model do not exist at present, data do exist that can give some insights into the model's usefulness, if some reasonable simplifying assumptions and approximations are made.

A Partial Test

Data on size of first clutch in *Urosaurus ornatus* from 3 locations during 3 years are presented in Table 11.2. The locations were (1) a lava outcrop near Animas, New Mexico, in the western Chihuahuan Desert where environmental conditions, particularly rainfall, vary greatly between and within years (see Michel, 1976; Ballinger, 1977); (2) the Edwards Plateau near Austin, Texas (see Martin, 1977), where environmental conditions are less variable between and within years and; (3) at

* It is easy to see that Y_i represents different populations or locations in this model; thus it would be possible to introduce additional variables to permit higher phylogenetic components to be evaluated.

† This would be typical, but a fixed-effects model is not required, as both Y_i and E_j could be chosen randomly.

Table 11.2 Clutch size variation between years in 3 populations of the lizard *Urosaurus ornatus.*

Population	Year of sample			
Animas, New Mexico	*1973*	*1974*	*1976*	*Average*
Mean clutch size	10.95	6.80	11.36	9.48
Variance	5.23	2.53	4.66	–
Sample size	57	49	25	131
Between-year variance				6.36
Austin, Texas	*1971*	*1972*	*1973*	*Average*
Mean clutch size	6.71	6.39	6.81	6.65
Variance	0.72	1.07	0.82	–
Sample size	32	33	42	107
Between-year variance				0.48
Portal, Arizona	*1973*	*1976*	*1978*	*Average*
Mean clutch size	9.72	7.89	10.07	9.41
Variance	2.89	2.88	2.58	–
Sample size	65	19	13	97
Between-year variance				1.37

NOTE: See text for explanation. Data from Austin, Texas, courtesy of Dr. R. F. Martin, University of Texas, Austin.

an altitude of 1700 m in the Chiricahua Mountains 8 miles south of Portal, Arizona, where environmental conditions are less variable between years but highly variable within years (cold winter, hot dry June, and wet July). The 3 locations represent 3 distinct, isolated populations within a single species. All have been recognized as different subspecies, *U. o. ornatus, U. o. linearis,* and *U. o. chiricahuae* respectively (Mittleman, 1942), even though the subspecies *chiricahuae* is now generally synonymized with *linearis.* The differences among the 3 populations represent the Y_i variance component in the ANOVA model.

Average clutch size data are available for 3 years. These years were not picked randomly nor purposefully for this analysis, nonetheless they represent 3 different E_j environments out of all the possible years that could have been chosen. These years also happen to represent similar environmental differences. Martin (1977) pointed out that 1971 was a year of extreme drought for *Urosaurus* in central Texas, as was 1974 in New Mexico (Ballinger, 1977). The variation in clutch size between years thus approximates the E variance component.

A random sample of individuals was taken and average clutch size obtained for each population and year. Because individuals were not the same in each year, the C component is not separable from the random (residual) variance associated with size of clutch produced with C, E, and Y held constant.

ANOVA components for these data using the model (Eq. 11.3) are presented in Table 11.3. The sums of squares, mean squares, and variance components have been computed according to methods of Scheffé (1959) for unbalanced samples. Because of the sampling scheme, no estimate of the residual (error) variance is possible. The error variance (V_R) is included in the variance among individuals, which represents 38.4 percent of the total variance. If error variance represents approximately 44 percent or less of this 38.4 percent then the variance (V_C) attributable to ecotypic adaptations of individuals is significant. The analysis shows that the variance among years is significant but the variance among populations is not. Due to the unbalanced data these tests are not exact but the probability levels are sufficiently striking to suggest that these approximate tests are meaningful. Tests are reinforced by inspection of the data. The populations are from a single species and are not expected to be sufficiently divergent to exhibit great differences in clutch size, due to phylogenetic or other design constraints. Similarly, there are conspicuous differences in clutch size between years in which rainfall patterns were drastically different (Ballinger, 1977; Martin, 1977). Such differences are expected to alter clutch size proximately through food availability. The variance due to proximate environmental factors, V_E (estimated among years within areas), represents 42.5 percent of the total variance whereas only 19.1 percent of the total variance is attributable to design constraints (V_Y). It is reasonable to speculate that if one were to look at species occupying less variable environments, the amount of variation attributable to E_j would be less, and if the divergence between Y_i components were increased a corresponding increase in that variance would be expected.

Conclusions

Variations in life-history traits in lizards result from 3 primary sources: design constraints, proximate environmental factors, and ecotypic adaptations. In an evolutionary context these sources result in identifiable components of total phenotypic variance. A simple nested analysis of variance can be used to determine the variance associated with each variance component. Data are insufficient for a full test of this approach, but data on clutch size in 3 different years from 3 populations of *Urosaurus ornatus* suggest that the approach is useful. In this partial test, less than 40 percent of the phenotypic variance was attributable to ecotypic adaptations, that is, to the evolution of clutch size in the usual sense of the adaptationist's paradigm. Over 60 percent of the variance was due to proximate environmental factors (approximately 43 percent) and constraints (approximately 19 percent) of phylogeny or design, both factors that are generally ignored in adaptationists' explanations of life-history variation. I recommend that questions regarding variation in

Table 11.3 Analysis of variance of *Urosaurus* clutch data presented in Table 11.2.

Source	df	Mean squares	Probability[a]	Expected mean squares[b]	Variance estimate	Total variance (percent)
Areas (or populations)	2	285.654	> 0.15	$\sigma_R^2 + \sigma_C^2 + 44.33\sigma_E^2 + 110.76\sigma_Y^2$	$1.352 = \hat{\sigma}_Y^2$	19.13
Years within areas	6	103.783	< 0.001	$\sigma_R^2 + \sigma_C^2 + 33.64\sigma_E^2$	$3.004 = \hat{\sigma}_E^2$	42.50
Individuals within years within areas	326	2.712	–	$\sigma_R^2 + \sigma_C^2$	$2.712 = \hat{\sigma}_R^2 + \hat{\sigma}_C^2$	38.37
Error	–					

a. Only approximate tests are possible due to unbalanced data.
b. Derived according to methods of Scheffé (1959).

life-history characteristics be addressed from a broader perspective than in the past. This parallels Gould and Lewontin's (1979) criticism of the excessive use of an adaptionist explanation for all observed biological phenomena. In the case of the variable clutch in *U. ornatus,* an explanation that incorporated factors other than adaptive ones would provide a model with more predictive power than would one that incorporated only adaptive factors.

It is easy to argue that the ability of an organism to respond variably to different proximate environmental factors is also adaptive and is expected to occur in highly variable environments, therefore the variation observed in life-history traits is totally adaptive. This argument may indeed be correct, but it results in the primary issue being overlooked, that issue being to identify the nature of phenotypic variations observed in life-history characteristics of lizards and to recommend a method for evaluating the most likely probable causes of these. That is, are observed life-history traits more predictable or best explained by a knowledge of phylogenetic, adaptive, or proximate factors? It seems certain that all factors are involved but in one example (clutch size in *Urosaurus ornatus*) phenotypic variation appears most sensitive to variations in proximate environmental factors. Similar evaluations of numerous life-history traits in many species are needed before the nature of life-history variations in lizards can be adequately understood.

12 Realized Niche Overlap, Resource Abundance, and Intensity of Interspecific Competition

Arthur E. Dunham

IN THE LAST 10 to 15 years students of community structure have amassed an impressive array of evidence which suggests that interspecific competition is a primary factor responsible for the observed structure of natural communities. Ecologists interested in the structure of lizard communities have provided many of these studies.

Most previous studies of interspecific competition in lizard communities have been comparative, and virtually all have provided support for the view that interspecific competition determines many of the patterns observed in lizard communities. Variation in patterns of morphology, resource allocation, microhabitat occupancy, food habits, foraging biology, niche overlap, and niche width in lizard communities has been related to variation in number and kinds of syntopic, potentially competing species in studies in Milstead (1957, 1961, 1965), Schoener (1968, 1969, 1970, 1974a, 1975, 1977), Schoener and Gorman (1968), Pianka (1973, 1974, 1975, 1976), Huey and Pianka (1974, 1977), Huey et al. (1974), Clover (1975), Lister (1976a,b) and Pianka, Huey, and Lawlor (1979) among others.

The other major approach to the study of interspecific competition in natural communities has been experimental. Direct density manipulations, either through removal or addition of individuals, can demonstrate the existence of competition in natural systems. This approach has been successfully applied to fresh-water aquatic communities (Reynoldson, 1964; Hall, Cooper, and Werner, 1970; Reynoldson and Bellamy, 1971; Wilbur, 1972; Werner and Hall, 1976), marine rocky intertidal communities (Dayton, 1971; Menge, 1972; Connell, 1974, 1975), plant communities (Müller, 1966), and a few terrestrial communities (Inger and Greenberg, 1966; Grant, 1969, 1970, 1971, 1972; Jaeger, 1970, 1971, 1972; McClure and Price, 1975; Schroder and Rosenzweig, 1975; Brown and Davidson, 1977; Hairston, 1980).

There have been, to my knowledge, only 4 attempts to study competition in lizard communities experimentally (Nevo et al., 1972; Smith, 1977; Dunham, 1980; Tinkle, 1982; but see also Roughgarden and associates, Chapter 16). Radovanovic (discussed by Nevo et al., 1972) studied 2 species of *Lacerta* that have mutually exclusive distributions on most small islands in the Adriatic, suggesting a history of competitive interaction. He introduced one or the other species onto 4 islands where only 1 of the species was previously known to occur. Nevo et al. (1972) followed up these experimental introductions. In only one case did density of the introduced species increase with apparent concomitant decline in the density of the other species. Thus, this study provided only weak evidence for interspecific competition between these 2 species. Smith (1977) carried out a replicated removal experiment in the Chiricahua Mountains of southeastern Arizona. He demonstrated the existence of competition between *Sceloporus virgatus* and *Urosaurus ornatus,* but only in the year following removal and in certain age classes. Tinkle (1982) removed *U. ornatus* and *Sceloporus magister* from a guild of 3 species of arboreal iguanid lizards and looked for effects on density, habitat utilization, and survivorship of the remaining species (*Sceloporus undulatus*). Tinkle found no significant experimental treatment effects in any of the 3 years of his study. None of these studies identified a mechanism for the competitive interactions they were attempting to document.

I experimentally studied interspecific competition in a simple guild of 2 abundant species of diurnal, saxicolous iguanid lizards, the canyon lizard (*Sceloporus merriami*) and the tree lizard (*Urosaurus ornatus*) at a location of ecological microsympatry in Big Bend National Park in southwestern Texas (Dunham, 1980). I demonstrated significant effects of the removal of *S. merriami* on the population density, individual foraging success, individual growth rates, survivorship, and prehibernation lipid storage and body weights of *U. ornatus.* These effects were found only during periods of reduced food availability, and the reciprocal experiment produced no convincing evidence of a competitive effect on *S. merriami* by the removal of *U. ornatus.*

Comparative studies of resource utilization and partitioning in natural communities have documented differences among syntopic ecologically similar species which are inferred to reduce competition and thereby facilitate stable coexistence. The limits that interspecific competition may place on the number of species that can stably coexist within a given community have been similarly analyzed. These studies have been reviewed by Schoener (1974a), Pianka (1976), and Dunham (1980). To the extent that the differences between species in morphology, foraging habitat, food-resource utilization, and the like revealed by these studies constitute "coexistence mechanisms" (Wiens, 1977) which facilitate coexistence by reducing the intensity of competition, the dynamic response of competitors to a reduction in resource availability (and hence an increase in the intensity of competition) should be divergence in resource use. This prediction was made explicitly by Svärdson (1949) and Lack (1946, 1971), both of whom studied resource partitioning in bird communities.

The formulation of quantitative niche theory (Hutchinson, 1957, 1975; Vandermeer, 1972; Pianka, 1976; Diamond, 1978) provided a theoretical background for the interpretation of empirical studies of resource partitioning. Currently, most students of resource partitioning view the frequency distributions of resource utilization along one or more resource axes as characterizing the niche. This analogy has allowed theoretical investigation of how similar 2 (or more) species can be in their resource utilization and still coexist. This is termed limiting similarity and was first investigated theoretically by MacArthur and Levins (1967) with deterministic models derived from the Lotka-Volterra competition equation; it has been widely studied since (for example, May and MacArthur, 1972; May, 1973; Roughgarden, 1974a,b).

To incorporate resource-utilization functions into the equations used to model competition, it was assumed that the overlap in resource-utilization functions for a *limiting* resource was proportional to the competition for that resource; that is, it was assumed that the interaction coefficients in the Lotka-Volterra model could be calculated directly from niche overlap. This approach has been recently questioned by Abrams (1976) and Levine (1976). Many ecologists studying resource partitioning in natural systems have, at least implicitly, made the assumption and employed measures of overlap in resource utilization functions as point estimates of the intensity of competition. A number of different techniques for estimating niche breadth and overlap have been proposed (Simpson, 1949; Morisita, 1959; Horn, 1966; MacArthur and Levins, 1967; Levins, 1968; Schoener, 1968, 1974b; Pianka, 1969, 1970, 1973, 1974, 1975, 1976; Colwell and Futuyma, 1971; Pielou, 1972, 1975; Roughgarden, 1972; Vandermeer, 1972; Hurlburt, 1978; Petraitis, 1979; Johnson, 1980; Lawlor, 1980). Several of these measures have been used (either explicitly or implicitly) to estimate the intensity of interspecific competition in lizard communities (for example Schoener, 1968, 1974a).

Optimal foraging theory predicts the optimal diet (range of types of prey consumed) and optimal foraging habitat (number and types of habitat patches in which an animal should forage) as a function of the distribution, abundance, and relative values of the available prey. Most theories of optimal diet are very similar to one another (review in Pyke, Pulliam, and Charnov, 1977) and consider a diet to be optimal if it maximizes net rate of energy (or other measure of prey value) intake per unit of foraging expenditure (usually energy or time). The predator is assumed to rank prey types according to this optimality criterion, with the most preferred prey type being the one with the highest value; items are added to the diet in order of decreasing values of net intake per unit foraging expenditure.

Most of these models predict that the inclusion of a particular prey type in the optimal diet depends only on the abundances of better (higher ranked) prey types. Any factor (for example, competition) that reduces the abundance of a prey type already included in the optimal diet will not result in the exclusion of that prey type but may result in the inclusion of less preferred types. The number of food types included in the optimal diet should stay the same or increase if a competitor which reduces the availability of one or more of the prey types included in the optimal diet is added to the system. Increasing the abundance of all potential prey types uniformly or of preferred prey types should lead to dietary specialization as less preferred (low-value) prey types are excluded from the optimal diet. MacArthur and Pianka (1966) used a graphical model to predict the optimal range of habitat types an animal should include in its foraging activities. In this model, if a competitor invades a particular habitat and reduces the food available there, an organism may more profitably spend its time visiting unaffected patches. This model predicts that competitors will be more likely to specialize on and partition particular habitat types than food types. This is termed the "compression hypothesis."

My experimental demonstration (Dunham, 1980) that competition between *S. merriami* and *U. ornatus* is a contemporary (although intermittent) phenomenon provided an unusual opportunity to test theoretical predictions of the effects of variation in resource abundance and intensity of interspecific competition on dietary breadth, foraging habitat, and resource partitioning in a system known to be competitive and in which temporal variation in resource abundance and the intensity of competition had been measured.

In this chapter I use data on resource abundance, stomach contents, individual foraging success, and foraging microhabitat utilization to test the prediction from Lack (1946, 1947, 1971) and Svärdson (1949) that the resource utilization of competing species should diverge as resources become more scarce and the intensity of competition increases. I also examine the usefulness of realized-niche overlap measures as point estimates of the intensity of competitive interaction. Finally, I attempt to test the prediction from optimal diet theory that, as the abundance of preferred prey increases, the number of types of prey included in the diet should decrease as a result of specialization on common, high-value food types.

Methods

Precipitation and Food Resource Abundance. Weather data for the study area were summarized by Dunham (1978, 1980, 1981). During this study daily precipitation data were available from a United States weather station located at Panther Junction, Big Bend National Park, at an elevation of 1140 m. Panther Junction is 9.3 km SSE of the study area. Two methods were used to estimate the abundance of potential prey for *S. merriami* and *U. ornatus* and are described in detail elsewhere (Dunham, 1978, 1980, 1981). In 1975, 1976, and 1977 masking tape squares (58.1 cm^2) were affixed to rock surfaces where these lizards normally fed. The exposed surface of each square was then coated with the adhesive Tanglefoot®. These "sticky traps" were used regularly from May through October of these years to sample arthropod abundance on the control plots of the experimental study of competition described by Dunham (1980). On each sampling date, "sticky traps" were randomly placed on each study site and the exact time each trap was set recorded. The total number of arthropods caught by each trap was then converted to a capture rate (captures $trap^{-1}$ h^{-1}).

In 1976, 1977, and 1978, a D-Vac® vacuum arthropod sampling device was also used to sample the arthropod abundance on each control plot. During each sampling period, open rock surfaces where *S. merriami* and *U. ornatus* normally fed were sampled for periods of about 2 minutes. The exact duration of each sample was determined with a stop-watch. On each sampling date, 2 samples were taken on each control site between 0930 and 1145, and the total number of arthropods captured per minute in each sample was determined.

Stomach Content Analysis. Samples for data on stomach contents were periodically taken from habitat similar to that on the study plots of the experimental study (Dunham, 1980). All such samples were taken within 2 km of the study plots and within the Grapevine Hills. All individuals destructively sampled were killed within 6 hours (usually within 3 hours). The stomach contents of these animals were removed and weighed to the nearest 0.01 g on a triple-beam balance. The wet weight of stomach contents was used as an estimate of individual foraging success (Dunham, 1978, 1980, 1981). There were 6 sampling periods (see Table 12.2) in 1976 and 1977 for which simultaneous sticky-trap estimates of food abundance and sufficiently large samples of stomach contents were available. For stomach content analysis I considered different arthropod morphospecies and/or developmental stages (larvae, pupae, adults) as distinct prey types. The total number of each prey OTU (operational taxonomic unit) found in each stomach was recorded. The greatest length (a) and width (b), exclusive of legs and antennae, of intact individuals of each prey OTU were measured to the nearest 0.01 mm with an ocular micrometer fitted to a dissecting microscope. The volume of each measured prey item was estimated using the equation for the volume of a prolate spheroid (Selby, 1965):

$$V = 4/3 \pi (a/2) (b/2)^2.$$

Mean lengths, widths, and volumes of each prey OTU were then computed. Many indices for quantifying niche overlap and breadth have been suggested and used (see reviews in Hurlburt, 1978; Petraitis, 1979; and Abrams, 1980). For comparison with other studies of lizard communities (for example, Pianka, 1969, 1970, 1973, 1975, 1976; Lister, 1976a), I used the inverse of Simpson's (1949) measure to quantify dietary diversity and niche breadth, B:

$$B = 1/\sum_{i=1}^{n} p_i^2, \tag{12.1}$$

where p_i is the proportion of the total diet made up of prey OTU i. Volume diversity (VD) was calculated using the same equation, where p_i is the proportion of the total diet volume made up of prey OTU i. The evenness component of diversity (Pielou, 1975) was calculated by dividing B and VD by the number of prey OTU's in the sample diet. These were termed standardized diet diversity (B/B_{max}) and standardized volume diversity (SVD), respectively.

To test the hypothesis from optimal foraging theory that an animal should become more selective in its choice of prey as food availability increases, I computed the likelihood measures of niche breadth suggested by Petraitis (1979) for each species based on the stomach content data for the sampling dates shown in Dunham (1980). For species 1 on a given sampling date, the standardized niche breadth is given by:

$$W_1 = (q_1)^{1/N_1} = rE, \tag{12.2}$$

where

$E = \log_r \lambda_1 q_1 / N_1$
$r =$ number of prey OTU's in the diet
$\lambda_1 = (q_j/p_{1j})^{N_{1j}}$
$q_j =$ the frequency of occurrence of prey OTU j in the environment
$N_{1j} =$ the number of items of prey OTU j in the diet of species 1

$$p_{ij} = N_{1j}/\sum_{j=1}^{r} N_{ij} = N_{1j}/N_1.$$

The combined frequency of use by both species was used to estimate the availability of each prey OTU on a given sampling date. The likelihood measures of niche breadth for each species were then compared across sampling dates and, hence, abundances using the statistical procedures recommended by Petraitis (1979). If the hypothesis of increasing dietary specialization with increasing food abundance is true, these measures of niche breadth should be a monotone decreasing function of prey abundance.

I calculated Pianka's (1973) symmetric estimate of volumetric overlap in food-resource utilization as

$$O_{jk} = O_{kj} = \frac{\sum_{i=1}^{n} p_{ij}p_{ik}}{\sqrt{\sum_{i=1}^{n} p_{ij}^2 \sum_{i=1}^{n} p_{ik}^2}}, \quad (12.3)$$

where n is the total number of prey OTU's, p_{ij} is the proportion of the total volume of the diet of species j made up of individuals of prey OTU i, and p_{ik} is the proportion of the total volume of the diet of species k consisting of individuals of prey OTU i. I also estimated volumetric overlap in food resources utilization using Levin's (1968) nonsymmetric measure

$$\alpha_{jk} = \frac{\sum_{i=1}^{n} p_{ij}\, p_{ik}}{\sum_{i=1}^{n} p_{ij}^2}, \quad (12.4)$$

where α_{jk} is the overlap of species k on species j and the other symbols are as in Eq. 12.2. Volumetric overlap was used for consistency with other workers (for example, Schoener and Gorman, 1968; Pianka, 1973) and because the energy content of an individual prey item is more accurately estimated by volume than by a linear measure such as length. Overlap in energetically important prey items is more likely to result in competition than is overlap in energetically trivial prey items.

During the lizards' active season in 1977 I carried out a series of observations designed to reveal the effects of changes in food resource abundance on breadth and overlap in foraging microhabitat utilization by *S. merriami* and *U. ornatus*. These observations were limited to 2 time periods and to adult males which were foraging. The first time period (July 12–19) followed the first substantial rains of the summer; 5.54 cm of rainfall had occurred in the week preceding this period. The second time period (August 10–15) followed 3 weeks without rain, and none fell during the observation period. During these 2 periods my assistants and

I observed individually marked lizards foraging on the control areas of the experimental study of interspecific competition described by Dunham (1980). The lizards were watched from a distance of at least 10 m with binoculars in order to minimize disturbance. I designated 5 qualitative and distinct habitat categories where these lizards were known to forage. These categories were exposed rock faces (R), bushes (B), rock crevices (C), rock faces covered with vegetation (RB), and on the ground (G). In addition, 2 categories of foraging tactic were designated. These were actively foraging (AF) and sit-and-wait foraging (SW). An animal was classified AF if it was actively moving through the environment and taking prey items as it moved; SW implied that the animal was stationary and attempted to capture prey that came near (to within 15 cm). When an animal was observed to make a foraging strike, the habitat type and foraging tactic were recorded. No more than 10 such observations were made of a given animal.

Results

Rainfall and Prey Abundance. The 4 years of this study differed greatly in the amount and distribution of precipitation (Dunham, 1980, 1981). The total annual precipitation was 48.0 cm in 1974, 21.4 cm in 1975, 44.7 cm in 1976, and 17.2 cm in 1977. The total active season (April–October) precipitation was 40.9 cm in 1974, 18.1 cm in 1975, 37.2 cm in 1976, and 15.5 cm in 1977. The mean annual precipitation based on 20 years of continuous data is 32.7 ± 4.2 (95 percent CI) cm; the mean active season precipitation for the same period is 26.8 ± 4.2 cm. Clearly 1975 and 1977 were much drier than 1974 and 1976 and both were significantly drier than the long-term average for both annual and active season precipitation; 1977 was the driest year since the establishment of the weather station at Panther Junction (1956).

Comparison of the arthropod abundance based on sticky-trap estimates suggests that saxicolous insectivores such as *S. merriami* and *U. ornatus* had more prey available in 1976 than in either 1975 or 1977 (fig. 2 in Dunham, 1980). In all cases the 1976 estimate was greater than that for the corresponding sampling period of 1975 or 1977. Pairwise Mann-Whitney U comparisons revealed that in all but 2 cases the differences were statistically significant ($P \leq 0.05$). The 1976 estimate for late June was greater than that for 1975, but the difference was not statistically significant ($P = 0.13$); the same was true for the comparison of the late August sample of 1976 with that of 1975 ($P = 0.08$). Pairwise Mann-Whitney U comparisons of the mean sticky-trap estimates for comparable sampling dates revealed that arthropod abundance was significantly greater in 1976 than in either 1975 ($U = 31$, $0.01 \leq P \leq 0.025$) or 1977 ($U = 33$, $0.005 \leq P \leq 0.01$) but that 1975 and 1977 did not differ significantly ($U = 15$, $P \geq 0.10$).

In the years for which there are adequate data (1976 and 1977), the highest correlation ($r = 0.65$, $P \leqslant 0.05$) between the D-Vac® estimate of arthropod abundance and precipitation was with the amount of precipitation during the preceding 2-week time interval. The 2 methods of sampling prey abundance are concordant in their estimates; there is a highly significant Spearman rank correlation ($r_s = 0.83$, $0.01 \leqslant P \leqslant 0.03$) between the sticky-trap and D-Vac® estimates of arthropod abundance for 1976 and 1977. Neither method of estimating arthropod abundance provides an unbiased quantitative estimate of food resources available to *S. merriami* and/or *U. ornatus* at any given time. However, both methods sample the same array of prey items and sizes as found in the stomachs of these lizards and thus provide satisfactory indices of relative abundance of prey between 2 time periods.

Individual Foraging Success. Elsewhere (Dunham, 1980, 1981) I have presented results of 3 attempts to measure individual foraging success in these lizards. The first estimate consisted of wet weights of stomach contents of adult male *S. merriami* and *U. ornatus* from animals taken in 1976 and 1977 (fig. 4 in Dunham, 1980). There was a significant Spearman rank correlation in both 1976 ($r_s = 0.93$; $0.01 \leqslant P \leqslant 0.02$) and 1977 ($r_s = 0.88$; $0.01 \leqslant P \leqslant 0.05$) between the weight of food in stomachs of adult male *U. ornatus* and *S. merriami*. There was also a significant Spearman rank correlation between weight of food in the stomachs of *U. ornatus* ($r_s = 0.83$; $P \leqslant 0.05$) and *S. merriami* ($r_s = 0.88$; $P \leqslant 0.05$) and the D-Vac® estimates of food abundance. There was a significant correlation between the D-Vac® estimate of food abundance and the mean mass of food in stomachs of both male tree lizards ($r_s = 0.91$; $P \leqslant 0.02$) and canyon lizards ($r_s = 0.89$; $P \leqslant 0.02$).

The second method consisted of direct field observation of foraging males of both species. My assistants and I observed individually marked lizards in 1976. These observations were carried out between 0800 and 1100 CDT, and only data from actively foraging lizards were counted in the analysis. Total length of time a lizard was observed was timed and the number of feeding strikes the animal made was counted and converted into a feeding rate. In order to have observations from an adequate number of lizards for analysis, these data were lumped into samples from 3 2-week intervals: late June, early August, and late August (Table 12.1). There are significant differences between the distributions of foraging success on these 3 dates as revealed by pairwise Mann-Whitney U comparisons. Observed feeding rates of both *U. ornatus* ($U = 596.0$; $P \leqslant 0.0001$) and *S. merriami* ($U = 1363.0$; $P \leqslant 0.0001$) were lower in late June than in early August. Feeding rates were also lower in late June than in late August in both *U. ornatus* ($U = 463.0$; $P \leqslant 0.0001$) and *S. merriami* ($U = 783.0$; $P \leqslant 0.0001$). The observed differences in foraging rates between early and late August were not significant in either spe-

Table 12.1 Foraging success of adult males as measured by observed feeding rates during 1976. (Seventy-five percent error bounds in parentheses.)

Species	Date	N	Mean feeding rate strikes h^{-1}
Urosaurus ornatus	Late June	24	5.47 (0.94)
	Early August	25	14.58 (0.98)
	Late August	21	14.57 (1.50)
Sceloporus merriami	Late June	29	6.57 (0.82)
	Early August	49	20.33 (0.85)
	Late August	27	21.45 (0.81)

cies. Thus, the lower prey abundance in late June was reflected in reduced foraging success by both species during that time period.

My final method of estimating foraging success involved quantifying prehibernation storage lipid levels in males and females of both species in a wet year (1976) and a dry year (1977). Using standard chloroform extraction techniques, I extracted and quantified lipids in the carcass, fat bodies, and liver from animals collected in late September and early October of 1976 and 1977. Total body lipid was calculated for each animal as the sum of dry weight of lipid stored in each of these depots. In males and females of both species, total prehibernation lipid depends on snout-vent length (SVL) and this relationship is nonlinear. I used analysis of covariance on logarithmically transformed variables to compare prehibernation lipid levels and body weights characteristic of animals collected in different years. Results of the covariance analysis of differences in body weight and prehibernation lipid levels are presented in tables 21 and 22 in Dunham (1981). In all cases, prehibernation lipid levels were significantly lower in the dry year (1977) than in the wet year (1976). For example, a prehibernation female canyon lizard which was 50 mm SVL in 1976 would be expected to weigh 3.94 g and to have about 0.28 g of stored lipid; in 1977 the predicted weight of the same female would be about 3.31 g and predicted total body lipid would be about 0.12 g, a weight difference of 0.63 g (16 percent) and a difference in stored lipid of 0.16 g (57 percent). Similar differences were apparent in male *S. merriami* and in both sexes of *U. ornatus.*

These data indicate that foraging success is related to arthropod abundance in both tree lizards and canyon lizards. Wet weights of stomach contents from adult males were positively correlated with D-Vac® estimates of food abundance on comparable sampling dates in 1976 and 1977. Foraging rates of individually marked lizards in 1976 were also positively correlated with the D-Vac® estimates of food abundance made during the same time periods. The most convincing data are those on prehibernation lipid levels and body weights in 1976 and 1977. Males and females of both species exhibited significantly lower (ANCOVA) prehibernation lipid levels and body weights in the dry year (1977). The comparisons of the sticky-trap estimates of arthropod abundance presented previously indicated that there was less food available to *S. merriami* and *U. ornatus* in 1977. The prehibernation weight and lipid data strongly suggest that these lizards were less successful in gathering food in 1977.

Resource Utilization and Foraging Tactics. Results of the stomach content analysis are presented in Table 12.2 and Figures 12.1 to 12.3. There were sufficient data available for adult males of both species to allow comparison on 6 sets of dates representing a range of estimated food abundance. Both species consumed a wide variety of arthropods. More than 2,950 individual prey items, representing 111 distinct prey OTUs, were found in stomachs of adult male *S. merriami* taken on these dates. More than 2,000 individual prey items of 78 distinct prey OTUs were found in stomachs of adult male *U. ornatus.* Diet diversity (B) of each species was calculated for each set of sampling dates using Simpson's (1949) index of diversity. Volume diversity (VD) of prey items in the diet was calculated using the same index and the proportion of the total volume of prey in the diet (represented by a given sample) contributed by each prey OTU. Both measures of niche breadth were standardized (Pielou, 1975) by dividing by the number of prey OTU's in the sample yielding standardized dietary diversity (B/B_{max}) and standardized volume diversity (SVD), respectively. There was considerable variation in B, B/B_{max}, VD, and SVD among sampling dates for both species (Table 12.2, Figs. 12.1 and 12.2). The same was true of the likelihood measures (Petraitis, 1979) of niche breadth (Table 12.2). There was no significant Spearman rank correlation between B, B/B_{max}, VD, or SVD with estimated prey abundance providing no evidence of dietary specialization with increasing prey abundance. However, of the measures of niche breadth employed here, only the likelihood measure (W) suggested by Petraitis (1979) includes the availability of resources, and only estimates based upon this measure may be statistically compared. There was no significant Spearman rank correlation of W with the sticky-trap estimate of prey abundance for either *S. merriami* or *U. ornatus.* However, likelihood estimates (W) of food-niche breadth for the 2 sampling dates with the highest estimated prey abundances were significantly lower (Fig. 12.2) than estimates for the other 4 sampling dates (F-tests, all $P \leq 0.05$) in both species. The estimates of W did not differ significantly among the four sampling dates with the lowest ranking estimates of prey abundance (F-tests, all $P \leq 0.05$). These results indicate that these two species may be exhibiting dietary specialization at high food abundances, but that the

Table 12.2 Estimates of food-niche breadth and overlap based on stomach content data.

Date	Estimated prey abundance (prey/trap h)	Species (N)	B	B/B_{max}	W	VD	SVD	$O_{ij} = O_{ji}$	α_{ij}
August 1–7, 1976	27.30	*Sceloporus merriami* (24)	4.134	0.172	0.485	5.39	0.225	0.506	0.402
		Urosaurus ornatus (27)	4.456	0.165	0.461	8.59	0.318	0.506	0.638
July 28–August 4, 1976	22.80	*Sceloporus merriami* (34)	2.360	0.069	0.342	4.21	0.124	0.391	0.372
		Urosaurus ornatus (26)	3.607	0.139	0.445	4.67	0.292	0.391	0.412
September 28–October 4, 1976	15.21	*Sceloporus merriami* (26)	6.851	0.264	0.745	5.37	0.207	0.147	0.211
		Urosaurus ornatus (20)	2.654	0.133	0.704	2.61	0.261	0.147	0.103
August 6–8, 1977	10.42	*Sceloporus merriami* (23)	5.908	0.257	0.734	4.50	0.196	0.141	0.169
		Urosaurus ornatus (9)	2.803	0.311	0.660	3.10	0.344	0.141	0.117
June 20–27, 1977	7.08	*Sceloporus merriami* (28)	4.493	0.161	0.614	7.81	0.279	0.148	0.169
		Urosaurus ornatus (28)	4.575	0.163	0.562	6.00	0.214	0.148	0.130
October 5–9, 1977	1.88	*Sceloporus merriami* (16)	2.137	0.134	0.622	5.18	0.324	0.043	0.045
		Urosaurus ornatus (10)	1.136	0.114	0.594	4.66	0.466	0.043	0.041

NOTE: Value for N is the number of prey OTUs in the diet on a particular sampling date. Estimated prey abundance is based on the sticky-trap estimate nearest the sampling dates.

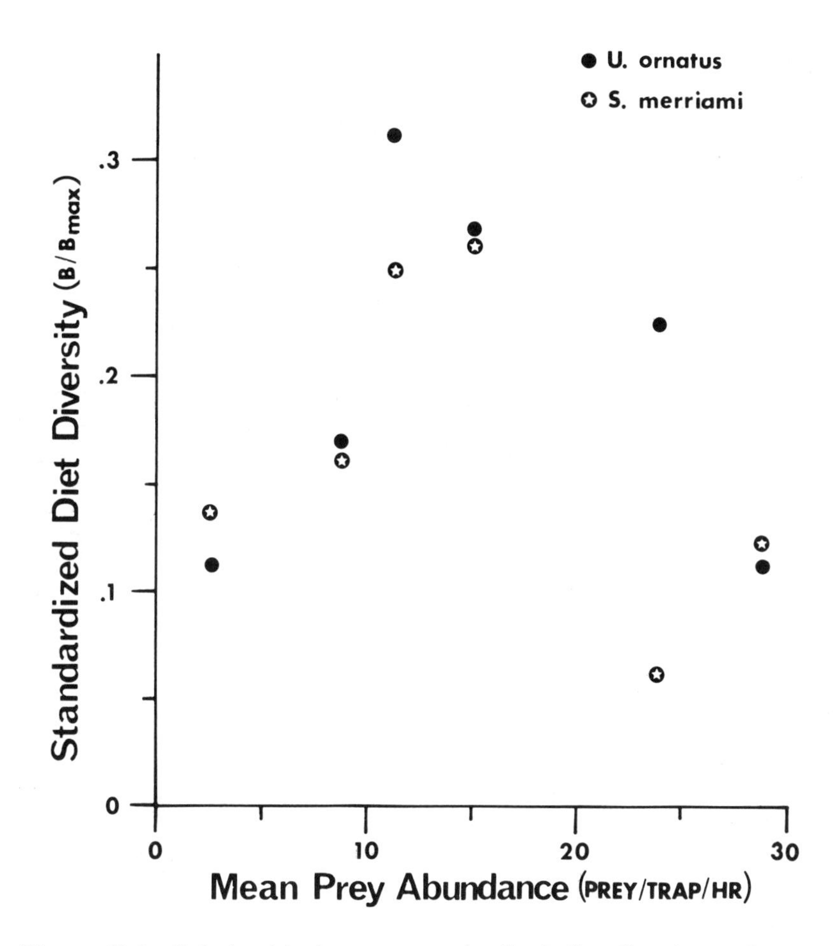

Figure 12.1 Relationship between standardized diet diversity and mean sticky-trap estimate of prey abundance for the 6 sampling dates given in Table 12.2.

relationship between dietary niche breadth and food abundance is complex.

Pianka's (1973) symmetrical overlap measure ($O_{ij} = O_{ji}$) and Levin's (1968) overlap measure (α_{ij}) were significantly correlated with the estimate of prey abundance ($r_s = 0.829$, $P \leqslant 0.05$ for O_{ij}; $r_s = 0.986$, $P \leqslant 0.01$ for α_{su}; and $r_s = 0.771$, $P \leqslant 0.05$ for α_{us}) indicating that realized niche overlap as estimated by these measures increases as abundance of prey increases (Fig. 12.3).

Foraging Microhabitat Utilization. Results of the observational study of foraging microhabitat utilization by individually marked adult males are given in Table 12.3. Proportional use of each of the 5 operational microhabitats varied greatly between the 2 observation periods for both

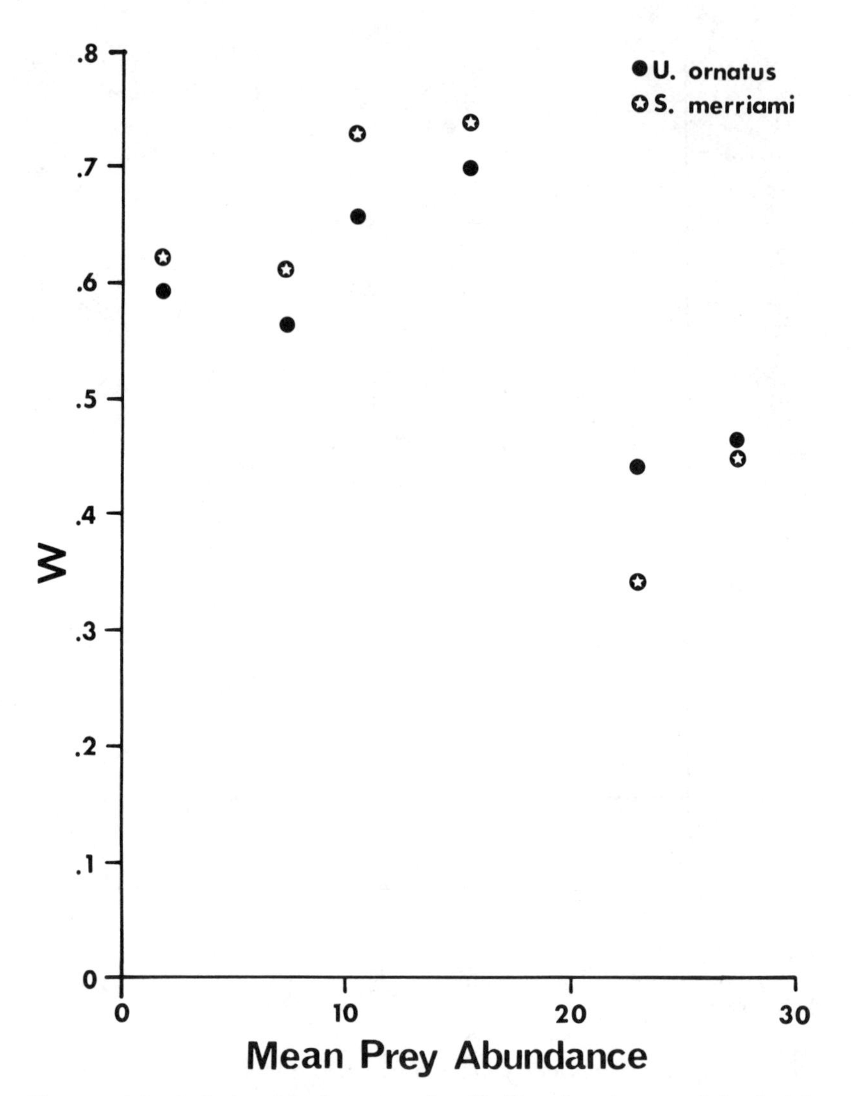

Figure 12.2 Relationship between the likelihood estimate of food-niche breadth and mean sticky-trap estimates of prey abundance for the 6 sampling dates given in Table 12.2.

species. Adult male canyon lizards decreased their proportional use of open rock faces as foraging sites from 0.701 during mid-July to 0.287 during mid-August ($\chi_1^2 = 50.93$; $P \leqslant 0.0001$). Concomitantly, use of rock faces covered with vegetation increased from 0.254 to 0.610 ($\chi_1^2 = 37.77$; $P \leqslant 0.0001$) and use of crevices increased from 0.037 to 0.104 ($\chi_1^2 = 4.75$; $P \leqslant 0.03$). Use of foraging sites on the ground declined from 0.007 to 0, but the change was not significant ($\chi_1^2 = 1.23$; $P > 0.2$). I

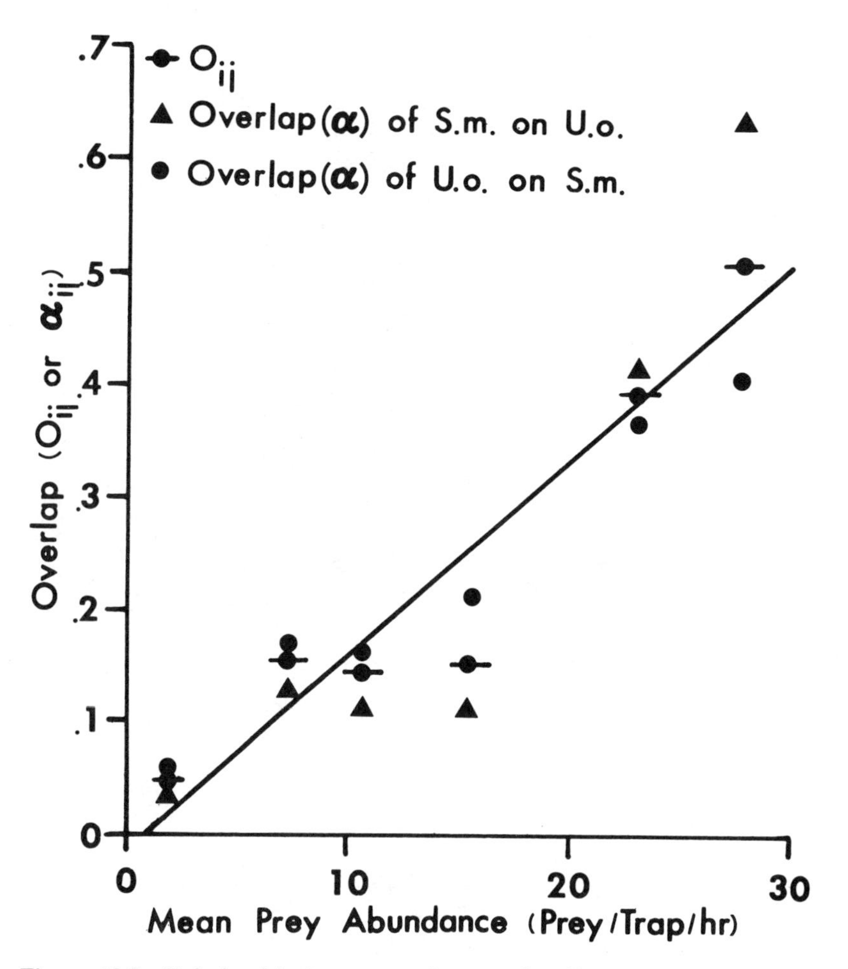

Figure 12.3 Relationship between estimates of realized food-niche overlap and the mean sticky-trap estimate of prey abundance for the 6 sampling dates given in Table 12.2.

have never observed adult male canyon lizards foraging in bushes. Adult male tree lizards also decreased their use of open rock faces from 0.342 to 0.102 ($\chi_1^2 = 29.77$; $P \leqslant 0.0001$) and their use of rock faces covered with vegetation from 0.385 to 0.267 ($\chi_1^2 = 5.49$; $P \leqslant 0.019$). Use of crevices increased slightly from 0.056 to 0.106 ($\chi_1^2 = 4.13$; $P \leqslant 0.05$). The proportional use of bushes as foraging sites increased dramatically from 0.186 to 0.604 ($\chi_1^2 = 62.43$; $P \leqslant 0.0001$). There was no significant change in the use of the ground as a foraging site ($\chi_1^2 = 1.82$; $P \leqslant 0.1$). During the first observation period (August 12–19) estimated food abundance was 25.0 arthropods trap^{-1}h^{-1}; during the second observation period

Table 12.3 Foraging microhabitat utilization of male *Sceloporus merriami* and *Urosaurus ornatus* under conditions of high and low prey abundance (1977).

Species	Date	Estimated prey abundance	Microhabitat: Rock faces	Microhabitat: Rock faces with vegetation	Microhabitat: Crevices	Microhabitat: Bushes	Microhabitat: Ground	B	B/B_{max}	O_{ij}	α_{ij}
Sceloporus merriami	July 12–19	25.0	94(0.701)	34(0.254)	5(0.037)	0(0)	1(0.007)	1.794	0.359	0.826	0.610
	August 10–15	5.8	47(0.287)	100(0.610)	17(0.104)	0(0)	0(0)	2.149	0.430	0.425	0.417
Urosaurus ornatus	July 12–19	25.0	55(0.342)	62(0.385)	9(0.056)	30(0.186)	5(0.031)	3.291	0.658	0.826	1.118
	August 10–15	5.8	19(0.102)	50(0.267)	3(0.106)	113(0.604)	2(0.011)	2.238	0.448	0.425	0.434

NOTE: Numbers in parentheses are proportions of foraging strikes made in each microhabitat. Variable B is Simpson's measure of niche breadth (1949); B/B_{max} is standardized niche breadth; O_{ij} is Pianka's symmetric measure of niche overlap (1973); and α_{ij} is Levin's measure of the niche overlap of species j with species i (1968).

(August 10–15), it was only 5.8 arthropods trap^{-1}h^{-1} (Table 12.3). During the period of high food abundance, adult male canyon lizards foraged more on rock faces (χ_1^2 = 20.42; $P \leq 0.0001$), less on rock faces covered by vegetation (χ_1^2 = 16.33; $P \leq 0.0001$), less in bushes (χ_1^2 = 60.00; $P \leq 0.0001$), and less on the ground (χ_1^2 = 5.33; $P \leq 0.021$) than did adult male tree lizards. There was no significant difference in proportional use of crevices (χ_1^2 = 22.29; $P \leq 0.13$). During the period of low food abundance, canyon lizards foraged more on open rock faces (χ_1^2 = 33.33; $P \leq 0.0001$), more in crevices (χ_1^2 = 19.60; $P \leq 0.0001$), and less in bushes (χ_1^2 = 226.00; $P \leq 0.0001$) than did tree lizards. Pianka's (1973) symmetrical overlap measure (O_{ij}) declined from 0.826 to 0.425 (Table 12.3) with the decline in prey abundance, indicating a reduction of realized overlap in foraging microhabitat utilization. Levin's (1968) measure (α_{ij}) behaved concordantly (Table 12.3). Both measures of the diversity of foraging microhabitats utilized by canyon lizards increased over the same period, indicating an increase in the evenness component of niche breadth. In contrast, both measures of the diversity of foraging microhabitats utilized by tree lizards decreased with reduction in food abundance, indicating specialization in foraging microhabitat use, and probably reflecting the greatly increased use of bushes as foraging sites during this time.

Foraging Tactics. Data on foraging tactics employed by both species during the observational study of foraging microhabitat use are summarized in Table 12.4. In both species there was a highly significant difference in the proportion of feeding strikes associated with each foraging tactic between the two observation periods. During the period of high food abundance (July 12–19) 81.8 percent of foraging strikes made by canyon lizards and 80.1 percent by tree lizards were classified as being

Table 12.4 Foraging tactics of male *Sceloporus merriami* and *Urosaurus ornatus* under conditions of high and low prey abundance (1977).

Species	Date	Estimated prey abundance	Foraging tactic: Sit and wait	Foraging tactic: Actively foraging	Contrast
Sceloporus merriami	July 12–19	25.0	25(0.182)	112(0.818)	65.039[a]
	August 10–15	5.8	108(0.643)	60(0.357)	
Urosaurus ornatus	July 12–19	25.0	32(0.199)	129(0.801)	117.434[a]
	August 10–15	5.8	162(0.753)	53(0.247)	

a. Attained significance level of $P < 0.0001$.

NOTE: Numbers in parentheses are proportions of total feeding strikes in which each tactic was observed. Prey abundance estimates are based on sticky-trap estimates. The value of the test statistics for the χ^2 test of independence of foraging success and prey abundance is also presented.

made while the lizard was actively foraging. In contrast, during the period of low prey abundance, only 35.7 percent of foraging strikes made by canyon lizards and 24.7 percent of those made by tree lizards were made while the lizard was actively foraging. The differences are highly significant ($P \leqslant 0.0001$) in both species (Table 12.4). The difference in proportion of active foraging between these two species was not significant ($\chi^2 = 0.127$; $P > 0.70$) during the period of high food availability. However, during the period of reduced food abundance, a significantly lower ($\chi_1^2 = 5.55$; $P < 0.02$) proportion of strikes by *U. ornatus* were made while actively foraging than was the case for *S. merriami*. Thus, *U. ornatus* may reduce active foraging more than *S. merriami* at low food abundances.

Discussion

Data presented here indicate that both seasonal and annual variation in arthropod abundance were correlated with variation in amount of precipitation. Further, I have demonstrated significant seasonal and annual variation in arthropod abundance and, hence, in food resources available to insectivorous lizards. Comparison of sticky-trap estimates of arthropod abundance for 1975, 1976, and 1977 demonstrated that arthropod abundance was greater in 1976 on comparable sampling dates. These differences were correlated with differences between these years in amount and distribution of precipitation. Correlation of weights of stomach contents, prehibernation size-specific body weights and total body lipids, and individual growth rates with estimates of food abundance indicates that successful foraging by these lizards depends, at least in part, on food abundance (Dunham, 1980, 1981). Results presented elsewhere (Dunham, 1980) argue strongly that interspecific competition is a contemporary phenomenon in this system, that this competition is probably for food resources and is probably exploitative in nature, that intensity of competitive interaction varies inversely as a function of food abundance, and that *S. merriami* exerts a much stronger competitive effect on *U. ornatus* than vice versa.

Results presented here indicate that during periods of reduced food abundance, *S. merriami* reduces its use of open rock faces as foraging sites and increases its use of rock faces covered with vegetation and of crevices. Under the same conditions *U. ornatus* reduces its use of open rock faces, rock faces covered with vegetation, and crevices, and greatly increases its use of bushes as foraging sites. Under conditions of reduced food abundance, overlap in foraging microhabitat use decreases, and both species reduce the proportion of active foraging and correspondingly increase sit-and-wait foraging. *S. merriami* exhibited increased foraging-microhabitat niche breadth and its use of available foraging microhabitats became more even under conditions of reduced food abundance. This is contrary to the response predicted by the compression hypothesis. *Urosaurus ornatus,* on the other hand, became more specialized in its use of available foraging microhabitats under conditions of low food abundance. This reflects its greatly increased use of bushes as foraging sites during periods of reduced food abundance. This shift is in the direction predicted by the compression hypothesis. Dietary overlap between these two species was much lower when food resources were scarce than was the case when they were abundant, and dietary overlap was positively correlated with estimates of food abundance. There was little indication that individuals of either species became more generalized in diet as food became scarce, contradicting the prediction to the contrary from optimal diet theory. Comparisons of likelihood measures of niche breadth suggested that both species may become more specialized, presumably on high-quality food items, at high food abundances as predicted by optimal diet theory. This is not a completely fair test, however, because standard models of optimal diet assume that the abundance of food types changes uniformly and that the same range of prey OTUs is present at all levels of food abundance.

These data do provide evidence for divergence of these two species in foraging-microhabitat utilization and diet as abundance of food decreases and competition presumably becomes more intense. Overlap in both diet and foraging microhabitat decreased with decreasing food abundance, indicating an overall divergence in foraging biology during periods of low abundance as hypothesized by Lack (1946, 1947, 1971) and Svärdson (1949). Prey OTUs included in the diet are probably not independent of foraging-microhabitat utilization, and most of the dietary differences between these two species may be due to differences in foraging site. A number of empirical studies of seasonal variation in diet or foraging habitats of vertebrates have produced results similar to those of this study; see review in Smith et al., 1978. For example, these authors, in their study of the effects of seasonal variation in food abundance on resource partitioning among Darwin's ground finches (*Geospiza*), found that diets of these finch species diverged and the range of food types taken was reduced in the dry season when food abundance was much reduced.

Both lizard species switched to a sit-and-wait foraging tactic during periods of low resource abundance, probably because this mode of foraging is energetically less costly than active foraging and when food abundance is low, net energetic return per unit of foraging expenditure is probably higher for the sit-and-wait tactic. Foraging in this manner may also lower the risk of mortality, especially since the foraging sites used during periods of resource scarcity (bushes and rocks covered by vegetation) seem to offer greater concealment. This evidence is anecdotal, however, and data on rates of predation in these habitats are needed.

The observed increase in overlap in both foraging-microhabitat utilization and diet with increasing prey abundance in a system in which intensity of interspecific competition has been shown experimentally to vary inversely with food resource abundance has several implications for comparative studies of the importance of interspecific competition in structuring communities. Studies which use measures of realized-niche overlap as point estimates of the intensity of competition seem likely to lead to incorrect conclusions about the competitive structure of communities. This study clearly demonstrates that overlap values may vary widely depending on variation in the availability of resources and, therefore, invites caution in uncritical acceptance of the results of studies of community structure which rely on overlap values measured in a single year in a fluctuating environment and/or without data on the availability of resources. Such studies of lizard community structure are common (Schoener and Gorman, 1968; Pianka, 1973, 1975; Lister, 1976a; reviews in Schoener, 1977). The dynamic behavior of the estimates of niche overlap used in this and other studies is opposite to that required of a point estimate of the intensity of competition. Several authors, including Sale (1974) and Schoener (1974b), have suggested weighting such overlap measures by the inverse of resource abundance to attempt to correct for this problem. However, the relationship of overlap in resource-utilization functions to intensity of interspecific competition is, at least potentially, system specific and unpredictable without knowledge of the underlying resource dynamics (Levine, 1976). Nonetheless, the effects of using data on the abundances of individual prey types remain to be critically examined empirically. At present, overlap values computed on the basis of weighted utilization data seem to be superior to those based on unweighted utilization data. Patterns of resource utilization may be due to processes other than competition, and the actual limiting resources (food, nesting sites, and the like) in a system may be different at different times. Therefore, additional observational studies which compare resource utilization and niche overlap between coexisting species seem unlikely to increase our knowledge of the importance of interspecific competition in determining limits to species packing or community structure. In natural communities, the intensity of interspecific (and intraspecific) competition is likely to be variable and, perhaps, absent much of the time. In communities containing vertebrates or other organisms whose generation times are lengthy relative to the rate of environmental fluctuation, long-term experimental studies combined with observation of the dynamics of resource utilization seem essential if the importance of competition in influencing species coexistence, community structure, and resource-utilization patterns is to be adequately assayed and current theory tested.

13 Temporal Separation of Activity and Interspecific Dietary Overlap

Raymond B. Huey
and Eric R. Pianka

SYMPATRIC PREDATORS are sometimes active at different times of day (Schoener, 1974). Lizards and raptors (owls and hawks) are conspicuous examples of groups that contain both diurnal and nocturnal species, and many other animal groups contain species that have more subtle differences in times of activity (Pianka, 1969b; Schoener, 1970; Pianka, Huey, and Lawlor, 1979). Interspecific nonsynchrony in activity can be related to the intensity of competition in two fundamental ways. First, if differences in times of activity lower frequencies of direct encounters between predators, interference competition may be reduced (Case and Gilpin, 1974). Second, if such temporal differences lower dietary overlap, exploitation competition for food can also be reduced (MacArthur and Levins, 1967; Levins, 1968).

Because of these direct and indirect effects of temporal separation on competition, time is often treated as a major niche dimension, along with space and food (Pianka, 1969b, 1973, 1974; Cody, 1974; Schoener, 1974). Time has sometimes been viewed as an indicator of dietary or spatial competition. Arguments supporting the use of temporal separation as an indicator of dietary separation rest, however, on the largely unverified assumption (Jaksić, Greene, and Yañez, 1981; Jaksić, 1982) that nonsynchronously active predators should be exposed to differing prey worlds (Lewis and Taylor, 1964) and should therefore encounter and eat different prey (Schoener, 1974). But if prey are rapidly renewed (MacArthur and Levins, 1967) or if differences in activity periods between consumers are major, such as between species active during different seasons, then renewal of prey itself could suffice as an alternative explanation for reduced exploitation competition (Pianka, Huey, and Lawlor, 1979; Jaksić, Greene, and Yañez, 1981).

Despite the intuitive appeal of these arguments, workers have long commented on the similarity of diets of nocturnal and diurnal raptors (Munro, 1929; Baumgartner and Baumgartner, 1944; Lack, 1946; Fitch, 1947; Jaksić, 1982). Indeed, prompted by evidence of a high overlap in diet between great horned owls and red-tailed hawks, Orians and Kuhlman (1956) suggested that a study of the interactions between diurnal and nocturnal predators would be worthwhile.

One indirect test of the assumption that differences in times of activity affect dietary overlap compares dietary overlaps of pairs of synchronous and nonsynchronous predators. If prey worlds are more similar for synchronous than for nonsynchronous pairs, then synchronous predators are more likely to have higher dietary overlaps than are nonsynchronous predators. Conversely, if time has no effect on foods eaten, then both synchronous and nonsynchronous pairs should show similar dietary overlap. Note that this prediction is statistical, not absolute; two synchronous predators that forage in different ways or in different microhabitats might actually be exposed to very different prey worlds (Huey and Pianka, 1981).

Lizards are ideal subjects for examining the influence of time of activity on diet. Here we present data on dietary overlaps among lizards that differ strikingly in short-term activity periods: nocturnal versus diurnal lizards from the Kalahari and Australian deserts. Specifically, we investigate whether pairs of synchronous species (nocturnal × nocturnal, diurnal × diurnal) tend to overlap more in diet than nonsynchronous pairs (nocturnal × diurnal). To determine the generality of our results, we supplement these data with examples from the literature on other diurnal/nocturnal predators (raptors, water snakes). Our analyses and those of Jaksić, Greene, and Yañez (1981) and Jaksić (1982) demonstrate that nonoverlapping activity periods are sometimes but not invariably associated with lower dietary overlaps. In fact, a few nonsynchronous pairs have nearly identical diets. The evolution of more subtle differences in activity periods (within a day or within a night) among closely related species is less likely to be related to reduced exploitation competition than to reductions in interference competition, risk of predation, or other factors.

Our analysis is based primarily on stomach contents of 4,214 lizards (mean ± standard error = 234 ± 45 stomachs/species, range = 19 to 688) of 18 species (12 diurnal, 6 nocturnal) collected in 10 study areas in the Kalahari semidesert of southern Africa (see Pianka, 1971) and from 3,376 lizards ($\bar{x}$ = 80 ± 14 stomachs/species, range = 10 to 511) of 42 species (28 diurnal, 14 nocturnal) collected in the Great Victoria Desert of Western Australia (Pianka, 1969a). We examined 94,915 prey items ($\bar{x}$ = 5,273 ± 1,183 prey/species of lizards, range = 64 to 33,216) from stomachs of Kalahari lizards and 107,820 prey items ($\bar{x}$ = 2,567 ± 1,145 prey/species, range 38 to 60,013) from stomachs of Australian lizards, measured volumes of each prey item, and classified prey into 46 (Kalahari) or 20 (Australia) taxonomic categories (Pianka, Huey, and Lawlor, 1979). After we determined the proportional volumetric representation of each prey category in the diets of all species, we computed dietary overlaps between all species pairs within deserts using a symmetrical formula (Pianka, 1973) that generates values between zero (no overlap) and one (complete overlap). Because certain broad prey categories (beetles, for example) undoubtedly include both nocturnal and diurnal species, we emphasize that these calculated values overestimate true dietary overlap. This bias is less severe in the Kalahari because of the larger number of prey categories and because termites, which constitute nearly half (41.3 percent by volume) of the total diet of all Kalahari lizards (Pianka, 1973), were identified to species and caste.

To reduce problems associated with species having small samples or unusual microhabitat associations, we do not consider data on 4 other Kalahari species and 19 other Australian species that were represented by fewer than 10 specimens, were fossorial, or had ambiguous activity periods. For the remaining species dietary-niche breadths computed with the diversity index of Simpson (1949) were not significantly correlated with either the number of stomachs per species of lizard (Kalahari $r_s = 0.10$, $P > 0.1$; Australia $r_s = 0.28$, $P > 0.1$) or with the number of prey per species of lizard (Kalahari $r_s = -0.18$, $P > 0.1$; Australia $r_s = 0.25$, $P > 0.1$), suggesting that sample sizes are adequate to characterize the diets of species. Moreover, proportions of species that are ground-dwelling versus arboreal among diurnal and nocturnal lizards do not differ significantly in either desert, suggesting that we have not inadvertently incorporated a habitat effect on dietary overlap.

The basic data contain an inherent statistical bias due to the transitive nature of overlap values (that is, if species A and species B have very high overlap, and if species B and species C also overlap substantially, then species A and species C will probably also have high overlap). The resulting lack of complete independence among overlap values violates assumptions of traditional statistical methods (Meagher and Burdick, 1980; Pimm, appendix to this chapter).

A second potential statistical problem arises when one uses a nearest-neighbor analysis (see below) to determine whether pairs of species that are synchronously active have higher dietary overlap than do nonsynchronous pairs of species. The relative number of synchronous and nonsynchronous pairs is equal only if the numbers of diurnal and nocturnal species are equal; if not, whichever group has the larger number of pairs will tend to have higher overlaps among nearest neighbors because of augmented sample size (R. K. Colwell, personal communication; Pimm, appendix). In our Kalahari and Australian samples synchronous pairs outnumber nonsynchronous pairs. Consequently, the bias favors the hypothesis that nearest neighbors that are synchronous should have higher overlap than do nearest neighbors that are not synchronous in activity.

To circumvent these statistical roadblocks, we exploit Monte Carlo computer simulation techniques (Pimm, appendix) and generate frequency distributions of similarity values obtained from randomly assigning the states "nocturnality" or "diurnality" to each of the observed species in the Kalahari and Australian samples. The resulting distributions allow direct computation of probabilities: thus, if the observed, actual, overlap value (or values more extreme) is encountered only twice in 200 such randomizations, its one-tailed probability of being different is 0.01, if it is encountered less than ten times in 200 runs, then $P < 0.05$, and so on (see Pimm, below).

Desert Lizards

In terms of relative dietary overlap, each species of lizard has a first, second, third, . . . , *n*th nearest neighbor that is synchronous for activity time (nocturnal × nocturnal pairs, diurnal × diurnal pairs) and a second set of neighbors that is not synchronous for activity time (diurnal × nocturnal pairs) (Inger and Colwell, 1977). Because our interest is in the possible effect of time of activity on diet of potentially competing species, we focus on overlaps among closest neighbors. Alternatively, one could compare average or median overlaps among all synchronous versus all nonsynchronous pairs, but this approach, which includes overlap values from many pairs with distinctive diets, obscures potential patterns of dietary differences.

We first compute the percentage of species whose nearest neighbor in diet is synchronous for activity time (Table 13.1). In both deserts significantly more first and second nearest neighbors are active synchronously than is expected using a random null hypothesis (Monte Carlo simulations), but the percentages of both third and fourth nearest neighbors that are synchronous do not differ from the null hypothesis. (We thank J. Felsenstein for helping us with these simulations.) Thus first and second nearest neighbors in diet are usually active at the same general times, but third and fourth nearest neighbors are as likely to be active at different times.

We next examine the magnitude of differences in dietary overlap by computing mean overlap values among all first, second, third, and fourth nearest neighbors for synchronous and nonsynchronous pairs (Table 13.2) and compare these observed averages with expected values that are generated from the Monte Carlo simulations (Fig. 13.1; see Pimm, below). In both deserts nearest neighbors with synchronous activity generally overlap significantly more than would be expected from a random null hypothesis. Moreover, this trend is highly significant (both

Table 13.1 Percentages of nearest neighbors (first through fourth) that are synchronously active in 3 independent communities.

	Nearest neighbor			
Population (diurnal, nocturnal)	First	Second	Third	Fourth
Kalahari lizards (12,6)				
Observed synchronous pairs	88.9	83.3	61.1	61.1
Predicted synchronous pairs	52.9	52.9	52.9	52.9
P	< 0.01	< 0.01	> 0.30	> 0.30
Australian lizards (28,14)				
Observed synchronous pairs	83.3	81.0	57.1	57.1
Predicted synchronous pairs	54.5	54.5	54.5	54.5
P	< 0.001	< 0.001	> 0.40	> 0.40
Michigan raptors (5,4)				
Observed synchronous pairs	66.7	55.6	55.6	55.6
Predicted synchronous pairs	44.4	44.4	44.4	44.4
P	> 0.25	> 0.35	> 0.35	> 0.35

NOTE: Observed values are derived from dietary comparisons. Probability levels (P) are derived from the results of 1,000 simulations (Monte Carlo) assuming a null hypothesis that the similarity of diet is independent of time of activity. The predicted percentages of synchronous pairs under a null hypothesis are calculated from a formula: $[D(D-1) + N(N-1)]/[(N+D)(N+D-1)]$, which is the probability that 2 species chosen at random without replacement will have the same activity time (J. Felsenstein, personal communication).

$P < 0.005$) when first through fourth nearest neighbors are examined together using Fisher combined probability tests. In these comparisons, dietary overlaps tend to be lower than expected among pairs with nonsynchronous activity (Table 13.2, Fig. 13.1), and overall trends are also significant nevertheless (Fisher combined probability tests: $P < 0.005$ in Kalahari, $P < 0.05$ in Australia). Clearly, time of activity significantly affects average dietary overlap among nearest neighbors of these lizards.

Both the above methods emphasize that time of activity influences dietary overlap. Nevertheless, the magnitude of this effect tends to decrease as nearness rank increases (Tables 13.1 and 13.2, Fig. 13.1). Average overlap values among first nearest neighbors that are nonsynchronous in activity times are comparable or even higher than the average overlap values for third and fourth nearest neighbors that are active synchronously (Fig. 13.1, Table 13.2). In other words, whereas the first and second most similar species are normally active at the same times, the third and fourth nearest neighbors are frequently active at different times (Table 13.1). The effect of time of activity on dietary overlap is thus conspicuous only among very close neighbors (Fig. 13.1).

Table 13.2 Average (± S.E.) dietary overlaps among first, second, third, fourth nearest neighbors that are synchronously or nonsynchronously active.

Population (number of species)	Average dietary overlap			
	First	Second	Third	Fourth
Kalahari lizards (18)				
Synchronous	0.84 ± 0.04	0.69 ± 0.04	0.58 ± 0.04	0.53 ± 0.4
Significance	< 0.001	0.010	< 0.001	< 0.001
Nonsynchronous	0.60 ± 0.05	0.54 ± 0.05	0.49 ± 0.05	0.41 ± 0.04
Significance	< 0.001	0.042	0.208	0.286
Australian lizards (42)				
Synchronous	0.87 ± 0.02	0.84 ± 0.02	0.79 ± 0.02	0.68 ± 0.03
Significance	0.070	< 0.001	0.005	0.231
Nonsynchronous	0.80 ± 0.03	0.71 ± 0.03	0.67 ± 0.04	0.64 ± 0.04
Significance	0.065	0.055	0.141	0.578
Michigan raptors (9)				
Synchronous	0.99 ± 0.01	0.88 ± 0.05	0.86 ± 0.05	
Nonsynchronous	0.95 ± 0.03	0.93 ± 0.04	0.78 ± 0.04	

NOTE: Significance levels for synchronous pairs indicate whether overlap is greater than expected on a random null hypothesis, whereas those for nonsynchronous pairs indicate whether observed overlap is less than expected.

Water Snakes and Raptors

Although our primary results concern the above analysis of our original Kalahari and Australian lizard data, we made a cursory search of the literature to determine whether the observed pattern has general application. Certain water snakes (*Nerodia* = *Natrix*) and raptors (hawks and owls), which differ in activity times and have well-studied diets, are suitable subjects from very different ecosystems.

Mushinsky and Hebrard (1977a,b) studied activity periods and diets of 4 species of *Nerodia* in Louisiana. Most important, they identified prey at least to genus, thereby minimizing the problem of overestimating dietary overlap. Mushinsky and Hebrard (1977a) observed some differences in diet between nocturnal and diurnal species (for example, the nocturnal *Nerodia rhombifera* ate more frogs and catfish, which are nocturnal, than comparable diurnal species). Although numbers of species are too few for meaningful statistical analysis (Pimm, below), average overlap between nonsynchronous pairs ($\bar{x}$ = 0.85, N = 4, range = 0.78–0.96) is actually slightly higher than that between synchronous pairs ($\bar{x}$ = 0.79, N = 2, range = 0.74–0.83). In particular, the nocturnal *N. rhombifera* and the diurnal *N. cyclopion* have nearly identical diets (0.96). Differences in times of activity certainly do not guarantee low dietary overlap.

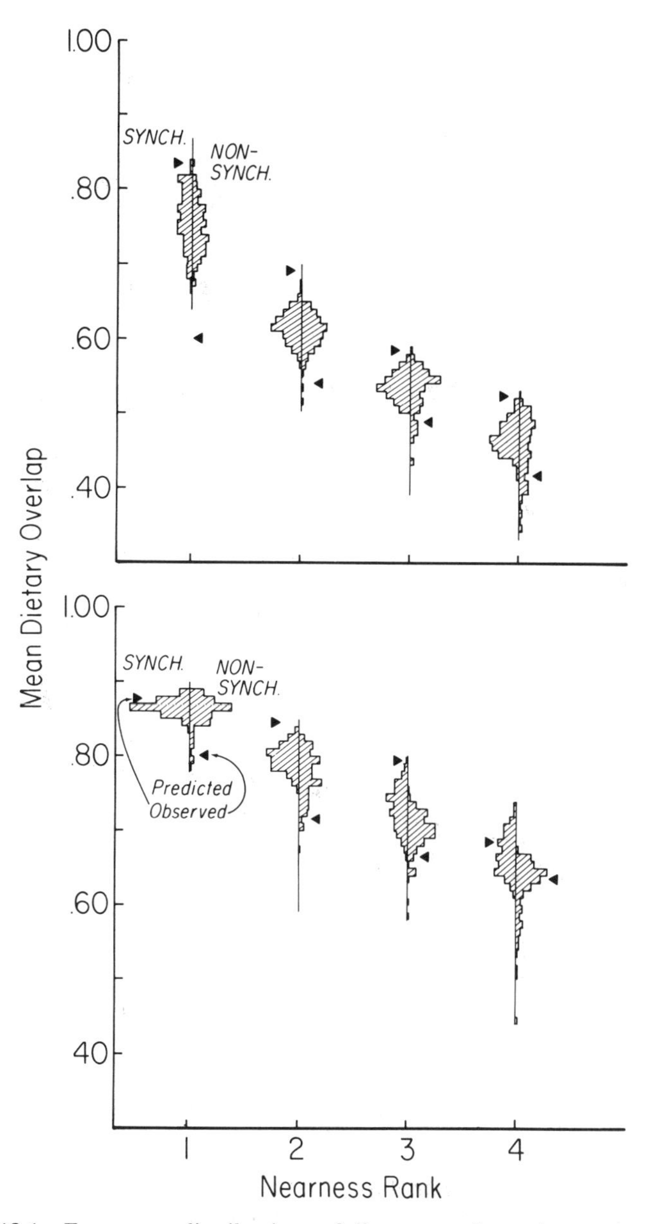

Figure 13.1 Frequency distributions of dietary overlap values generated by Monte Carlo simulations for the first 4 nearness ranks in niche space for lizards in the Kalahari semidesert (*upper panel*) and for lizards from the Great Victoria Desert in Western Australia (*lower panel*). Monte Carlo histograms are cross-hatched: those on the left side of the vertical line represent synchronous pairs (N = 200), whereas those to the right represent nonsynchronous pairs (N = 200). Observed average overlap values are depicted by solid triangles outside the histograms.

Diurnal and nocturnal raptors are an especially rich source of comparative information (see Jaksić, 1982). Samples are usually large, and prey are often identified as to species. For example, Craighead and Craighead (1956) presented extensive data on diets of nocturnal owls and diurnal hawks in Superior Township, Michigan. We used these data to calculate dietary overlaps among all species pairs during the fall and winter. Percentages of first through fourth nearest neighbors that were synchronously active were not significantly higher than expected on a random null hypothesis (Table 13.1). Moreover, this lack of an effect of time on diet holds even if the probabilities are combined by a Fisher exact text ($P > 0.10$). As is the case for the water snakes, time of activity thus seems to exert little effect on dietary overlap among raptors during these seasons.

Korschgen and Stuart (1972) presented 20 years of data on diets of red-tailed hawks, great horned owls, and barred owls from Missouri. Year-to-year changes in the proportional utilization of prey species are generally parallel among these raptors. Between-year Spearman rank correlations in percentage (by volume) of *Sigmodon hispidus,* of rabbits (primarily *Sylvilagus*), and of *Microtus* in the diets are 0.77, 0.56, and 0.42, respectively (all $P < 0.05$), for red-tails versus great horned owls; 0.82, 0.50, and 0.58, respectively (all $P < 0.05$), for red-tails versus barred owls; and 0.88, 0.48, and 0.67, respectively (all $P < 0.05$) for great horned owls versus barred owls. Thus yearly shifts in diet are similar between raptors with different activity periods (Korschgen and Stuart, 1972).

Time of Activity and Dietary Overlap

Our analysis of diets of nocturnal versus diurnal predators demonstrates that differences in time of activity are sometimes associated with significantly lower dietary overlaps, but only among species with very similar diets. Even so, the magnitude of this effect is not great: moreover, nocturnal × diurnal pairs sometimes have nearly identical diets. Therefore, degree of synchrony in activity periods is evidently an unreliable index of dietary overlap.

These results, which refute expectations of a strong relationship between activity time and diet, are surprising. Undoubtedly, the crudeness of many prey taxa categories obscures some real dietary differences. Nevertheless, even with the data on raptors and water snakes, where prey were identified as to genus or species, time of activity has no dramatic effect on diet. If these patterns are real, we need to question our initial assumptions rather than the data base.

The first assumption is that prey are either nocturnal or diurnal. However, some prey species are probably crepuscular (Lewis and Taylor, 1964) and might be eaten by both diurnal and nocturnal predators (Jaksić, Greene, and Yañez, 1981). Some predators, including certain owls, are also crepuscular. Moreover, other prey (and some predators) may switch from diurnality to nocturnality on a daily (Wilson and Clark 1977) or a seasonal basis (Jaksić, Greene, and Yañez, 1981) and thus be eaten by both types of predators.

A second implicit assumption, that prey are vulnerable to predation only while active, is certainly not universally true. In particular, few inactive arthropods enjoy inviolate sanctuaries. In the Kalahari, for example, nocturnal scorpions spend daylight hours in underground burrows, but are nevertheless the dominant prey of *Nucras tessellata,* a strictly diurnal lizard which searches for inactive scorpions in burrows (Pianka, Huey, and Lawlor, 1979). An Australian lizard species, the legless and snake-like *Pygopus nigriceps,* is also a scorpion specialist, but in this case the predator is nocturnal and captures its prey above ground during the latter's period of activity at night. Interestingly, no North American desert lizard species is a scorpion specialist, even though these large arachnids are abundant. Perhaps the small snake *Chionactis occipitalis* has usurped this ecological role (Norris and Kavanau, 1966).

Similarly, predators such as the diurnal lizard *Cnemidophorus tigris* that dig or search for inactive prey (termites) can have high dietary overlap with other predators like the sympatric nocturnal gecko *Coleonyx variegatus* which eats the same prey but captures them when they are active (dietary overlap between this species pair is 0.897 compared with an overall community-wide average overlap in diet of only 0.43). In short, diurnal predators can often find nocturnal prey in their diurnal retreats, whereas nocturnal predators may frequently be able to do the equivalent with diurnally active prey. For these reasons, time of activity may be of limited significance in reducing dietary overlap; attempts to analyze predator-prey interactions from measured or predicted overlaps in activity times (for example, Porter et al., 1973) are potentially risky.

In an analysis of vertebrate predators in Chile (hawks, falcons, owls, foxes, and snakes), Jaksić, Greene, and Yañez (1981) found only subtle differences in diet between diurnal and nocturnal predators. They argue that temporal separation of activity is an inefficient mechanism for reducing dietary overlap and propose that predatory guilds should be recognized "solely on food-niche overlap patterns, because activity times and habitat selection of both predators and prey are thereby implicitly included."

Evolution of Thermal Preferences

The times that a lizard can be active depend primarily on interactions among the physical environment, thermoregulatory repertoire, morphology, physiology, and thermal preference (Porter et al., 1973). Typically, lizards with high thermal preferences are active during warm periods, both on daily and seasonal bases (Huey, Pianka, and Hoffman, 1977; Pianka and Huey, 1978). Consequently, thermal preferences represent one potential mechanism by which temporal overlap can be adjusted.

Nevertheless, for several reasons we doubt that the evolution of thermal preferences is often related to exploitation competition for food. Thermal preferences seem to evolve very slowly in most ectotherms (Bogert, 1949; Huey and Slatkin, 1976; but see Hirshfield, Feldmeth, and Soltz, 1980; Huey, 1982): thus a particular competitive interaction would probably have to persist for a long period of time to influence thermal preferences. Also, in thermally heterogeneous habitats, even species with nonoverlapping thermal preferences can still overlap extensively in times of activity (see Huey, 1982).

These arguments suggest that the evolution of thermal preferences in lizards should be weakly related to exploitation competition for food. However, a divergence in thermal preference could be influenced by interference competition (Case and Gilpin, 1974), by divergence in habitat associations, or even by risks of predation. Indeed, perhaps reductions in temporal overlap with predators were the major reason for evolution of activity at low body temperatures among nocturnal lizards such as geckos and some skinks, as well as at very high body temperatures (for example, *Dipsosaurus dorsalis, Nucras tessellata,* and *Ctenotus leae*).

Our results and those of Jaksić, Greene, and Yañez (1981) and Jaksić (1982) demonstrate that major differences in time of activity, such as those between diurnal and nocturnal predators, do not invariably result in low dietary overlap. These findings challenge the widely accepted assumption that temporal separation of activity is invariably effective in lowering dietary overlap and invite caution in using time as a niche dimension. Minor differences in activity periods (within a day), which occur commonly in lizards (Pianka, 1971; Schoener, 1977), would appear to be even less effective in reducing dietary overlap. We encourage more empirical studies on this general problem as well as additional theoretical analyses (Case and Gilpin, 1974) on time of activity as a mechanism of coexistence of species. Both would be timely.

Appendix: Monte Carlo Analyses in Ecology

Stuart L. Pimm

The central feature of a statistical test is conceptually simple. We measure a quantity and then determine the probability of the appearance of a value as extreme as or more extreme than the one measured. This probability is stated in terms of a mathematically formulated hypothesis (the null hypothesis). If this probability is sufficiently small (usually < 0.05), then we reject the null hypothesis and the biological hypothesis that it implies. We usually obtain the probability from a well-known model of how the values of the quantity should be distributed and more directly, from a set of tables (F, t, χ^2, etc). Although the majority of distributions we encounter fit or can be made to fit one of a very limited number of simple distributions, not all do. In these exceptional cases we must resort to other techniques to find distributions and, from these, the required probabilities. Sometimes these special distributions can be derived analytically. More often, this is difficult or impossible due to the special constraints biology places on the distributions. But we can often approximate the distributions using numerical methods on a computer. Such is the case with the problem posed above by Huey and Pianka's study of the influence time of activity has on dietary overlap.

Their data involve the similarity in diet between lizard species that are either nocturnal or diurnal. To simplify discussion of the problems the data pose, I shall consider the dissimilarity between species' diets. These values can be obtained simply by subtracting the similarity values from unity. And the dissimilarity values can be represented graphically by the distances between points (which represent species). Though the dimensionality of the data is large—it requires up to $n - 1$ dimensions to represent all the distances between n species—a two-dimensional representation (Fig. 13.2) of the problem is adequate for my purposes.

The figure shows a hypothetical array of 3 nocturnal and 3 diurnal species. We need to ask: how should the distances between these 6 points be distributed? The answer is "not simply" for several reasons:

Suppose the nearest diurnal species to diurnal species D_1 is D_2; the distance between them is x. And further suppose that the nearest nocturnal species to D_1 is N_1, the distance between them being y. Now, a requirement of all simple statistical tests is that each quantity involved (say the individual observations that contribute to a mean) be independent of all the other quantities in the test. This requirement is clearly violated in our example. The distance between D_2 and N_l—call it z—is limited by simple geometry thus: $z < x + y$, and $z > y - x$. These and similar conditions mean that distances between species that are active at similar times of day (either both nocturnal or both diurnal) and those distances between species active at different times of day will be interrelated. (For convenience, I shall call the first set of distances nocturnal-nocturnal or diurnal-diurnal distances, matched, and the other distances, nocturnal-diurnal and vice versa, mixed.)

Other constraints exist on the distances too. Because the distance between two species cannot be greater than unity, even if D_1 is maximally distant from D_2, D_2 from D_3, and so on, D_n still cannot be further away from D_1 than unity—the points lie in a hypersphere with unit diameter.

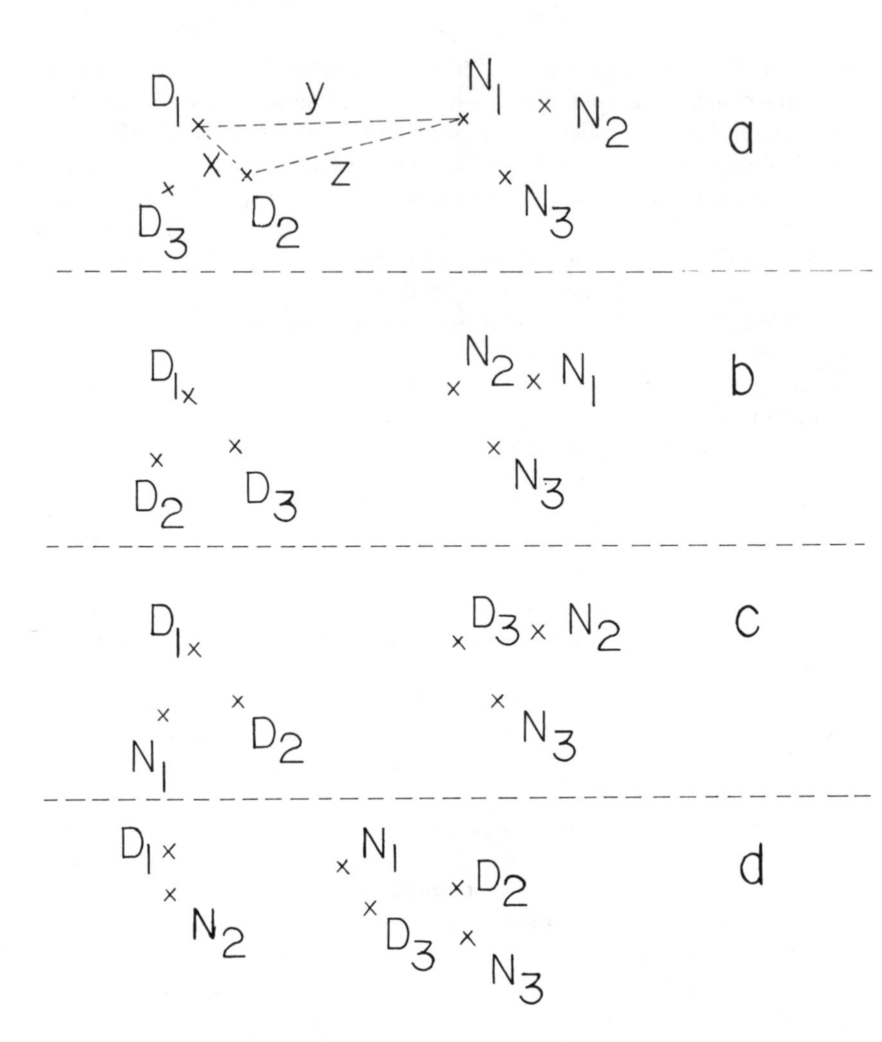

Figure 13.2 Four examples to illustrate the methods used to calculate statistics on dietary similarities. There are 6 species: 3 diurnal (D_1,D_2,D_3) and 3 nocturnal (N_1,N_2,N_3). The arrangement of their positions reflects their dietary similarities such that the dietary dissimilarities are represented by distances. Thus, the dietary dissimilarity between D_1 and D_2 is x; alternatively the dietary similarity between D_1 and D_2 would be $1 - x$. (*a*) The hypothetical observed data. (*b*) A randomization of names that preserves the same arrangement of diurnal and nocturnal species. (*c*) A randomization that does not preserve the same arrangement of diurnal and nocturnal species. (*d*) A different hypothetical observed arrangement. In (*a*) diurnal species are always most similar to diurnal species and nocturnal species to nocturnal species. No other arrangement of diurnal and nocturnal labels—for example, (*c*)—has this feature. In (*d*) diurnal species are always most similar to noctural species and vice versa. No other arrangement of labels for these data would have this feature.

Another reason to use numerical techniques with this particular problem is a common one in many ecological applications: the statistics are strongly sample-size dependent. In this case, the nearest species distances depend on the proportions of species that are diurnal and nocturnal. Because the distances between species must all fall within a hypersphere of unit diameter there will be a tendency for species to be, on average, closer to other species the more species there are involved in the comparison. Suppose there are n nocturnal and m diurnal species. There are $\frac{1}{2}[n(n-1) + m(m-1)]$ matched comparisons and mn mixed comparisons. Unless the numbers of comparisons in each set (mixed, matched) are equal, we should expect, by chance alone, for the smallest distance to be usually in the set that involves the greater number of comparisons.

In sum, the distances between species and hence the similarities between species are interrelated and have properties that prevent them from being the normally and independently distributed variables required for most statistical tests. But this need not prevent our developing and testing hypotheses using them.

Huey and Pianka are interested in such questions as: are diets of species that are matched for time of activity more or less similar than those that are mixed? The question "more or less similar" can be addressed in terms of a null hypothesis. Consider this null hypothesis: suppose each species takes its array of prey species without regard to whether it is diurnal or nocturnal. Then this hypothesis would imply, for the data in Figure 13.2, that the 6 points (whose relative positions reflect diets) would have arbitrary designations (D_1, D_2, . . . , N_3). In the example, there are 6 species and thus 6! (equals 720) ways of arranging the names to the points. Some of these, however, involve the same arrangement of diurnal and nocturnal species—Figure 13.2b is an example. Others represent distinct arrangements (Fig. 13.2c). There are 36 ways of arranging the names that maintain each particular configuration of nocturnal and diurnal species. So, if we were interested in the distribution of mixed and matched distances there would be 20 different configurations (720/36). Each one of these 20 possibilities would occur equally probably under the null hypothesis that time of activity did not influence diet.

Finally, we ask the crucial question: how unusual is the observed arrangement of diets shown in Figure 13.2a? It is clear from the figure that each diurnal species is always more similar in diet to a diurnal species than to any nocturnal species and each nocturnal species is always more similar to a nocturnal species than to a diurnal species. In short, matched distances are smaller than mixed distances in each case. Only 2 of the 20 possible configurations are this extreme: Figure 13.2a represents one possibility; the patterns produced by interchanging each diurnal and nocturnal species would be the other. In all the other 18 sets of arrangements, one or more mixed distances would be smaller than a matched distance. Thus, the chance of finding an arrangement this extreme is 0.1 (2 out of 20).

Finally, consider the arrangement of Figure 13.2d. Here each mixed distance is always smaller than a matched distance. In biological terms, each diurnal species has a nocturnal replacement with very similar diet. This also is an extreme arrangement and has an identical probability of occurrence of 0.1.

Depending on our prior biological knowledge we may wish to formulate 1 of 3 pairs of hypotheses. The first pair involves a null hypothesis and an alternative that supposes that matched distances will be less than mixed distances. In the second pair the alternative supposes that matched distances will be greater than mixed differences, and the alternative for the third pair supposes only that matched and mixed differences will differ. The structure of these hypotheses should be familiar. The first two pairs are considered one-sided: of the two possible kinds of extreme arrangements only one kind will reject the null hypothesis in each case. The third pair of hypotheses is two-sided: either of the two extreme arrangements would reject the null hypotheses. For the first pair, we ask how likely is it that the arrangement we observe (Fig. 13.2a) or some more extreme arrangement will occur, given the null hypothesis. In this case there are no more extreme arrangements (matched distances are less than mixed in each case). The chance of obtaining this arrangement is 0.1, as I have already discussed. If the data were as in Figure 13.2d and we were testing the second pair of hypotheses, then the chances of this arrangement would be 0.1 yet again. Of course, if we were testing the third pair of hypotheses (with either the data of Fig. 13.2a or 13.2d), then either extreme arrangement would satisfy us and the chance of getting one of them would be 0.2 (4 out of 20).

Because we usually choose to reject hypotheses if their probabilities fall below 0.05, we would not reject the null hypothesis in any of these cases even though the data may be the most extreme arrangement possible. Simply, for the data shown in the figure, rejecting the null hypothesis is not possible. The probability of accepting the null hypothesis, when the truth is that the alternative hypothesis is correct, is assigned a value, β; we call $1 - \beta$ the *power* of a statistical test and hope our tests will be powerful. In this example, the test is not and the lack of power is one problem inherent to analyses of this kind when data are few. I shall return to this problem later, but it does not cause any difficulties in the case of Huey and Pianka's data.

Analyses

Nothing in the analyses of Huey and Pianka's data is conceptually different from the example I have just discussed. Dietary similarities between each pair of species were calculated and for each species the species most similar in diet identified for species active at the same time of day (matched values) and for species active at a different time of day (mixed values). I then calculated the mean of the differences between matched and mixed values: I call this mean value $D_{1,\text{obs}}$ (the observed difference between the dietary similarities of species active at the same and different times of day averaged over all the $n + m$ species).

In the above example, the number of possible arrangements was only 20. Each arrangement could be explored. With the real data, the number of possible arrangements is often vast. Some authors, notably Schoener (1982), using their own minicomputers and having adequate time, have chosen in these circumstances to investigate all possibilities. For those of us with limited computer time, a satisfactory alternative is to select randomly a sufficient number of arrangements (200 in this case) from the total array of possibilities. For each random arrangement I used a program that assigned $m + n$ labels randomly to the $m + n$ species: n of the labels implied a species was nocturnal, m, diurnal. From this point the calculations were identical to those performed on the actual data and yielded means I call $D_{1,i}$ (i = 1 to 200). The final stage was to compare $D_{1,\text{obs}}$ with the statistical distribution of values of D_1 under the null hypothesis, the $D_{1,i}$. If, for a one-sided test, less than 10 (5 percent) of the simulated means, $D_{1,i}$, were greater than the observed mean $D_{1,\text{obs}}$, then one would conclude that dietary similarities were greater between species active at the same time of day than one would expect by chance.

The extension of these analyses to the other tests follows directly. I calculated the dietary similarities not only of the nearest but also of the second, third, and fourth nearest species both for matched (call these S_1, S_2, S_3, and S_4) and mixed (call these M_1, M_2, M_3, M_4) for time of activity. From the mean of $S_1 - M_1$ I calculated D_1 as described above and also D_2 (the mean of $S_2 - M_2$; that is, the mean of the differences between the second most similar species active at the same time of day and the second most similar species active at a different time of day), D_3, and D_4 defined analogously. Also calculated were cross comparisons; for example $M_3{-}D_1$—the mean difference between the third nearest species active at a different time of day and the nearest species active at the same time of day. In each case, calculations were performed on the actual data and then repeated 200 times on randomized data to generate distributions of these means under the null hypotheses.

Several comments need to be made. First, the sample size of 200 is chosen as a compromise between accuracy and computer time. The values of interest are proportions and, therefore, are binomially distributed. The critical value is 0.05 and the standard error for such a proportion is $[(0.05)(0.95)/\text{N}]^{1/2}$, where N is the sample size. For N = 200, the standard error is 0.015. For small proportions confidence intervals are F-distributed and the upper-95-percent confidence interval approaches

3/N for large N. Simply, with a sample size of 200 we can be certain that the true level of significance will not be more than a few percentage points from our assertion.

Second, these methods can often lack power because the number of distinct arrangements under the null hypothesis is small. In such cases it still may be possible to obtain biologically useful results by repeating the analyses on many different sets of data. Proportions from each of these analyses can be combined to give an overall test of the null hypotheses. Examples of this are given in Pimm (1980).

Finally, we note that Monte Carlo methods, of which this paper describes a special case, are becoming popular in ecology. (Some examples are Connor and Simberloff, 1979; Pimm, 1980; Pimm and Lawton, 1980; Schoener, 1982). Yet it is my impression that they are underused. If this impression is correct, it may be because the problems that require one to resort to such methods—interdependence of data, sample-size-dependent biases, etc.—are common in ecological problems but are usually overlooked. This may reflect an antipathy toward statistics or a reluctance to compute. Such a reluctance would have been unfortunate in this case. Over 80 percent of the computer code required by this analysis involved the calculations of dietary similarities between species and the extraction from these data of the first through fourth mixed and matched values. These calculations were required for analysis of the observed data; performing the simulations required only the addition of a small subroutine which randomly assigned diurnal or nocturnal status and a loop to repeat the process. Such minor modification of data analysis routines to permit statistical inferences to be drawn is, in my experience, quite typical.

14 Sympatry and Size Similarity in *Cnemidophorus*

Ted J. Case

THE DISTRIBUTIONAL RANGES of any closely related group of animals often appear as a collage; it is not at all obvious what factors determine species' distributions and overlaps. Certainly, the distribution of any species is influenced by climatic barriers (Chapter 3). Yet if some species are able to flourish under particular stressful physical conditions, why have not more species evolved similar capacities? MacArthur (1972) believed that the ultimate explanation for species' range boundaries must lie in the distributions of other species—competitors, predators, mutualists, or parasites. Over evolutionary time species will coevolve, adjusting their ranges and niches in the process, as in a multiplayer chess game (Chapter 16). Does the collage of species' ranges that we see today represent some sort of multilateral ecological "checkmate"?

In this chapter I develop some methodology to analyze geographic patterns and apply this methodology to the distribution of *Cnemidophorus* lizards. In particular, I address the following questions: (1) Do dissimilar-sized lizards cohabit more commonly than chance would predict? (2) Indeed, what is an appropriate null model for species' size relations?

The theory and methodology I develop to answer these questions is general, and although I have *Cnemidophorus* lizards in mind when I think about these problems, my arguments are not constrained within these taxonomic bounds. *Cnemidophorus* lizards are ideal for such an analysis for a number of reasons. *Cnemidophorus* is a relatively species-rich genus, with a great deal of interspecific variability in body size. *Cnemidophorus* lizards display relatively little intra- and interspecific agonistics, so that predictions based on resource overlap and exploitative competition are not confounded by overt interference interactions. Finally, the genus is a distinct ecological guild, in the sense of Root (1967), characterized by diurnal, insectivorous lizards which are wide-ranging foragers, relying mostly on smell rather than sight to detect their prey. All other sympatric lizards are ecologically distinct.

Null Hypothesis for Community Structure

Before we can properly analyze geographic patterns of sympatry and size similarity in *Cnemidophorus* lizards, we must first find a nonbiased analysis to apply. In recent years a number of papers have attempted to provide null hypotheses or null models of community structure (Sale, 1974; Caswell, 1976; Simberloff, 1978; Connor and Simberloff, 1979; Poole and Rathcke, 1979; Pianka, Huey, and Lawlor, 1979; Lawlor, 1980). One recent attempt was that of Strong, Szyska, and Simberloff (1979) who attempted to determine if the body and bill sizes of sympatric species in insular bird communities were more different than chance might predict.

Their analysis of the Galapagos finches is particularly germane since these birds serve as one of our textbook examples of character displacement. Strong, Szyska, and Simberloff asked if the observed pattern of bill sizes in the Galapagos finches could be adequately explained without invoking character displacement. Specifically they asked if the observed bill sizes of sympatric finch populations are more dissimilar than the bill sizes of randomly formed communities of these same species. With the Galapagos finches, the absence of a recognizable mainland species pool hinders the formation of a null community to serve as a yardstick against which to measure any emergent patterns in community organization on the islands. The course adopted by Strong, Szyska, and Simberloff (1979) to get around this problem was to treat the entire archipelago as a universe from which individual island populations were selected at random. To populate each null island they randomly drew species and populations of each species from the real archipelago. From this analysis Strong, Szyska, and Simberloff concluded that the pattern of bill sizes among Galapagos finches can be explained solely by random assortment; character displacement linked to competitive interactions between species need not be invoked.

Suppose one suspects that a particular pattern in nature arises from a particular structuring force. A proper null model to test this effect should incorporate all other structuring forces but the force in question. The null model must then pass two tests. On the one hand, a pattern produced at random with respect to the process in question should, upon comparison with the null model, be judged as random. Failure to pass this test is indicative of Type I statistical error. That is, the null model is rejected when in fact it is correct. On the other hand, if a pattern contains within it the very structuring force that the null model lacks, then this pattern must, upon comparison with the null model, be recognized as being nonrandom. Failure to pass this test is indicative of Type II statistical error. The null model is accepted when in fact it is wrong. Grant and Abbott (1980) criticized the Strong study on a number of grounds;

one argument they raised is that of Type II statistical error. They complain that "the two samples tested for similarity (the real data is one sample and the set of randomly generated faunas is the other) are not independent. In fact the real data for any given island are a subset of the randomized data. The statistical bias that this kind of nonindependence causes is not random; it consistently favors the acceptance of the null hypothesis."

Case and Sidell (1982) asked whether the Strong, Szyska, and Simberloff procedure for forming null models could detect competition even if it exists; that is, can it pass the second test referred to in the previous paragraph. Case and Sidell formed model communities where the size similarity of populations was the only nonrandom factor determining sympatry or allopatry of species on islands. If the hypothetical size-similarity barrier was not a factor then chance colonization was the only determinant of community structure. In these cases, the Strong approach succeeds—random communities are recognized as exactly that. On the other hand, in other simulations the size-similarity barrier between species was made so severe that, on average, 40 percent of the communities formed by random colonization could not remain as such. In these cases, the application of the Strong procedure would fail to recognize this pattern and wrongly conclude that these communities were randomly formed with respect to body size. This occurs in spite of the fact that in these simulations, communities were concocted to be highly nonrandom.

In nature, size similarity certainly does not affect community structure quite as severely as in these counterexample models, nor is it the only force affecting community structure. My aim, however, is not to mimic nature in these simulations, but rather to pose this important question: suppose communities are structured in this strict and stringent manner; even then, could observed communities be statistically distinguished from null communities formed in the manner of Strong, Szyska, and Simberloff?

To understand how these problems can be avoided in an analysis of *Cnemidophorus,* we must understand how the bias arises. As we shall see, two distinct and complementary tests may be applied to questions of size-similarity patterns in natural communities. Neither of these tests (which I call size-assortment and size-adjustment tests) suffers from the same inherent Type II statistical bias that plagues the Strong approach. These tests are able to detect size structure in the same model communities where the Strong test fails, and at the same time recognize random communities as being random. I begin with a description of one of the Case and Sidell counterexample models. I then show how and why the Strong group's procedure fails to detect the built-in structure. Next I show how size assortment and size adjustment may confound one another to produce this bias. I illustrate all this here with one of Case and Sidell's (1982) 3-species examples. Finally, I adopt these new size-assortment and size-adjustment tests in my analysis of the size structure of *Cnemidophorus* lizard guilds.

A Model of Size Similarity

Consider 3 hypothetical species in a mainland source pool with the overlapping and rectangular body-size distribution plotted in the top of Figure 14.1. Suppose that all 3 species can colonize all the islands in some hypothetical archipelago. Next, imagine that some limiting size similarity s is set by ecological interactions. That is, if we rank the 3 populations on each island according to their body size, and take the absolute size difference between each contiguously ranked species pair, this difference must exceed the value s for both species to persist. (Strong and his coworkers actually confined most of their statistics to size ratios rather than size differences. This presents no serious complication since one may consider the size axis in my computer models as being on a log scale. Because $\log (X_1/X_2) = \log X_1 - \log X_2$, the results are identical. I prefer to work with size differences here because their variance is smaller than the corresponding variance for size ratios.) An initial island community of species A, B, C may only remain as such if both size differences between contiguous ranked species exceed s. Otherwise, the community will collapse to one of the 3 pairwise sets A,B or B,C or A,C or perhaps one of the 3 single-species sets. For most of my analyses, I discarded all single-species populations to mimic the analysis of Strong, Szyska, and Simberloff (1979), but inclusion of single-species sets does not qualitatively alter these results. This process of community-wide selection is equivalent to one in which 2 of the 3 species are chosen at random and then a colonizing individual is chosen from each species already chosen. If the size difference between the two exceeds the value s, a colonizing individual from the third missing species is chosen at random to invade. The invasion is successful only if each of the 2 resulting size differences between the 3 size-ranked species exceeds s. If not, the invasion fails and the community remains a 2-species set. Hence, the mechanism of community-wide selection may be viewed either as extinctions due to size similarity or unsuccessful invasions for the same reason; the result will be the same.

Case and Sidell (1982) explore various alterations of these assumptions by varying the number of species from 2 to 4; by changing the distribution of sizes from rectangular to gaussian and thereby, in effect, allowing multiple colonizations of each species; by allowing island heterogeneity with respect to maximum species number; and by not discarding single species islands. None of these variations seriously affects the qualitative conclusions discussed below.

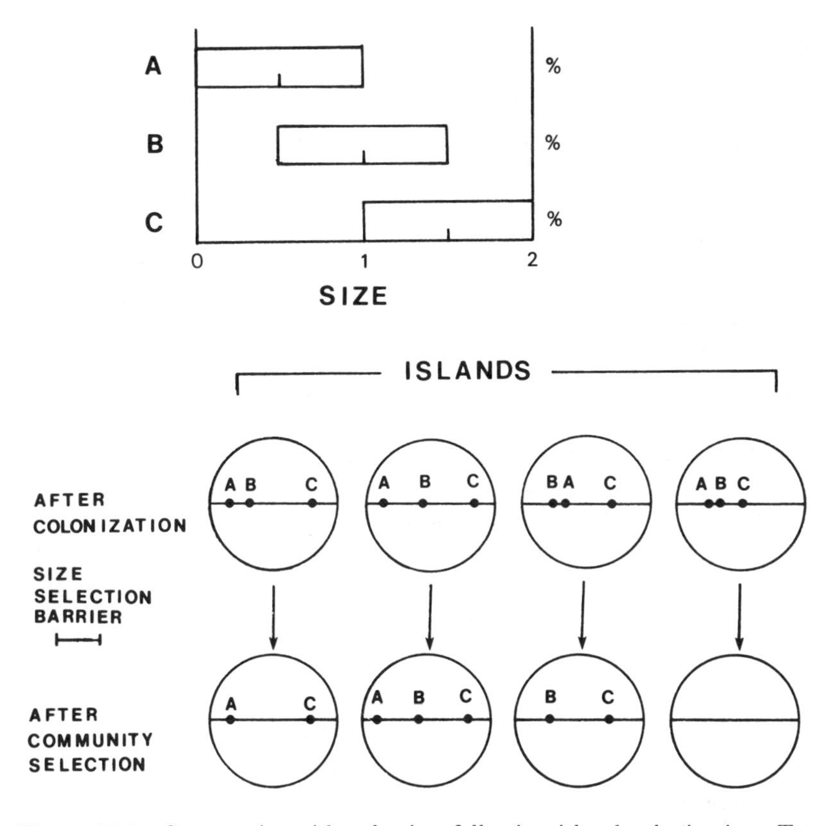

Figure 14.1 Community-wide selection following island colonization. *Top:* the uniform body-size distributions for the 3 species A, B, and C in a mainland source pool. *Bottom:* 4 hypothetical islands shown as circles, after colonization and after community-wide selection. The position of the point along the diameter of the circle indicates the size of the particular colonist individual drawn at random for each species (the mean body size of the ensuing population on the island is assumed to equal the size of the colonist from that species). If the absolute size difference between any 2 contiguous species is less than the size similarity barrier (*s;* shown by the length of the bar on the left), then the interior species of the pair goes extinct on that island. If the size difference between the 2 remaining species also does not exceed *s,* then they both go extinct as on the fourth island. This latter rule seems unrealistic, but there is no nonarbitrary way of deciding which of the 2 species should remain. Moreover, Strong, Szyska, and Simberloff (1979) excluded single-species islands in their analysis.

Recall that the size distributions of the 3 species in these models are staggered but overlapping (Fig. 14.1). Thus even though species A has a larger average size than species B, the particular B present on some islands may be larger than the particular A, purely by random colonization (Fig. 14.1, third island). If *s* is zero (that is, there is no limiting similarity barrier) all islands will contain A, B, C, and each species averaged over all islands will have the size distribution that it has in the colonizing pool shown in Figure 14.1. However, as *s* increases, the size distributions of species after selection may become skewed. The skew occurs even in 2-species models and is illustrated in Case and Sidell (1982). Moreover, a smaller proportion of islands contains species triplets; pairs or single-species islands predominate. Because of the symmetry (Fig. 14.1), communities A,B and B,C will be represented equally and will occur less frequently than the pair A,C. Based on 1,200 computer simulations for each value of *s,* I obtained the community representations shown in Figure 14.2. As expected the frequency of A,C sets increases and the frequency of A,B,C sets decreases as *s* increases.

For moderate to large *s,* the observed size difference between A and B on A,B islands is greater than that found by sampling the colonizing pools of A and B; compare the values marked by the open circles in Figure 14.3a to the average size difference of 0.5 between these species in the colonizing pool (Fig. 14.1). Similarly, if we compare these A,B islands to a random sample of all populations of A and B throughout the archipelago (closed circles in Fig. 14.3a), we again find support for community-wide character displacement, at least for this species set. This same argument applies to B,C islands; on average the B's will be smaller and the C's larger than in the colonizing source pools of these species.

On A,C islands (and these are the most frequent category when *s* is large), the average observed size difference between A and C (Fig. 14.3b, open circles) is usually smaller than it is between these same species in the source pool (1.0; Fig. 14.1) and it is always smaller than the null community of this same species set (closed circles; Fig. 14.3b). This is because, on average, those communities that had a larger size separation

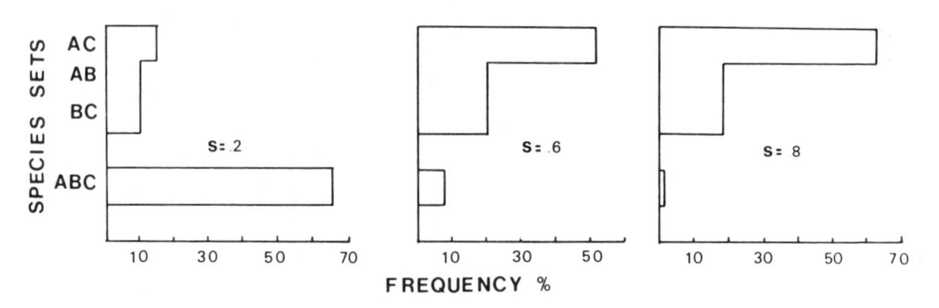

Figure 14.2 Frequency of various species sets in a simulated archipelago after community-wide selection. Colonists are drawn from the size distributions shown in Figure 14.1. The magnitude of the size-similarity barrier (*s*) is 0.2, 0.6, or 0.8. As the similarity barrier increases, fewer 3-species sets are represented in the archipelago and among the 2-species sets, the particular set AC predominates. (Results based on 1,200 computer-generated islands.)

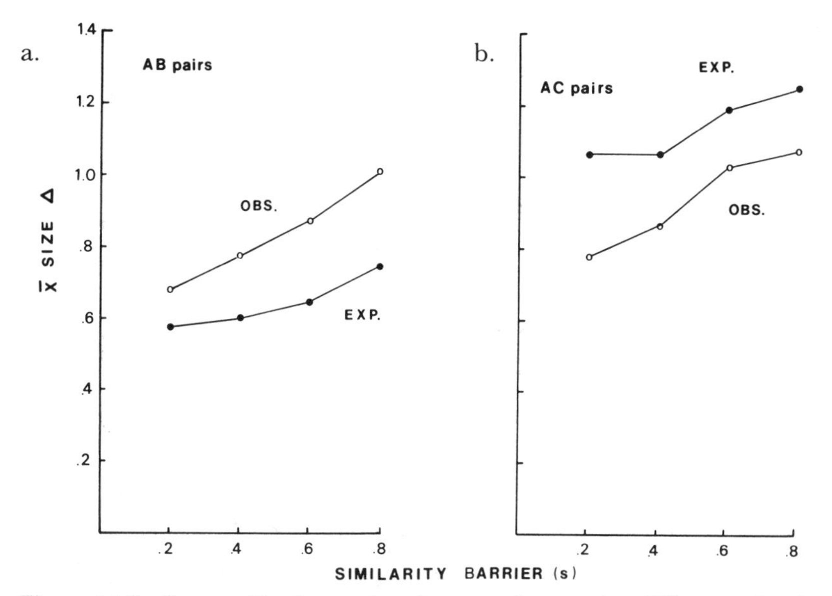

Figure 14.3 Set-specific observed and expected mean size differences for the computer simulations of Figures 14.1 and 14.2. For islands inhabited only by species A and B (*left*) the mean observed size difference after community-wide selection exceeds the mean expected value for a null community composed of these same species; moreover, the discrepancy increases as *s*, the similarity barrier, increases. For the species set A,C observed values are greater than expected (*right*). (Results are based on about 300 islands of each set.)

between A and C were also able to support species B in the middle of the size axis and therefore did not persist as A,C communities. The result is that A,C communities do not display community-wide character displacement in the sense employed in Strong; in fact, they display size convergence. This occurs in spite of the fact that our model has been strictly contrived so that species coexistence is based on size dissimilarity.

The final step in this procedure is to pool the observed computer communities (those left on islands after community-wide selection) and redraw null communities from this pool for various species numbers. The exact details of Strong and coworkers' randomization protocol, however, are somewhat ambiguous. On page 900 they explain that "random draws are taken by first choosing a species at random, then a population at random for each species already chosen, until the appropriate number of species populations is obtained (2 species populations for islands with 2 species, 3 for islands with 3 species, etc)." Here and elsewhere they do not specify whether species were chosen at random from a uniform distribution or from the archipelago at large, nor do they state whether or not they allow multiple draws of the *same* species for a given island; although in fact they do not. Thus 2 interpretations of their procedure are possible:

(A) Species are chosen at random from a uniform distribution until the appropriate number of species population is obtained. For each selected species, a population of that species is selected with equal probability.

(B) Species are chosen at random from a distribution weighted by each species' frequency of occurrence in the archipelago until the appropriate number of species is obtained. Populations of each species already chosen are selected with equal probability. This procedure is equivalent to simply choosing populations at random from the archipelago (without duplicating species) until the appropriate number of species populations is obtained.

Notice that with either protocol both species and populations are chosen at random. Thus, community-wide character displacement as measured by this test might result from 2 very different phenomena: (1) species may simply assort nonrandomly on islands with respect to their bill or body sizes, and (2) bill or body sizes of each species might evolve differently on different islands with different species composition.

Naively, either protocol A or B might be considered appropriate. Protocol B, however, produces a null archipelago that more closely resembles the actual archipelago. To the extent that actual species occurrences have already been altered by competitive interactions, Protocol B will incorporate more of this structure into its composition relative to Protocol A. This creates the risk of Type II statistical error.

Figure 14.4 plots the grand average size differences (and their standard deviations) for observed and redrawn (null) 2- and 3-species communities. Null communities were redrawn according to both protocols A and B. When $s = 0.2$, about 40 percent of the randomly formed 3-species islands cannot exist as such. In spite of this moderately strong selection, the observed and null 2-species communities produced by either protocol are nearly indistinguishable on the basis of their mean size differences. For 3-species sets (at $s = 0.2$) the inherent bias is almost as bad (Fig. 14.4; right-hand side).

As s increases, the intensity of community-wide selection increases, and the mean size difference between species in actual communities diverges more and more from those predicted by the null model. Note, however, that both types of null communities tend to track the actual communities and themselves diverge increasingly from the expected size difference for 2- and 3-species communities sampled directly from the mainland pool (dashed line). As expected, this deviation is more extreme under protocol B than protocol A. If one plots the actual and predicted minimum size differences (rather than the mean) versus s, the pattern

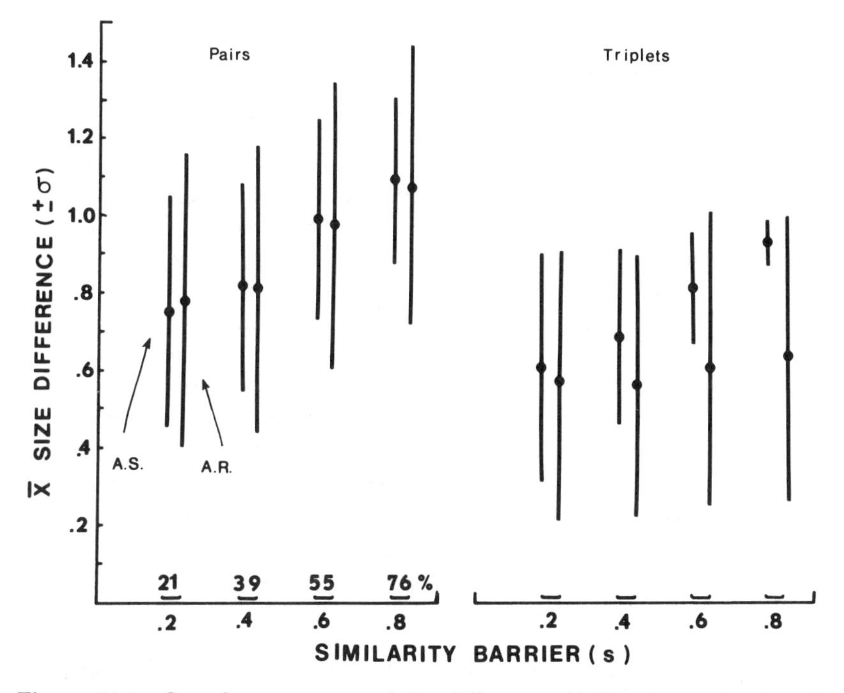

Figure 14.4 Grand mean expected size differences (A.R., after redraw) and observed size differences (A.S., after selection) for island pairs (*left*) and triplets (*right*) as a function of s. The mean is represented by a dot and the line shows one standard deviation. The values above the abcissa on the left are the average percentages of randomly drawn pairs of individuals which violate s. For pairs, the observed and expected size differences are indistinguishable on the basis of mean size differences; for triplets they can be distinguished given adequate sample size, particularly for large values of s. For both pairs and triplets the observed and null communities are more easily distinguished (that is, with smaller sample sizes) on the basis of their different variances rather than their different means.

produced is naturally identical for pairs, and roughly equivalent for triplets.

The observed 2- and 3-species communities are characterized by a sharply smaller variance in size differences when s is large. Overall, the more evenly spaced sizes of species within the observed communities compared to the null communities make an analysis of the variance of within-community size differences a more revealing statistical test than one simply comparing mean size differences.

When either protocol A or B is modified so that the size distributions of the 3 colonizing species are gaussian rather than uniform, these results are barely altered. Incorporating 4 or even 5 species into the source pool also does not erase this problem.

Size Assortment and Size Adjustment

In these computer simulations of communities formed under a limiting size-similarity barrier, several patterns emerge. Certain species combinations occur more frequently in the archipelago than others. The more frequently represented species sets are those which, on average, have greater mean size differences between species in the colonizing pool and smaller within-community variances of size differences. For example, with a 3-species pool, the species pair A,C predominates over the pairs A,B and B,C and this discrepancy in representation increases as s increases. Communities which obey this pattern will be called size assorted. Size assortment in communities arises from the greater persistence of some species combinations over others; in particular those sets which are competitively compatible and at the same time resistant to invasion.

Another pattern typifying these computer communities selected by a limiting similarity barrier is the tendency for the observed size differences for specific species combinations (for example, pair A,B) to deviate from the null community of these very same species. The direction of this deviation, whether it is positive or negative, and its magnitude, depends upon which species combination is considered. For example (see Fig. 14.3), for A,C communities (the species pair with the greatest average size difference of all 3 possible pairs) the observed size differences are actually smaller than the expected size differences. On the other hand, pairs that are more similar sized in the source pool like A,B and B,C will exhibit size differences larger than expected. Size adjustment is used to denote this pattern.

How the Bias Arises and How It Can Be Avoided

If community-wide selection operates according to a limiting similarity barrier, then those species that overlap in size with most other species in the colonizing pool will be less frequently represented over all islands (and species sets) than those species with less average overlap. For the 3-species example considered earlier, species B is intermediate in size to A and C and its representation in the archipelago at large decreases with increases in the threshold of size similarity s (Fig. 14.5).

This distortion in species representation is a major factor along with the skew induced in the size distributions of species after selection that can produce a Type II error, using Strong, Szyska, and Simberloff's methods for analyzing community-wide character displacement. These null communities are based on pooling all existing populations and redrawing populations from each species at random from this pool. Popu-

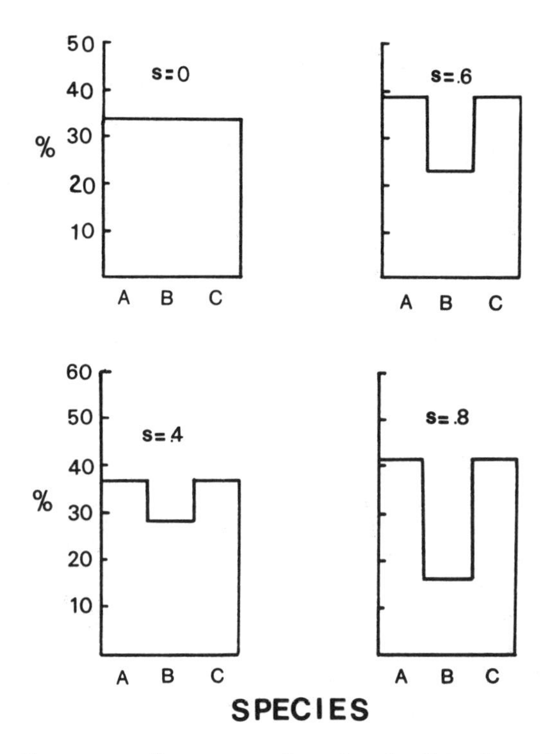

Figure 14.5 Frequency of representation of each of the 3 species within an archipelago as a function of the size-similarity barrier (s). As s increases, the frequency of the intermediate-sized species B decreases on islands after community-wide selection.

lations of different species are put into the pool in the frequency in which they exist. To the extent that competition has already influenced species representations, then the resulting null community contains within it a residue of the very competitive process whose presence it seeks to test.

In the Strong approach the questions of which species combinations are more common in the archipelago and why are not considered. As we have already seen for the 3-species example considered earlier (and this is equally true for 4 and more species), if community-wide selection did occur in the manner outlined here, then certain subsets of species that are dissimilar in size before selection actually will exhibit closer and more even size packing than the null community of that set after selection.

If the statistics for each particular species set are not pooled as in the Strong null model but rather compared separately to expectations, much of this bias can be avoided. Two separate questions arise. First, are size-dissimilar sets more common on islands than chance would predict (is there evidence of size assortment)? Second, for each specific set how do the observed and expected size differences compare? Specifically, for size-similar sets are the observed size differences between species significantly larger than the expected values for that set? For size-dissimilar sets is the reverse true? A positive answer to these questions indicates size adjustment. By teasing apart size assortment from size adjustment, these two effects cannot confound one another in the null model to produce a Type II statistical bias. In the foregoing computer simulations, it is much easier to detect the presence of a size-similarity barrier by examining size assortment (Fig. 14.2) and size adjustment (Fig. 14.3) than by comparing the grand mean expected and observed size differences in Figure 14.4.

Testing for Size Assortment and Size Adjustment

Are species sets which are more size-dissimilar more prevalent in the archipelago than chance would predict? A simple test for the presence of size assortment in natural communities is based on the application of the binomial distribution. For each set size (pairs, triplets, and so on) rank the species sets according to their mean (or minimum) size differences, as in Fig. 14.2 or 14.6. If a limiting size-similarity barrier has played no role in the formation or persistence of natural communities, then the frequency distribution of the observed species sets plotted along these ranked axes should not be biased or skewed in any particular direction. This expectation serves as the null hypothesis.

This approach assumes that, a priori, each of the possible species sets is equally probable. One may argue, however, that if some species are more ubiquitous in the archipelago than others, then certain combinations involving these species should occur more frequently than combinations involving species rare in the archipelago. It is a simple matter to modify the null hypothesis to account for different expectations for each specific species combination by weighting probabilities along these lines. However, as we have seen, the commonness or rarity of a species may, in part, be determined by its competitive compatibility with the other species, and this, in turn, is determined by its size and the sizes of the other species. Therefore, it can become dangerously tautological to take the existing species frequencies as givens to input into the null model. Nevertheless, I will consider both weighted and unweighted null models in my analysis of *Cnemidophorus*.

The expectations for size adjustment are (1) that among high-ranked species pairs (those dissimilar in size) observed size differences on islands should be smaller than the expected size differences obtained from randomly drawing populations of each species from the existing species pool and (2) that the reverse will hold for the mid- and low-ranked species pairs. To test these expectations, we plot the percent deviation between observed (O) and expected (E) size differences as $(O - E)/E$ on the y-axis against the ranked species pairs (ranked from high to low average size difference) on the x-axis. Each point plotted in this grid represents a single observed 2-species island community. If the cluster of points falls mostly above the $y = 0$ line, except for those points representing the highest-ranked species pairs, then the community is size adjusted. This can be tested statistically by calculating the binomial probability for having at least m out of n points fall above the zero line and second by determining if there is a significant positive parametric or nonparametric correlation coefficient for the points. Since the rankings of species pairs are independent of the rankings of triplets, quads, etc., these larger community subsets may be combined in the same analysis. This, however, requires that the scale of rankings for all communities be adjusted to conform to the same range. Recall that n species sets (where $n > 2$) may be ranked according to either the average of the $n - 1$ within-community size differences or according to the average minimum size differences (minimum of the $n - 1$ size differences between the n-ranked species averaged over all islands with that species). These 2 rankings cannot be combined in an analysis since they are not independent.

Case and Sidell (1982) applied the analysis of size assortment and size adjustment to the Galapagos finches studied by Lack (1947), Abbott et al. (1977), and Strong et al. (1979). We found that in both the ground finches (*Geospiza*) and the tree finches (*Camarhynchus*), size-dissimilar species sets are more common in the archipelago than chance would predict. We found no evidence of size adjustment in either group. However, the number of island populations with complete bill-size date is so small that the latter negative results may be meaningless.

Before moving on to the analysis of *Cnemidophorus* lizards, I stress that the mere detection of a geographic pattern (no matter how statistically significant) does not shed any light on the process producing this pattern. In particular, even if we observe size assortment, we cannot conclude that interspecific competition is the responsible agent. Size-discrepant species sets may be more common than size-similar sets for any number of biological reasons: similar-sized species may hybridize more readily, spread pathogens to one another more easily, or enable predators more easily to form search images. An association of large species with small species may be due to an association of certain habitat types across islands. The methodology presented here cannot distinguish between these and other alternatives.

Tests with *Cnemidophorus*

The analysis described in the previous section is not necessarily restricted to true island systems. Any set of disjunct localities within the colonizing range of a consistent set of species is appropriate. Continental situations, however, usually have higher rates of immigration and emigration between sites. This movement of individuals and species will tend to blur the development of *in situ* size-assortment and size-adjustment patterns. With continental localities, there is an additional problem of avoiding any bias in the selection of sites that might favor particular species sets over others. With islands, the natural ocean boundaries nicely delimit the number and position of all sites. On continents, there may be an almost infinite number of sites a priori, and the selection of particular study sites from this entire array presents severe difficulties. A final proviso is that the analysis of size assortment and adjustment is limited in practice to cases where the total species pool under consideration involves only about 3 to 6 species and roughly 10 times as many localities. Otherwise, the number of potential species subsets is many times greater than the actual number of observed communities, thus spreading the data too thin and introducing huge sampling errors.

To avoid these problems, I considered 3 separate geographic regions and the *Cnemidophorus* therein. I included all sites in these regions that I have made a census of over the past 6 years as well as reports in the literature. In each case these sites were selected (by me or others) without regard to the analysis which I am now applying. Hence, although they were not chosen at random, they were selected in an essentially blind manner. Region 1 consists of all localities adjacent to the Gulf of California and the islands therein. Included is eastern Baja California, southern California, southwest Arizona, western Sonora, and Sinaloa, Mexico, and the Gulf islands. Five *Cnemidophorus* species inhabit this region: *C. tigris, C. hyperythrus, C. sonorae, C. burti,* and *C. costatus.* The upper-decile snout-vent length (UDSV) of the *Cnemidophorus* at each site was determined from my own samples (Case, 1979), from museum specimens, or from data gathered from the literature. (I believe UDSV is the metric of choice for reptiles or other species with continuous growth because it is not as greatly influenced by changes in age structure as is mean snout-vent length, and UDSV is not as sharply influenced by sample size as is the maximum snout-vent length for a population.) The localities of Region 1 are listed in Table 14.1 along with the UDSV of each species present at each site.

The second region consists of southern Arizona east of the Dragoon Mountains, through New Mexico as far north as Bernalillo (Table 14.2, and see Cuellar, 1979, for his site locations). The *Cnemidophorus* here are *C. tigris, C. tesselatus, C. inornatus, C. neomexicanus, C. uniparens,* and a member of the *C. exanguis* group (either *C. sonorae* in the west or *C. exanguis* in the east). Region 3 consists of only Texas localities (Table 14.3, and see Schall, 1976, for his site locations) and all but one of the sites are drawn from the literature. No site was considered, however, if less than 6 lizards were observed and no species was considered as a true member of the community at that locality if less than 3 percent of the lizards sighted consisted of that particular species.

For the last 2 regions it was impossible to get complete body-size records from each species at each site. In lieu of accurate site-specific size data, I calculated the average maximum snout-vent length of each species and considered it uniform across all sites (Tables 14.2 and 14.3). This simplification still allows tests for size assortment in Regions 2 and 3, but prevents any analysis of size adjustment except in Region 1.

Results for the analysis of Region 1 are shown in Figures 14.6 and 14.7. In Figure 14.6, there is no significant tendency toward size assortment: 3 pair-wise sets lie above the midpoint line and 11 lie below. There is only one 3-species set. However, a trend toward size adjustment is visible in Figure 14.7. Twelve out of 14 observed communities are more size-discrepant than the null communities for those very same species sets. (A deviation this extreme or more occurs at $P < 0.05$ from a binomial expectation of equal frequencies.) The correlation coefficient, however, is not significantly positive, mainly due to the single outlying point represented by the very similar-sized species pair, *C. sonorae* and *C.*

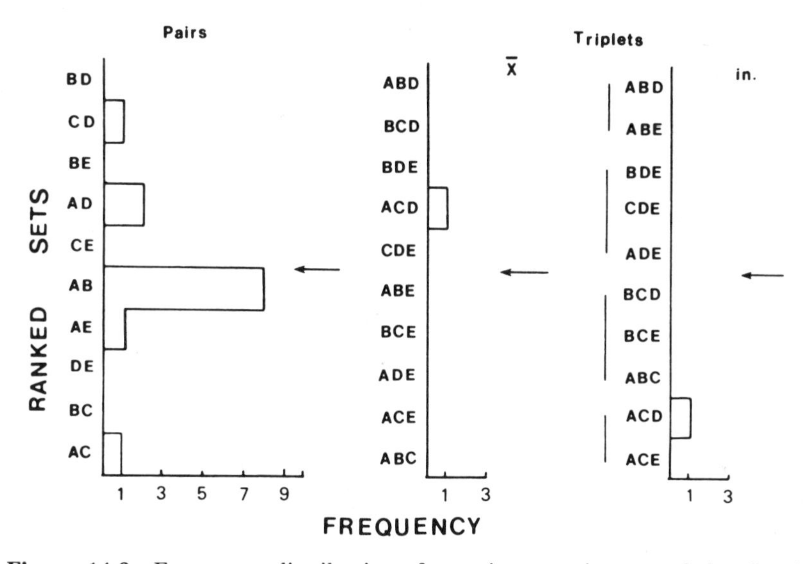

Figure 14.6 Frequency distributions for various species sets of the *Cnemidophorus* of Region 1 plotted in the manner of Figure 14.2. Each alphabetic symbol represents a particular *Cnemidophorus* species (see Table 14.1). The arrow in each subfigure represents the null point; on average, 50 percent of the observed communities should fall above this point, 50 percent below. There are 2 possible ways to rank triplets (3-species communities), by the mean expected size difference ($\bar{X}$) or by the expected minimum (min) size difference between member species. Sets that yield tie values are connected by a vertical line to the left of the set names. There were no observed 4- or 5-species sets. For Region 1 no trend of size assortment is evident in this figure.

Table 14.1 Upper decile snout-vent lengths of *Cnemidophorus* from Baja California, Sonora, Mexico, and southwest Arizona and southeast California.

Location	*C. tigris* (A)	*C. hyperythrus* (B)	*C. sonorae* (C)	*C. burti* (D)	*C. costatus* (E)	Number of species
Mainland						
Sonora-Sinaloa border	66	–	–	–	105	2
San Carolos, Sonora	88	–	–	106	–	2
Baja Norte, Mexico	100	62	–	–	–	2
Loreto, Baja, Mexico	99	62	–	–	–	2
Baja Cape, Mexico	134	62	–	–	–	2
Kino, Sonora	85	–	–	–	–	1
Lower Madera Canyon	83	–	83	–	–	2
Madera Base, Arizona	85	–	–	–	–	1
Vekol Road, Arizona	82	–	–	–	–	1
Upper Sabino Canyon, Arizona	–	–	78	130	–	2
Lower Sabino Canyon, Arizona	81	–	–	125	–	2
Lower Bear Canyon, Arizona	82	–	–	–	–	1
Molino Basin, Arizona	–	–	80	–	–	1
Campbell Ave., Tucson, Arizona	78	–	78	130	–	3
Tucson, Arizona	82	–	–	–	–	1
Cochise Stronghold, Arizona	–	–	80	–	–	1
Ajo, Arizona	76	–	–	–	–	1
Sells, Arizona	83	–	–	–	–	1
Kitt Peak, Arizona	–	–	82	–	–	1
Gila Bend, Arizona	83	–	–	–	–	1
Palm Springs, California	100	–	–	–	–	1
Mojave Desert, California	100	–	–	–	–	1

Islands, Gulf of California						
Angel de la Guarda	87	–	–	–	–	1
Carmen	84	64	–	–	–	2
Danzante	82	–	–	–	–	1
San Lorenzo Sur	75	–	–	–	–	1
San Marcos	86	61	–	–	–	2
San Pedro Martir	76	–	–	–	–	1
Partida Norte	91	–	–	–	–	1
Tiburon	100	–	–	–	–	1
Salsipuedes	71	–	–	–	–	1
San Esteban	79	–	–	–	–	1
San Francisco	94	63	–	–	–	2
San Lorenzo Norte	76	–	–	–	–	1
Santa Catalina	82	–	–	–	–	1
Espiritu Santo	111	61	–	–	–	2
San Pedro Nolasco	75	–	–	–	–	1
San José	108	65	–	–	–	2
Monserrate	–	61	–	–	–	1
Cerralvo	–	86[a]	–	–	–	1
Totals	34	10	6	4	1	

a. Size given for the insular endemic species *C. ceralbensis.*

Table 14.2 Typical maximum snout-vent lengths for *Cnemidophorus* from southeast Arizona and southern New Mexico.

Source and location	*C. tigris* (A)	Exsanguis group[a] (B)	*C. tesselatus* (C)	*C. inornatus* (D)	*C. neomexicanus* (E)	*C. uniparens* (F)	Number of species
Case (personal observations)							
Grassland east of Dragoon Peak, Arizona	–	83	–	–	–	70	2
Chiricahua National Monument, Arizona	–	83	–	–	–	–	1
Fort Bowie, Arizona	90	–	–	–	–	–	1
30 mi south of Socorro, New Mexico	90	83	–	65	–	–	3
Albuquerque, New Mexico	–	–	–	–	75	–	1
Gage, New Mexico	90	–	–	–	–	70	2
Las Cruces, New Mexico	90	–	–	–	–	–	1
Corralis, New Mexico, 5,700 ft	–	–	–	–	75	–	1
Corralis Hills, New Mexico, 6,050 ft	–	–	96	–	75	–	2
Wilcox, Arizona	–	–	–	65	–	70	2
Dixon and Medica (1966)							
White Sands, New Mexico 1967	90	–	–	65	75	–	3
Medica (1967)							
Rio Grande of South Central New Mexico	–	–	–	–	–	–	–
Site I	–	83	–	65	–	–	2
Site II wet year	90	83	–	–	–	–	2
Site II dry year	–	83	–	–	–	–	1
Site III wet year	90	–	–	–	75	–	2
Site III dry year	90	–	–	65	75	–	3

Cuellar (1979)							
Conchas Reservoir	–	–	96	–	–	–	1
Cedro Canyon	–	83	–	–	75	–	2
Tijeras Canyon	–	–	–	–	75	–	1
San Antonio riparian	–	83	96	–	75	70	4
Bosque uplands	90	83	–	65	75	–	4
Bosque riparian	–	83	–	65	–	70	3
La Joya	90	–	–	–	75	–	2
Elephant Butte, riparian	–	83	96	–	–	70	3
Elephant Butte, Bajadas	90	–	96	–	–	70	3
Caballo Dam	90	–	–	–	–	70	2
Totals	12	11	5	7	11	8	

a. Either *C. sonorae* in the west or *C. exsanguis* in the east.

Table 14.3 Typical maximum snout-vent length of Texas *Cnemidophorus*.

Source and location	*C. tigris* (A)	*C. septemvittatus* (B)	*C. exsanguis* (C)	*C. tesselatus* (D)	*C. gularis* (E)	*C. inornatus* (F)	Number of species
Schall (1976)							
Burnt House Canyon	–	–	91	–	90	70	3
Red Pens Ranch	–	–	91	107	90	–	3
Sibley Ranch	95	–	–	–	90	70	3
Blakemore Ranch	–	–	–	107	–	70	2
Miller Ranch	–	–	91	107	–	70	3
Balmorhea Lake	95	–	–	–	90	–	2
Case (personal observations)							
Palo Duro	–	–	–	107	90	–	2
Guadalupe Mountain	–	–	–	107	–	70	2
Milstead (1957a,b)							
Black Gap							
M-H[a]	95	–	–	–	–	–	1
H-C	95	–	–	–	–	70	2
S-L	95	107	–	–	–	70	3
Sierra Vieja							
B-C	–	–	–	–	–	70	1
C-T	–	–	91	107	–	70	3
T-G	–	–	91	–	–	70	2
M-H-B	–	–	91	–	–	70	2
C-C-B	–	–	–	107	–	70	2
C-C	–	–	91	–	–	70	2
S-B	–	–	91	–	–	–	1
C-G	–	–	91	107	–	–	2
L-B	–	–	91	–	–	–	1

La Mota Mountains							
O-C	95	–	–	107	–	70	3
C-O	95	–	–	107	–	–	2
M-C	95	–	–	107	–	–	2
M-M-C	95	–	–	107	–	–	2
Stockton Plateau							
C-S	–	–	–	–	90	70	2
C-O	–	–	–	107	90	–	2
P-S	–	–	–	107	90	–	2
M-C	–	–	–	107	90	70	3
W-O-W	–	–	–	107	90	–	2
Degenhardt (1966)							
Tornillo flat	95	–	–	–	–	–	1
Grapevine Hill	95	107	–	–	–	–	2
Burnham flat	95	–	–	–	–	–	1
Green Gulch, 4,200 ft	95	107	–	–	–	–	2
Green Gulch, 4,650 ft	–	107	–	–	–	–	1
Green Gulch, 5,250 ft	–	107	–	–	–	–	1
Totals	13	5	10	16	10	16	

a. These initials stand for the dominant plant species that Milstead used to characterize each site. Milstead (1957a) was unaware that he was dealing with 6 rather than 4 species of *Cnemidophorus*.

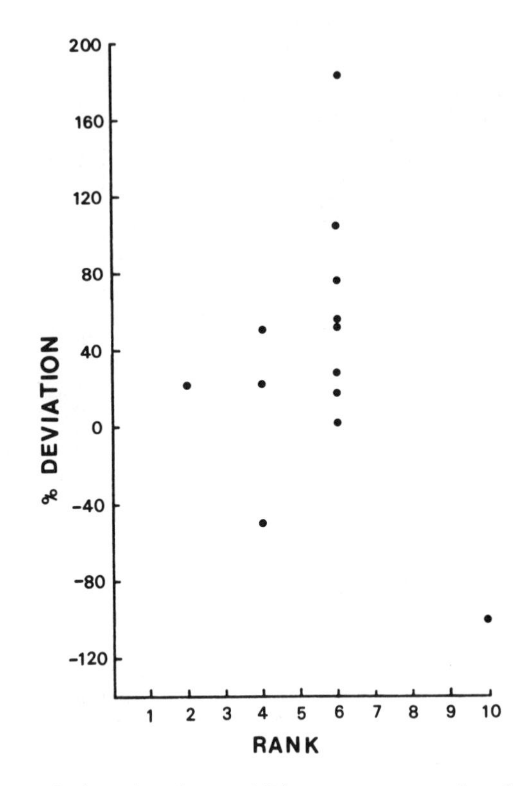

Figure 14.7 Deviations (on the y-axis) between expected and observed mean size differences for the various *Cnemidophorus* species of Region 1. Significantly more points lie above the $y = 0$ line than below (supporting a hypothesis of size displacement) but contrary to the complete fulfillment of this hypothesis, there is no positive slope to these points. The most deviant point from the standpoint of fulfilling a size-displacement pattern is that in the lower-right-hand corner, which represents the species pair *C. sonorae–C. tigris*. (See text and Figures 14.8 and 14.9 for a possible explanation.)

tigris, represented as the point in the lower-right-hand corner of Figure 14.7. These species are sympatric at a number of sites in southern Arizona yet display no size adjustment relative to allopatric populations.

A closer examination of this single exception, however, reveals that ecological interactions (perhaps competitive) are affecting these species but not on a niche axis related to body size. The geographic ranges of *C. tigris* and *C. sonorae* are broadly overlapping but the 2 species only co-occur along a narrow altitudinal band approximately 500 feet wide in the mountains of southern Arizona. *Cnemidophorus tigris* occurs only below about 3,500 feet, and *C. sonorae* extends from slightly below this altitude to between 6,000 and 8,000 feet at the peaks (Fig. 14.8). In the narrow zone of sympatry, both species are relatively rare (Fig. 14.8)

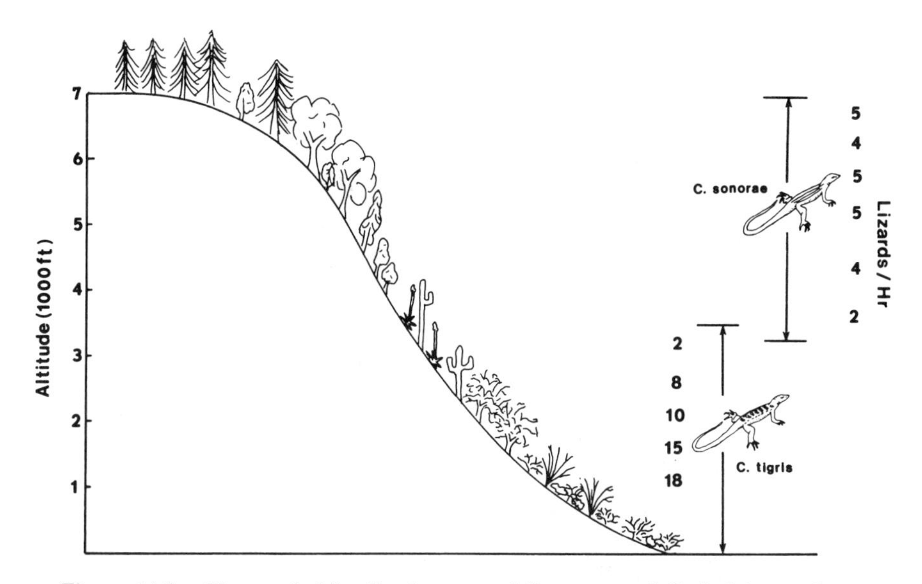

Figure 14.8 Observed altitudinal ranges of *C. sonorae* and *C. tigris* in southern Arizona. The 2 species' geographic ranges overlap broadly, but their altitudinal overlap is very narrow. The relative density of the 2 species along a altitudinal transect on Mt. Lemmon is shown on the right as the average number of lizards of each species seen per hour. Both species are rare in the zone of overlap.

based on my timed lizard counts (see Case, 1975, for techniques). The altitudinal point of species turnover is somewhat different in different mountain ranges (Fig. 14.9), presumably because of varying exposures, climate, latitude, and the like. Usually the turnover point occurs coincident with the entry of *Nolina* spp. into the plant community. At altitudes where junipers and piñon pines are common *C. tigris* is completely absent.

In areas outside the geographic range of sympatry, both species extend their altitudinal ranges in opposite directions. In the absence of *C. tigris* in the Sierra Opunto (east face) *C. sonorae* are present all the way down to 2,000 feet into the Rio Bavispe valley (also see Wright, 1968). One must go considerably farther west and north to the Mojave's New York Mountains in eastern California to find extensive altitudes above 4,000 feet which are outside the range of *C. sonorae.* Because of this latitudinal and longitudinal change, the altitudinal zones containing pine, oaks, and junipers begin at lower elevations than those in southern Arizona. Nevertheless, *C. tigris* (although a different subspecies, *C. t. septentrionalis*) extends up to 5,800 feet in the New York Mountains (Fig. 14.9) where it inhabits plant communities dominated by oaks, piñon pines,

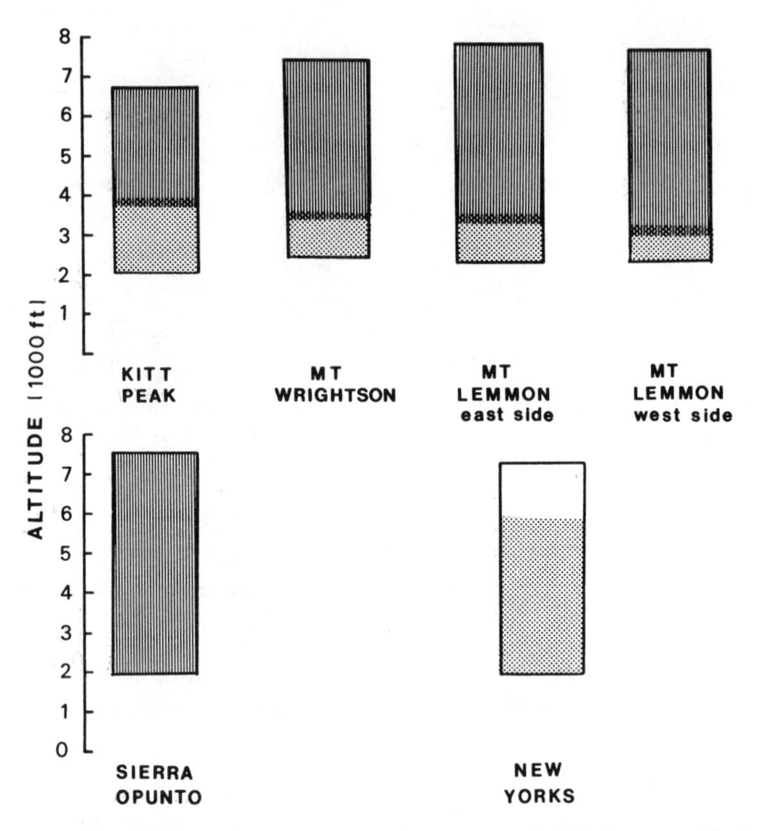

Figure 14.9 Actual altitudinal ranges of *C. tigris* (*dots*) and *C. sonorae* (*lines*) on some mountains in the Southwest. When sympatric, the altitudinal turnover point occurs at about 3,500 ft (*top row of figures*). In the Sierra Opunto, in the absence of *C. tigris, C. sonorae* extends down to about 2,000 ft. In the absence of *C. sonorae* in the New York Mountains of California, *C. tigris* extends up to about 5,800 ft.

and even ponderosa pines. Miller and Stebbins (1964) have recorded *C. tigris* at 5,000 feet in Joshua Tree National Monument. Johnson and associates (1948) recorded *C. tigris* from 6,200 feet in the Clark Mountains of California, and Johnson (1969) reports *C. tigris* at 6,000 feet in the Sierra Nevada. In short, there appears to be niche displacement in this species pair that occurs along a habitat dimension rather than a body-size dimension. If we then exclude *C. sonorae* from the analysis on the basis of its rather different habitat occupation, the pattern of size adjustment among the other 4 *Cnemidophorus* becomes more apparent. The following picture emerges (Fig. 14.10). In the southernmost Cape region of Baja California, the very large *C. tigris maximus* (synonym, *C. maximus*)

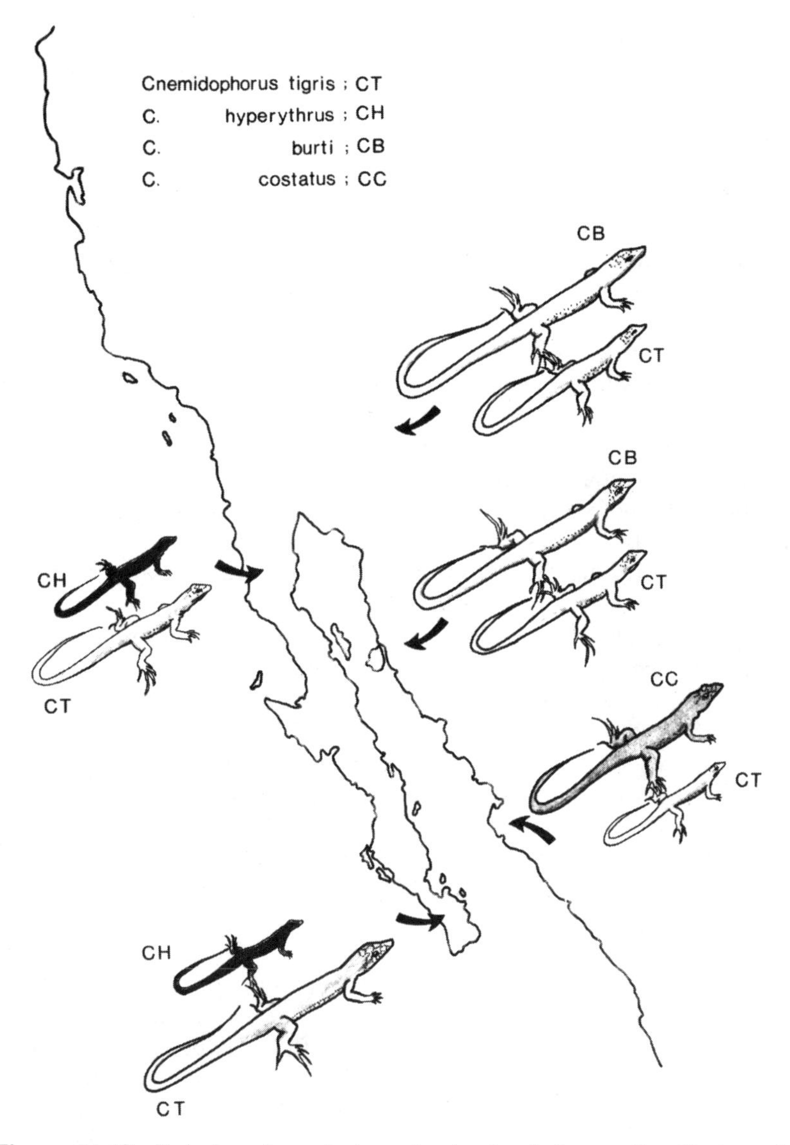

Figure 14.10 Relative size relations for lowland desert *Cnemidophorus* in Region 1.

occurs sympatrically with the much smaller *C. hyperythrus.* Proceeding north up Baja California, *C. tigris* becomes somewhat smaller in the form of *C. t. rubidus* and *C. t. tigris,* also sympatric with *C. hyperythrus.* In southern Arizona and northern Sonora, Mexico, *C. tigris gracilis* is smaller yet and is now sympatric with a larger *C. burti.* In southern Sonora *C. tigris aethiops* is also relatively small and sympatric with a different subspecies of *C. burti.* Finally, proceeding south into northern Sinaloa, Mexico, *C. burti* drops out and a new large species *C. costatus* overlaps the range of *C. tigris* along the Sonora-Sinaloa border. At the point of overlap *C. tigris* is quite small, as small as the *C. hyperythrus* on the Baja California Cape directly across the Gulf of California.

Northern Sinaloa and the Cape of Baja California are remarkably similar in vegetative structure, having a short, semideciduous, subtropical thorn forest. Both places receive similar weather patterns and are at similar latitudes (Shreve, 1951; Hastings and Turner, 1969; Hastings et al., 1972), yet in one place (the Baja Cape), *C. tigris* has the largest body size and in the other (northern Sinaloa) the smallest body size of any locality throughout its geographic range (Case, 1979). In the Cape *C. tigris* is sympatric with the much smaller *C. hyperythrus.* In Sinaloa it is sympatric with the much larger *C. costatus.*

The evidence for character displacement (that is, size adjustment) is strengthened by comparing the body sizes of *C. tigris* and *C. hyperythrus* on islands in the Sea of Cortez or off the west coast of Baja California. On the 6 oceanic island banks where the smaller *C. hyperythrus* is absent, *C. tigris* is consistently smaller in size relative to locations (island or mainland) where the 2 species occur sympatrically (Case, 1979). On Angel de la Guarda and Partida Norte, this size difference is small and insignificant, but on islands like San Pedro Martir, Santa Catalina, the San Lorenzos, or San Pedro Nolasco it is very great indeed. *Cnemidophorus hyperythrus* or its insular endemic derivative occur allopatrically on only 2 islands (both oceanic), Monserrate and Cerralvo. On Monserrate *C. hyperythrus* reaches a maximum size 5 mm greater than mainland relatives. On Cerralvo, the endemic *C. ceralbensis* is nearly 20 mm larger in snout-vent, thus approximating the size of *C. tigris* on oceanic, allopatric islands. *Cnemidophorus tigris* also occurs allopatrically on a few land-bridge islands (Danzante, Smith, and various land-bridge islands off the west coast of Baja California). On these islands, no marked reduction in body size is apparent (Case, 1979). Thus the evidence for size adjustment in Region 1 becomes compelling when only syntopic species are included in the analysis and when proper attention is paid to the very different size relationships on oceanic versus land-bridge islands.

Region 2 includes the *Cnemidophorus* in southeastern Arizona and New Mexico. There are only 18 multispecies sites in this region, making any statistical comparison difficult. A statistically significant pattern of size assortment for only 18 sites would have to yield 13 sites above the midpoint and this is not the case (Fig. 14.11). For the 2 rankings, 9 ($\bar{x}$) and 11 (min.) communities fall above the midline.

In Region 3 (Texas) there is a richer array of *Cnemidophorus* associations, and a definite pattern of size assortment is evident both for 2- and

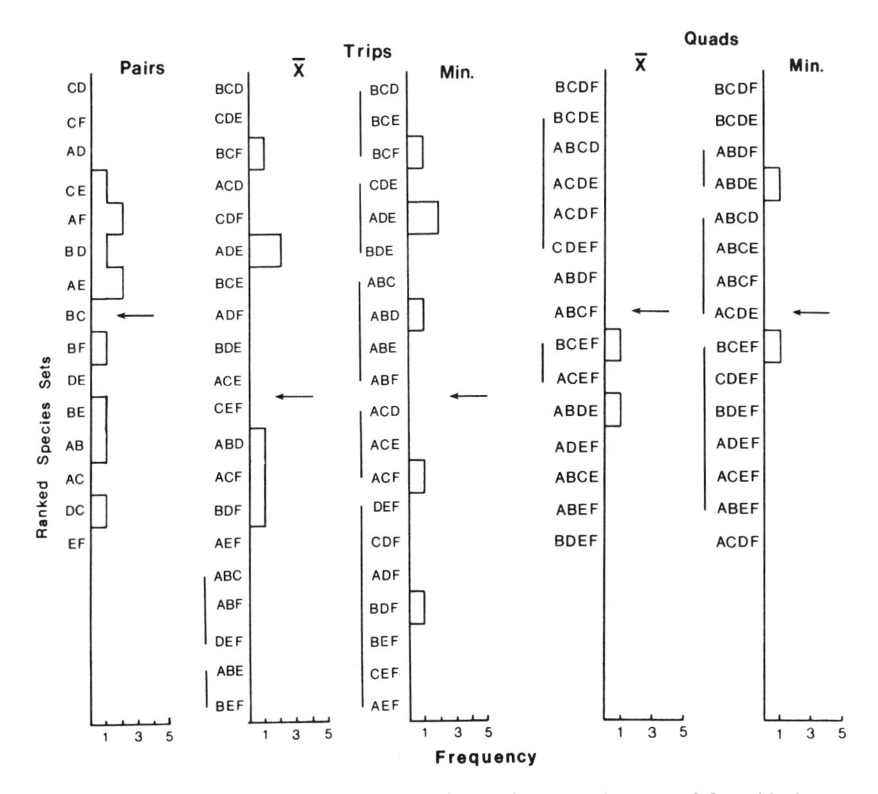

Figure 14.11 Frequency distributions for various species sets of *Cnemidophorus* in Region 2 (eastern Arizona and New Mexico) plotted as in Figure 14.6. No significant size assortment is evident.

3-species subsets. Combining the rankings based on mean size differences for pairs and triplets, we find that 19 out of a total of 27 fall above the half-way mark (Fig. 14.12). Based on a binomial expectation of equal frequencies a deviation at least this extreme in this direction (that is, one tail of the distribution) occurs at $P \leqslant 0.05$. A ranking based on minimum (rather than mean) size differences yields exactly the same result.

These conclusions are based on the assumption that each community subset of a given species number is equiprobable. This assumption may be changed to reflect the different representations of various species across the sites. The probability of each specific species combination occurring may be calculated by using the individual frequencies of each species across all sites. For example, if there is a total of 100 species occurrences and species A accounts for 50 of these, the individual probability for A is 0.5. Once these probabilities are calculated for all species, it is

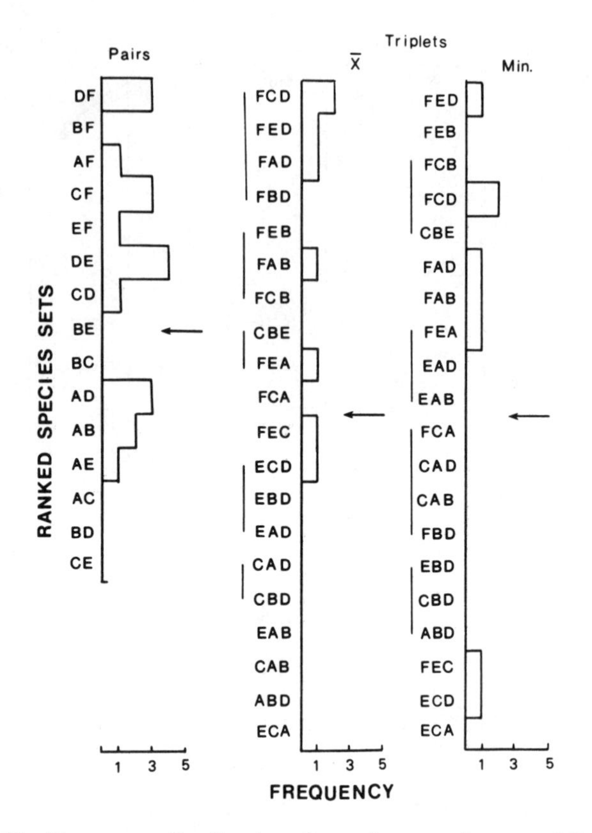

Figure 14.12 Frequency distributions for various species sets of *Cnemidophorus* in Region 3 (Texas) plotted as in Figure 14.6. Significant size assortment is evident.

a simple matter to calculate the probability of each specific combination given that it occurs in an *i*-species set. After this modification, the dividing line separating equal expectations for Region 3 shifts to a position approximately between pairs D,E and C,D and between triplets F,E,A and F,C,A. Now 18 communities out of 27 fall above the null expectation for equal frequency, and this is only significant at $P < 0.08$. On the other hand, for Region 2 the null line for pairs shifts downward by about 0.63 of a pair and is unaltered for triplets and quads, so that now 10 out of 18 species sets fall above the new midpoint (for $\bar{x}$). This upward-biased ratio still does not differ significantly from 0.5 at $P < 0.05$.

As discussed earlier, a modification of the null expectations in the above manner verges on tautology if indeed size similarity influences species coexistence. The commonness or rarity of a species will be, in part, determined by its size and not independent of it. Therefore, any null expectation that weights probabilities according to the commonness or rarity of individual species will a priori be biased in favor of accepting the null hypothesis.

In conclusion, for at least 1 of the 3 geographic regions studied, there is a significant trend of size assortment in community structure. Only Region 1 allows an analysis of size adjustment, and for this region the null hypothesis of no size adjustment must be rejected. Moreover it is very rare to find more than 3 sympatric species of *Cnemidophorus* at a single site in this region; many localities contain only a single species. These observations plus the altitudinal separation of the similar-sized *C. tigris* and *C. sonorae* support the hypothesis that ecological interactions, perhaps competitive, mediated through body-size similarity are one set of factors determining the distribution and local sympatry of some *Cnemidophorus* lizards.

It is significant in these analyses that the region with the least evidence for size assortment is Region 2 (eastern Arizona and New Mexico). For this region 4 out of the 6 species considered are all-female, parthenogenetic species probably not more than 13,000 years old (Brown and Wright, 1979). The geographic areas occupied have undergone rapid ecological transitions in the last few thousand years (Wright and Lowe, 1968), and niche separation between these species is more often aligned along habitat and microhabitat differences than morphological differences.

15 | Ecomorphs, Faunas, Island Size, and Diverse End Points in Island Radiations of *Anolis*

Ernest E. Williams

IN 1972 I CHOSE Puerto Rico, because it was relatively well known and had a fauna of moderate complexity, for a test analysis of the evolutionary radiation of anoline lizards within an island. It seemed reasonable to regard the Puerto Rican *Anolis* as a readily analyzable stage in the evolution of faunal complexity in relation to island size and topographic diversity. I expressed the hope that this analysis could be extended to the faunas of the other Antillean islands once sufficient data became available.

Unfortunately more data have not permitted fulfillment of this hope—certainly not in the simple fashion that was then expected. It is now apparent that an increase or decrease in area has effects that are much more than simple additions and subtractions. There are qualitatively differing end points and even—as we shall see for Hispaniola below—differing faunal end points on the same island.

The ecomorph concept that was introduced in the 1972 paper remains crucial to the new analysis. The concept is basically the familiar one of convergent evolution—a set of animals showing correlations among morphology, ecology, and behavior, but not lineage—a concept usually applied to widely divergent taxa (for example, the birds of different continents; Karr and James, 1975) but here seen in the radiations of a single genus within a single archipelago.

The phenomenon that I have called ecomorph is obtrusively evident in the Greater Antillean islands. Anyone who visits more than one of these islands is struck at once not only by the abundance and diversity of the anole faunas and the corresponding diversity of the microhabitats in which they occur but also by the conspicuous presence of highly similar species from one island to another, always occupying very similar habitats. The similarity in each case extends to color, size, body proportions, perch, and foraging and escape behavior. Further information, however, always makes it quite clear that these are ecological analogues, not closest relatives. On the contrary, the several very distinct types on one island may be much more closely related to one another than any of them is to analogous species on another island.

Figure 15.1 shows the major ecomorph categories found in the Greater Antilles and indicates that both the major structural categories within *Anolis,* alpha and beta of Etheridge (1960), are represented by species in each of these categories.

It is noteworthy that this crossing of phyletic boundaries by ecomorphs does not depend upon any specific classification. Currently, Etheridge's major subdivisions are under attack (Gorman, Buth, and Wyles, 1980; Shochat and Dessauer, 1981), but the proposed rearrangement, while it switches the presumed relationship of whole lineages, still finds the same ecomorph occurring in two or more major groupings.

As in the many other cases, the ecomorph concept began to take shape

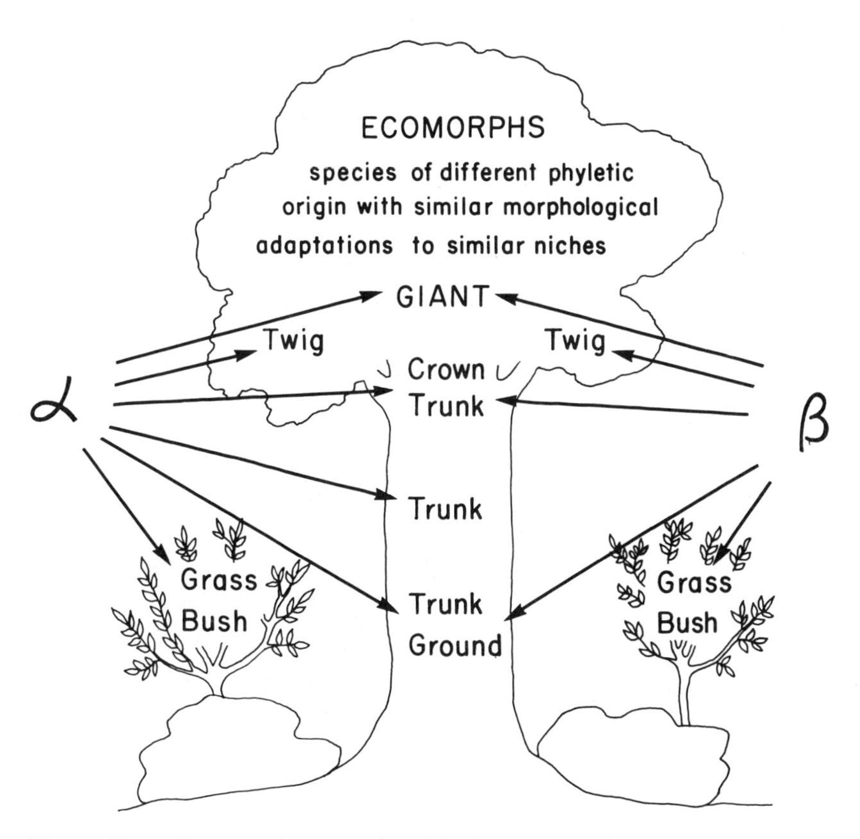

Figure 15.1 Six named ecomorphs with the perch position characteristic of each.

as soon as a close look was taken at the relationships of the compared taxa. The more conspicuous ecological types of Greater Antillean anoles were first recognized by A. S. Rand (1964) in Puerto Rico. It was soon realized that similar types existed on the other Greater Antilles but that not all of these could possibly belong to similar phyletic stocks. In 1969 Rand and I found it useful to define the characters and to provide simple names for these ecological types. Table 15.1 details some of the defining characteristics of the major ecomorphs more fully than has been done before. I coined the term *ecomorph* in 1972 to emphasize the morphological aspect of the similarities between types and to avoid confusion with the significantly different term *ecotype* proposed by Turesson (1922) and much used by botanists.

Unfortunately the ecomorph categories that Rand and I erected have their strict application only for the anoles of the Greater Antilles, not even for the Lesser Antilles, and not consistently for mainland anoles, let alone other lizards. Even for the anoles of the Greater Antilles other than Puerto Rico we have had to make emendations and additions to the concepts that had their origin and primary base in Puerto Rico.

Before I go further, it will provide perspective to look in broad terms at the niche and mode of radiation of *Anolis*. The background facts are that *Anolis* are primarily but not exclusively arboreal, are exclusively diurnal, and include both thermoregulators and thermoconformers, that is, species that bask and those that do not bask. Figure 15.2 describes the niche of *Anolis* in terms of 3 axes: size, perch, and climate, and their attendant morphological and behavioral correlates. These 3 axes seem to have general descriptive utility and validity.

However, the ways in which these axes determine the species characters of *Anolis* are clearly different in different areas (Chapter 16). Size, perch, and climate all differ in the sympatric species pairs that occur in the Lesser Antilles, as Schoener and Gorman (1968) first announced for the 2 species of Grenada, but differences in morphology (other than size) are muted as compared with the stronger specialization that occurs on the Greater Antilles. The Lesser Antillean pairs have clearly not reached the level of differentiation that I called ecomorph in Puerto Rico and of which 5 examples typical in size, perch, and shape are given in Figure 15.2.

In the Greater Antilles size and perch sort out the ecomorphs. These two factors are correlated with foraging and defense behavior (Table 15.1). Position, whether on tree, bush, or grass, determines foraging opportunities and defensive possibilities. Size has an influence on size of prey and mode of defense. Climate subdivides ecomorphs and by permitting some spatial separation provides room for more species. Body form, as a defining feature of ecomorph, relates to the constraints and

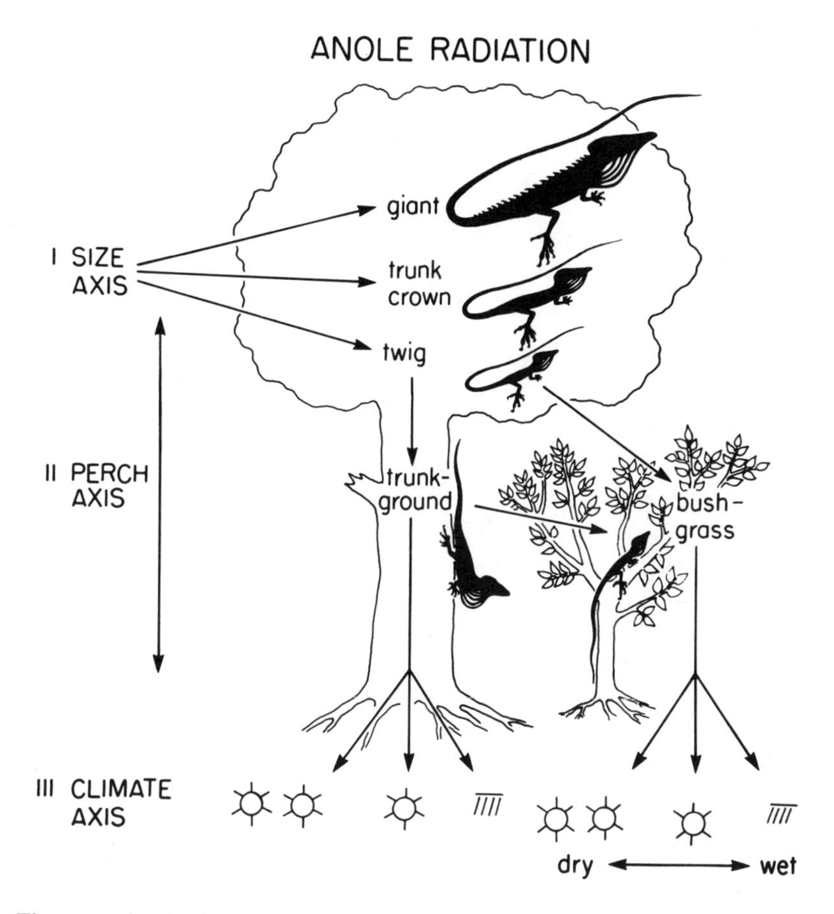

Figure 15.2 Ecological axes that appear to be involved in the evolution of *Anolis*. Five of the 6 ecomorphs of Figure 15.1 are shown in characteristic sizes as well as perches.

opportunities imposed by the vegetational matrix (Moermond, 1979). Color may promote crypsis or may be related primarily to mate choice. Color and particularly changes in color may also be related to thermoregulation. The scale differences cited in Table 15.1 are less obviously adaptive, but they contribute to the very visible sharp distinctions between the ecomorphs. The contrast between this Greater Antillean diversity and the severely limited between-species variation found in the Lesser Antilles is very striking.

Size, perch, and climate are clearly important also in mainland anoles, and one might naively expect the mainland to have anole faunas much more diverse than those of the largest islands. We must confess at once that we know much less about mainland anoles than those of the

Table 15.1 Defining characteristics of major ecomorphs.

Characteristic	Ecomorph					
	Crown giant	Twig dwarf	Trunk-crown	Trunk	Trunk-ground	Grass-bush
Size:	> 100 mm	< 50 mm	> 70 mm	< 50 mm	> 60 mm	Usually < 50 mm
Color:	Green, patterned or not	Gray, lichenate	Green, sometimes grayish	Variable, green, grayish, or brownish	Brown with variable pattern, more rarely green	With distinct lateral or dorsal stripe in both sexes
Modal perch:	Typically high in the crown	Twigs of canopy	Canopy and upper trunk	On trunk between trunk-crown and trunk-ground	On lower trunk	On grasses or bushes
Body proportions:	Head large, massive, often casqued	Long head, short body, short legs	Large head, body tending to be long, short legs	Head and body short	Head relatively short, body short and stocky, limbs long	Head moderately long, body slender, tail long
Scales:	A vertebral crest present	Uniform dorsal scales	Uniform dorsal scales	Dorsal scales usually uniform	Middorsal scales abruptly (in 2 rows) or gradually enlarged	A zone of few to many rows of dorsal scales
Foraging behavior:	Primarily a canopy forager	A *slow* searcher on twigs	A searcher on leaves and branches	Primarily a forager on its trunk perch	Sit-and-wait predator on ground prey	Primarily a grass-bush forager
Defensive behavior:	Primarily aggressive	Primarily crypsis	Flight upward	Squirreling	Flight downward	Flight downward

NOTE: Moermond (1979, 1981) describes twig and trunk-crown ecomorphs as "crawlers," the trunk ecomorph as a "runner," and the trunk-ground and grass-bush ecomorphs as "jumpers." He cites different types of prey attack behavior as associated with these ecomorphs also—that is, "stalk-strike" with the twig ecomorph, "jump-strike" primarily with grass-bush anoles, whereas "stationary-strike" and "approach-pause-strike" were used by all ecomorphs some of the time.

islands, too little even of their taxonomy, less about their ecology, and almost nothing of their evolution. What information we do have, however, does not show them impressively more diverse ecologically or morphologically than those of the islands. There is, in fact, a difficulty in comparability, and it is evident that the Greater Antillean ecomorphs are only imperfectly paralleled on the mainland.

It is facile to explain the special features of mainland anoles by the fact that, in contrast to the island species, they are parts of truly complex faunas of which they are not the most conspicuous or important components. It is, for example, vividly and immediately evident to the observer coming from the islands that mainland anoles have far less dense populations than their island congeners, so much less dense that any competitive interactions must be much more with distantly related taxa (even other classes or phyla) than with congeners.

The high divisibility of the *Anolis* niche is a major feature of the genus; it is the other side of the phenomenon of radiation that is so much an anoline characteristic. My concern in this chapter is with the effects of area on this niche divisibility and hence on faunal complexity, an interaction that was only partly explored in my 1972 paper. But this new look at the phenomenon which is here proposed is necessarily limited by the availability of data. The mainland cases must be put aside for the reasons mentioned just above. Cuba likewise cannot be adequately treated. Although it is far from a terra incognita, it is still relatively poorly studied, even at the alpha taxonomic level, and, unique among the islands, species of *Anolis* have been described for Cuba that I do not know even as preserved specimens. (See Ruibal, 1964, for a useful but outdated review of Cuban *Anolis*. Schwartz and Thomas, 1975, and Schwartz, Thomas, and Ober, 1978, provide a more current list of names.)

Omission of mainland anoles is not a problem: the factors that have impinged upon them seem clearly different from those that have affected the island anoles. The omission of Cuba may also not seriously flaw our analysis. It is now very evident that the Hispaniolan anole fauna may be even more complex than that of Cuba and may therefore quite plausibly serve as the final term in a series of West Indian anole faunas grading upward in size and complexity.

Such a series of West Indian islands and faunas cannot, however, show an even gradation. To speak of emergent islands only, the smallest of the Greater Antilles (the main island of Puerto Rico, 3,421 mi^2) is about 5 times larger than the largest of the small islands (Guadeloupe, 687 mi^2), and Hispaniola is 40 times larger than Guadeloupe. A logical first grouping is into large and small islands, but the small islands must immediately be broken into two categories: the *small old islands,* old enough and high enough to have been emergent for a long time with the consequence that the evolutionary processes that we are interested in have played their appointed role, and the *small low islands,* which are so recently emergent that they are better studied in terms of ecological rather than evolutionary time. As for the large islands, it will be a major point of this study that they are each quite different and that the substantial differences in area between them may be, in great part, the explanation of their faunal differences.

Where published documentation of the habits and habitat of the anoles discussed below is available, I have cited it. However, much detail is still unpublished or, worse, unknown. Some published data—quantitative though they may be—are based on observations at restricted localities and for limited periods. Most of the species I have seen alive myself, and I have not hesitated to reinterpret the observations or interpretations of others in the light of my own observations.

All determinations about relationship and phylogeny are my own. I crystallized these judgments into a formal system in an earlier paper (1976). Unfortunately not all the reasoning behind those assignments is yet in print. In extenuation I can only report that a series intended to summarize information on all the West Indian anoles is in preparation, of which Williams (1976) was the first paper.

The Small Old Islands

The small old islands of the Caribbean include, in addition to the Lesser Antilles of geographers, 3 islands adjacent to the mainland of Venezuela: Curaçao, Bonaire, and Blanquilla. All of these islands, although 3 distinctive lineages of *Anolis* are represented on them, show only 1 or 2 anoles per island.

In 1972 I used the so-called solitary (named by Schoener, 1970) anoles and species pairs of these islands as models of the first stages in the evolution of the coadaptation of faunas. At that time I emphasized size. Solitary species—so named because they occur without any congener—were found to belong somewhere in the middle range of all sizes exhibited by *Anolis*, whereas pairs of species were usually skewed away from that middle range, one of each species pair being distinctly below the middle range and the other distinctly above it (Schoener, 1970; Williams, 1972).

Grenada (Schoener and Gorman, 1968), as the ecologically first studied of the Lesser Antilles, may serve as an example of all the species pairs (Fig. 15.3).

It must be emphasized that the size difference between members of species pairs is considerable, up to nearly 2 times in snout-vent length. It is an impressive difference, but the point I want to stress is that size is the

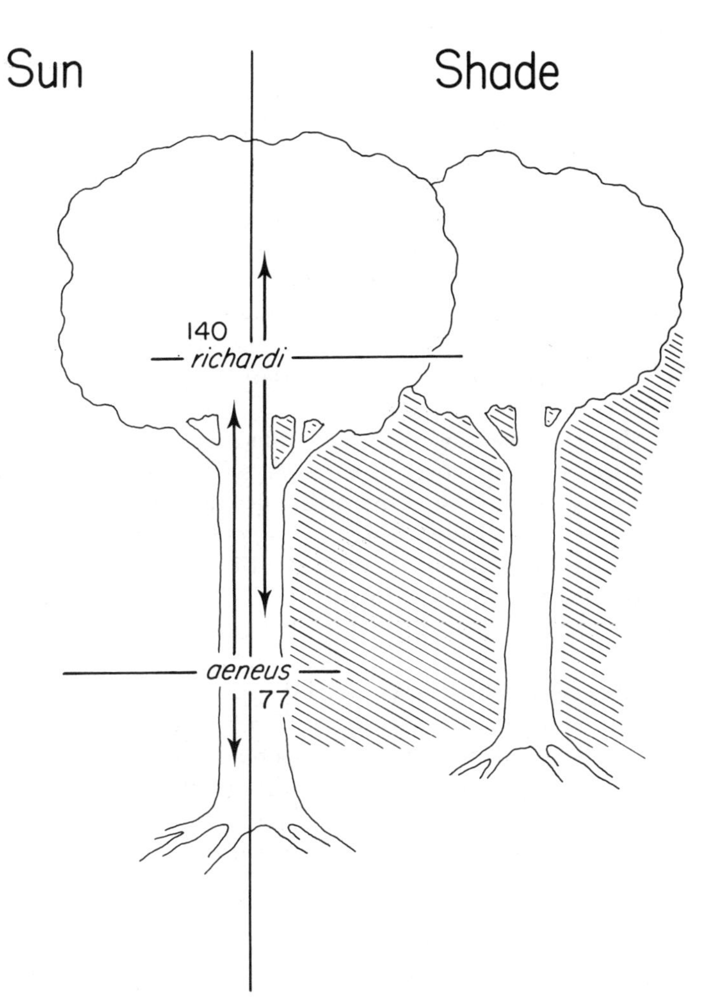

Figure 15.3 Perch and climatic preferences of the 2 anoles of Grenada. Numbers are maximum adult male snout-vent sizes in mm. Horizontal lines indicate preferential positioning of *A. richardi* in the shade, of *A. aeneus* in the sun. Vertical arrows indicate the preference of *A. richardi* for higher perches than those favored by *A. aeneus.*

only impressive difference between these species pairs. The differences in scalation between members of the species pairs are not impressive in terms of the differences between species on the larger islands, and, similarly, the climatic preferences between species, while real, allow major overlap (Schoener and Gorman, 1968). Perch does differ in that the larger species tend to perch higher and prefer bigger trees. Where the habitat does not permit big trees, the larger species tend not to occur. Within this limitation, species pairs in the Lesser Antilles are almost ubiquitously syntopic, occurring in the same trees, only at different heights. See, however, the species of St. Martin (Williams, 1972; Roughgarden et al., Chapter 16).

The Lesser Antillean pairs thus have some striking features in common: (1) there is always at least partial sympatry and, when there is sympatry, there is often near total syntopy; (2) the major differentiation is in size axes, and there is minimal differentiation as regards other morphology; (3) no species is wholly montane, although one species may extend altitudinally higher than the other; (4) neither member of any species pair shows any geographic variation comparable to that of certain solitary species.

Morphological and ecological differences can be greater within solitary species than between species pairs. The most striking example is *A. marmoratus* within the island of Guadeloupe, where scale size and color change vary radically within the species, and there is one differentiated population that is montane (Lazell, 1964, 1972). There are parallels in *A. oculatus* and *A. roquet* (Lazell, 1962, 1972) that are somewhat less extreme and less clearly differentiated geographically (personal observation). In the species pairs of any island bank, there is nothing quite comparable either within or between the species. Although, as Lazell (1972) has emphasized, there is very striking variation within *A. aeneus,* it is a phenomenon of individuals, not populations. There may be considerable nongeographic or erratically geographic variation. One of the 2 species may climb higher into the mountains than the other, or farther out into scrubby vegetation, but strong modifications are not seen.

The Small Low Islands

The small low islands in the West Indian region may show only 1 or 2 anoles per island or in the exceptional case, the emergent islands of the Great Bahama bank, 4 species.

The 4-anole fauna of the Great Bahama bank, first studied on Bimini by Schoener (1968), might on the basis of numbers be considered one of the terms in a series of more and more complex faunas. In fact, however, it does not fit into a series in complexity, conceived as an evolutionary sequence. The fauna of the Great Bahama bank, like that of the other low West Indian islands, is recent and derivative, compounded from the specialized endemics of adjacent large islands (Williams, 1969). As I have personally observed, 4 strongly marked ecomorphs do exist on Bimini, but they have not evolved there. Instead, they have come as fully differentiated species, preadapted to coexist (see Case, Chapter 14).

The situation is comparable to that occurring on a newly emergent portion of a large island just opened to colonization. The new area is colonized by that portion of the fauna of the major island that (1) can reach it and (2) is suited to depauperate areas of relatively uniform ecology.

The animals that arrive in such situations are not just fractions of larger faunas but highly selected fractions. Three of the Great Bahaman anoles—*A. carolinensis, A. angusticeps, A. sagrei*—are Cuban in origin and all are edge animals; they have no readjustments to make in order to coadapt ecologically. They were already coadapted on Cuba. The fourth animal—*A. distichus*—is more interesting; it is a Hispaniolan animal that has inserted itself within a set of Cuban animals. Note that it has never invaded Cuba, although the overwater distance to the Great Bahama bank is greater than that to Cuba. It is, however, preadapted; it has the ideal ecology to insinuate itself into the depauperate fauna of the Bahamas because its ecological relationships to the other species parallel its native niche in Hispaniola: between, on the one hand, the trunk-crown ecomorph—in the Bahamas, *A. carolinensis* (in Hispaniola it was *A. chlorocyanus* or *A. coelestinus*)—and, on the other hand, the trunk-ground ecomorph—in the Bahamas, *A. sagrei* (in Hispaniola it was *A. cybotes* or a close relative). The forms on Cuba that are ecologically analogous to *A. distichus* are more mesic than the colonizing populations of *A. distichus* and may have therefore been unable to follow their 3 successful Cuban colonizers of the Bahamas on to these low, relatively dry, and open islands. (On these points see Williams, 1969.)

The Large Old Islands: The Greater Antilles

In 1972 the anole fauna of Puerto Rico was very elaborately described and discussed. It will not be desirable to describe Jamaica and Hispaniola in equivalent detail, nor will it be possible to develop a phylogeny for the anoles of the other 2 islands as well documented and as apparently tidy as that which was presented for Puerto Rico. What I intend here instead is a relatively gross comparison, with emphasis on the distribution of ecomorphs over the 3 islands.

Table 15.2 makes a comparison solely in terms of number of species per ecomorph present on each island. I have ordered the island faunas from left to right in an order of complexity which is the same as the order of size of island banks. (The Puerto Rican bank, which includes the Virgin Islands, is much larger than the Jamaican bank which has very little offshore extent.) Table 15.2 in conjunction with Table 15.3a and b, which adds species names, size, and some other details, allows an analysis in terms of three topics: (1) what is regularly present, (2) what is added, (3) what is left out.

(1) What is *regularly* present in the three islands?

The ecomorphs are, of course, by definition the set of categories that

Table 15.2 Number of species per ecomorph.

	Jamaica	Puerto Rico	Hispaniola
Giant	1	1 + 1[a]	3
Twig I	1	–	1
Twig II	–	1	2
Trunk-crown I	1	1	2
Trunk-crown II	1	1	2 +
Trunk-ground	1 (subspecies) + 1[b]	3 (1 with subspecies)	5 + 1[c]
Grass	–	–	3 + (2)
Bush	–	3	3
Trunk	–	–	5 + (1)
Other	1	–	7
Total	6 + 1[b]	10 + 1[a]	36 + 1[c]

a. Probably extinct.
b. Possible invader (*A. sagrei,* see text).
c. Invader (*A. cristatellus:* Roughgarden, 1974; Williams, 1977).

regularly recur and that in color, habitus, squamation, ecology, and behavior are dramatically similar, in spite of the fact that they may not be at all closely related.

For Puerto Rico I recognized in 1972 only 5 ecomorph categories: crown giant, twig, trunk-crown, trunk-ground, and grass-bush anoles. For the whole set of large islands I now (Tables 15.2 and 15.3) recognize 9; 2 of these are subdivisions of an older category, that is, I now recognize 2 sizes of twig anoles (I, giant, mentioned above, and II, dwarf) and 2 of trunk-crown anoles (similarly I, large and II, small). I recognize also a trunk ecomorph which does not occur in Puerto Rico and I subdivide the grass-bush category. (The last is the weakest of my new decisions, since the subdivision occurs only in Hispaniola.) I shall call these standard-sequence ecomorphs.

In Tables 15.2 and 15.3 there is a category "Other," in which I list animals that in one way or another do not fit the ecomorph concept, the essence of which is convergence and stereotypy. In the table I have accommodated some species that I would formerly have listed as other by subdividing the twig and the trunk-crown ecomorphs into 2 sizes of categories.

Discrepant taxa aside, however, the morphological, ecological, and behavioral congruence of the species that are listed in any one ecomorphic category is amazing, the more so since (a) closest relatives may belong to different ecomorphs, for example, *A. gundlachi* and *A. krugi* on Puerto Rico (see discussion in Williams, 1972), and, especially, all the native Jamaican anoles which, despite their rather full roster of ecomorphs, are closer phyletically to each other than to other anoles anywhere (Underwood and Williams, 1959; Williams, 1976), and since (b)

Table 15.3a Anole ecomorphs on three Greater Antillean islands.

Island	Crown giant	Twig I	Twig II	Trunk-crown I	Trunk-crown II
Jamaica N = 6 (7) (1 invader)	*A. garmani*^{124} FW	*A. valencienni*^{86}		*A. grahami*^{72}	*A. opalinus*^{56}
Hispaniola N = 36 (37) (1 invader)	[*A. ricordii*^{160} FW *A. baleatus*^{180} FW *A. barahonae*^{158} FW]	*A. darlingtoni*^{72} FW–M *A. fowleri*^{77} FW–M	*A. insolitus*^{47} FW–M *A. sheplani*^{41} M	[*A. chlorocyanus*^{80} *A. coelestinus*^{84}]	[*A. aliniger*^{60} M *A. singularis*^{52}]
Puerto Rico N = 10 (11) (1 extinct?)	[*A. cuvieri*^{137} FW *A. roosevelti*^{157}]		*A. occultus*^{40} FW	*A. evermanni*^{78}	*A. stratulus*^{50}

NOTES:
N = species number per island or island bank.
FW = forest/wet habitat.
M = montane.
Superior number = reported maximum male size, snout-vent, mm.
Allospecies are bracketed.

Table 15.3b Additional anole ecomorphs on three Greater Antillean islands.

Island	Trunk-ground	Grass	Bush	Trunk	Other
Jamaica	*A. lineatopus*^{73} FW/OA				*A. reconditus*^{88} FW
	(*A. sagrei*)^{56}				
Hispaniola	[*A. cybotes*^{81}	[*A. olssoni*^{48} OA	[*A. hendersoni*^{49} FW–M	[*A. distichus*^{58}	[*A. armouri*^{59} M
	A. marcanoi^{65}	*A. semilineatus*^{47}	*A. bahorucoensis*^{51} FW–M	*A. brevirostris*^{51}	*A. shrevei*^{56} M]
	A. longitibialis^{72}	*A. alumina*^{40}]	*A. dolichocephalus*^{51} FW–M]	*A. marron*^{47}	*A. rimarum*^{45} FW–M
	A. strahmi^{79}	*A. etheridgei*^{43} FW–M		*A. websteri*^{47}	*A. monticola*^{48} FW–M
	A. whitemani^{62} OA]	*A. koopmani*^{39} FW–M		*A. caudalis*^{48}]	*A. rupinae*^{56} FW–M
	(*A. cristatellus*)			*A. christophei*^{49} FW–M	*A. eugenegrahami*^{72} FW
Puerto Rico	[*A. gundlachi*^{72} FW		[*A. krugi*^{55} FW		
	A. cristatellus^{74}		*A. pulchellus*^{51}		
	A. cooki^{62} OA]		*A. poncensis*^{48} OA]		

NOTES:

FW = forest/wet habitat.

OA = open/arid habitat.

M = montane; in Hispaniola there is also a montane genus, *Chamaelinorops*.

Superior number = reported maximum male size, snout-vent, mm.

Allospecies are bracketed; invader species are enclosed in parentheses.

All doubly underlined species are in the *monticola* species group.

very similar ecomorphs may be phyletically quite distant, for instance, *A. cybotes* on Hispaniola and *A. lineatopus* on Jamaica.

Clearly, selective pressures on the several islands have been powerful enough and similar enough to call into existence on 4 islands (Cuba also has the standard-sequence ecomorphs) strongly convergent ecological types, so strongly convergent that the cross matches are ecologically nearly perfect.

(2) What is differentially added on each of the islands?

Within the series Jamaica to Puerto Rico to Hispaniola there are both (a) additions within the standard sequence, that is, climatic vicariants and allospecies, and (b) additions outside the standard sequence, such as montane faunas and specialist species using some niche unknown in the standard sequence. Table 15.4 lists examples of these additions.

Both climatic vicariants and allospecies multiply the species count within ecomorphs but they do so in differing and interesting ways. *Allospecies* by definition are closely related species with allopatric or parapatric distributions. However, though they may achieve parapatry, they are characteristically ecologically close enough to exclude one an-

Table 15.4 Examples of additions to the standard ecomorphs.

Within the standard sequence
- Climatic vicariants (closely related species with strongly differing modal climatic associations *and* marked morphological differences)
  - Among Hispaniolan trunk-ground ecomorphs (Williams, 1963b)
    - *A. cybotes* (mesic, large, ventrals smooth)
    - *A. whitemani* (arid, smaller, ventrals keeled)
  - Among bush ecomorphs of Puerto Rico (Williams, 1972)
    - *A. krugi* (mesic, few dorsal rows enlarged)
    - *A. pulchellus* (less mesic, several dorsal rows enlarged)
    - *A. poncensis* (arid, many dorsal rows enlarged)
- Allospecies (closely related species, parapatric or allopatric, differing little in climatic preference or in morphology)
  - Among Hispaniolan trunk-ground ecomorphs
    - *A. cybotes/marcanoi* (Williams, 1975)
  - Among Hispaniolan grass ecomorphs
    - *A. semilineatus/alumina* (Hertz, 1976)

Outside the standard sequence (morphologically and behaviorally distinct)
- Montane
  - In Jamaica
    - *A. reconditus*, a montane generalist
- Specialist
  - In Hispaniola
    - *A. eugenegrahami*, a semiaquatic species

other and are not even locally syntopic except in very narrow zones or occasionally (apparently by accidental transport) very temporarily, even in terms of ecological time.

Climatic vicariants, on the other hand, while they may be in large measure allopatric, if climatic conditions are different over wide areas, are often in broad sympatry and may in intermediate or edge situations be locally syntopic, even exhibiting the phenomenon that Schoener (1970) called "nonsynchronous spatial overlap"—literal occurrence in identical places but at different times of day.

Anolis cristatellus and *A. cooki* are such climatic vicariants. In southwest Puerto Rico, they do co-occur and are even intimately interspersed, but there is good evidence that they utilize slightly different climatic microhabitats within the same general habitat (Huey and Webster, 1976; Lister, 1976; personal observation).

Climatic vicariants commonly show significant differences in scale characters, for example, scale size, degree of keeling, as well as in thermal behavior. They may have overlapping temperature tolerances but with different maxima and minima.

It can be inferred that all climatic vicariants were once allospecies but have become further differentiated. What is implied here is a spectrum of climatic specialization; it follows that the distinction between allospecies and climatic vicariants will sometimes be subtle or equivocal. Probably all allospecies differ somewhat in climatic preference.

Montane faunas, at least in the examples before us, are to the species of the classic ecomorph sequence as allospecies to climatic vicariants. Some montane species (for example, *A. koopmani* on the mountains of southwest Haiti is a parallel to the lowland grass-bush species) approach the morphological and behavioral characters of the lowland ecomorphs, but they do so incompletely. They are certainly quite independent of their lowland parallels and they are also more or less divergent. I shall regard them below as representing different end points alternative to the classic sequence.

Specialist species: I have in mind here especially the aquatic species—*A. eugenegrahami* in Hispaniola with its parallel in Cuba, *A. vermiculatus*—phyletically quite distant, morphologically dissimilar, analogous only in their semiaquatic habits. The species pair in Cuba, *A. lucius* and *A. argenteolus,* restricted to the trunks of huge complex trees with special hiding places, also qualify as specialists—in this case without known parallels on other islands. *A. bartschi,* a cliff species replacing the *A. lucius-argenteolus* species pair in western Cuba is another specialist.

(3) What is differentially absent in the three islands?

An absence is invisible unless it has left a trace or unless it is suggested by a presence elsewhere (Williams, 1969).

If, as I argued in 1972, there is a natural ecological sequence in ecological radiation, then some absences can be accounted for by the stage in any sequence reached in any island or subfauna. The absence of a grass-bush ecomorph in Jamaica might be accounted for in this fashion. Similarly, the absence of a trunk ecomorph in both Jamaica and Puerto Rico, while this ecomorph occurs in both Hispaniola and Cuba, could be accounted for as an earlier stage in the sequence. Both trunk and grass-bush ecomorphs in my 1972 scheme are late in the ecological sequence and might never have evolved on the smaller islands.

But another possible explanation for the absence of ecomorphs is extinction. This would seem unlikely for the major ecomorphs in the standard sequence, since the niches that they occupy are present nearly everywhere. However, there are cases in front of us that point to the real possibility of extinction for certain categories. One of the 2 large-twig anoles of Hispaniola, *A. darlingtoni,* is known from a single specimen. It is certainly local; it might already be extinct or on the verge of extinction. The apparently demonstrated rarity of *A. fowleri,* known from only 7 specimens, which I interpret as the north island representative of *A. darlingtoni,* may be a stronger case. (The exact habitat of *A. darlingtoni* has not been penetrated since its discovery.) *A. fowleri* occurred in a region of Hispaniola which has been devastated by the recent hurricanes. The absence of a large-twig anole in Puerto Rico or the absence of a small-twig anole in Jamaica might be explained by an analogy with these examples.

The second giant anole of Puerto Rico, *A. roosevelti,* is known from 2 specimens. It has been vigorously looked for and is probably extinct. *Anolis eugenegrahami,* the semiaquatic anole very recently discovered in Haiti, is known from a single locality. It may or may not be genuinely rare, but its habitat requirements appear to be such that its extinction is not unlikely. I mentioned in 1972 *A. poncensis* and *A. cooki,* confined to the most arid areas of southwest Puerto Rico, as species that might be threatened by climatic change and that *A. cooki* may also be threatened by competition with the more eurytopic *A. cristatellus.*

The species confined to the broadleaf forests of montane Hispaniola—*A. christophei, A. etheridgei, A. insolitus*—although they have been abundant, will surely suffer the fate of those forests, if they do not disappear before the forests do.

Note also the restricted range of the *monticola* group on the south island, the Massif de la Hotte in farthest western Haiti. Clearly, the *monticola* group, if it is a natural unit as it appears to be, must have had a distribution well to the east, up to and across the Cul de Sac plain. There must at some time have been a very severe contraction of range. This area (cf. *A. darlingtoni* above) has not been well explored and new taxa and new localities for those already known are to be expected. The rarity or commonness of individual taxa cannot be assessed until the region is better canvassed.

From these cases of apparently or genuinely vulnerable species, it is necessary to extrapolate to the possibility of unobserved species that went extinct before collections were ever made, perhaps normally as part of the fauna-building process, perhaps as a result of secular climatic change, perhaps as a result of post-Columbian, particularly recent, destruction of habitats.

Extinction is clearly an undefined—worse, indefinable—term in our assessment of possible anole communities. The anole communities we see are very possibly a residue only. On the worst view we see only the animals commensal with or tolerant of man; on the best view we see climax communities, the end result of a long history of species-species, species-environment interactions in which many nascent and perhaps also many fully evolved species have gone to extinction. On this point, the discovery in 1977 of an animal so distinct as the semiaquatic *A. eugenegrahami* should give skeptics pause. There are still unknown areas and unknown animals in the West Indies.

Components of the anole faunas of the 3 large islands and the individual species were categorized above. Differences between the 3 islands and especially the relationship of fauna to area will, however, be better appreciated if the faunas are compared as assemblages.

Jamaica

The 7-anole fauna of Jamaica (of which I have seen all species alive and in the field many times) is the simplest of the Greater Antilles. We can be sure that 6 of the 7 evolved in place; the seventh, *A. sagrei,* is conspicuously a late, perhaps human assisted, invader from Cuba that is still confined to the western half of the island. The native Jamaican anoles are all clearly their own closest relatives, an unquestioned intra-island radiation.

This may be our only example of a complex anole fauna evolved within an island. I formerly (1972) tried to portray Puerto Rico as such an island. Since the discovery of *A. sheplani,* more primitive than its relative, Puerto Rican *A. occultus,* (Schwartz, 1974a), on Hispaniola, there is good reason to reject that hypothesis and to assume that Puerto Rico has had at least 3 invasions, one ancestral to *A. occultus,* one ancestral to *A. cuvieri,* and a third for the stock ancestral to all the remaining Puerto Rican anoles. Cuba certainly has had 2, while Hispaniola has been invaded and back-invaded (Williams, 1969, also Table 15.5 and Figure 15.4).

Ecomorphs are clearly present on Jamaica: (1) the active and aggressive *A. garmani* is the crown giant, tending to occur high; (2) a somewhat smaller species, *A. valencienni,* a slow searcher, tending also to be high but preferring smaller perches, even twigs, represents what I now call the twig-giant ecomorph; (3) *A. grahami* is the larger of the 2 trunk-crown

Table 15.5 Invasions into and within the West Indies, and the resulting faunas.

Target	Invasion number	Faunal number
Cuba (43,036 mi^2)	2	> 35[a]
Hispaniola (28,242 mi^2)	4	> 35[a]
Puerto Rico (3,421 mi^2)	3	11
Northern Lesser Antilles	1	1 or 2 per island bank
Southern Lesser Antilles	1	1 or 2 per island bank
Jamaica (4,450 mi^2)	1 + 1[b]	7

NOTE: Island area according to Rand, 1969.

a. Species numbers for Cuba and Hispaniola are already 2 to 3 species greater than 35, with more species to be described.

b. *Anolis sagrei* remains equivocal—probably imported by man, but not demonstrated to be so.

ecomorphs that I now recognize, foraging in the crown and on the upper trunk; (4) *A. opalinus* I now regard as a second smaller trunk-crown ecomorph, on this island somewhat more shade loving than its larger counterpart; (5) *A. lineatopus,* representing the trunk-ground ecomorph, typically perches head downward, foraging for prey on the ground. Different races of this last species (Fig. 15.5) prefer sun (*A. l. lineatopus*) or shade (*A. l. neckeri*).

Anolis sagrei (separated as a nonnative species in Fig. 15.5) is a second trunk-ground ecomorph. In western Jamaica its perch is lower than that of neighboring *A. lineatopus* and it is sun loving in contrast to western shade-loving *A. lineatopus neckeri.*

The last Jamaican species, *A. reconditus,* is a montane isolate in eastern Jamaica. It is outside the ecomorph series, a solitary generalist of high wet forest.

This is a summary of data reported in Underwood and Williams (1959), Rand (1967b), Schoener and Schoener (1971a) and Hicks (1973) as well as personal observations. One caveat is necessary: the Jamaican ecomorphs are most distinctive in size and color, less so in body proportions or scales, and least so in behavior, which depends substantially on the presence or absence of other species (see Jenssen, 1973).

Modal perch and climatic preferences are diagrammed in Figure 15.5. Maximum snout-vent length in males is indicated opposite each name. An altitudinal profile (Fig. 15.6) is intended to indicate that lowland diversity tends to attenuate (in reality only slightly) until only *A. garmani* and *A. opalinus* are adjacent to the wet montane forest that *A. reconditus* inhabits alone (Hicks, 1973). Note in Figures 15.5 and 15.6 that in Jamaica the bush habitat has no species specifically adapted to it, that there is no dwarf-twig ecomorph, and that the trunk-crown ecomorph is divided by size.

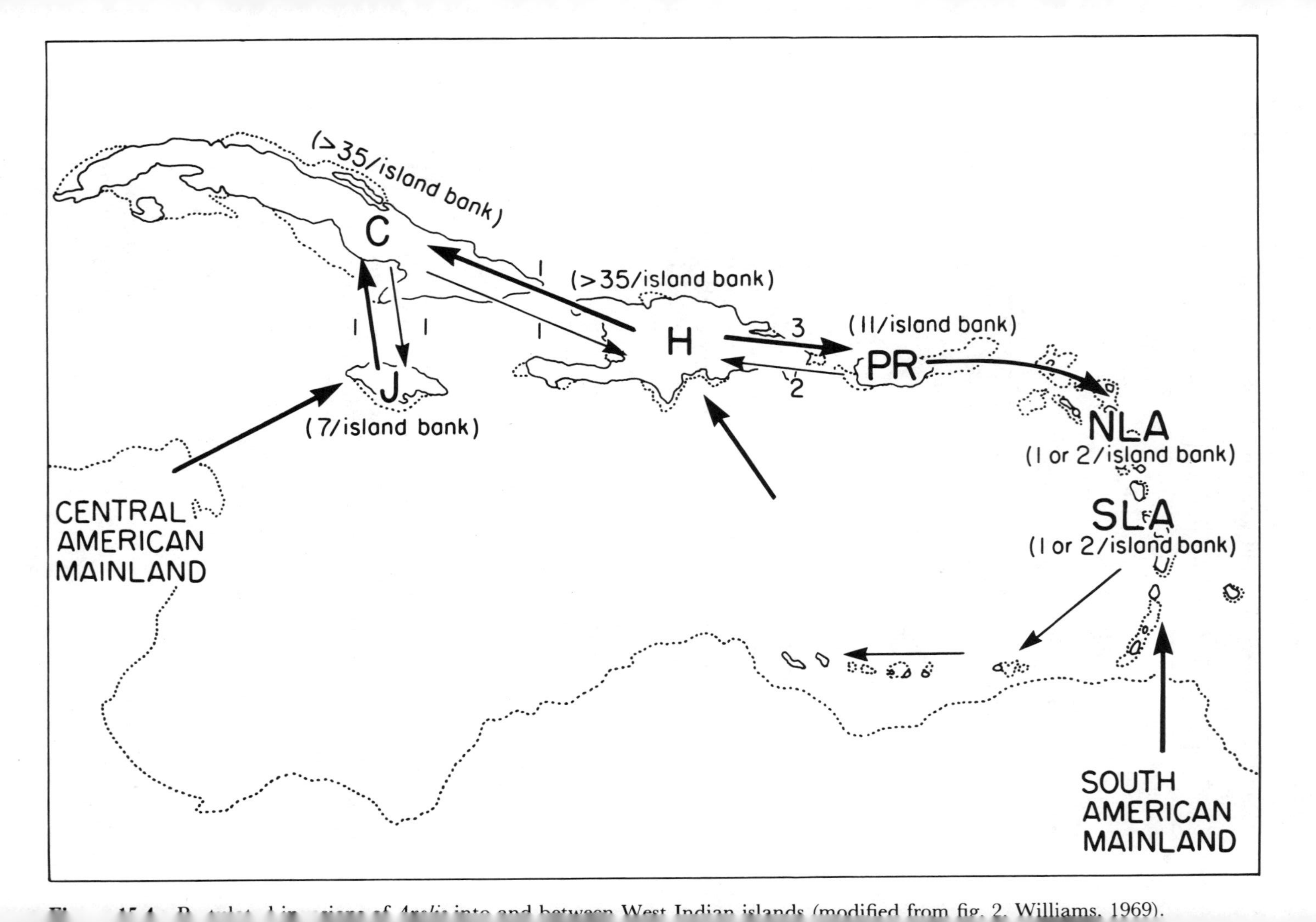

Figure 15.4. Postulated invasions of *Anolis* into and between West Indian islands (modified from fig. 2, Williams, 1969).

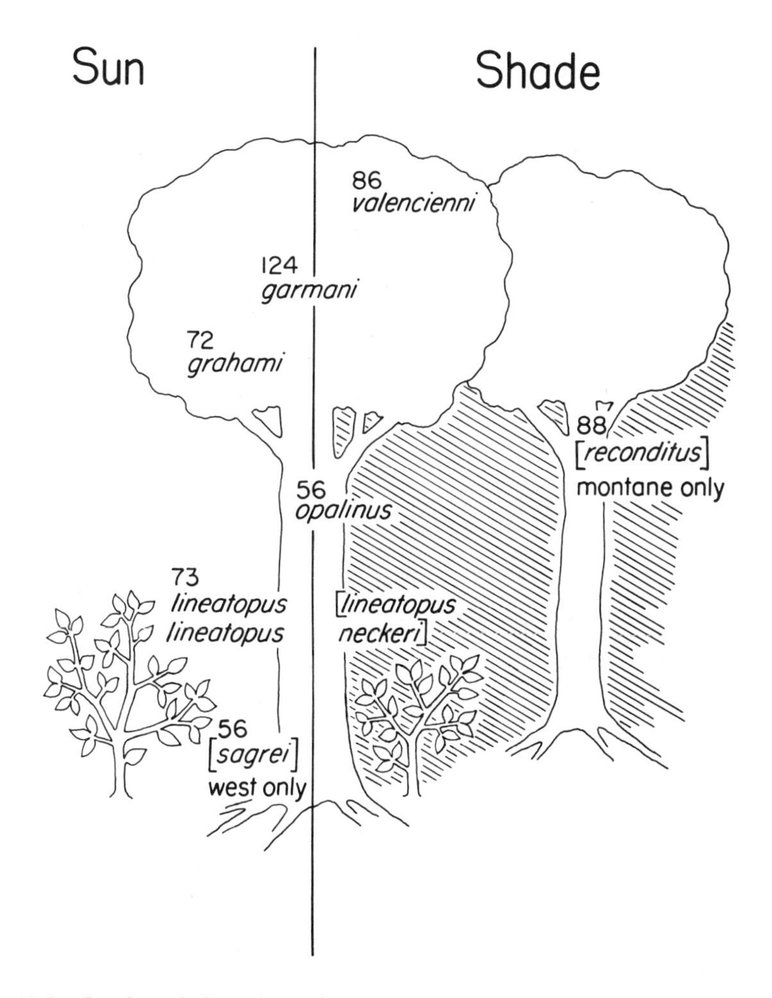

Figure 15.5 Perch and climatic preference of Jamaican species. Numbers are maximum adult male snout-vent size in mm. Note that there are no species specialized to the grass-bush niche. The ecological relationships of the 5 anoles not bracketed are as shown on the campus of the University of the West Indies near Kingston (Rand, 1967).

In Jamaica we have neither climatic vicariants nor allospecies. Climatic adaptation occurs within species—conspicuously within *A. lineatopus* where climatic variants are sharply distinguished by color, including dewlap color, and are recognized as subspecies. A lesser climatic adaptation exists within *A. grahami,* identifiable by color and taxonomically recognized as subspecific.

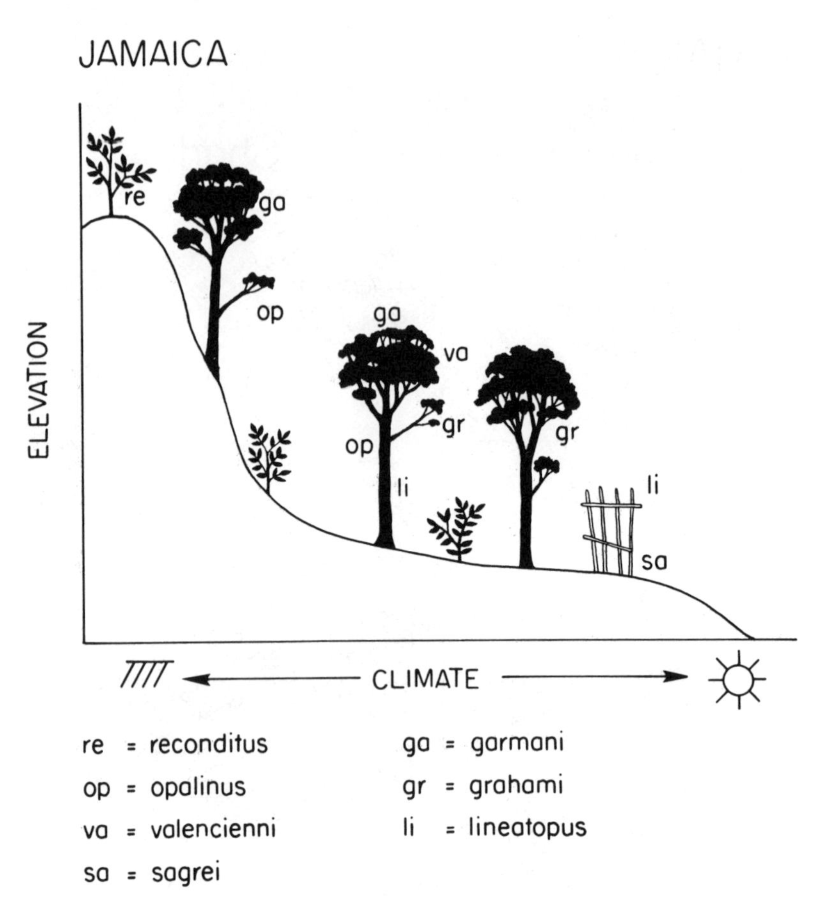

Figure 15.6 Diagrammatic profile of Jamaica showing variation of community structure with elevation and climate. There is considerable behavioral and some morphological variation within species with respect to climate, for example, *A. lineatopus* would not typically be found above *A. sagrei* on fence posts because in western Jamaica, where *A. sagrei* occurs, *A. lineatopus* is a relatively shade-loving species. In the Kingston area, in contrast, *A. lineatopus* is the characteristic lizard of sunny fence posts.

All Jamaican species are sympatric, often syntopic, except *A. reconditus,* which minimally overlaps with *A. garmani* and *A. opalinus. Anolis reconditus* is the only real addition in Jamaica and one, as mentioned above, that is outside the classic sequence.

Puerto Rico

In Puerto Rico there are only standard-sequence ecomorphs (Fig. 15.7): 2 giants, *A. cuvieri* and *A. roosevelti,* a cryptic dwarf-twig *A. occultus,* a large green anole (trunk-crown ecomorph I) *A. evermanni,* and a smaller

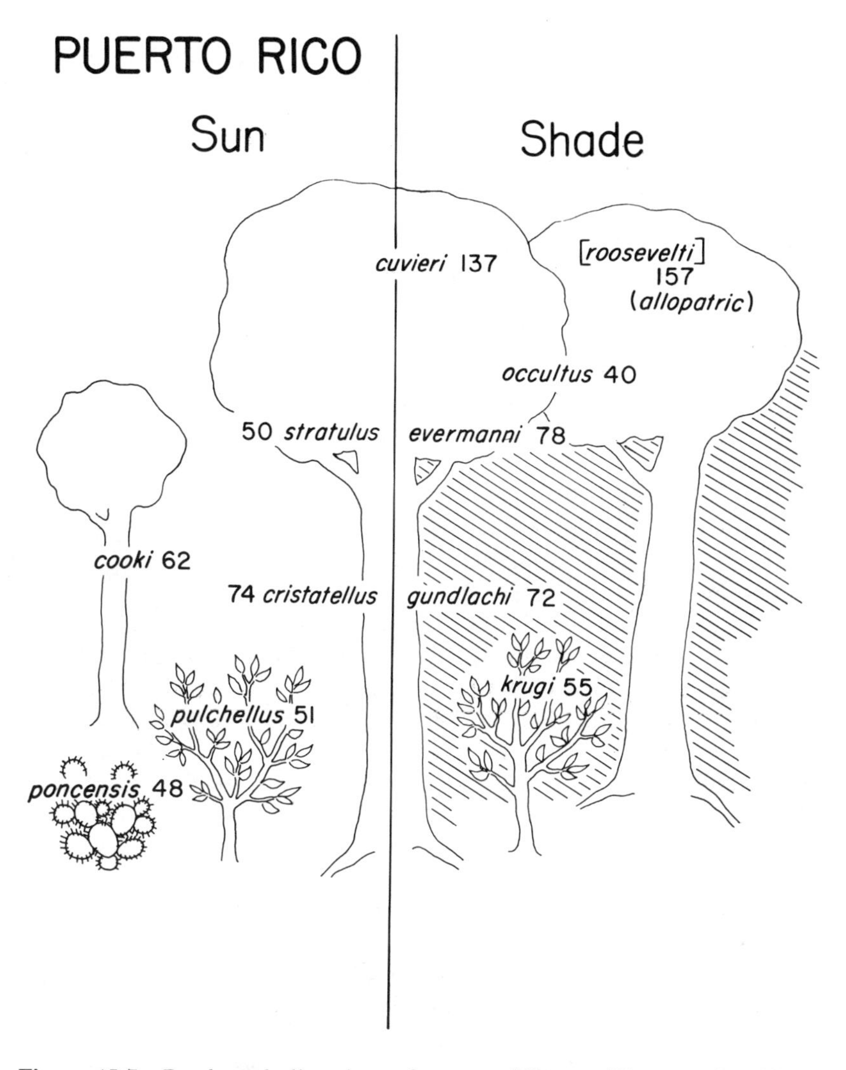

Figure 15.7 Perch and climatic preferences of Puerto Rican anoles. Numbers are maximum adult male snout-vent sizes in mm. *Anolis roosevelti* is bracketed because it is known only from Culebra Island and may be extinct (modified from Williams, 1972).

greyish species (trunk-crown ecomorph II) *A. stratulus,* 3 trunk-ground species, *A. gundlachi, A. cristatellus* and *A. cooki,* and, in addition, a type missing on Jamaica, the grass-bush ecomorph, here represented by 3 species (relatively heavy-bodied for this ecomorph), best called bush ecomorphs, *A. krugi, A. pulchellus* and *A. poncensis.* I have seen all species alive and in the field with the sole exception of *A. roosevelti.* Named from

2 specimens, this species has not been found again in spite of determined search.

Missing in Puerto Rico is the large-twig anole. (There appears to be some complementarity—see below for the Hispaniolan fauna—between giant and dwarf-twig species.) Missing also is the trunk ecomorph. In contrast to Jamaica, also, there is no distinctive montane species. Even the dwarf-twig anole, *A. occultus,* formerly Puerto Rico's last candidate for an exclusively montane species, has been found at elevations of only 200 m in northwestern Puerto Rico (Thomas and Thomas, 1977). (Note that in Puerto Rico the mountains rise only to 4,000 feet, while those in Jamaica go to 7,000.)

The striking new additions in Puerto Rico are by climatic vicariance (Fig. 15.8). In contrast to Jamaica, where climatic divergence is infraspecific, visible in behavior or at the subspecific level, in Puerto Rico climatic divergence shows itself at the specific level. Even the 2 giants appear to be such: *A. cuvieri* on mainland Puerto Rico occurs characteristically in more mesic situations than are (or were) available to *A. roosevelti* in the comparatively dry forest of Culebra.

Quite classic cases of climatic vicariance are the trio of trunk-ground species, *A. gundlachi, A. cristatellus, A. cooki,* which range in that order from habitats that are quite wet to those that are very arid. These are paralleled by the trio of grass-bush species, *A. krugi, A. pulchellus, A. poncensis,* which have the same range of climatic preference and hence quite parallel distributions. The members of neither trio are allospecies. All of these species are quite distinct in many ways and at least interdigitate. *A. cristatellus,* in fact, is sympatric, even syntopic, with *A. cooki* over most of the latter's range.

I emphasize that there are no allospecies in Puerto Rico. This, as we shall see, is in stark contrast to the situation on Hispaniola.

It is an interesting point that the smaller of the two trunk-crown species of Puerto Rico, *A. stratulus,* is believed to represent the exact stock that, when it colonized Hispaniola from Puerto Rico, evolved into the *distichus* complex in Hispaniola, which are the classic trunk ecomorphs.

As might be expected in such an ancestor-descendant relationship there are similarities between *A. stratulus* and the distichoids. They are similar in size and even in shape. In behavior, however, they differ strongly. *Anolis stratulus,* like all trunk-crown anoles, do spend much time on the trunks of the trees on which they are resident, but *A. stratulus* is found with higher frequency in the canopy. Above all, *A. stratulus* is never inserted between the larger trunk-crown species and a trunk-ground species; it is never inserted between *A. gundlachi* and *A. evermanni.* In relation to *A. cristatellus,* it behaves as a trunk-crown species should and there is no species above it. In fact, because it is a less shade-loving species than *A. evermanni,* it is often found at the outer margins of forests

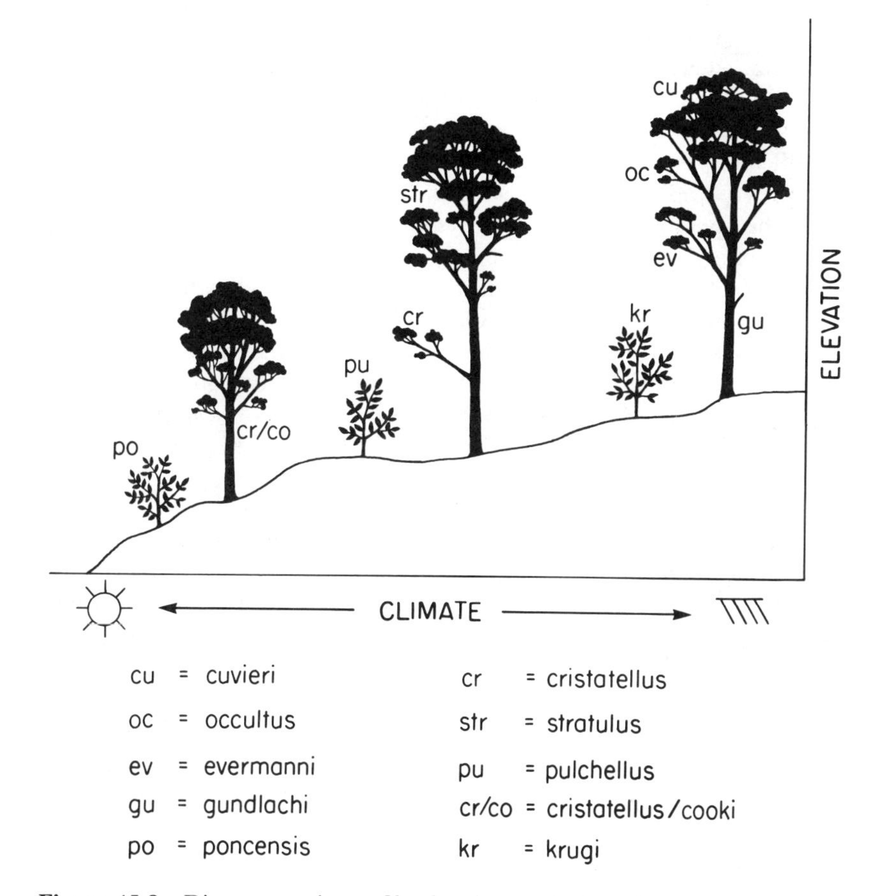

Figure 15.8 Diagrammatic profile showing variation of community structure in Puerto Rico with elevation and climate.

inhabited by *A. evermanni,* including the top of the canopy, thus higher than *A. evermanni.*

Hispaniola

Hispaniola is very different from the 2 smaller islands in much more than mere faunal number. The increased number is achieved in a distinctive way: there is an extraordinary proliferation of allospecies (see Tables 15.3a and b where they are named and marked off in brackets). Each of the standard sequence ecomorphs is represented by two or more allo- or parapatric species. In 2 of the more complex cases—the trunk-ground and grass ecomorphs—there are additionally in each case examples of the climatic vicariance first found in Puerto Rico: in the trunk-ground series *A. whitemani* is in the very arid lowlands and *A. cybotes* in adjacent more mesic situations, and in the grass series *A. olssoni* is in arid lowland, whereas *A. semilineatus* is typically in more mesic situations in the lowlands and at high elevations. In the trunk ecomorph—this is the first level of complexity at which this ecomorph is found—the situation is still more complicated: there is a series of allospecies, the *brevirostris* complex (Arnold, 1980)—*A. websteri, A. caudalis, A. marron, A. brevirostris*—the members of which are sometimes allopatric to, sometimes parapatric to, sometimes complexly interdigitated with or intervening between subspecies of a sister species that is infraspecifically divided climatically—*A. distichus* (cf. Schwartz, 1968).

In this island, while I have seen most of the 36 described species repeatedly in the field, I have seen *A. eugenegrahami* alive only in captivity and *A. darlingtoni, A. dolichocephalus, A. haitianus, A. koopmani, A. marron, A. monticola, A. rimarum, A. rupinae,* and *A. sheplani* only as preserved specimens.

The allospecies of Hispaniola are only partly correlated with the major physiographic subdivision of the island—a trough, the Cul de Sac–Valle de Neiba plain, partly below sea level—that divides Hispaniola into north and south islands (Williams, 1961). This trough, which was at times in the Pleistocene a real physiographic barrier, must have played a part in the origin of the trunk-crown I allospecies, *A. chlorocyanus* and *A. coelestinus;* it still coincides with the boundary between them. It is astonishing, however, that this is the only instance in which the below-sea-level trough and the consequent north and south islands are a sufficient explanation of Hispaniolan allospecies. In other cases dispersal across the boundary is the minimum explanation required, while in still other cases no obvious explanation of allospecies is currently available. *Anolis marcanoi* (Williams, 1975) is an especially provocative example. It is a close relative of *A. cybotes,* and its range in the south central Dominican Republic is entirely surrounded by that species. Schwartz and Thomas (1977) have pointed to the parallel occurrence of a *Sphaerodactylus* with a distribution entirely coincident with that of *A. marcanoi.* Yet neither for *A. marcanoi* nor for the *Sphaerodactylus* has it been possible to point to any physiographic or vegetational or climatic factor that can in any way isolate these 2 taxa or explain their origin and persistence in the area they occupy.

The proliferation of allospecies in Hispaniola becomes especially interesting when it is realized what it implies: allospecies are sets of populations that have begun the speciation process but have not carried it beyond its first stages; they have been unable to achieve sympatry with their sister populations. Thus, in contrast to Jamaica and Puerto Rico, where all populations were either clearly species in the fullest sense of the term or were as clearly below that level, it is a characteristic of Hispaniola that the majority of named species-level taxa appear to have at best only recently crossed the species threshold. The situation could be described in terms of permissiveness and restraint: Hispaniola, presumably because of its size and physiographic complexity, has been permissive of differentiation but has also, for whatever reason, held most differentiation to a relatively low level.

Extraordinary also in Hispaniola is the existence of not one but several montane faunas. The exclusively montane species are marked by a special symbol "M" in Tables 15.3a and b. There is first a highly distinctive fauna (evolved mostly from a single-species group, the *monticola* species group, doubly underlined in Table 15.3b) in montane broadleaf forest.

In addition, there is a second montane fauna made up of taxa either directly ascending from the lowlands, for example, *A. cybotes, A. chlorocyanus, A. distichus,* the *ricordii* allospecies (*A. ricordii, A. baleatus, A. barahonae*), or the more or less modified descendants of such lowland dispersers, *A. shrevei* and *A. armouri* (Williams, 1963b).

This is a far cry from the single montane species of Jamaica and the absence of distinctively montane species in Puerto Rico. Furthermore, the greater part of the montane faunas either do not fit the classic sequence and are as clearly separately evolved as compared with the lowland members of the standard sequence as if they had evolved on a different island.

Montane *A. cybotes, A. chlorocyanus, A. coelestinus, A. distichus, A. semilineatus,* and the *ricordii* allospecies, it is true, are changed little or not at all from their corresponding lowland populations. But these are not, except perhaps the *ricordii* allospecies, inhabitants of undisturbed montane broadleaf forest and appear to do relatively poorly or fail entirely in montane pine forest. These montane representatives of lowland species appear to be, as Schwartz (1974b) has suggested, very recent immigrants into open areas, especially the areas opened by roads and the disturbance associated with man.

The truly montane species include species closely related to the characteristic lowland species as well as the others that are quite separate lineages. These montane vicariants of lowland species were not seen on the smaller islands. They include the trunk-crown II species, the smaller subdivision of that ecomorph. There are 3 species: *A. aliniger* and *A. singularis* (Williams, 1965) and an undescribed blue-dewlapped taxon known only from the Sierra Martin Garcia. The first 2 have distributions parallel to those of their close lowland relatives, *A. aliniger* primarily on the north island, *A. singularis* solely on the south island, but *A. aliniger* has somehow gained a foothold south of the Cul de Sac on the Massif de la Selle. In scales also they quite parallel their lowland relatives: *A. singularis* is distinguishable only minimally except in size; *A. aliniger* has one unique peculiarity—its partly pigmented but wholly scaleless axilla. For all 3 Hispaniolan trunk-crown II species, however, their restriction to high elevations is presumably related to a physiological difference from the primarily lowland Hispaniolan trunk-crown I species.

(There is clearly a tendency for differences other than size to distinguish the 2 subdivisions of the trunk-crown ecomorph, but this ecological divergence appears to differ opportunistically on the different islands; thus on Jamaica *A. opalinus* prefers shade or cooler situations than does *A. grahami.* It tends therefore to be found in open situations in the mountains only. In contrast, Puerto Rican *A. stratulus* accepts sun more readily than *A. evermanni.* In Hispaniola the 2 sizes of trunk-crown ecomorphs are separated by still another device, lowland-highland vicariance.)

There are montane vicariants also in the trunk-ground ecomorph. *A. armouri* (Massif de la Selle) and *A. shrevei* (Cordillera Central) are obviously closely related to *A. cybotes.* (In 1963, I considered *A. armouri* a subspecies of *A. cybotes.*) Animals of the montane pine forests, they live primarily on the ground and rarely perch on pine. Smaller than *A. cybotes* (and *A. shrevei,* unlike *A. cybotes,* heavily keeled), they have, however, diverged from their lowland relative more in behavior than in structure.

The bush niche in Hispaniola is montane. It is occupied by the *hendersoni* (Williams, 1963a; Schwartz, 1977) set of allospecies (*A. hendersoni, A. bahorucoensis, A. dolichocephalus*) limited to the mountains of the south island—south central and western Hispaniola. They are in habit and habitat bush anoles but distinctive in morphology: their extraordinarily long heads, reduced dewlaps, and spectacular coloration put them outside the standard set of bush-grass anoles and require that they be recognized as another of the unique montane components of the Hispaniolan fauna, in this case probably rather distantly related to the Hispaniolan trunk-crown anoles and not nearly as closely to the Hispaniolan grass anoles. As bush animals they are inhabitants of the edge rather than the interior of montane broadleaf forest (Moermond, 1979; personal observation).

Montane *A. insolitus* and *A. sheplani* (Williams and Rand, 1969; Schwartz, 1974a) belong in the standard sequence as twig dwarfs and as such closely resemble the twig dwarf of Puerto Rico, *A. occultus,* in both morphology and behavior.

Anolis darlingtoni (Cochran, 1935) is known from a single specimen from the Massif de la Hotte in southwest Haiti; *A. fowleri* (Schwartz, 1973) is known from the Cordillera Central. Nothing at all is known of the habits of the first; of the second nothing is known except where it was found sleeping, "on twigs" and "across branches." On the basis of the same close resemblance that made Cochran (1935) place *A. darlingtoni* in the same genus (*Xiphocercus*) that Jamaican *A. valencienni* was then relegated to, I infer that the habits of *A. darlingtoni,* when known, will be very like those of *A. valencienni,* those of a twig giant. (These 2 species are quite distinct phyletically; the resemblance is clearly ecological convergence.) *Anolis fowleri* has never previously been compared with *A. darlingtoni.* It is clearly a quite distinct species, but I again infer from its morphology that it is a second giant twig species of Hispaniola—a geographic representative of *A. darlingtoni* in the Cordillera Central and phyletically related as well as ecologically equivalent.

It is noteworthy that twig anoles are curiously spotty in their distribution; only Jamaican *A. valencienni* is genuinely widespread within its island. All 4 Hispaniolan species are not only strictly montane, they are local within this restriction. *Anolis insolitus* is locally common (personal observation) and known from several localities but all are in the Cordillera Central. *Anolis sheplani,* while known from fewer localities, does occur on both sides of the Cul-de-Sac–Valle de Neiba plain, south of it on the Sierra de Baoruco, north of it on the Sierra de Neiba. *Anolis fowleri,* known from 7 specimens from 2 localities, in the Cordillera Central, may be genuinely rare. *Anolis darlingtoni,* although nothing can be said about its local abundance in its unvisited type locality in the Massif de la Hotte, appears to be really absent from the relatively well-collected mountains of the eastern portion of the Hispaniolan south island.

Close to the same basal stock as the twig anoles is the *A. monticola* series—the one really distinctive montane radiation in Hispaniola, as diverse as it is distinctive. There are 2 widely disjunct subgroups within it, the more primitive on the 2 northernmost of the 3 mountain ridges that dominate the Hispaniolan north island—the Cordillera Central and the Cordillera Septentrional, the second, highly specialized, again, like *A. darlingtoni,* confined to the Massif de la Hotte in the extreme southwest.

Anolis christophei (northern group) is much the most primitive of these. It has been classified in Table 15.3b as a trunk anole. It fits in size but not very well in any other way. In montane broadleaf forest it has the tree trunks and the branches of adjacent bushes and banks and rock ledges to itself. It has neither a trunk-crown anole above it nor a trunk-ground anole below it. Its habitus is quite unlike that of *A. distichus.* It is something of an analogue but not at all a homologue of a standard trunk anole.

Anolis etheridgei is in the undergrowth under the trees of the northern montane broadleaf forest. In morphology it resembles neither the grass anole of the standard sequence nor the *hendersoni* allospecies that has the same niche in the south island of Hispaniola. It is again an analogue rather than a homologue.

Anolis rimarum (again northern group) is an anole of rocky fields but not in any obvious way specialized for this habitat. It has no parallel in Jamaica or Puerto Rico and does not belong in the standard sequence, does not in fact make any approach to any of the standard ecomorphs.

Anolis monticola and *A. rupinae* (southern group) are again two rock anoles, the second larger than the first, with which it is syntopic. They are both gaudily colored, obviously closer to each other than to anything else, and not especially similar to *A. rimarum.* They clearly do not belong to the standard sequence. (For a comparison of *A. monticola* with the standard sequence see Moermond, 1979.)

Anolis koopmani is phyletically closest to *A. monticola* and *A. rupinae* (Thomas and Schwartz, 1967; Webster, Hall, and Williams, 1972) but behaviorally and morphologically it is the most similar of any of the *monticola* series to a standard ecomorph, specifically the grass ecomorph (Williams and Webster, 1974; Moermond, 1979). In a disturbed area in southwest Haiti where Moermond observed it, it exists alongside *A. semilineatus* like a climatic vicariant of the latter, although it is conspicuous that the resemblance is not phyletic.

The anole fauna of the Hispaniolan montane broadleaf forest, in fact, looks as though it evolved quite separately from the lowland fauna, as though it were on an island within an island. More interesting than that, however, it may present us with an example of an incomplete or an *alternative sequence* (see below).

The montane faunas contribute most of the ecomorph category "Other" in Hispaniola. There is, however, one *outré* species that is lowland: recently described *A. eugenegrahami* (Schwartz, 1978). Ecologically it fits very well in the semiaquatic niche well known from several species in South and Central America and from one species in Cuba. But this is a niche that apparently does not impose any rigid morphological constraints. *Anolis eugenegrahami* does not closely resemble any other anole, not Cuban *A. vermiculatus,* not the *lionotus* group, nor *A. barkeri* nor *A. aquaticus* of the mainland. On this point see Schwartz (1978), who has made all the necessary comparisons. *Anolis eugenegrahami* is not an ecomorph in my sense; except in its behavior, it is idiosyncratic. If its distribution is as limited as present information might indicate, it would fit, classically, a category of "local survivor"—if, in fact, it has long to survive.

The allospecies and the montane faunas so complicate the distributional/ecological patterns within Hispaniola that figures fully comparable to those for the other islands cannot tell the full story. Figures 15.9–15.12, however, should assist.

It is necessary to treat the lowlands and montane areas separately, and the north and south island faunas must be distinguished also. Figure 15.9 shows size, perch, and climatic preferences for species from the north island lowland faunas, and Figure 15.10 does the same for the north island montane faunas. Figure 15.11 provides an altitudinal/climatic profile for all north island species; arrows indicate the altitudinal placement but not the perch of two species recently discovered and still poorly known. Figure 15.12 is a south island altitudinal and climatic profile. Comparison with Figure 15.11 will show obvious general parallels despite dissimilarity in the species represented. As in Figure 15.11, species doubly underlined belong to the *monticola* series. The figure is composite in that several of the species are localized even within the south island, for example, *A. armouri* is known only from the central south island range, the Massif de la Selle, *A. sheplani* only from the Sierra de Baoruco, the 3 *monticola* series species only from the western Massif de la Hotte, *A. alumina, A. longitibialis,* and *A. strahmi* only from the Barahona Peninsula. *Anolis brevirostris,* although isolated populations or related species occur on other parts of the south island, is itself known only from the Barahona Peninsula and the adjacent Cul-de-Sac–Valle de Neiba plain. The three allopatric species of the *hendersoni* superspecies, spread across the whole of the south island, are represented in Figure 15.12 by the single symbol "h."

As the comment on the last figure implies, these figures understate the complexity, which is at a completely different level from that of the smaller islands.

Discussion

Faunas—Ephemeral or Persistent

In an evolutionary sense no faunas are permanent. At best they are more or less persistent. In this chapter I have treated present faunas as persistent, and even local situations as persistent—not without occasional cautionary phrases about the past, present, and future effect of human action. Roughgarden and his colleagues (Chapter 16) treat even the simplest of my faunas, the Lesser Antillean species pairs, as evanescent, a temporary phase in a dynamic ineluctable process. I do not argue the case, but I take the opportunity here to warn that the details that I have reported here are quite possibly labile and unstable, even if they have been stated with total accuracy—something more to be hoped for than achieved—but I am confident that large patterns here merit attention, whether persistent or dynamic, and whether I have presented them with entire fairness or not. For a critique of generalization from faunal data, see Sih and Dixon (1981) and the immediately following reply by Fretwell (1981).

Syntopy, Interdigitation, and Parapatry

Complex faunas here described are made complex by both spatial overlap and spatial separation of their contained species.

Table 15.6 lists for selected localities in each of the 3 large islands the number of anole species that might be found within a few feet of one

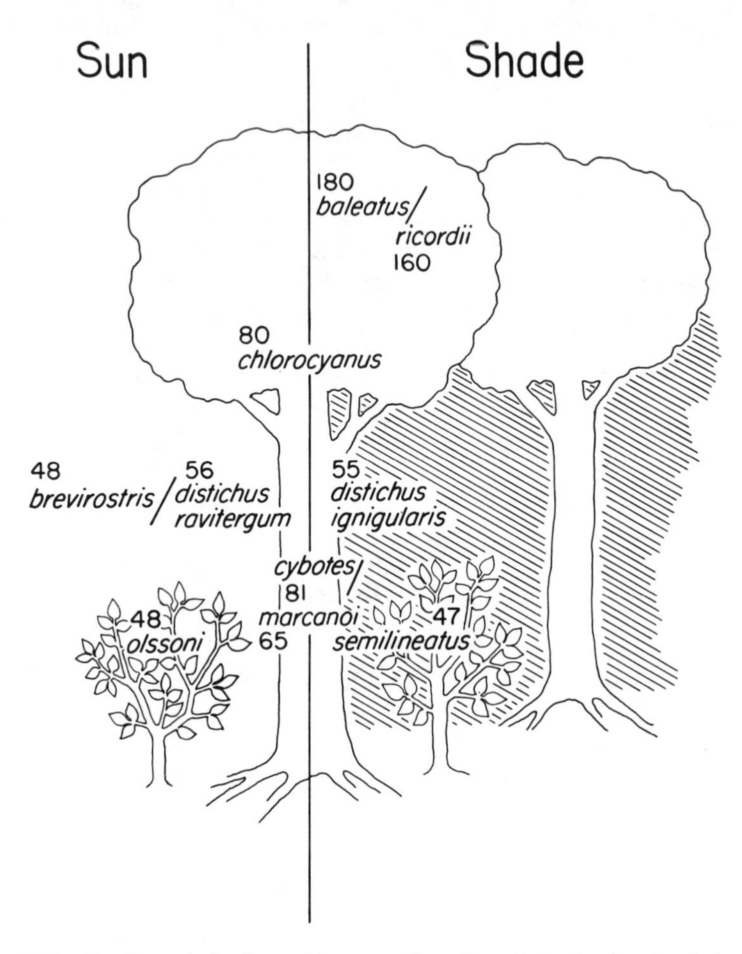

Figure 15.9 Perch and shade preferences of anoles of the lowlands of the north island of Hispaniola. All are standard sequence ecomorphs. Numbers are maximum adult male snout-vent sizes in mm. Species separated by a solidus (/) are allospecies.

another, that is, those that might be found in one tree or a few and their immediately adjacent bushes. These animals may usefully be regarded as syntopic. Three is a very usual count; the extreme count possible might be 6.

Where there are climatic vicariants interdigitation of habitats will sometimes permit species usually well separated to be seen within yards of one another. Table 15.7 lists some localities at which this situation

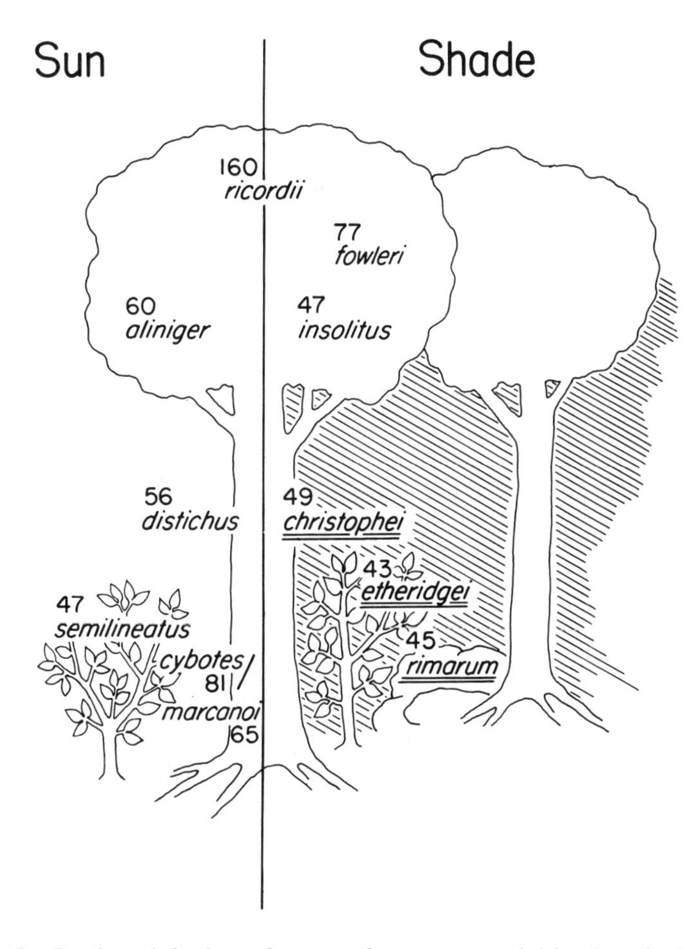

Figure 15.10 Perch and shade preferences of montane north island anoles in Hispaniola below the pine zone. Numbers are maximum adult male snout-vent sizes in mm. Species doubly underlined belong to the strictly montane *monticola* group. *Anolis shrevei* occur in the pine zone above these species.

would be seen. The extreme count of such species in Hispaniola or Cuba might be 10.

The tables cite maxima for specific localities. It should be remembered, however, that there are also minima, and that even on the largest islands in the densest faunas there are local situations and times in which faunas seem depauperate, despite apparent adequacy of habitat, and in which close observations find some to many fewer species than

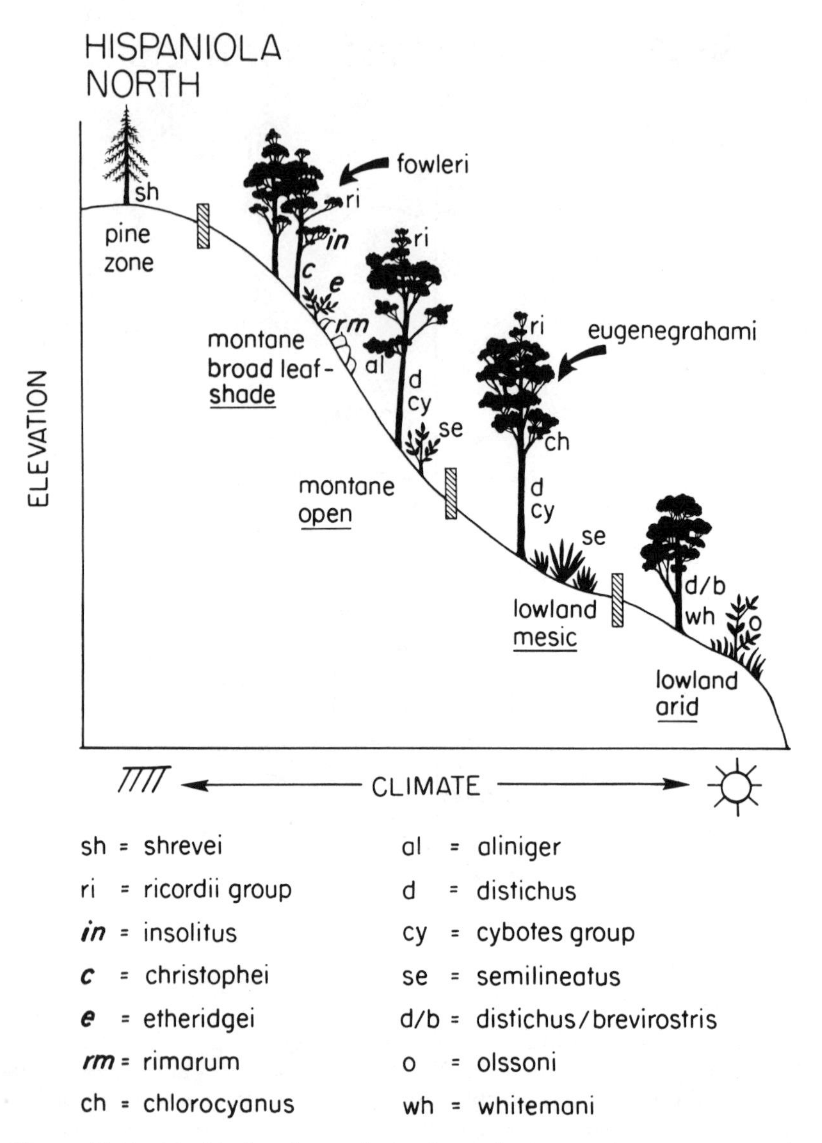

Figure 15.11 Profile showing variation in anole community structure on the north island of Hispaniola. Montane broadleaf and montane open communities are for the most part not closely related. Except in the case of *A. brevirostris* and *A. distichus,* which are complexly interdigitated, allospecies are not shown. Arrows indicate the occurrence of 2 recently discovered species that are poorly known.

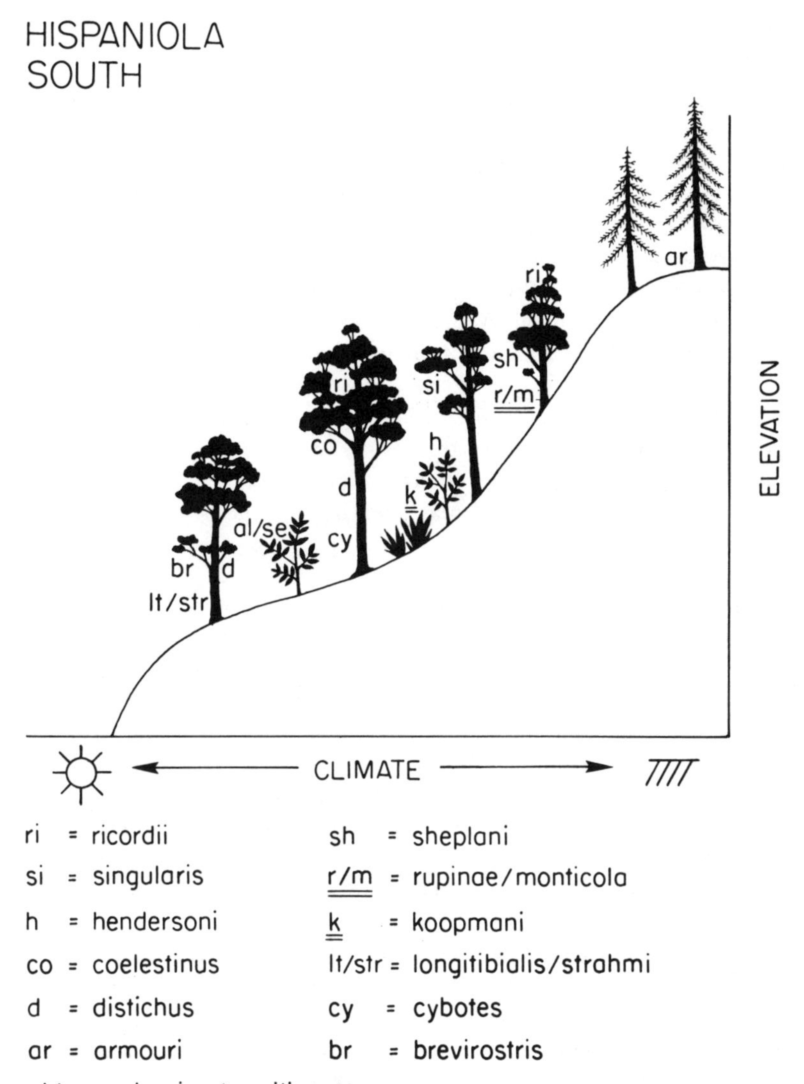

Figure 15.12 Diagrammatic profile of the anole communities of the south island of Hispaniola. Doubly underlined species belong to the *monticola* series.

might be expected. This may be a temporary phenomenon or an artifact of observation (or the individual observer), but such situations are sufficiently often confronted that they must be considered part of expected faunal variance, whether or not any ad hoc explanation can be devised. The maximal numbers of coexistent anoles given above for the larger

West Indian islands are not rare or even unusual, but they are not nearly modal.

Another local condition is not cited in any of the lists of Tables 15.6 and 15.7. This is the occasional doubling of species at the contact zones of allospecies, sibling species, or semispecies. There is sometimes a gap rather than overlap and then there is no doubling; one condition may be as frequent as the other. Thus on Hispaniola, *A. marcanoi* sometimes overlaps *A. cybotes, A. brevirostris* may be separated by an apparent gap from *A. distichus* or may contact that species (and may then hybridize or not, depending on locality), and Schwartz and Thomas (1975) report the overlap of *A. cybotes* with *A. armouri* and of *A. hendersoni* with *A. bahorucoensis.*

The lists in Tables 15.6 and 15.7 represent essentially coadapted communities; they correspond to the results of the 3 axes of adaptation of *Anolis* in Figure 15.2 (size, perch, and climate). The contacts of allo-, sibling, and semispecies are not adaptive; they are the byproducts of an incomplete speciation process.

Thus 2 components to the size and complexity of anole faunas exist: (1) the high divisibility of the anole niche (speciation completed) and (2) the phenomenon of *species nascendi.*

Modes of Speciation, Modes of Colonization, Faunal Buildup

Most discussions of faunal buildup (Lack 1947, 1976; MacArthur and Wilson, 1967; Diamond, 1975) have used birds as their empirical examples. Classically faunal buildup in these models has been by multiple invasions—colonizations separated by time intervals and occurring across some discrete barrier, such as an oceanic strait. Invaders are considered to have already reached the level of species. The process so envisaged is one of accretion or accumulation.

This does seem to be the process by which island avifaunas originate. But Diamond (1977), who distinguishes 3 modes of allopatric speciation for Pacific land birds (continental, intra-archipelagal, and inter-archipelagal), goes on to make an observation very pertinent for anoline lizards: "For taxa whose dispersal ability and population density differ from those of birds, the relative contributions of the three modes may be very different. What is an island to one group of taxa may be a continent to another."

This is very much to the point. Clearly, to *Anolis* the Greater Antilles have been (relatively speaking) continents. In this regard the contrast between birds and lizards is very great.

Lack (1976:193) comments: "In the land birds, there has been virtually no adaptive radiation within the West Indies, though on Hispaniola the two species of todies and the two of endemic warblers represent a first step. Instead, as is usual in oceanic archipelagos elsewhere, each

Table 15.6 Syntopic or nearly syntopic anoles at selected localities on the three large islands.

Jamaica	Puerto Rico	Hispaniola
Mona forest, eastern Jamaica	El Yunque forest, eastern Puerto Rico	Vicinity of Santo Domingo, Dominican Republic
A. garmani	*A. cuvieri*	*A. baleatus scelestus*
A. valencienni	*A. occultus*	*A. chlorocyanus*
A. grahami grahami	*A. evermanni*	*A. cybotes*
A. opalinus	*A. gundlachi*	*A. distichus ignigularis*
A. lineatopus lineatopus	*A. krugi*	*A. semilineatus*
Whitehouse, western Jamaica	San Juan	La Palma, Dominican Republic (broadleaf forest)
A. garmani	*A. stratulus*	*A. baleatus sublimis*
A. valencienni	*A. cristatellus*	*A. insolitus*
A. grahami grahami	*A. pulchellus*	*A. christophei*
A. opalinus	Cabo Rojo, western Puerto Rico	*A. etheridgei*
A. sagrei	*A. cristatellus*	La Palma, Dominican Republic (open)
	A. cooki	*A. aliniger*
	A. poncensis	*A. cybotes*
	Culebra Island, Virgin Islands	*A. distichus ignigularis*
	A. stratulus	Barahona City, Dominican Republic
	A. cristatellus	*A. coelestinus*
	A. pulchellus	*A. cybotes*
		A. brevirostris wetmorei
		A. alumina
		Ducis, near Aux Cayes, Haiti
		A. coelestinus
		A. cybotes
		A. distichus vinosus

NOTE: Data from Rand and Williams (1969), Schoener and Schoener (1971a), Moermond (1979), and personal observation.

Table 15.7 Some localities on the three large islands at which sympatric (interdigitating) climatic vicariants occur.

Jamaica	Puerto Rico	Hispaniola
Mandeville (grounds of the Mandeville Hotel), central Jamaica *A. garmani* *A. valencienni* *A. grahami grahami* *A. opalinus* *A. lineatopus neckeri* *A. sagrei*	Maricao forest, western Puerto Rico *A. occultus* *A. evermanni* *A. stratulus* *A. gundlachi* *A. cristatellus* *A. krugi*	Oasis edge, Manneville, Haiti *A. chlorocyanus* *A. cybotes* *A. whitemani* *A. brevirostris brevirostris* *A. distichus dominicensis* *A. olssoni*
	El Verde Field Station, eastern Puerto Rico *A. cuvieri* *A. evermanni* *A. stratulus* *A. gundlachi* *A. cristatellus* *A. krugi* *A. pulchellus*	Ravine at ca. 1000 ft on road to Jarabacoa, Dominican Republic *A. chlorocyanus* *A. cybotes* *A. distichus ignigularis* *A. semilineatus* *A. christophei*
		Boutillier Road, south of Port-au-Prince, Haiti *A. coelestinus* *A. cybotes* *A. distichus dominicensis* *A. semilineatus* *A. olssoni* *A. hendersoni*

Les Platons, north of Aux Cayes, Haiti
A. coelestinus
A. cybotes
A. distichus vinosus
A. semilineatus
A. dolichocephalus
A. koopmani
A. monticola

NOTE: Data from Schoener and Schoener (1971b), Moermond (1979), and personal observation.

ecological niche is filled by a different colonist from the mainland. The Galapagos finches and the Hawaiian sicklebills are quite exceptional in this respect."

Lack notices the striking difference with *Anolis* and he suggests "that these lizards spread much more infrequently from one island to another than do land birds, so that it may have been possible for some of the colonists to become adapted to what, for *Anolis,* are unusual niches, without this being prevented by the arrival of a more efficient occupant of such a niche on the mainland."

The relative slowness of lizard colonization compared with that of birds is certainly part of the story. Beyond this, however, there are the correlates of this slowness, its difficulty and its selectivity: few lizard invaders ever came from the mainlands, and these were apparently moderate generalists. In lizards, the specialists are not, in general, good colonizers (Williams, 1969). Specialization—the differentiation of ecomorphs—occurred in situ on the islands.

Given slowness, difficulty, and selectivity of colonization, *Anolis* speciation on islands has, for the most part, been not by accumulation but by radiation. This requires a different use of space. Space between islands not on the same bank slows the process too much to play a major role in the formation of the larger *Anolis* faunas. Space within islands, at least within island banks, which are, sometimes, for *Anolis* demicontinents, serves instead.

However, some islands do not show internal speciation. All the Lesser Antilles seem to belong to this class. Although 2 quite different phyletic stocks inhabit the northern and southern islands and although there are a great many islands from the very small to many times larger, there are 1 to 2 anoles per island, never more, and, in all cases (see Lazell, 1972, for one version of the possible histories) the 2 species faunas appear to have been built up by cross-water invasion.

Yet 4 species of *Anolis* can clearly live in sympatry on islands far smaller than the major islands of the Lesser Antilles. Four species coexist, for example, on Bimini, in numbers that should resist extinction except by major geological, meteorological, or climatic catastrophes. The failure of the Lesser Antilles to have more than 2 anoles per island must depend therefore on 2 factors: (1) the difficulty, almost to the point of impossibility, of colonization in the face of filled faunas or broadly adapted species; (2) the similar apparent impossibility or near impossibility of speciation, that is, the full genetic partition of species, within islands below a certain size.

Note, however, there is not a failure to respond to environmental pressures. Geographic differentiation can be sharp and spectacular (for example, in *A. marmoratus* cited above) and is at least in part a response to climatic factors. But in the Lesser Antilles strong differentiation except in size is always infraspecific and involves usually color, much more rarely scales.

Apart from size, indeed, differences within species, either solitary or a member of a species pair, may be stronger and more conspicuous than between species. Between species pairs there are, as was said above, ecological differences—thermal preference, perch—but these are not, as was also said, very strong, nor are the morphological correlates of thermal preference and perch very marked. Differences exist—the members are displaced ecologically and morphologically with respect to one another, the species have moved apart—but except in size not very much.

In contrast to the Lesser Antillean anoles, the faunas of the Greater Antilles all have utilized for the origin of at least a major part of their species number intraisland space, and the differences between their several island faunas is probably best understood in terms of the space available for speciation.

Although on the Greater Antilles intraisland speciation has clearly occurred, much of its mode and detail remains obscured by the "fog of time" (Williams, 1969). The speciation process has gone to completion for all members of the faunas on 2 of the islands, Jamaica and Puerto Rico, but for Hispaniola (and Cuba) this is untrue.

All species of Jamaica, except *A. reconditus,* are fully sympatric. But, except for *A. sagrei,* an obvious invader, which is not part of the problem, their mode of speciation is not understood. For geographic speciation, as commonly understood, there are obvious difficulties. Jamaica is very much a unit island, with only a few small and close-in satellite islands on its bank. Physiographic barriers within the island are not obvious, and vegetational fragmentation sharp enough to permit the evolution of full species status is not obvious either, certainly not at the present time. It is, indeed, easier to explain the absence of some members of the standard sequence—the absence of bush-grass or trunk ecomorphs—than to explain the existence of 5 full species plus 1 quite distinct montane isolate. Quite obviously the failure of Jamaica to develop climatic vicariants or allospecies is, given its topographic unity and simplicity, the least of problems.

The happiest solution may be to suppose that in Jamaica the operational areas for allopatric speciation were vegetational islands—the products of the very sharp climatic transitions that can occur on islands—possibly assisted by near-inshore islands separated by narrow water gaps (cf. Lazell, 1966). The mechanism—capture of a fragment of a species by the alternative environment—proposed by Vanzolini and Williams (1981) was suggested as an explanation of climatically vicariant species. In modified form it may assist in explaining the evolution of ecomorphs. Ruled out apparently is chromosomal speciation; the karyotypes of the Jamaican anoles are too similar. There is intraspecific polymorphism, rampant in *A. grahami* (Judith Blake, in preparation), but there is a single karyotype present in common in all the native Jamaican species.

But the absence of climatic vicariants at the species level in Jamaica does not imply a failure to respond genetically to local environmental pressures. Again the genetic response is infraspecific; the named subspecies of *A. lineatopus*—all of them color races with or without minimal scale differences (compare Puerto Rican vicariant species below)—have a clear correlation with the major climatic regions within Jamaica: *A. lineatopus lineatopus* with the dry south central region, *A. l. neckeri* with the wetter central mountains and the west, *A. l. merope* with the dry north coast, *A. l. ahenobarbus* with the very wet eastern region. In fact, the climatic differentiation of *A. lineatopus* is subtler than the named taxa imply (personal observation) and local populations may differ from valley to valley. Of the two recognized subspecies of *A. grahami, A. g. aquarum* applies to the population of the wet eastern region that borders the John Crow Mountains. It is sharply distinct in color and in 1 scale character from *A. grahami grahami,* which occupies the remainder of the island. There are differences within the latter (see Jenssen, 1981) but these are not as clearly associated with ecological aspects of the habitat. The others of the Jamaican species (*A. garmani, A. opalinus*) show geographic variation but, like that within *A. grahami grahami,* it is neither sharp nor well bounded.

For all that, the native anoles of Jamaica are distinct full species, all but *A. reconditus* sympatric, all fully isolated reproductively. Despite their differentiation of true ecomorphs, they are, except for *A. valencienni,* morphologically more alike and behaviorally less constrained than are their parallels on the 3 other large islands. Within the spectrum of differentiation and specialization from the smallest to the largest islands, these animals clearly fit below the level of those of the other large islands, although well above those of the Lesser Antilles.

In terms of the usual models of allopatric speciation, Puerto Rico is, at least at a superficial level, an easier case than Jamaica. Its total bank is much larger than that of Jamaica and, more important, it divides into 2 distinct sections, the main island of Puerto Rico and the remainder of its bank, the almost linear series of small islands, the Virgins. Most of the main island is mesic (although the extreme southwest is very arid), while the Virgins are relatively xeric but not maximally so. As I wrote in 1972, there is clearly enough geographic complexity for the allopatric model of speciation, although its details are not easy to make out.

Judged on the results, the Puerto Rican bank has been optimally fragmented for the speciation process. Alone of the 3 large islands, all its species are both fully reproductively isolated and ecologically well constrained. Not only the ecomorphs but the climatic vicariants are all also morphologically well separated. Puerto Rico presents a tidy picture, all clean lines and no shadows.

Climatically vicariant species first occur on this island and within 3 of the 5 ecomorphs. The color and scale differences between these climatic vicariants in each case show interesting parallels. The shade-wet-preferring *A. evermanni* of the trunk-crown ecomorph is green and large; the more open-dry-adapted *A. stratulus* is gray-brown and smaller. Both have uniform dorsal scales. In the trunk-ground and grass-bush climatic series there are in each case 3 species ranging the spectrum from wet to drier to arid habitats. In neither series is there any important size difference between the species, although in each case the arid-adapted species tends to be smallest. There is in each case a series in body color from wet to dry—dark brown to lighter brown to gray-brown—and in each an increase in scale size from wet to arid.

The curious fact here is that, although involving different parameters, the species difference between climatic vicariants is as sharp as that between ecomorphs, that is to say, the species level does appear to have the potential to heighten ecological adaptation.

Hispaniola is quite another case, being a huge island divided into 2 parts by a trough that is partly below sea level. It has a great deal of physiographic complexity, but in no way can it be said to be fragmented in the sense that the Puerto Rican bank is fragmented. As the description of its anole fauna above has plainly indicated, the north-south island division is of very limited aid in the explanation of its faunal complexity. In fact, the absence of more than one major physiographic barrier in so large an area may be, in major part, the explanation for the species level untidiness that is characteristic of the island. In a number of cases the speciation process in Hispaniola has not been able to advance as far toward completion as have all the species of Puerto Rico.

Hispaniola has, indeed, by the usual method of counting, more than 3 times (37) the number of species of anoles that the Puerto Rican bank has (11). But only some of these 37 species are genuine equivalents of Puerto Rico's clearly defined 11. Twenty-four of the named species of Hispaniola are allospecies or semispecies. Only 13 are so distinct or show their distinctness so clearly by extensive sympatry with their closest relatives that they can be called species in quite the same sense that the species of Puerto Rico are so called.

The phenomenon we confront here is, of course, much more common than just the anoles of Hispaniola. It was for this sort of situation that Mayr and Short (1970) invented the term "zoogeographical species," which they succinctly explain as follows: "The basic units of our analysis are the 'zoogeographical species.' These are superspecies (Mayr, 1963; Amadon, 1966) or individual species not belonging to a superspecies. When several species comprise a superspecies they are counted as *one* zoogeographical species just as is each species not belonging to a superspecies."

A count of zoogeographical species shows Hispaniola with 20 anoles (13 species not in superspecies and 7 superspecies). Twenty is not a great increase beyond 11 and not impressive at all when the areas of Hispaniola and Puerto Rico (even including its bank) are compared. Clearly Hispaniola has been permissive of the initiation of the species process, but, by the same token, it has not forced completion of that process.

Forced may be a key word. Sympatry is clearly less easily achieved than reproductive isolation. Looking back now at Jamaica and Puerto Rico, we can see these islands as large enough to permit speciation but small enough to impose intense competition for successful completion of speciation.

Any speciation process will include aborted examples: incipient and nascent species that in the end are left by the waysides of faunal history. These are not visible in the transect of Recent time available to us and because of their low level of differentiation are unlikely to be discoverable in the fossil record.

I believe it safe to assume that every *Anolis* species sympatric with others now was once—sometimes, perhaps, a number of speciation events back—an allospecies. We may also assume that over the course of faunal history some allospecies (and some full species) have lost out. These are part of that "invisible history" that I have commented on elsewhere (Williams, 1969). We can track them, if at all, only by their effect upon the surviving faunas. This is not an issue that can be dwelt on here, but it should always be kept in mind.

There is another phenomenon in Hispaniola that requires comment, the montane faunas. I have remarked above that there are not one but several montane faunas in Hispaniola. The montane faunas of the Cordillera Central and of the Sierra de Baoruco and of the Massif de la Hotte have elements in common with each other and with the lowland faunas but the differences are more impressive. The montane faunas are islands within islands and their resemblances can be charged to past connections between them and present connections within the lowlands, their differences to periods of isolation like the present.

The sizes of the present montane islands must have varied greatly in the past and there is a further factor to be considered: the montane areas of Hispaniola have 2 types of forest, pine forest (on the highest and least favorable areas) and montane broadleaf forest. The pine forests are relatively unfavorable for *Anolis; A. shrevei* in the Cordillera Central and *A. armouri* in the Massif de La Selle are known inhabitants of pine forests. It is therefore faunas of the broadleaf forests that we particularly think of when the term montane faunas is used and the area of broadleaf forest must, under all climatic regimes, have been significantly smaller than the total montane area.

It is possible therefore that the montane broadleaf forests of Hispaniola, or some of them, correspond to the missing intermediate term in the island sizes between the largest of the Lesser Antilles and the smallest of the Greater Antilles. Is it not possible then that the seemingly anomalous montane anole fauna of Hispaniola—an alternative sequence, as I have called it above—at least that of the broadleaf forest, corresponds to some stage in the evolution of the main sequence ecomorphs so well represented in the lowlands of this and the other Greater Antilles? If we suppose that this has been true, an enlarged view of the evolution of faunas may be necessary. The intermediate evolutionary stages may have been more experimental than I supposed in 1972; the strict sequence that I propounded in 1972 from giant ecomorph to twig ecomorph to trunk-crown ecomorph to grass-bush ecomorph may quite underestimate the ecological variability of the transitional stages of faunal evolution. The main sequence may be just the survivors of more varied and variable faunas. On the smaller islands they would be the only survivors, while on the large islands the permissiveness of greater area may have allowed the survival of such specialists as *A. eugenegrahami* on Hispaniola (or its analogue *A. vermiculatus* on Cuba).

The untidiness of Hispaniola as compared with the tidiness of Puerto Rico does suggest that tidiness is a secondary phenomenon, the end result of a severe weeding out process. There is another suggestion: Puerto Rico (and perhaps Jamaica) may be at its faunal climax, with only further loss a probability. Hispaniola, on the other hand (and presumably Cuba), may still be in a formative phase with active coadjustment going on—some species still forming, others now dying.

Of course, whatever the evolutionary process in the West Indies may have been before human contact, the hand of man has been heavy on these islands. All of them correspond to gardens in which those patches of "natural" vegetation that survive are only those that have been unfavorable for gardening. We are looking at a scarred and ruined picture, the subject and lesson of which may have been difficult to interpret when pristine but are far more difficult to discover now. In attempting to assess the history of West Indian faunas, we contend not only with the fog of time but with human-imposed distortions.

Coda

This has obviously been a close rather than a distant view of faunal radiations in the sense I have earlier used (1969) (see also Lack, 1976). Similarly, my paper on the origin of faunas (1972) was a close rather than a distant view. In presenting these detailed synopses of lizard biogeography in the West Indies, I have not at all intended to denigrate efforts (MacArthur and Wilson, 1967, and others) to obtain generality by a distant view. However, for my part, I am very conscious of a need to assure myself of the homogeneity of the data that are being used for generalization. The distant view is, it appears to me, best reinforced or emended by a close view.

In this regard, the present case appears to provide a very significant warning against any too simple attempts at estimating the meaning of area for species number. (I point out that in 1972 I was guilty of just this simplification.) Counting of species assumes as a first premise the equivalence of the entities counted. In the series of faunas we have just examined, this equivalence is not present. The allospecies of Hispaniola and the local and specialist species of all the large islands are not biologically or evolutionarily equivalent to the fully sympatric species of Puerto Rico or Jamaica or the Lesser Antilles. We obscure a part of biological reality when we count all of these together.

This failure of species to be equivalent should be less apparent in island birds than in island lizards. The very factors that permit intra-island radiation in lizards (and presumably in other slow colonizers) have the consequence that some species are caught *in statu nascendi.*

Diamond's 1977 warning is very pertinent: differences in dispersal ability and population density may mean very different end results. In this study we have seen substantive differences between similar faunas that seem to have resulted from the different areas and topographies of different island banks. But even the largest islands are several grades in area and complexity below the continental mainlands. Even for lizards Hispaniola is not really quite a continent.

Island anoles, both those of the large and the small islands, are abundant, conspicuous, and dominant in island ecology. Precisely because they are so major a feature of the areas they inhabit, they can be treated as if evolving alone (Williams, 1969). They can be treated, in fact, as if the only factors important in their evolution were two: (1) interactions within the genus and (2) constraints that physical habitat, in particular areal size, place upon these interactions.

These phenomena, which tend to be characteristic of many island species, make the biota of islands easy systems to study and make a close view of whole sections of faunas far more feasible and attractive than parallel studies on mainland. But by the same token, there is imperfect correspondence between the histories and the effective factors in the evolution of biotas in the 2 classes of areas. Partial correspondence there is, but a simple transference of concepts from one frame to the other is not possible. As we have found in this study, when areas of differing magnitudes are compared, there is as much change as there is congruence.

16 Coevolutionary Theory and the Biogeography and Community Structure of *Anolis*

Jonathan Roughgarden,
David Heckel, and
Eduardo R. Fuentes

IN THIS CHAPTER we present an account of the biogeography and community structure of the *Anolis* lizard populations from the Lesser Antillean islands of the Caribbean. These islands comprise a model system for the study of the coevolution of two competing populations. Also, we propose an explanation for the empirical findings in this system based on the mathematical theory of evolutionary community ecology. What is new is both the account of the phenomena and the appreciation that the theory of evolutionary community ecology yields predictions which coincide to a remarkable degree with these empirical findings.

In brief, we show that independent replicated instances of the coevolution of 2 competing populations have yielded outcomes that are qualitatively different from one another, and that the coevolution of 2 competing populations has not produced a parallel community structure in every instance. The islands differ in the magnitude of the present-day competition that is occurring between 2 species on an island, in the identity of the resource axis used to partition the space used for territories, and in the existence of species replacement along environmental gradients. Also, we show that the theory of evolutionary community ecology predicts (a) a very slow nonequilibrium turnover of species over coevolutionary time, resulting from the coevolutionary process; (b) multiple simultaneously stable equilibrium patterns of resource partitioning involving several resource axes; and (c) a threshold, determined by an island's topography, that controls whether competing species evolve habitat segregation along geographical transects on that island. The coevolutionary turnover, the multiple equilibria, and the threshold related to island topography, all expected by theory, appear to explain the community structure and biogeography of the populations in this system.

Coevolutionary theory predicts the outcome of the simultaneous evolution of ecologically interacting species (reviews in Slatkin and Maynard Smith, 1979; Roughgarden, 1979). The present results of this theory are of general importance in population biology and include theorems implying that coevolution in a community often destabilizes a community and causes the extinction of species in a community. Other theorems imply that coevolution generally leads to an adaptational mismatch between interacting species. Nonetheless, coevolutionary theory has not been oriented toward explaining known phenomena except in the most general terms. It has not been very helpful for understanding particular instances of the coevolutionary process. The principal aims of this study are to describe a system in which coevolution occurs with an emphasis on those properties of the system to which theory can be related and to specialize coevolutionary theory so that it speaks to the facts observed in this system. Thus, this study is principally intended to exhibit the explanatory power of coevolutionary theory for a particular situation where coevolution has occurred.

A secondary aim of the study is to provide an alternative theory of island biogeography for groups that do not possess the features explained by the now classical turnover theory of island biogeography developed by MacArthur and Wilson (1967). This theory was originally intended to explain two regularities in the number of species on small oceanic islands. The first regularity, called the "area effect," is that the number of species on an island increases with the island's area, often according to a power law. The second regularity, called the "distance effect," is that the number of species on an island is a decreasing function of the distance from the continental source of the island's fauna. The central concept to the turnover theory of island biogeography is that the number of species on an oceanic island is the steady-state value at which the continual arrival of new species by immigration equals the continual loss of established species by extinction. According to this concept the number of species on an island is approximately constant through time, whereas the identity of the species is continually changing as a result of the steady-state turnover of species. This idea was expressed initially with a simple mathematical model (MacArthur and Wilson, 1967), has been successfully tested with insects (Simberloff and Wilson, 1970) and birds (Diamond, 1973), and has been receiving elegant refinement as more is learned about the basic immigration and extinction processes involved (Diamond et al., 1976).

However, the pretheoretical facts that motivate the turnover theory of island biogeography are absent in the *Anolis* system of the Lesser Antilles. Specifically, there is no area effect or distance effect. The number of species on an island is not related to its area, its maximum elevation, its rainfall, or its distance from either Puerto Rico or Venezuela. These points can be noted in Figure 16.1 and will be discussed in more detail below.

Furthermore, turnover, as envisaged in the turnover theory of island biogeography, is almost surely absent in this system. The turnover theory refers to turnover in "ecological time," that is, to a short average residence time for a species on an island, say, on the order of 10 to 10^3 generations. If the average residence time is long, say, 10^3 to 10^5 or more generations, then substantial evolutionary modifications of the species are likely to occur. In fact, in the *Anolis* system of the Lesser Antilles each population from every island bank is taxonomically distinct, at least at the subspecific level and usually at the specific level. Moreover, the total abundance on an island for an *Anolis* population in this system is typically on the order of 10^8. This figure is obtained by extrapolating the value of 25 lizards per 100 m^2 (see Table 16.2 below) to a total island area of 400 km^2. The extinction of such a huge population as a result of ordinary fluctuations in population size is an event of practically zero probability.

In the *Anolis* system of the Lesser Antilles the process of faunal buildup is qualitatively different from that modeled in the ecological turnover theory of island biogeography. A theory is required that explicitly incorporates the coevolutionary process. The aim of this new theory is to predict not only the number of species on an island but also the phenotypes of those species.

This new theory of island biogeography is not a rival to the ecological turnover theory of island biogeography; instead it applies to groups and situations for which the ecological turnover theory was not intended. Such groups might include snakes, mammals, and others with poor dispersal powers. The new theory may also apply to the *continental* biogeography of birds and insects, even though the turnover theory is an excellent theory of *island* biogeography for these groups.

This chapter summarizes a 5-year collaborative project that has combined empirical research in community and physiological ecology with research in theoretical ecology. We have also developed new techniques during the course of this work.

The System

The system consists of the *Anolis* populations on the islands of the Lesser Antilles beginning with St. Croix of the United States Virgin Islands, and terminating with Curaçao of the Netherlands Antilles (Lazell, 1972; see Fig. 16.1). It includes the British and French West Indies.

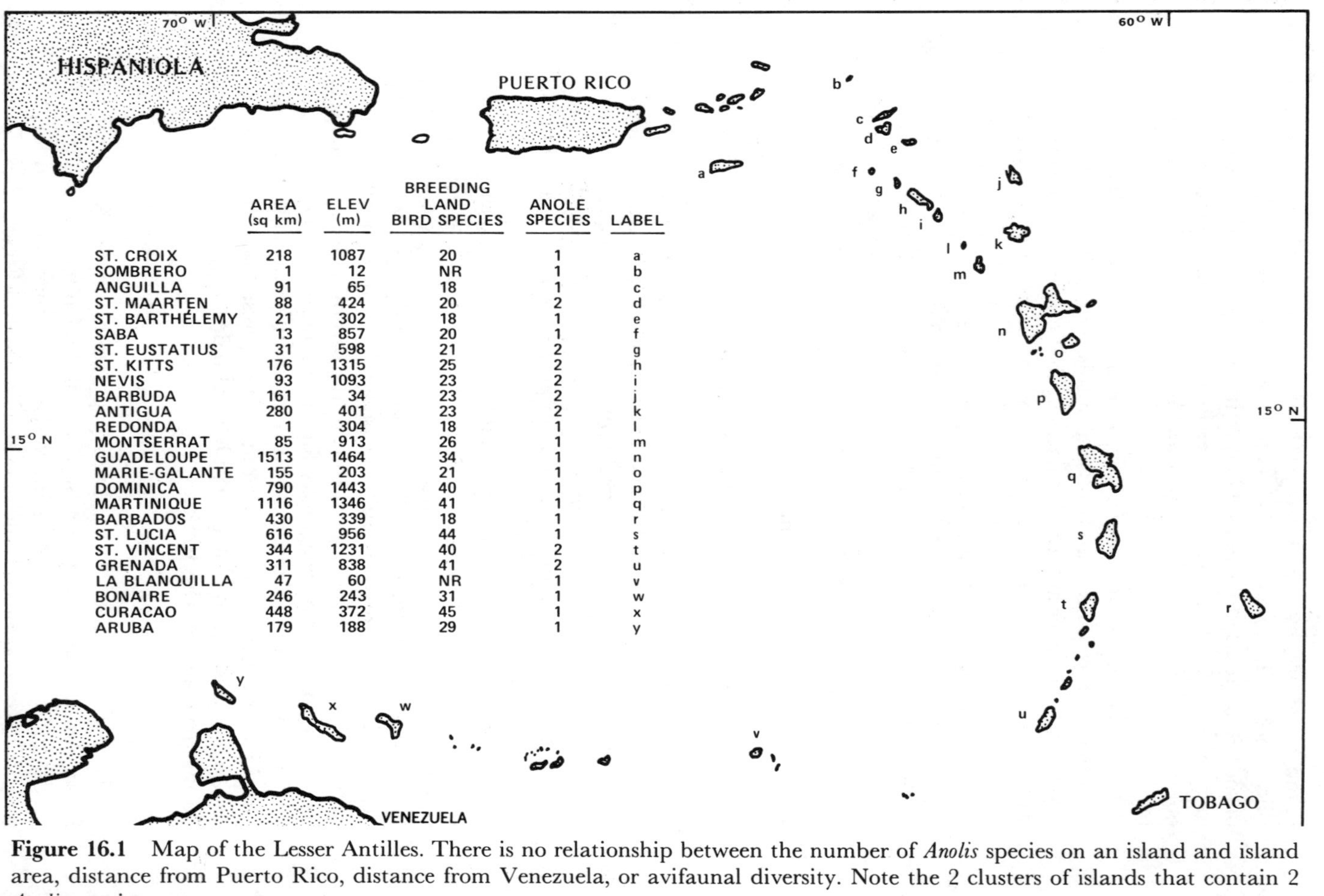

	AREA (sq km)	ELEV (m)	BREEDING LAND BIRD SPECIES	ANOLE SPECIES	LABEL
ST. CROIX	218	1087	20	1	a
SOMBRERO	1	12	NR	1	b
ANGUILLA	91	65	18	1	c
ST. MAARTEN	88	424	20	2	d
ST. BARTHÉLEMY	21	302	18	1	e
SABA	13	857	20	1	f
ST. EUSTATIUS	31	598	21	2	g
ST. KITTS	176	1315	25	2	h
NEVIS	93	1093	23	2	i
BARBUDA	161	34	23	2	j
ANTIGUA	280	401	23	2	k
REDONDA	1	304	18	1	l
MONTSERRAT	85	913	26	1	m
GUADELOUPE	1513	1464	34	1	n
MARIE-GALANTE	155	203	21	1	o
DOMINICA	790	1443	40	1	p
MARTINIQUE	1116	1346	41	1	q
BARBADOS	430	339	18	1	r
ST. LUCIA	616	956	44	1	s
ST. VINCENT	344	1231	40	2	t
GRENADA	311	838	41	2	u
LA BLANQUILLA	47	60	NR	1	v
BONAIRE	246	243	31	1	w
CURACAO	448	372	45	1	x
ARUBA	179	188	29	1	y

Figure 16.1 Map of the Lesser Antilles. There is no relationship between the number of *Anolis* species on an island and island area, distance from Puerto Rico, distance from Venezuela, or avifaunal diversity. Note the 2 clusters of islands that contain 2 *Anolis* species.

The lizards are mainly insectivorous and forage during the day. They tend to arboreality, and adults of both sexes are typically territorial. This system is a part of the radiation of the genus *Anolis* that has occurred throughout Central America, northern South America, and the Caribbean islands including the Bahamas. The defining feature of the Lesser Antillean system is that it consists of islands each of which has exactly 1 or exactly 2 native populations of *Anolis* lizards. There is *no* island with 3 or more species, and the species on each island bank are native to it; typically, endemic at the specific level, otherwise at the subspecific level. There are 8 islands with 2 species, which provide replicated instances of the coevolution of 2 competing species, and 16 islands with 1 species which provide replicated instances of the evolution of a solitary population. Furthermore, there has been no speciation within these small islands. The nearest relative of a species from an island with 2 species is not the other species on that island but a similarly sized species from a nearby island.

The islands toward the north contain anoles that are subtly different from those of the south. The northern islands, beginning with St. Croix and ending with Dominica, contain species that belong to a group known, informally, as the *bimaculatus* group, and those to the south, beginning with Martinique and ending with Bonaire, contain species of the *roquet* group. The island of Curaçao contains a solitary species of South American affinity. Among the bimaculatus-group islands, those to the east, St. Croix, Anguilla, St. Maarten, St. Barths, Barbuda, Antigua, the eastern half of Guadeloupe, and Marie Galante, are old, flat, and dry. The islands to the west, Saba, St. Eustatius, St. Kitts, Nevis, Redonda, Montserrat, the western half of Guadeloupe and Dominica, are new, mountainous (volcanic), and wet. The roquet-group islands are all new, mountainous, and wet, except for Barbados and Bonaire.

This study emphasizes the islands with 2 species, but before proceeding to a detailed examination of the 2-species islands we review the principal fact concerning the 1-species islands (Schoener, 1969; Williams, 1972; Roughgarden and Fuentes, 1977): on 15 of the 16 1-species islands, lizard body size has evolved to approximately the same value, although there is great divergence in other traits including body coloration, physiological tolerances to heat and water stress, and courtship behavior. A useful summary statistic is that the average jaw length for adult males on a single-species island is between 10 mm and 12 mm. The one exception is the population on the island of Marie Galante near Guadeloupe, where the average jaw length is 16 mm (Roughgarden, 1974). The important point is that the exception consists of a population in which the body size is too large, relative to the standard established by the other solitary populations. There is no exception wherein the body size is too small, yet much smaller anoles do occur on the 2-species islands in this system and on the Greater Antilles and the continent.

Figure 16.1 also presents a summary of the number of breeding land bird species for most of the islands that we compiled from Voous (1957) and Bond (1971). The islands from Grenada through Dominica each have about 40 species. Proceeding north, Guadeloupe and Montserrat have about 30 species apiece, and beyond these the islands from Antigua up to St. Croix each have about 20 species. This pattern does not correspond with the distributional pattern of *Anolis* populations in any respect. Specifically, the "knee" to the species diversity curve for birds that occurs near Guadeloupe and Montserrat does not match the location of the separation between the roquet and bimaculatus communities; the bird species diversity of the islands with 1 *Anolis* species as a group, do not differ from the group of islands with 2 *Anolis* species; and the exceptional cases in the biogeographic pattern of *Anolis* (Marie Galante and St. Maarten) do not correspond with exceptional features to the pattern of bird species diversity. We have also compiled checklists for the total herpetofauna, the bat fauna, the butterfly fauna, and the flora of each island, though the latter 2 have serious inaccuracies. These additional checklists also fail to show any obvious relationship to the biogeographic pattern of community structure of *Anolis* populations.

Project Description and Techniques

On islands with 2 species we established several study sites chosen to span the range of site productivities at sea level and to provide an altitudinal gradient. The descriptive field work involved a team of 3 people during the month of August in the years 1976 through 1979. The total effort represented approximately 12 man-months of field work. The peak of the wet season is typically in October and the peak of the dry season in March to April, so the field work is approximately midway between the seasons, heading into the wet season. The study began with Barbuda and Grenada in 1976, then St. Maarten in 1977, St. Kitts in 1978, and St. Eustatius in 1979. As our technique for microclimate analysis was developed during the project, we returned to previously studied islands to obtain additional data as needed. Also, we were offered unpublished data of G. Gorman and J. Lynch from St. Vincent, taken in 1975, which we have confirmed and supplemented as necessary.

For this project we developed 2 new techniques. The first is a fast and accurate method of determining the population size of *Anolis* lizards (Heckel and Roughgarden, 1979). The technique consists of making 3 visits to the site, and marking all lizards seen on each visit with a color that is specific to that visit. The animals marked on a previous visit are re-marked with an additional color at each subsequent sighting. Marks are applied with a paint spray gun. At the end of 3 visits it is possible to summarize the sighting data in the form: number seen on day 1, day 2, and day 3; number not seen on day 1 but seen on day 2 and day 3; and so on for all the remaining possibilities except one. One possibility, the number not seen on any of the days, is, of course, necessarily unknown. We fit a statistical model to the known categories and use it to project the number in this unknown category. The estimated total population size is then the sum over all categories, including the estimated category. To determine the statistical model to use in each projection, we begin with the simplest statistical model, namely one assuming independence of the sighting events on different visits. Then, if this model is rejected by a χ^2 goodness-of-fit test, we use a model with an interaction term for 2 of the visits. If this model is also rejected, we include still an additional interaction term. By this procedure we eventually settle on the simplest model that is consistent with the data. This method is an improvement over the familiar 2-census Lincoln index for 2 reasons. First, in the Lincoln index, the statistical model of independence between the 2 visits is assumed but cannot be checked against the data. Second, even when the independence model is acceptable, as it frequently is, the 2 censuses of the Lincoln index give less accuracy than the 3 censuses used in this method. Thus, this method provides a statistically justified estimate that is also typically more accurate than the Lincoln index.

The second technique developed for the project consists of a method for characterizing the microclimate where a lizard perches (Roughgarden, Porter, and Heckel, 1981). The technique begins with the definition of an inanimate reference object that weighs 5g, has the shape of a lizard, and is gray in color. We characterize the microclimate where a lizard perches as the equilibrium temperature that our reference object would attain when placed where the lizard perches. The temperature of the reference object at some place is called the "gray-body temperature index" of that place, or GBTI, for short. The GBTI integrates the effects of convection, incident illumination, and black-body radiation. We have never actually had to manufacture the reference object because the formulas of environmental biophysics can be used to predict the temperature of the reference object directly from the measurements of solar radiation, wind speed, and air temperature at the spot where the lizard was perching. This technique is preferable to assessing a lizard's microclimate by determining the lizard's body temperature because the relationship between the body temperature and the microclimate of the space it is occupying is itself species-specific. The relationship depends on traits like the body mass, absorptivity, orientation behavior, tendency to thermoregulate, and physiological parameters such as evaporative water loss through the skin.

The objective of our study within each site is to determine the population sizes of the lizard species and their niche positions. As with many organisms, the potentially limiting resources consist of food and space. We find it useful to consider these resources as having separate resource axes. In this genus it has been established that food, primarily insect prey, is partitioned with respect to prey size and that jaw length and prey length are closely correlated (Schoener and Gorman, 1968; Roughgarden, 1974). In relating jaw length to prey length it is important to use the distribution of prey mass with respect to prey length from stomach contents as the basis for statistical analysis and not the distribution of prey number with respect to prey length. Thus the principal resource axis for food resources relative to which the niche position is scored is the prey-length axis, and jaw length is used as an indirect indicator of position on this axis.

For space resources there are two axes which are known to be important in this genus in the Lesser Antilles (Rand, 1964): perch height and perch microclimate. In previous studies of the genus, the microclimate of the perch was indirectly indicated by the lizard's body temperature, but in this study we have measured the perch microclimate directly with the perch GBTI discussed above. The perch height is ascertained by direct inspection of where the lizards perch in their habitat. Thus, for space resources, there are two axes, perch height and perch GBTI, relative to which the niche position is scored. We should add that in the Greater Antilles, where the local species diversity of anoles is higher, the dimensionality required to describe the niche positions is also higher.

In addition to the measurements of population size and niche position of the lizards in a site, data are taken on the site itself. The amount of insects available at ground level is assayed with insect traps prepared from Tanglefoot coated paper plates. The height of the vegetation and its density is measured. In some cases we have also measured the availability of microsites indexed by their GBTI throughout the day.

The use of these techniques within each site, together with the network of sites established on an island, allow us to determine (1) whether there is any local resource partitioning of food and space, and, for space, what axes are involved; and (2) whether there are different regions of the island where each species characteristically occurs or whether both species are spread throughout all habitats together.

Results

We derived 3 principal empirical findings from this study. First, bimaculatus-group communities partition space mainly with respect to perch height and only slightly or not at all with respect to perch microclimate (GBTI). In contrast, roquet-group communities partition space mainly with respect to perch microclimate (GBTI) and only slightly or not at all with respect to perch height. This phenomenon is referred to as "niche complementarity" (Schoener, 1974). Second, the bimaculatus-group communities on St. Kitts and Barbuda each consist of 2 species that are both found in approximately the same relative abundance throughout virtually all habitats on the island. In contrast, the roquet-group communities each consist of 2 species such that one is the most abundant in certain habitats and the other is the most abundant in other habitats. This phenomenon is referred to as "habitat segregation." There is habitat segregation in the roquet-group 2-species islands and none in the bimaculatus-group islands of St. Kitts and Barbuda. Third, in the bimaculatus-group community on St. Maarten, and to a lesser extent in the community on St. Eustatius, there is evidence of very strong present-day competition, strong enough to cause actual competitive exclusion of one species by the other from certain habitats on the basis of present-day ecological competition.

Niche Complementarity

The data on the niche positions relative to the axes involved in describing the use of space are summarized in Table 16.1. The essential

Table 16.1 Niche positions on resource axes pertaining to the use of space.

	Perch height (feet)			Perch GBTI (degrees C)		
	N	$\bar{X}$ (S.E.)	*d*	N	$\bar{X}$ (S.E.)	*d*
Bimaculatus group						
St. Kitts						
A. wattsi	47	0.4 (0.2)		17	34.0 (1.3)	
A. bimaculatus	77	3.0 (0.3)	3	30	32.5 (0.7)	0*
Barbuda						
A. wattsi	143	2.0 (0.2)		11	33.7 (0.6)	
A. bimaculatus	87	5.8 (0.2)	4	11	34.1 (0.8)	0*
St. Maarten						
A. wattsi	250	2.6 (0.1)		19	28.2 (0.5)	
A. gingivinus	111	4.8 (0.2)	2	18	29.4 (1.2)	0*
St. Eustatius						
A. wattsi	112	0.1 (0.1)		15	30.4 (0.2)	
A. bimaculatus	50	5.0 (0.4)	5	16	30.1 (0.2)	0*
Roquet group						
Grenada						
A. aeneus	126	4.5 (0.2)		41	30.1 (0.4)	
A. richardi	181	4.5 (0.2)	0*	39	28.3 (0.3)	2
St. Vincent						
A. trinitatis	286	3.7 (0.2)		6	31.0 (0.8)	
A. griseus	146	4.1 (0.2)	0*	7	29.0 (0.5)	2

NOTE: N is number of instances; *d* is difference in perch height or GBTI (gray-body temperature index) between species pairs. An asterisk indicates that measured difference was not significant.

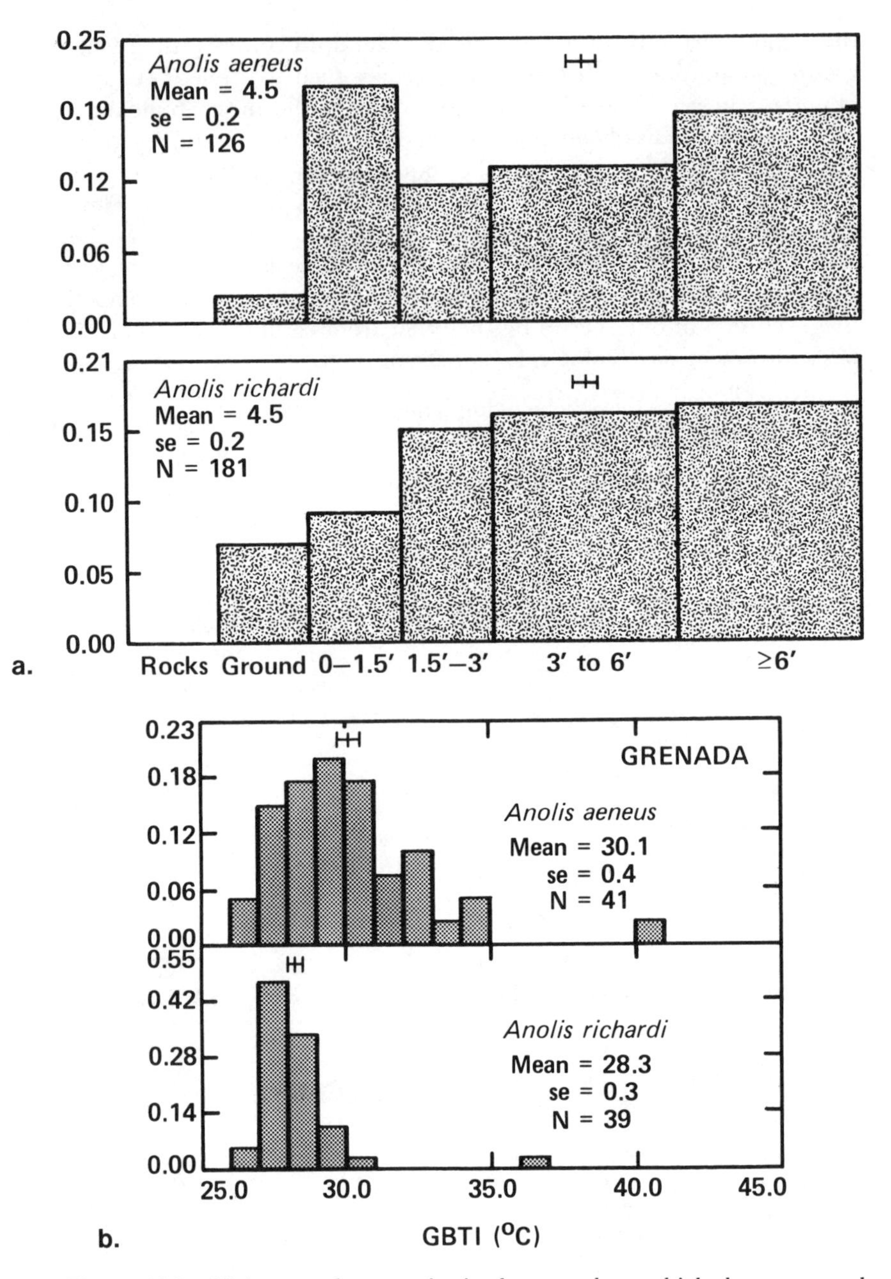

Figure 16.2 Niche complementarity in the axes along which the space used for territories may be partitioned. In Grenada perch height (*a*) is not an axis of separation, whereas there is large separation along the thermal-microclimate axis of GBTI (*b*). In St. Kitts there is large separation along the perch-

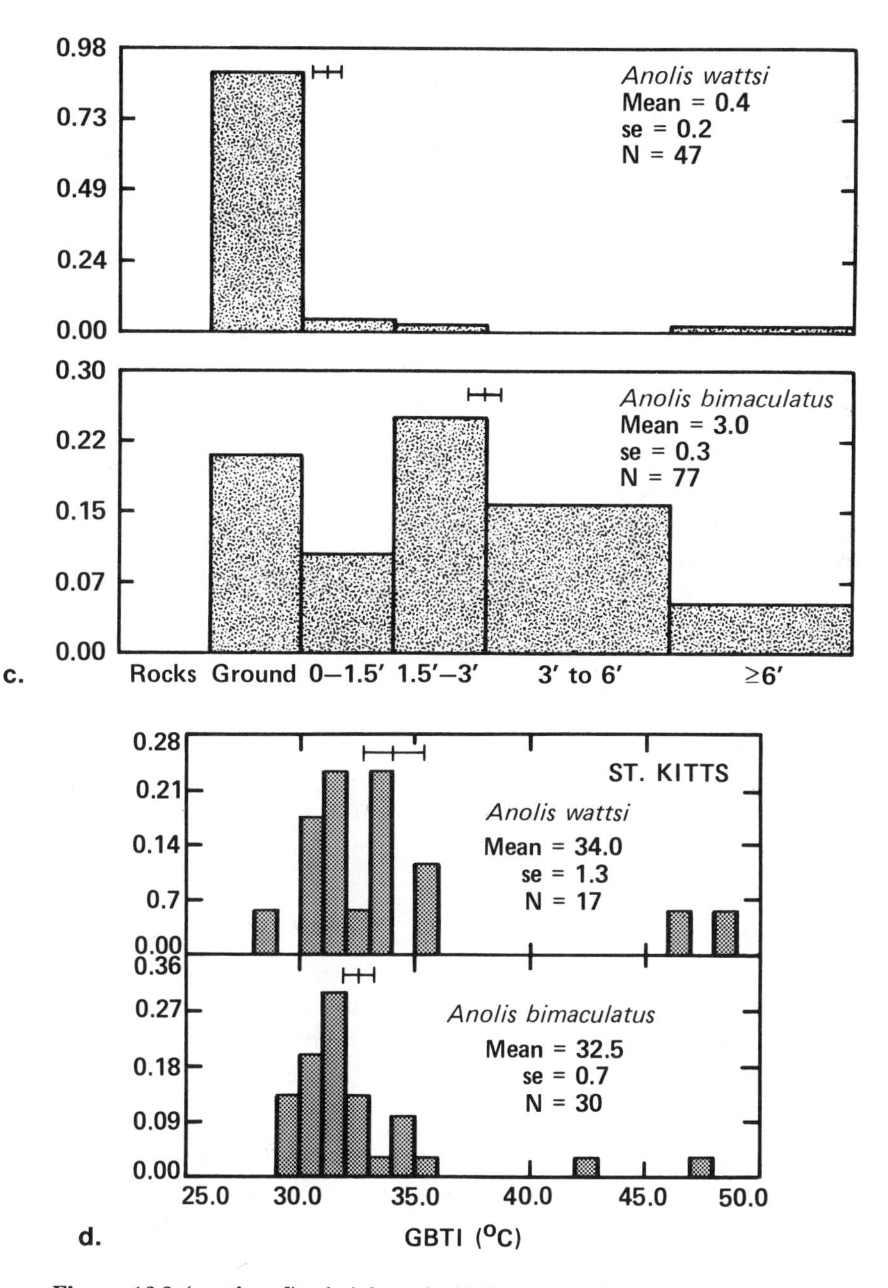

Figure 16.2 (*continued*) height axis (*c*) but none along the thermal-microclimate axis of GBTI (*d*). In all figures the vertical axis is the fraction of observations and the horizontal axis is either perch height in feet or temperature in degrees Celsius, the units of the GBTI. The figures presenting GBTI data are adapted from Roughgarden, Porter, and Heckel (1981).

point is that the bimaculatus communities all exhibit a statistically significant niche separation in perch height, typically 3 ft, and no statistically significant or consistent difference in perch GBTI. In contrast, the roquet communities all exhibit a statistically significant niche separation in perch GBTI, typically 2 ° C, and no statistically significant or consistent difference in perch height. Figure 16.2 illustrates the difference between the utilization curves for these kinds of communities. We now record further details concerning these findings.

The GBTI data refer to observations between 0930 and 1500 hours, the time during which it is physically possible for differences to be manifested. The GBTI for perches outside this time interval are largely the same for both species. The body temperature of the animals whose GBTIs were determined was also statistically significantly different between species in the roquet communities and not in the bimaculatus communities. Hence, body-temperature data would lead to the same qualitative finding. However, the numerical relationship between the perch GBTI and the body temperature realized by a lizard has proven to be species specific in a way not explained only by species differences in body mass.

An interesting incidental finding is that the distribution across perch GBTI in the field is not closely related to the maximum tolerable temperature as ascertained under artificial conditions. On St. Eustatius we determined the panting temperatures of the 2 species by experimentally exposing lizards to direct sun until panting began. The results appear in Figure 16.3. It is clear that *A. bimaculatus* can withstand higher temperatures before panting then *A. wattsi.* Nonetheless, in sites in which they both occur, these species do not differ either in perch GBTI or body temperature.

The perch data were analyzed numerically by assigning values of 0 and −1 to sightings on the ground and on rocks respectively. This convention tends to lead to an underestimation of the true niche separation, as Figure 16.2 illustrates, and thus is a convention that is conservative relative to the detection of statistically significant resource partitioning along the perch-height axis.

There may be statistically significant differences in the perch heights at certain sites in roquet communities. Such differences are not consistent from site to site and are caused by a strongly nonuniform distribution in the availability of perch GBTI throughout the site. The species that uses hot sunny perch microsites is concentrated at whatever height such perches are available, and conversely for the species using cool shady microsites. Such site-specific perch-height differences average out when many sites are considered.

Table 16.2 Number of lizards per 100 m^2 in bimaculatus and roquet communities.

Bimaculatus communities		
St. Kitts		
	A. wattsi	*A. bimaculatus*
Top Antenna Road (woods)	49.7 (4.5)	Rare
Med. Antenna Road (forest)	29.2 (4)	Present
W. Farm Crest (forest)	76.3 (8.8)	Rare
Monkey Site (scrub)	29.1 (1.9)	Present
Little Hill (scrub)	Present	Present
Barbuda		
All-in-Well (woods)	35.1 (3.5)	5.9 (1.1)
Junction (woods)	18.8 (1.4)	3.1 (0.6)
Cross Path (scrub)	4.6 (0.8)	1.2 (0.2)
Coral Slope (scrub)	2.7 (0.7)	0
Roquet communities		
Grenada		
	A. aeneus	*A. richardi*
Les Advocate (forest)	1.3 (0)	16.3 (2.5)
Beach Forest (forest)	10.8 (2.6)	18.0 (2.6)
Junction (woods)	4.1 (0.9) adult	0
	29.0 (8.6) hatch	
Coral Slope (scrub)	Rare	0
St. Vincent		
	A. trinitatis	*A. griseus*
Kings Hill (forest)	5 (3)	55 (17)
Layou (woods)	50 (10)	Rare

NOTE: Rare = sighted adjacent to quadrat, or only 1 marked in quadrat; present = 2 or more marked inside quadrat. Standard error of estimate is given in parentheses.

Habitat Segregation

The data on the abundance of lizards per 100 m^2 for the bimaculatus communities of St. Kitts and Barbuda and for the roquet communities appear in Table 16.2. The essential point is that in the bimaculatus communities, the smaller lizard is always more abundant than the larger lizard, regardless of habitat (sizes in Table 16.3). On Barbuda, where the crown is sufficiently low for us to obtain very accurate estimates of the *A. bimaculatus* population, the ratio is approximately 6 to 1, small to large animals. On this island both species are more abundant in mesic habitats than in xeric habitats. In contrast, in the roquet communities, the smaller species is numerically dominant in open xeric habitats, and the

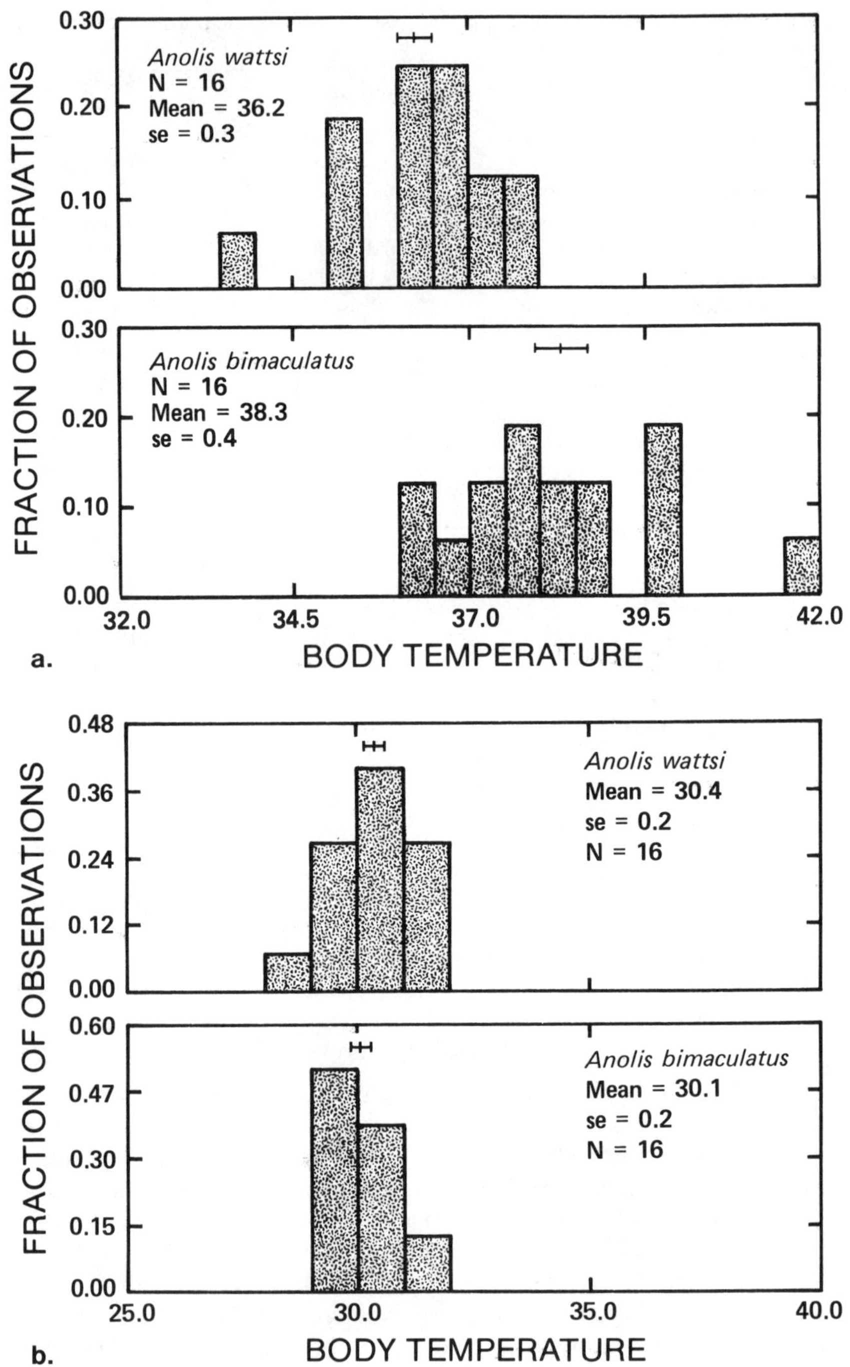

FRACTION OF OBSERVATIONS
0.30
0.20
0.10
0.00
Anolis wattsi
N = 16
Mean = 36.2
se = 0.3
0.30
0.20
0.10
0.00
Anolis bimaculatus
N = 16
Mean = 38.3
se = 0.4
32.0
34.5
37.0
39.5
42.0
a.
BODY TEMPERATURE
FRACTION OF OBSERVATIONS
0.48
0.36
0.24
0.12
0.00
Anolis wattsi
Mean = 30.4
se = 0.2
N = 16
0.60
0.47
0.30
0.15
0.00
Anolis bimaculatus
Mean = 30.1
se = 0.2
N = 16
25.0
30.0
35.0
40.0
b.
BODY TEMPERATURE

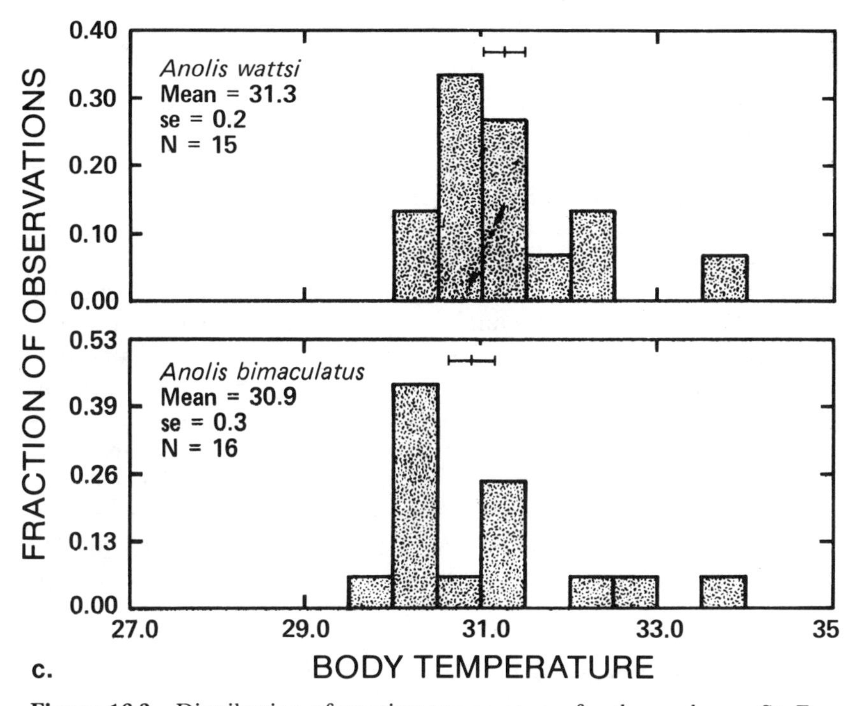

Figure 16.3 Distribution of panting temperatures for the anoles on St. Eustatius (*a*). Vertical axis is the fraction of observations, horizontal axis is degrees Celsius, the units of body temperature. The panting temperature is the body temperature at which the animal begins panting. *Anolis bimaculatus* has a higher panting temperature than *A. wattsi.* In the distribution of GBTI for the anoles on St. Eustatius (*b*), the 2 species do not differ in perch GBTI. Furthermore, the 2 species do not differ in body temperatures (*c*). Thus, the differences in panting temperatures in (*a*) do not imply the existence of differences in the distribution of the lizards in a site with respect to microclimate.

larger species is numerically dominant in forests with deep shade. These facts mean that there is habitat segregation in the roquet communities and virtually none in the bimaculatus communities of St. Kitts and Barbuda.

Present-Day Competition

There are 4 lines of evidence that point to the existence of strong present-day competition on St. Maarten. First, the niche positions of the 2 species on St. Maarten are closer than on any other island with 2 species. Table 16.3 presents data on jaw lengths. Note that the sizes for St. Maarten and St. Eustatius are quite close. Recall from Table 16.1 that the perch heights are also very close on St. Maarten, although not on St. Eustatius. The high niche overlap implied by this closeness of the niche positions does involve overlap in the use of resources that appear to be limiting. Concerning food, Table 16.4 presents the population sizes of the anoles on St. Maarten. Note that there is a perfect rank ordering between the insect biomass and the density of lizards at a site, for sites at which only *A. gingivinus* is present. This relationship between food abundance and lizard abundance would be expected if food were limiting. The overlap in the consumption of prey implied by the closeness of the jaw lengths is genuine; our observations show that both species consume the same general taxa of prey and forage in the same general places and with the same general foraging behavior. Concerning space, we observe reciprocal interspecific territorial interaction, that is to say, interactions where *A. wattsi* chase *A. gingivinus* out of a territory, and also interactions where *A. gingivinus* chase *A. wattsi.* Interspecific territoriality would be expected if the same types of space were used by both species. Thus this first line of evidence for strong present-day competition on St. Maarten lies in the close niche positions and high

Table 16.3 Jaw lengths, adult males (in millimeters).

	N	$\bar{X}$ (S.E.)	*d*
Bimaculatus group			
St. Kitts			
A. wattsi	49	8.4 (0.1)	
A. bimaculatus	62	14.6 (0.3)	6
Barbuda			
A. wattsi	42	8.9 (0.1)	
A. bimaculatus	21	14.7 (0.6)	6
St. Maarten			
A. wattsi	53	8.5 (0.1)	
A. gingivinus	169	11.7 (0.0)	3
St. Eustatius			
A. wattsi	15	9.2 (0.1)	
A. bimaculatus	8	12.6 (0.7)	3
Roquet group			
Grenada			
A. aeneus	27	11.1 (0.1)	
A. richardi	13	15.0 (0.6)	4
St. Vincent			
A. trinitatis	41	10.8 (0.2)	
A. griseus	13	14.8 (1.2)	4

NOTE: N is number of specimens; *d* is difference in length.

overlap for apparently limiting food and space resources.

Second, the smaller lizard, *A. wattsi*, is restricted in its distribution to the low hills in the center of St. Maarten. These hills are slightly more wooded and mesic than the scrub at sea level. The distribution of *A. wattsi* on St. Maarten is important because its relatives on Barbuda and on St. Kitts are present in all sea level habitats, including arid scrub.

Third, the distribution of perch positions of *A. gingivinus* shifts substantially in the presence of *A. wattsi* from the distribution observed where it is solitary. Figure 16.4 presents the perch-height distributions of *A. gingivinus* from the sites where it is solitary (excluding Point Blanc), and from sites where *A. wattsi* is present. Notice that *A. gingivinus* shifts upward in perch position in the presence of *A. wattsi*. This shift is not explained by any obvious lack of high perches at the site where *A. gingivinus* is solitary. (Point Blanc is excluded for this reason; the scrub there rarely exceeds 6 ft in height.)

Fourth, the abundance of *A. gingivinus* is substantially lower in the presence of *A. wattsi* than where it is solitary (see Table 16.4). Point Blanc is an exceptionally arid and exposed site and, apart from this site, *A. gingivinus* is approximately twice as abundant in the absence of *A. wattsi* than it is in the presence of *A. wattsi*. Moreover, we have measured the abundance at 3 of the sites in Table 16.4 (Pic du Paradis, Boundary, and Naked Boy Mountain) at a time near the peak of both the dry and wet seasons each year since our initial study in 1977. The rank order of

Table 16.4 Number of lizards per 100 m^2 on St. Maarten and index of insect biomass.

	A. gingivinus	*A. wattsi*	Index of insect biomass[a]
Sites with *A. gingivinus* alone			
Boundary[b] (scrub)	50.1 (2.2)	–	50.0 (14.3)
Well (woods)	42.4 (1.9)	–	33.7 (3.7)
Naked Boy Mountain (scrub)	43.2 (3.4)	–	7.6 (3.1)
Point Blanc (scrub)	6.4 (0.5)	–	3.4 (0.9)
Sites with both species			
Pic du Paradis (woods)	8.4 (2.6)	10.2 (0.6)	5.8 (1.3)
Medium Pic (woods)	10.5 (0.7)	17.7 (0.7)	10.2 (3.8)
St. Peter Mountain (woods)	19.7 (1.4)	16.9 (0.8)	9.2 (2.3)

a. Captured in Tanglefoot traps.

b. At Boundary, 2 *A. wattsi* were seen but not resighted. During the same time 91 distinct *A. gingivinus* were seen, most of which were sighted 2 or more times.

NOTE: Standard error of estimate is given in parentheses.

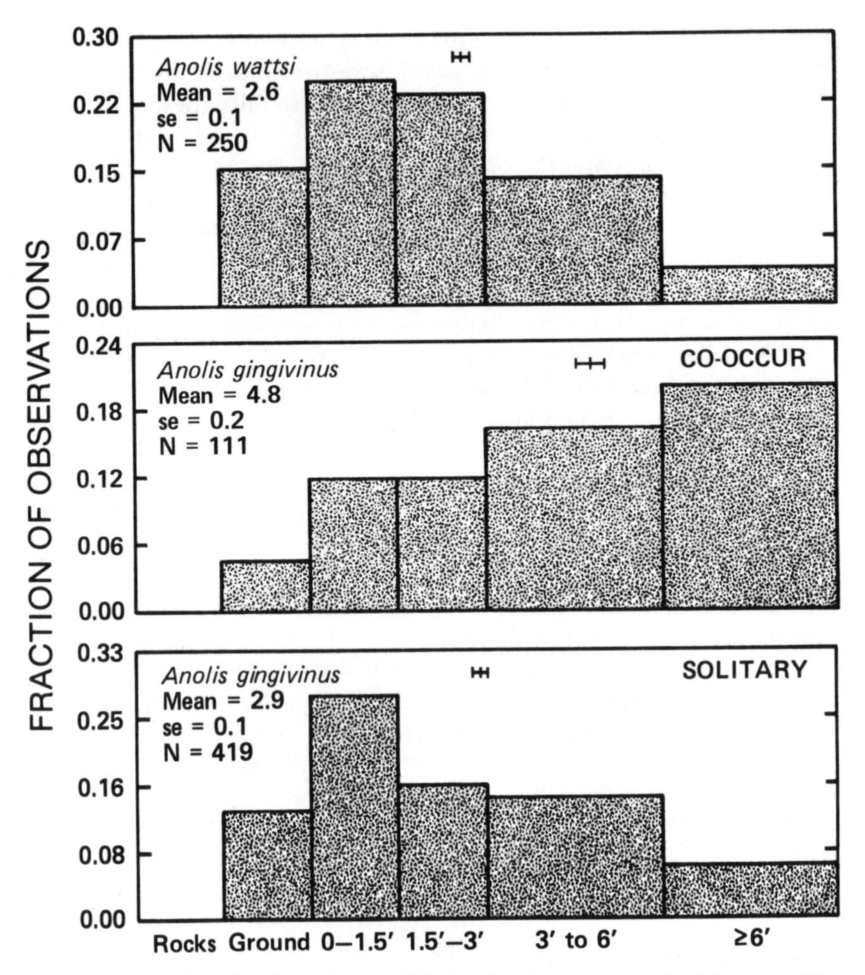

Figure 16.4 Distribution of perch heights for the anoles on St. Maarten. The perch height of *A. gingivinus* shifts upward in the presence of *A. wattsi.*

abundance of the species at these sites has remained the same during this time. Further long-term study is continuing at these sites in St. Maarten.

Each of these lines of evidence for present-day competition is susceptible to ad hoc alternative interpretation, and even taken together they do not conclusively demonstrate strong present-day competition. Nonetheless, the weight of this evidence certainly strengthens the proposition that there is strong present-day competition on St. Maarten, and that *A. wattsi* is absent from the habitats at sea level as a result of present-day competitive exclusion by *A. gingivinus.*

We have also begun experimental studies to confirm the presence of strong competition and to measure its magnitude. A preliminary report of the finding appears in Roughgarden, Rummel, and Pacala (1982). We located a cay near St. Maarten which contains resident *A. gingivinus.* We introduced *A. wattsi* from St. Maarten to the cay at 2 study sites. At one of these sites we removed half of the resident *A. gingivinus* prior to introducing *A. wattsi,* and at the other site we left the residents undisturbed. We obtained twice the survivorship of *A. wattsi* during the first 6 months at the site where the residents were removed as compared to the sites where the residents were undisturbed. Also, we have adult *A. wattsi* individuals of both sexes that have survived in the cay for over 14 months, and we have observed *A. wattsi* hatchlings on the cay. These preliminary results already establish that *A. wattsi* can physically survive and breed in the sea-level scrub habitat from which it is absent on St. Maarten, and that there is a strong negative effect of *A. gingivinus* on the survivorship of *A. wattsi* in this habitat. Furthermore, Pacala has just completed as yet unpublished studies that demonstrate a strong effect of the presence of *A. wattsi* on the rate of body growth of individuals of *A. gingivinus.* These experiments further strengthen the proposition that *A. wattsi*'s absence from sea-level locations on St. Maarten is due to competitive exclusion in the sense of the Lotka-Volterra competition equations.

St. Eustatius is somewhat intermediate in its ecology between St. Maarten and the other bimaculatus-group communities of St. Kitts and Barbuda. The size separation between the species is low, but the perch height separation is high. Both species co-occur throughout almost all of the island with *A. wattsi* being slightly more common than *A. bimaculatus* (see Table 16.5). However, extreme scrub habitats on St. Eustatius contain only *A. bimaculatus* in contrast to the distribution pattern on St. Kitts and Barbuda. Furthermore, *A. bimaculatus* is rare in the forest at the top of the island in a volcanic crater called the Quill. This distribution pattern is one of habitat segregation that contrasts with the absence of such segregation on St. Kitts and Barbuda. However, the distribution pattern on St. Eustatius also contrasts with the habitat segregation of the roquet communities because the larger lizard predominates in open scrub habitats on St. Eustatius, whereas the reverse occurs in the roquet communities. In addition, *A. bimaculatus* shifts upward in perch height where it co-occurs with *A. wattsi* as compared to its perch position at sites where it is solitary. But this finding is compromised by the comparative lack of high perch positions in the open habitat where *A. bimaculatus* is solitary as compared to the sites where it co-ccurs with *A. wattsi.* Thus St. Eustatius exhibits some of the characteristics of St. Maarten—close niche positions, tendency for *A. wattsi* to be restricted to central hills, suggestion of present-day competition—but these characteristics are vaguely formed when compared to St. Maarten.

Table 16.5 Number of lizards per 100 m² on St. Eustatius.

	A. wattsi	*A. bimaculatus*
Quill Top (woods)	25.3 (1.4) adult 39.3 (11.9) hatch	Rare
Central Hill (woods)	32.7 (2.4)	21.3 (5.9)
Cannon Point (scrub)	0	11.9 (2.0)

NOTE: Standard error of estimate is given in parentheses.

The Problem

The major features of the island biogeography and community structure of the *Anolis* lizard populations in the Lesser Antilles based on our field studies are summarized below. The problem is to explain the entire set of facts listed below on the basis of the processes of competition, coevolution, and invasion hypothesized to be occurring in this system.

(1) Sixteen islands contain 1 native species.
  (a) On 15 of these there is convergent evolution to a characteristic body size referred to as the "solitary size."
  (b) On 1 island there is a species whose body size is larger than the solitary size.
  (c) No single-species island has a species whose body size is smaller than the solitary size.
(2) Eight islands contain 2 native species.
  (a) On 6 of these islands 1 species is smaller than or equal to the solitary size, and 1 species is much larger than the solitary size. In these communities there is relatively low overlap in the use of food and space by the 2 species.
  (b) On St. Maarten:
    (i) The larger lizard is of solitary size while its relatives on all nearby islands are larger than solitary size.
    (ii) The smaller lizard has a distribution wherein it is present in the center of the island and absent from suitable habitat around the circumference of the island.
    (iii) There is high overlap in the use of both food and space.
    (iv) There is strong present-day competition.
  (c) St. Eustatius is intermediate between the island of St. Maarten and the remaining 6 islands with 2 species.
(3) No island in this system has 3 or more species.
(4) No island has 2 species of the same size. Specifically, no island has two species of solitary size even though there are many potential colonists of this size.
(5) There is niche complementarity in the axes governing the partitioning of space.

(a) The 2-species communities of the bimaculatus group use the axis of perch height.
(b) The 2-species communities of the roquet group use microclimate as restricted by the axis of the gray-body temperature index (GBTI)

(6) There is no habitat segregation in the bimaculatus-group 2-species communities whereas there is substantial habitat segregation in the roquet-group 2-species communities.

A Coevolutionary Theory of Faunal Buildup

To explain the biogeography of the Lesser Antillean anoles we need a theory that offers a simple conceptualization of the process of faunal buildup on an island. The theory should begin with a statement of what resources are present on the island and of what resources are used by any member of the colonizing genus. It should conclude with representation of all the possible pathways of faunal buildup together with the community structure at key points along each pathway.

The theory that we introduce for this purpose may be understood as a mapping from 4 numbers to a faunal buildup graph. Two of the 4 numbers describe the resources of the island, the other 2 characterize the resource use by members of the colonizing taxon. The representation of the possible pathways of faunal buildup is accomplished by the use of a special graph. A graph is a set of nodes with lines connecting them. In the faunal buildup graph each node represents the community structure that occurs at a key point during the faunal buildup process. The transition from one key point to another key point is marked by a line that runs from one node to another node. The theory consists of an *algorithm* that takes as input the 4 numbers that characterize the island to be colonized and the genus of colonists and constructs the unique faunal buildup graph corresponding to those 4 numbers.

The mathematical framework for the theory of faunal buildup originates from niche theory in community ecology (review in Roughgarden, 1979, chap. 24). The resources on the island are characterized by a univariate Gaussian carrying capacity function, $K(x)$, given by

$$K(x) = K_{\max} e^{-(\frac{1}{2})x^2/\sigma_k^2}. \qquad (16.1)$$

By definition $K(x)$ is the population size of a solitary population with niche position x, and $K_{\max}$ is the maximum population size attainable on the island, achieved at a niche position of $x = 0$. The value of $K_{\max}$ is an increasing function of both an island's productivity and area, and σ_k^2 measures the variety of resource types.

The members of the genus are assumed to share the same competition function. The competition coefficient for the effect of an individual whose niche position is at x_1 from an individual whose position is at x_2 is

$$\alpha(x_1 - x_2) = e^{w^2\kappa^2} e^{-(x_1 - x_2 + 2w^2\kappa)^2/(2w)^2} \tag{16.2}$$

The niche width is measured by w, and there is asymmetry in the competition function due to κ. If κ is positive then an animal with a high x has a larger effect against an animal with a small x than vice versa. The history of this competition function is that it is obtained from the MacArthur-Levins overlap formula, if the utilization curves are Gaussian with standard deviation, w, and if there is exploitative asymmetry. (See Roughgarden, 1979: 532, for illustrations and more detail.)

There are 3 reasons for the incorporation of asymmetry in the competition function if the resource axis is related to body size. First, there is exploitative asymmetry in the sense that large animals consume more total food than small animals so that a given amount of overlap between normalized utilization curves implies that more food is being taken away from a smaller animal by a large animal than vice versa. Furthermore, there are 2 kinds of interference asymmetry. Large animals more often succeed in securing space from small animals than vice versa. Finally, large animals occasionally capture and eat small animals.

Thus, the 4 numbers for input to the algorithm are σ_k, K_{max}, w, and κ. The first 2 describe the island, the second 2 describe the genus.

The community structure at any key point in the buildup process is described by a list of niche positions, ordered from left to right. The number of species in the community is determined by counting the number of entries in the list. The population size of each member is obtained by forming the α-matrix using the competition function, forming the K-vector using the carrying capacity function, and solving for the vector of equilibrium population sizes based on the equilibrium solution of the Lotka-Volterra competition equations.

There are 2 key points in the process of faunal buildup at which it is of interest to know the community structure: (a) after a community has undergone coevolution among its members; (b) after a community has been invaded.

Furthermore, there are several interesting variants on the 2 kinds of points. First, coevolution leads to 1 of 3 outcomes: (i) coevolutionary equilibrium among all the members, (ii) the extinction of a member, or (iii) the fusion of 2 members with one another. Second, the invasion of a community by a new species leads to 1 of 4 outcomes: (i) faunal enrichment in which the invader is added to the community, (ii) faunal substitution in which one of the residents is replaced by the invader, (iii) faunal decline in which 2 or more residents are replaced by the invader, or (iv) failure of the invasion itself.

The faunal buildup graph will contain a node listing the niche positions in the community whenever one of the 7 conditions discussed above occurs. A community will always be identified as one of these 7 variants. The transition of one community to another is indicated by a line pointing from one community to another. Adjacent to the line appears a number signifying the time used to make the transition.

The algorithm that constructs the faunal-buildup graph includes instructions for how a transition is made from one key point to another, instructions for how to determine an invader, and instructions for when and how to stop. The mathematical framework is based on the Lotka-Volterra competition equations. The density-dependent fitness function for an individual whose niche position is at x_i, according to the Lotka-Volterra model and assuming the intrinsic rate of increase equals 1, is

$$W(x_i) = 2 - \frac{1}{K(x_i)} \sum_j \alpha(x_i - x_j)\, N_j \qquad (j = 1, \ldots, S). \qquad (16.3)$$

The equations that determine how a transition is made are based on this function.

The transition of a community from a state where it is not at coevolutionary equilibrium (for example, a community subsequent to an invasion) to a community where it is at coevolutionary equilibrium is accomplished by iterating the following system of equations:

$$\Delta x_i = (\text{speed}) \left. \frac{\partial W_i}{\partial x_i} \right|_{\{N_i = \hat{N}_i\, (x_1, \ldots, x_S)\}} \qquad (i = 1, \ldots, S), \qquad (16.4)$$

where x_i is the niche position of species i. The system changes by a constant, called speed, times the gradient in the fitness function. The population sizes in Eq. 16.3 are evaluated at equilibrium population sizes associated with the current set of niche positions. Thus it is assumed that the ecological equilibrium population size tracks the coevolution of the niche positions. A coevolutionary equilibrium is attained if

$$|\Delta x_i| < \text{accuracy} \qquad (i = 1, \ldots, S) \qquad (16.5)$$

There are two global constants in the algorithm, speed and accuracy. Conventionally, speed equals 1 and accuracy equals 10^{-6}. An extinction occurs if one of the $\hat{N}_i$ becomes less than 1 during the course of iterating Eq. 16.4. A fusion occurs if the α-matrix becomes singular during the iteration. The criterion for singularity is referenced to the constant, accuracy. If either event occurs a node is generated, and the coevolution continues with the remaining community. Eventually a community is obtained that is at a stable coevolutionary equilibrium. The formation of a node representing a coevolutionary stable community terminates a coevolutionary phase.

The background to the formula for Δx_i lies in the theory of quantitative inheritance with weak selection. According to this theory, the change in the mean value of a character, $\bar{x}$, is given as

$$\Delta \bar{x} = h^2(\bar{x}_W - \bar{x}),$$

where h^2 is the heritability of the character, and $\bar{x}_W$ is the mean after selection. This quantity is defined as

$$\bar{x}_W = \frac{\int x \; W(x) \; p(x) \; dx}{\int W(x) \; p(x) \; dx},$$

where $W(x)$ is the fitness of an individual whose character value is x and $p(x)$ is the density function for the distribution of the character in the population. If we expand $W(x)$ to first order about $\bar{x}$, we obtain

$$\bar{x}_W = \bar{x} + \frac{\sigma^2}{W(\bar{x})} \frac{dW(\bar{x})}{dx},$$

where σ^2 is the variance of the character. Furthermore, according to the theory of natural selection on a quantitative character, the population variance approaches a limit, independent of the trajectory of the population mean, given by

$$\sigma^2 \rightarrow \frac{\sigma_L{}^2}{1 - (h^2)^2/2}.$$

In this expression $\sigma_L{}^2$ is called the variance of the segregation kernel. It is a constant that describes the variability of offspring phenotypes resulting from parental crosses. (See review in Roughgarden, 1979, chap. 9.) Hence, the expression for the change in the mean value of a character under weak selection becomes

$$\Delta \bar{x} = \frac{\text{const}}{W(\bar{x})} \frac{dW(\bar{x})}{dx},$$

where const equals $h^2\sigma_L{}^2/[1 - (h^2)^2/2]$. This result motivates Eq. 16.4 for the evolution of the niche position of a species. Moreover, if the population sizes are their equilibrium values for the current niche positions, then $W(\bar{x})$ equals 1, thus producing an equation of the form of Eq. 16.4.

The transition of a community from a state where it is at a coevolutionarily stable equilibrium to a state where it contains an invader is accomplished by iterating the Lotka-Volterra system itself:

$$\Delta N_i = (\text{speed})\ (W_i - 1)N_i \qquad (i = 1, \ldots, S), \tag{16.6}$$

where speed is the global constant also appearing in Eq. 16.4. As the initial condition for the iteration, the residents are assumed to be at the equilibrium abundances they had before the arrival of the invader, and the invader is assumed to have an initial population size of 1. An ecological equilibrium is attained if

$$|\Delta N_i| < \text{accuracy} \qquad (i = 1, \ldots, S), \tag{16.7}$$

where accuracy is the same constant appearing in Eq. 16.5. An extinction occurs if one of the N_i becomes less than 1. That species is then removed and the iteration proceeds with the remaining species until eventually all extinctions have occurred and equilibrium is attained. The formation of a node at which there is ecological equilibrium terminates an invasion phase.

The instruction for determining the identity of an invader is based on finding the locations within a coevolutionarily stable community where an invasion is most likely to occur. The most likely places along the resource axis are places where $W(x)$ is a local maximum given the positions of the residents. Furthermore, for the invasion to be feasible at a place where $W(x)$ is a local maximum, the equilibrium population of the invader, after its invasion, must exceed 1. The instruction for determining the identity of an invader consists of finding all places that are local maxima to $W(x)$ and testing for feasibility. Then, exactly 1 invading population is introduced, separately, at each qualifying position.

The faunal buildup graph branches at any node that represents a community where invasion can occur at 2 or more places. This branching represents the basis for multiple pathways of faunal buildup.

There are 2 criteria for ending a path in the buildup graph. First, a pathway terminates in a coevolutionarily stable community if it has no feasible place of invasion. This kind of termination is represented by a node with no line pointing out from it. Second, a pathway terminates if a community at some point in a pathway points to a coevolutionarily stable community that occurred earlier in that same pathway. In this case, the pathway cycles back on itself.

Convergence in pathways occurs if a community is produced in one pathway that is identical to that in another pathway. If convergence occurs the pathways are merged.

The algorithm begins with an island containing one species located at

$x = 0$. This is the state that any invader on an empty island evolves, independent of the 4 input numbers, by ordinary single-species K-selection. The node with one species at $x = 0$ is the "root" for the faunal buildup graph.

Results

Figure 16.5 illustrates the faunal buildup graphs for 3 choices of input. The figure contains, for purposes of illustration, more information about the community at each key point than is required on a buildup graph. The simplest graph is illustrated in Figure 16.5*a* and results from the parameters $\sigma_k = 1.25$, $K_{max} = 100$, $w = 1$, and $\kappa = 0$. In this case the island is invaded by a population whose niche position evolves to $x = 0$. In the graph $\pm 2\sigma_k$ is indicated by the end points on the horizontal axis. The niche position of a species is indicated by a heavy dot on this axis at $x = 0$. The population size, which equals 100, is indicated by the height of the horizontal bar centered on the niche position at $x = 0$. The length of the bar is $\pm 2w$. For these input values the algorithm reveals that there are no points along the horizontal axis where an invader has a positive rate of increase. Hence, there are no branches from this node. For these input values the entire tree consists of only one node.

In Figure 16.5*b* we see a more complex tree. Here σ_k is made wider, equal to 1.50. Again $\pm 2\sigma_k$ is indicated by the end points on the horizontal axes. As before, the initial invasion leads to a root node that represents a solitary population whose niche position is at $x = 0$ and whose population size equals K_{max}, that is, 100, as a result. However, this community is invadable, and the niche positions representing local maxima for an invader's rate of increase are at -2.06 and 2.06. (The invasion points are symmetric about 0 since $\kappa = 0$.) The presence of 2 invasion points means the graph must branch from this node. The left branch illustrates the node that results after invasion at -2.06. The population size of the invader is 5, and the size of the resident drops to 98. Similarly, the right branch illustrates the node resulting from invasion at 2.06. Returning now to the left branch, the community consisting initially of a species at -2.06 and another at 0 with abundances of 5 and 98 respectively coevolve in niche position until coevolutionary equilibrium is attained. The result is a community of 2 species, one at -0.47 and the other at 0.47, each with an abundance of 53. Also, the community in the right branch converges through coevolution to the same result even though the initial condition is different. The 2 pathways thus merge at the node at the bottom of Figure 16.5*b*. This coevolutionarily stable node has 3 places where the rate of an invader is a local maximum, -2.94, 0, and 2.94. However, the equilibrium population size for an invader at any of these 3 points is less than 1. Hence this node is declared closed to further invasion and the buildup graph is complete.

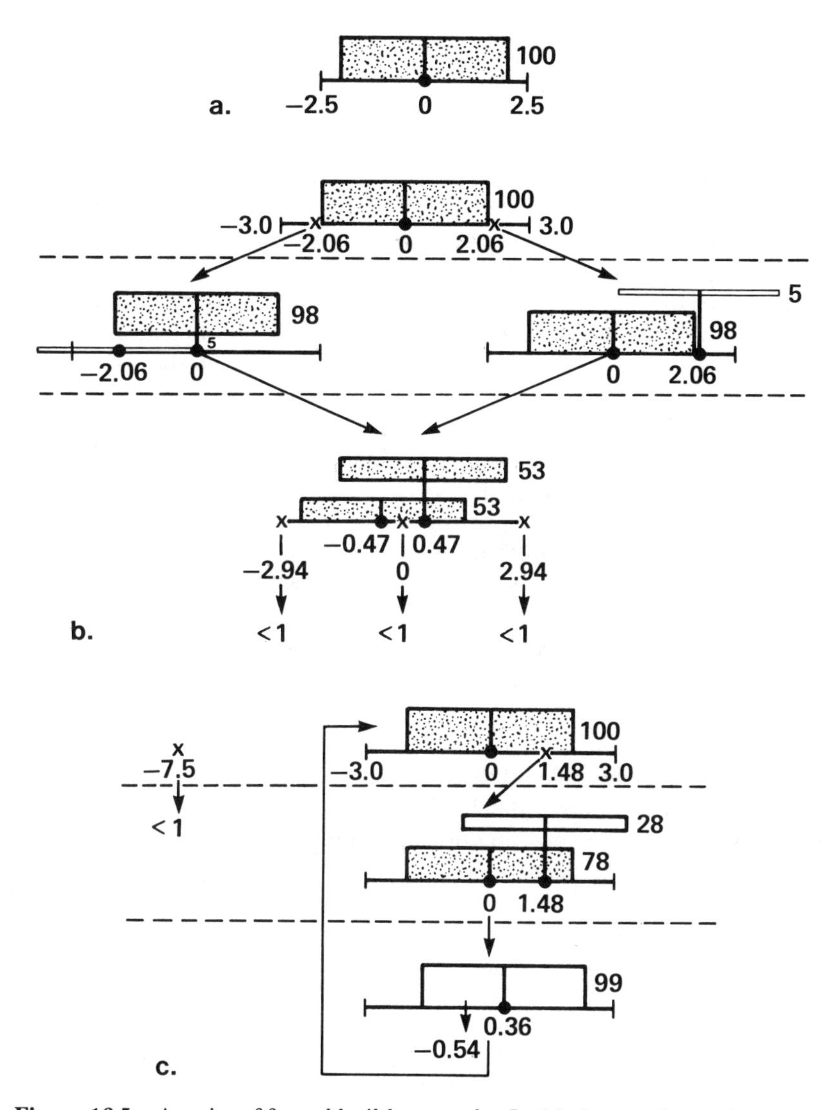

Figure 16.5 A series of faunal buildup graphs. In (*a*) the graph terminates in an island with only 1 species whose niche position lies under the peak of the carrying capacity function. There are no positions at which the rate of increase of an invading propagule is positive. Here $\sigma_k = 1.25$, $K_{max} = 100$, $w = 1$, and $\kappa = 0$ (note $\sigma_k < \sqrt{2}\,w$). In (*b*) the graph terminates in an island with 2 species, each displaced symmetrically to one side of the peak of the carrying capacity function. The terminal community has 3 places where an invasion is likely, but the equilibrium population size attained by an invader at any of these points is too small. Here $\sigma_k = 1.50$, $K_{max} = 100$, $w = 1$, and $\kappa = 0$ (note $\sigma_k > \sqrt{2}w$). Graph (*c*) shows cycling between communities of 1 and 2 species. Here $\sigma_k = 1.50$, $K_{max} = 100$, $w = 1$, and $\kappa = 0.2$.

Figure 16.5*c* illustrates the most important graph from the standpoint of explaining the biogeography of *Anolis*. The input values are the same as in Figure 16.5*b* except that asymmetry is introduced by setting $\kappa = 0.2$. There are again 2 places in the root community where the rate of increase of an invader is a local maximum, but they are now asymmetrically located about the origin, one at -7.5, the other at 1.48. However, invasion at $x = -7.5$ results in a population size for the invader less than 1 so this point is ruled out. Invasion at the other point leads to a community where the resident's abundance is 78 and the invader's is 28. This community now coevolves, but *during the course of the coevolution the original resident species becomes extinct.* At the time of the extinction the resident has evolved a position at -0.54 and the invader is at 0.36 with a population size of 99. This community, now consisting of a single species at 0.36, then evolves back into the root community. Although the root community is reestablished through this cyclic process, there has been a species substitution. This graph represents coevolutionary turnover. It is turnover caused by the coevolution. Without coevolution the turnover would not occur because the initial community that contains both the resident and the invader is ecologically stable.

A graph of the form in Figure 16.5*c* can account for the biogeographic patterns of body size and species number for *Anolis* in the Lesser Antilles in the following sense. If there is actually a coevolutionary turnover process occurring among the islands of the Lesser Antilles as diagrammed in Figure 16.5*c,* then we would expect to see the biogeographic pattern that is actually observed. Specifically, beginning with the top node in Figure 16.5*c,* we would expect a majority of the islands to contain exactly 1 species with a characteristic body size. Any exception to this rule for solitary populations should be a solitary population that is too large, like that in the bottom node of Figure 16.5*c*. The only exception to the rule for solitary populations is the population on Marie Galante and indeed it is too large. A minority of the islands should contain 2 species and typically 1 of the species should be smaller than the solitary size and the other larger than the solitary size. The 2-species islands should represent transition states between the middle node in Figure 16.5*c* and the bottom node in the figure. An island with 2 species that has nearly attained the state of the bottom node in Figure 16.5*c* should resemble St. Maarten in that it should contain a larger species that has evolved to a size near the solitary size and that has, as a result, driven the smaller species nearly to extinction through what would count as strong present-day ecological competition if it had been observed in the past. Other 2-species islands should be intergrades between the middle node and bottom node in Figure 16.5*c* as apparently occurs with St. Eustatius.

Also, according to Figure 16.5*c* there should be no exception to the solitary-size rule consisting of a species that is too small, there should be no island with 3 species, and there should be no island with 2 species of the same size including no island with 2 species both of solitary size even though there are many potential invaders of this size. Indeed, none of these situations is observed. Thus, the predictions of Figure 16.5*c* coincide almost exactly with the observations on body size and species number for *Anolis* in the Lesser Antilles.

The clustering of the 2-species islands is a natural consequence of any realization of the process in Figure 16.5*c*. A cycle from a solitary state through a 2-species state back to a solitary state is initiated when a solitary island is invaded by a large species. Hence, the availability of a larger species to serve as an invader becomes the factor which limits where and how often the cycle occurs. The more isolated an island, the more likely it is to reside in the 1-species node. Islands that are physically close to one another, like the 2-species Leeward Islands, and St. Vincent, and the Grenada Bank in the Windward Islands, can provide invaders for one another and thus reside in the 2-species state for a longer time. Ultimately, however, the 1-species state is absorbing in Figure 16.5*c*.

Thus, it is an important consequence of Figure 16.5*c* that all 2-species islands are transient. But this point is delicate. Are there any islands in the Lesser Antilles where the 2-species state is permanent? The evolution of a habitat segregation in St. Vincent and Grenada is a complication affecting these islands. The habitat segregation there reduces the exposure of the species to one another and should tend, as a result, to stabilize the presence of both species on those islands. Moreover, in the faunal buildup theory itself, there exist input values which do lead to 2-species communities that are not transient. For example, if $\sigma_k = 1.75$, instead of 1.5 as in Figure 16.5*c*, then a stable 2-species node exists as well as transient 3-species nodes. Thus, it is possible under the faunal buildup theory to obtain stable 2-species communities on large productive islands (with the higher σ_k). But if so, the absence in the Lesser Antilles of any 3-species islands must be explained by the absence in that system of any suitably sized invaders to invade the 2-species islands in order to bring about transient 3-species states. In view of these points, it seems reasonable to suggest that the 2 species of St. Vincent and Grenada are relatively stable because of the stabilizing effect of the habitat segregation there. In the Leeward Islands, only St. Kitts has high rainfall and productivity, and this island might exhibit a stable 2-species state. For the remaining islands of St. Eustatius, Antigua, and Barbuda as well as St. Maarten the eventual extinction of the smaller species is predicted as a consequence of the coevolutionary turnover hypothesized according to Figure 16.5*c*.

Another delicate point is that the body size of *A. ferreus* on Marie Galante is presently larger than the body size of *A. gingivinus* on St. Maarten. Marie Galante is a low island (128 m in altitude). We have no solid explanation for why the small anole hypothesized to have occurred on Marie Galante became extinct before *A. ferreus* attained the size of *A. gingivinus* on St. Maarten. Possibly Marie Galante is analogous to Anguilla, also a low island, with extensive human disturbance. The smaller species formerly of Anguilla, *A. wattsi,* has become extinct during the last few decades due to human destruction of habitat (Lazell, 1972).

If this coevolutionary turnover is occurring in the Lesser Antilles, then it is possible that the single-species islands may exhibit some effects of this process. Some of the single-species islands may have completed an entire cycle whereas others may have yet to begin one. If so, we would expect the lineage relationships of the populations on single-species islands that have not yet cycled to lie with the smaller species of the nearby 2-species islands, and those that have completed a cycle to be with the larger species of the nearby 2-species islands. Thus the single-species islands may be divisible into 2 groups, precycle and postcycle. We have observed that the habits of some solitary populations in the Leeward Islands are trunk-ground in character, ecologically similar to *A. wattsi,* and that others are more arboreal, ecologically similar to *A. bimaculatus* and *A. gingivinus*. Perhaps the trunk-ground solitary anoles are precycle populations with actual genetic affinities to the *A. wattsi* radiation, and the more arboreal solitary anoles are postcycle populations with genetic affinities to the *A. bimaculatus* radiation. If so, it is also predicted that the precycle solitary populations, being older, should be more genetically diversified *inter se* than the postcycle solitary populations.

In coevolutionary turnover theory, the principal character that exhibits evolution is the niche position. The niche width is assumed to be comparatively constant through time. Also, the niche width is assumed to consist almost entirely of the within-phenotype component. The asymmetrical selection pressure that shifts niche position may also produce within-species resource partitioning (a nonzero between-phenotype component to the niche width). If so, the mean of the species will be higher than predicted in the theory above (see Roughgarden, 1979: 536). For a solitary island the consequence would be that the invasion of large species would be less likely than anticipated in the theory above. Indeed, if a solitary species attains the equilibrium between-phenotype component prescribed by selection pressure, then no species can invade (see Roughgarden, 1979: 562). The possibility of faunal buildup beyond 1 species requires that a second species arrive before the first species has had time to thoroughly diversify within its habitat. The apparent conservativeness of the between-phenotype component (Roughgarden, 1974) is potentially of great biogeographic importance.

Coevolution and Niche Complementarity

To explain the phenomenon of niche complementarity we need a theory that predicts the pattern of resource partitioning which evolves for resources that may be partitioned with respect to 2 or more resource axes. We quote here some unpublished results of Pacala and Roughgarden on this issue.

The model is familiar; it is the multivariate extension of the niche theory formulation already introduced previously in the theory of faunal buildup:

$$W_1 = 1 + r_1 - \frac{r_1}{K(\underline{x}_1)} N_1 - \frac{r_1}{K(\underline{x}_1)} N_2 \alpha(\underline{x}_1, \underline{x}_2);$$

$$W_2 = 1 + r_2 - \frac{r_2}{K(\underline{x}_2)} N_2 - \frac{r_2}{K(\underline{x}_2)} N_1 \alpha(\underline{x}_2, \underline{x}_1). \tag{16.8}$$

Here W_i is the fitness of a member of species i; N_i is the population size of species i; r_i is the intrinsic rate of increase in species i; and $K(\underline{x}_i)$ is the carrying capacity of species i. The niche position of species i, $\underline{x}_i$, is a vector. Each of the p elements of $\underline{x}_i$ represents a position along a different resource axis. There are p resource axes. The carrying capacity function, $K(\underline{x}_i)$, is assumed to be multivariate Gaussian

$$K(\underline{x}_i) = K^* \exp\left\{-\frac{1}{2}\left[\frac{x_{i1}^2}{\sigma_1^2} + \frac{x_{i2}^2}{\sigma_2^2} + \ldots \frac{x_{ip}^2}{\sigma_p^2}\right]\right\}. \tag{16.9}$$

In Eq. 16.9, there is zero covariance between all axes. However, this assumption of statistical independence does not restrict the generality of the model because there is an algorithm that transforms the general problem to this case.

The competition coefficient $\alpha(\underline{x}_i, \underline{x}_j)$ is assumed given by the MacArthur-Levins (1967) overlap index,

$$\alpha(\underline{x}_i, \underline{x}_j) = \frac{\int_{-\infty}^{\infty} u_i(\underline{z}) u_j(\underline{z}) d\underline{z}}{\int_{-\infty}^{\infty} u_i^2(\underline{z}) d\underline{z}}, \tag{16.10}$$

where $u_i(\underline{z})$ is the resource utilization function of the ith species and z_i is a location on the ith resource axis. The $u_i(\underline{z})$ are assumed multivariate Gaussian and identical except for the position of their centroids which are the niche positions $\underline{x}_i$.

Equation 16.10, assuming statistical independence between axes, leads to

$$\alpha(\underline{x}_i, \underline{x}_j) = \exp\left\{-\frac{1}{2}\left[\frac{(x_{i1} - x_{j1})^2}{2w_1^2} + \frac{(x_{i2} - x_{j2})^2}{2w_2^2} + \ldots + \frac{(x_{ip} - x_{jp})^2}{2w_p^2}\right]\right\}, \quad (16.11)$$

where w_i^2 is the variance of the utilization function (niche width) along the *i*th resource axis. Notice that competition is symmetrical so that $\alpha(\underline{x}_i, \underline{x}_j) = \alpha(\underline{x}_j, \underline{x}_i)$.

The problem is to maximize the fitness of each species with respect to its own niche position. The optimization involves solving the system $\partial W_i/\partial x_{ij} = 0$ for all species i and axes j under the constraints that $W_i = 1$ and $N_i > 0$ and checking that the solution is a maximum. (The N_i are found by solving the system $W_i = 1$ for all i.) The first constraint ensures that the optimal niche positions are consistent with population dynamic equilibrium and the second ensures feasibility.

Results

The principal results derived from this model for 2 species are as follows:

(1) The only coevolutionary equilibrium niche positions lie on axes. Figure 16.6 illustrates examples with 2 resource axes. In the figure, although 2 axes are available for resource partitioning, only 1 axis is used in any coevolutionary equilibrium. There is exactly 1 coevolutionary equilibrium per axis.

(2) All of these equilibria, or some subset of them, may be simultaneously stable in a coevolutionary sense. Figure 16.6*a* illustrates the typical situation with 2 axes in which only 1 of the equilibria is stable. The axis along which substantially farther separation is possible is the axis that contains the stable equilibrium. However, if the separation achieved on each axis is about the same, by a factor on the order of 2, then both equilibria are simultaneously stable, as illustrated in Figure 16.6*b*. In this case, the axis which is involved in the resource partitioning observed in a particular system depends on the initial condition for that system.

Application to *Anolis* Community Structure

The result quoted above provides a hypothesis to explain the phenomenon of niche complementarity observed in the Lesser Antillean anoles with respect to the axes that govern the partitioning of space. It is possible that the partitioning of space with respect to microclimate (the GBTI) and with respect to perch height are alternative, simultaneously stable coevolutionary equilibria. According to this hypothesis, the pres-

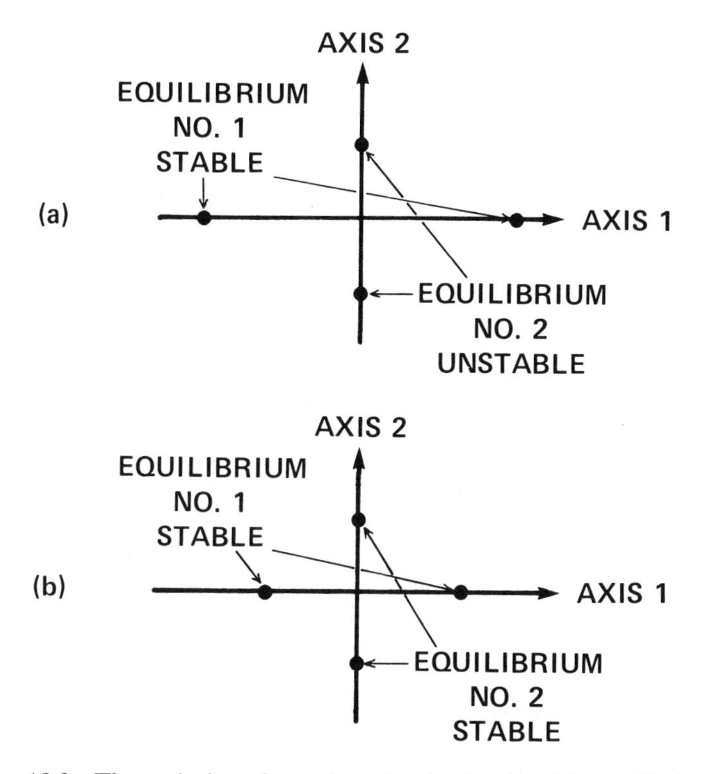

Figure 16.6 The typical configurations that lead to 1 stable equilibrium, or to 2 simultaneously stable equilibria involving the partitioning of a resource that is indexed along 2 or more dimensions. For the application to *Anolis,* the resource is the space used for territories, and the dimensions are perch height and thermal microclimate (GBTI). A dot marks the centroid of a species' multivariate utilization curve.

ence of partitioning with respect to one or another axis on an island is an accident of the initial dispositions of the colonizing lineages.

Why would the bimaculatus group be disposed to partitioning in the perch-height dimension and the roquet group in the microclimate dimension? Possibly there is a disposition toward local resource partitioning of space with respect to microclimate in the roquet group that is caused by the prior evolution of the habitat segregation that exists in the roquet-group islands. Given the habitat differentiation (one species numerically predominating in open habitats, the other numerically predominating in forests), then the physiological differences associated with these habitat differences can predispose these groups to the evolution of microclimate as the basis for the local resource partitioning of space within the habitats where they co-occur in approximately equal abundance. In contrast, the colonists in the bimaculatus-group islands may have originally employed perch height as one of the major axes to partition space as is currently observed in the Puerto Rican anole fauna. These colonists may have brought this disposition with them. In any case these matters are pure speculation. According to the hypothesis, it is an historical accident, no more, that accounts for the particular identity of the axis used for the partitioning of space. However, it is an ecological proposition that multiple simultaneously stable equilibria exist, one per axis, even though which equilibrium is attained on a particular island is traced to a historical accident.

It may seem surprising that the axes of perch height and GBTI can be taken as independent. Yet data in Roughgarden, Porter, and Heckel (1981) support this proposition. High perches generally do have a higher illumination, but these environments also have steady breezes. Hence, the air temperature profile is not large. The higher wind speed at high perches compensates for the higher illumination thereby yielding approximately the same distribution of GBTI as at lower perches.

Coevolution and Habitat Segregation

To explain the evolution of habitat segregation, we need a theory that predicts the conditions under which a population that is potentially distributed throughout an entire geographic region evolves so as to be restricted in its distribution to a subset of the whole geographic region that is available. We quote here some unpublished results by Heckel and Roughgarden on this topic.

The model studied is the beginning of an extension of the theory of density-dependent selection to a spatially varying environment. Let $P(x,t)dx$ be the number of "A" alleles at a locus belonging to those organisms located between x and $x + dx$ at time t, and similarly for $Q(x,t)dx$ with respect to the number of "a" alleles. The number of individuals between x and $x + dx$ at time t is $\frac{1}{2}[P(x,t) + Q(x,t)]dx$. The model is

$$\frac{\partial P(x,t)}{\partial t} = \frac{l^2}{2}\frac{\partial^2 P(x,t)}{\partial x^2} + \frac{P(x,t)}{P(x,t) + Q(x,t)}$$
$$\times \left\{ P(x,t) r_{AA}(x) \left[1 - \frac{P(x,t) + Q(x,t)}{K_{AA}(x)} \right] + Q(x,t) r_{Aa}(x) \left[1 - \frac{P(x,t) + Q(x,t)}{K_{Aa}(x)} \right] \right\}; \tag{16.12a}$$

$$\frac{\partial Q(x,t)}{\partial t} = \frac{l^2}{2}\frac{\partial^2 Q(x,t)}{\partial x^2} + \frac{Q(x,t)}{P(x,t) + Q(x,t)}$$
$$\times \left\{ P(x,t) r_{Aa}(x) \left[1 - \frac{P(x,t) + Q(x,t)}{K_{Aa}(x)} \right] + Q(x,t) r_{aa}(x) \left[1 - \frac{P(x,t) + Q(x,t)}{K_{aa}(x)} \right] \right\}. \tag{16.12b}$$

The first term on the right-hand side of each equation is the contribution to the change in $P(x,t)$ or $Q(x,t)$ from dispersal. The parameter l is the standard deviation of the dispersal function, a probability density function describing the distance an animal disperses from its place of birth. The parameter l describes the typical dispersal distance of an animal. The next term is the contribution to the change in $P(x,t)$ or $Q(x,t)$ from selection. The fitness at location x is taken to be of the logistic form with r and K dependent on both genotype and location. Since P and Q are numbers of genes, for convenience we let $K_{ij}(x)$ be twice the number of individuals of type A_iA_j that are supportable between x and $x + dx$.

We further assume that organisms do not leave the region. Hence, there is no flux of organisms across the boundaries of the region, and this condition implies that the slopes of all solutions are flat at the boundaries of the region. Finally, we assume that the flux of organisms exists and is continuous everywhere within the region; this condition implies that all solutions and their first derivatives must be continuous.

We have studied a special example of this model as a first step toward understanding the evolution of habitat segregation. We assume the population is initially fixed for an allele, "A," whose carrying capacity is constant throughout the whole geographical region, O to M, as illustrated in Figure 16.7. Then an allele, "a," is introduced that has a higher

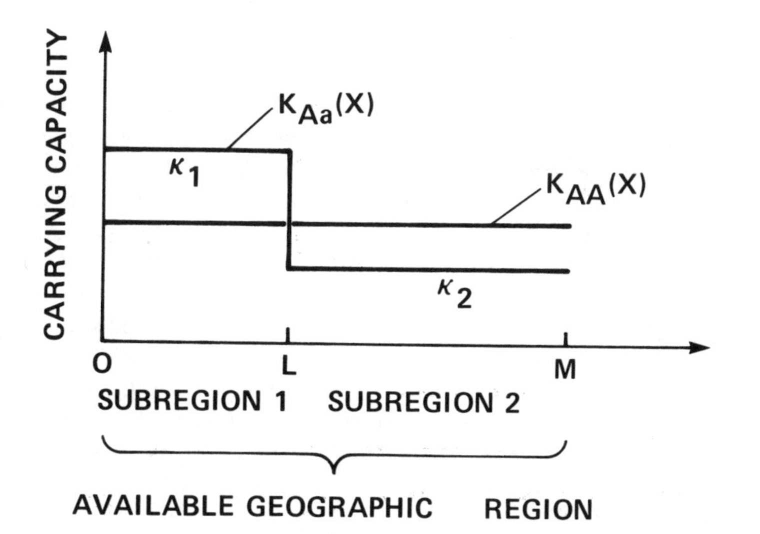

Figure 16.7 Setup for the model that relates to the evolution of habitat segregation. The horizontal axis represents location within the geographic range. The vertical axis represents the carrying capacity of a genotype as a function of location. The population is initially fixed for a genotype with a uniform carrying capacity function, and the problem is to determine the condition under which a mutant allele can increase, given that it has a nonuniform carrying capacity function.

carrying capacity in a subregion, O to L, at the expense of a lower carrying capacity in the remainder of the region L to M. (Thus we assume $\kappa_1 > K_{AA} > \kappa_2$.) Also, the intrinsic rate of increase of Aa may be different in the 2 subregions (r_{Aa} is ρ_1 in region 1 and ρ_2 in region 2). If this gene enters the population, then the consequence is that the population will depart from its pattern of a constant abundance throughout the region and attain, instead, a pattern where it is concentrated in the subregion O to L and is comparatively rare in L to M. Thus the entrance of the allele, "a," marks the first step in the evolution of habitat segregation. The problem is to determine the mathematical condition that leads to the spread of the "a" allele.

Results

The results may be expressed as a decision tree, illustrated in Figure 16.8. There are 2 levels in the decision tree. If region 1 is good enough in an absolute sense, then "a" enters independent of its performance in region 2. This first branch point depends solely on properties of region 1, namely, its size L and quality Z_1. If region 1 is not good enough absolutely to ensure the entry of "a," then it is still possible for "a" to enter if region 1 is good enough *relative* to region 2. The number B_1 is computed from properties of region 1 and B_2 from properties of region 2. If B_1 is greater than B_2 then "a" enters; if B_1 is less than B_2 then "a" cannot enter; and in the case where B_1 exactly equals B_2 (an event of zero probability) the outcome has not been determined.

The important point concerning the results of this theory is that the conditions which control whether habitat segregation evolves depend on topographical features of the region. Of particular importance is the size of the subregion within which the "a" allele is able to confer an advantage. In Figure 16.8, if L is sufficiently large, that is, $L \geqslant (\pi/2)/\sqrt{Z_1/D}$, then habitat segregation evolves absolutely. As L is lowered below this value, habitat segregation evolves, but now based on the fact that $B_1 > B_2$ when L is less than but near $(\pi/2)\sqrt{Z_1/D}$. As L is lowered still further eventually B_1 becomes less than B_2 and habitat segregation does not evolve. Indeed, there is a threshold size for the region, say L_o, such that if $L > L_o$ then habitat segregation evolves, and if $L < L_o$ then it does not evolve. The threshold size for L is obtained by setting $B_1 = B_2$ and solving for L (graphically or with a root-finding computer program).

Application to *Anolis*

The 2-species *Anolis* communities on the bimaculatus-group islands do not exhibit segregation, whereas those in the roquet group do. In the roquet group, one species predominates in shade forest and the other predominates in open woods and scrub. Why? For some of the bimaculatus-group islands there is trivially little variation in habitat to begin

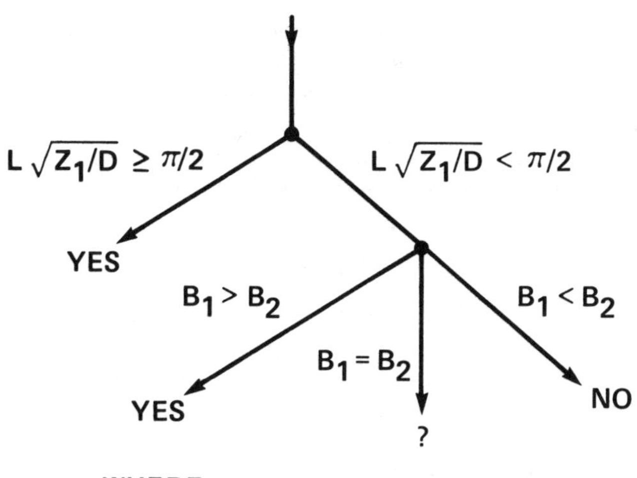

WHERE

$$D = \ell^2/2$$
$$Z_i = \rho_i(1 - K_{AA}/\kappa_i)$$
$$B_1 = \sqrt{Z_1}\tan[L\sqrt{Z_1/D}]$$
$$B_2 = \sqrt{-Z_2}\tanh[(M-L)\sqrt{-Z_2/D}]$$

Figure 16.8 Decision tree to determine if the mutant allele whose spatial carrying capacity function, $K_{Aa}(x)$, as depicted in Figure 16.7, can increase.

with. Barbuda, Antigua, St. Maarten, and St. Eustatius all are quite flat and arid with little other than open woods and scrub. But St. Kitts in the bimaculatus-group islands does possess a mountain as tall as those in the roquet-group islands of St. Vincent and Grenada. It is this island that poses the issue, for there indeed is a nontrivial amount of deep-shade forest on St. Kitts as well as open woods and scrub.

If the theory presented above is relevant, then the principal factor controlling the likelihood that a region of habitat will evolve a habitat specialist to it is the product of the habitat's size with a measure of its favorability. Now the principal topographic difference between St. Kitts and the roquet islands of St. Vincent and Grenada is that the fraction of the island's area that is mountainous on St. Kitts is much lower than on St. Vincent and Grenada. If we assume that the inherent favorability of the corresponding habitats on these islands are comparable (Grenada forest comparable to that on St. Kitts), then the problem becomes one of the areas involved in the various habitat types. So we suggest that even on St. Kitts, and obviously more so for the remaining bimaculatus-group islands, there is not a large enough area of deep-shade forest to permit the evolution of a specialist to this habitat type. In contrast, we suggest that the islands of Grenada and St. Vincent do possess large enough areas of deep-shade forest to permit a specialist to this habitat type to evolve. Presumably, the evolution of habitat specialization in 1 of 2 competing species then promotes the reciprocal evolution by the competitor of specialization to the complementary habitat type.

Summary

In this chapter we have detailed the biogeography and community structure of the *Anolis* populations present on the islands of the Lesser Antilles. We have also demonstrated that coevolutionary theory for competing populations appears to account for these empirical findings with remarkable accuracy.

Specifically, we have shown that there is a biogeographic rule pertaining to body size in this system such that 15 of the 16 solitary populations have converged in body size to a characteristic value, and yet there is one exception consisting of a species that is larger than this solitary size. Furthermore, the biogeographic rule for 6 of 8 2-species islands is that there is a high separation in body size with one species being smaller than the solitary size and the other being much larger than the solitary. Yet there is one clear exception consisting of a community with a very low separation in body size and evidence of very strong present-day competition between the 2 species. Moreover, the slightly larger species in this community has apparently evolved down toward the solitary size from the very large size possessed by its relatives on nearby islands. The smaller species is now excluded by competition from sea-level habitat where it is physically able to survive and breed. This smaller species has a relictual distribution in the central hills where there is slightly less water stress than at sea level. Still another community appears to be an intergrade between this clear exception and the remainder of the 2-species communities. Also, we have noted that the 2-species islands occur in clusters. All these findings can be explained by a hypothesis that there is a slow turnover of species on islands. This turnover results from extinction of the smaller species as a consequence of coevolution with a larger species, followed by a convergence to solitary size of the remaining species, and ending with subsequent reinvasion by a larger species, reestablishing a 2-species community. During this cycle most islands would reside in the 1-species state awaiting colonization by a larger form supplied by some external faunal source. This hypothesis is improbable, and yet it is predicted by mathematical theory for the coevolution and invasion of competing species based on the conceptualization of niche theory.

We have also demonstrated that there is a qualitative difference in the partitioning of the space used for territories in 2 groups of the 2-species islands. In one group an axis pertaining to microclimate is the dimension used for partitioning space and in the other group an axis of perch height is used. This *qualitative* difference is explained by a hypothesis that there are alternative simultaneously stable equilibrium patterns for resource partitioning between 2 competing species. If so, historical considerations would be relevant to explaining which of these equilibria is attained by a particular island. This hypothesis is also predicted by mathematical theory for the coevolution of 2 competing species based on a niche-theory conceptualization.

Finally, we have demonstrated that the same 2 groups of islands that differ in the axis used for the partitioning of space also differ qualitatively with regard to the existence of habitat segregation. In one group both species are present in approximately the same relative abundance in all habitats, whereas in the other group one species is numerically dominant in one habitat (forest) while the other is numerically dominant in an alternative habitat (open scrub). The mathematical theory of density-dependent selection in a spatially varying environment predicts that there is a minimum size for a habitat patch that will support the evolution of a specialist to that habitat patch. We conjecture that the absence of habitat segregation in one group of islands is explained by the absence of a sufficiently large area of forest habitat, thereby preventing the evolution of habitat segregation.

The comprehensiveness of our theory for these empirical findings in the Lesser Antillean islands makes it difficult at this time to develop an equally informative alternative theory. Nonetheless, we can suggest an approach.

To begin, we suggest that one type of alternative, a completely non-interactionist hypothesis, is already known to be false in this system. It is already clear that there is always the potential for competition between Lesser Antillean *Anolis* populations and that in some circumstances very strong present-day competition has been realized.

A more interesting alternative, we suggest, would be a hypothesis providing invasion and competition but without coevolution. This alternative hypothesis can also be used to generate a faunal buildup graph. We can imagine, as in the theory above, that invasion occurs one species at a time, at places on the resource axis where the rate of increase for a propagule is a local maximum. Following an invasion, niche positions do not change. However, an invasion may cause extinction of a resident according to the Lotka-Volterra competition equations. After each invasion the axis is again searched for the most likely invasion points and invasion is repeated. A pathway stops when it has produced a community that is not invadable or when it is cycling back to a community that appeared earlier on the pathway. This latter possibility requires that there be an invasion at some point which causes the extinction of a resident. This alternative theory may be quite useful because it potentially predicts the same kind of information as the theory developed for *Anolis* but without the assumption of coevolution. It would predict a faunal buildup graph and the relationship between phenotype and species number on an island. In some cases species turnover will be predicted, but without assuming either random extinction, as in the MacArthur-Wilson theory, or coevolution, as in the *Anolis* theory.

In conclusion, we have demonstrated that the mathematical theory of coevolution is, in a special case, extremely relevant to understanding natural processes and we have developed a new theory of island biogeography for a situation to which the MacArthur-Wilson theory does not apply.

Conclusion: Lizard Ecology, Viewed at a Short Distance

I SHALL CONCLUDE this book by discussing the state of lizard ecology from the point of view of one who has studied the ecology of birds and mammals. Several contributors have already shown in their chapters that they are well aware of the similarities and differences between the ecology of lizards and the ecology of homeotherms—much more aware, I believe, than are bird and mammal ecologists. Nevertheless, I shall try to assess the current strengths of lizard ecology and conclude by making some suggestions for future research.

But first some general remarks about evolution. Ecology has been transformed by evolutionary thinking in the last 20 years. At its best it has resulted in a deeper understanding of behavioral, physiological, and ecological properties of animals. This probably needs no elaboration. At its worst it has spawned untestable hypotheses and unwarranted generalizations. Lizard ecology has been spared the worst, and so that this may continue I offer the following deliberate parody of sociobiological explanation. From it I draw an important conclusion.

The Tongue-in-Cheek Release Hypothesis

In an impressively thorough discussion Simon (Chapter 6) raised, but did not completely answer, the question of why lizards stick out their tongues. I have an answer, based upon the work of Stamps (Chapter 9), Crews (Chapter 10), and my own studies of a mammal which I will identify by its scientific name only, *Homo sapiens*. The answer is that the behavior has been strongly favored by sexual selection.* My evolutionary theory is as follows.

Once upon a time lizards did not stick out their tongues. They walked

* C. Simon (personal communication) informs me that sexual selection has been invoked to explain tongue extrusion in snakes. This supports at least 3 of my points.

around tongue-in-cheek. One day a mutant arose. When he grew up he had the disturbing habit of sticking out his tongue. A neighboring female, when she saw this, quivered with excitement without knowing why. He liked her reaction and continued. Soon she too was sticking out her tongue and waving it in the air. Their mutual excitement rose to a point at which, in Crews's words, the inevitable happened, and 9 months later a family of baby lizards was born. This family comprised sons and daughters who stuck out their tongues when they grew up, and mated without suffering inbreeding depression (in fact, they rather enjoyed it). And so on, generation after generation. The tongue-sticking-out gene rapidly increased to fixation.

It can easily be seen how the trait is maintained now in the population. Surrounded by males all sticking out their tongues and waggling them, a female chooses the male with the longest and most rapidly moving tongue. She discriminates by measuring (really estimating) the length of the male's tongue in relation to his snout-to-vent length. The task is simple because she only looks at the largest one-third of the males. Male-male competition is simultaneously resolved by the tongue extrusion behavior. Males stick out their tongues at each other, and the infallible rule is that he who has the longest tongue becomes the dominant by striking terror into the heart of his rivals, and secures the most copulations, all without the necessity of fighting. The queen of the females dispatches her rivals with a flick of the tongue in like manner. Thus through intrasexual and intersexual competition, selection has run away, and has been brought to a halt only by the tendency of males with the longest tongues to trip over them, which lowers their fitness.

I believe my theory can explain all different degrees of tongue-sticking-out among lizard species by suitable intuitive modification. In fact I am so convinced I am right I would prefer to call it a law, rule, or principle; but go ahead and falsify it if possible. It could be argued that an olfactory function is performed by the extruded tongue, but this is more prosaic, less interesting, and anyway incidental or subsidiary to my theory and therefore unnecessary.

My aim has not been simply to ridicule sociobiology, which is a fascinating and important area of current research that is underrepresented in this volume. The caricature has been a vehicle for several ideas, the most important being this. It is easy to think up an explanation for any set of observations, and then extend it to fit some new facts. The problem is that several explanations of varying degrees of plausibility can usually be constructed for a given set of observations, and it is a really difficult task in ecology to design and carry out tests that will distinguish unambiguously between alternative explanations. This difficulty is not restricted to lizard ecology, of course. But the degree to which it is overcome is one measure of progress of lizard ecologists.

Interspecific Competition among Lizards

I once published a paper on island lizards in *Copeia* (Grant, 1967), which has never been cited and for which I have never received a reprint request. But I possess a more meaningful qualification that permits me to conclude this book: I attended the 1965 lizard symposium in Kansas City, Missouri.

I had studied birds on islands with a view to discriminating between a competitive-release hypothesis and a climatic-selection hypothesis to account for certain morphological features they possessed (Grant, 1965). The evidence came down in favor of the competition hypothesis. I then tested further predictions of the hypothesis with new data, and results were generally in accordance with predictions (Grant, 1966, 1968). But I felt what was really needed at that time was an experimental test of competition between species which in their allopatric populations had undergone an apparent competitive release in one form or another. William W. Milstead, who had done some interesting studies implicating competition among species of *Cnemidophorus* lizards (Milstead, 1961), suggested I should come to the 1965 symposium and speak to Donald W. Tinkle, and this I did. We decided a *Holbrookia* and *Uta* pair of species would be suitable to work with. But I then took a job in a lizard-free part of Canada, switched to small mammals, and successfully carried out the competition experiments with them (Grant, 1972).

It would be distressing indeed if I had to say that more than 15 years after the symposium we still do not have a really good experimental demonstration of competition between lizard species. Fortunately this is not so. Two recent Ph.D. dissertations from the University of Michigan, by D. C. Smith (1977) and A. E. Dunham (1978), have provided that demonstration (see also Chapters 12 and 16). They were supervised by Tinkle, and like the work of his other students they reflect his legacy to the field of lizard ecology: long-term field studies carried out with exemplary thoroughness. It is very sad, however, that Tinkle did not live to see the completion of his own long-term study of interspecific competition between lizards.

Things Done Particularly Well in Lizard Ecology

Reflecting on the 1965 symposium, what immediately comes to mind is the high standard of work in the area of physiological ecology. The present book shows that subdiscipline to be in excellent health. I am impressed by the thoroughness, rigor, precision, and abundance of measurements of animal activity. Population ecologists can never hope to reach these high technical standards but should be, and I believe have been, inspired by them. The characteristic of lizard physiological ecology which stands out to me is something I think of as ecological realism, a careful attention to the life-historical features of the species being studied. In this I perceive the influence of George A. Bartholomew—somewhat like the presence of Alfred Hitchcock in his own films, felt more than seen.

A second area that deserves emphasis is the subdiscipline of evolutionary ecology because it has made such enormous progress since the 1965 symposium, much more than physiological ecology because it had much further to go. Probably the most widely read textbook on the subject is by one of the editors of this book, Eric R. Pianka (1978). There has been a substantial development of testable theory in evolutionary ecology and island biogeography, and actual testing of that theory with lizard data, most notably by Pianka, Schoener, Roughgarden, and Case, all of whom are contributors here. I should mention that their work far transcends the boundaries of lizard ecology, and in fact they would doubtless consider themselves ecologists first and lizard ecologists second.

The third characteristic which strikes me is that stimulating links have been established between the various subdisciplines of lizard ecology. Consider for example the cooperative ventures of Porter and Roughgarden (reviewed in Chapter 16). Porter seeks to understand how the physiological machinery of lizards can operate under different microclimatic conditions and how those conditions dictate activity and distribution of lizards. By contrast, Roughgarden seeks to account in coevolutionary terms for regularities in morphology and distribution of lizards on islands of the West Indies. From such different starting points they have converged on problems of common interest, and the fruits of such cooperation have yet to be fully realized. Another example is Dunham's work, which combines two previously separate fields: community structure, through analyses of niche overlap and inferred interspecific competition (Chapter 12), and demography, with an actual demonstration of interspecific competition. His work is important because he has been able to show, contrary to earlier interpretations, that niche overlap is not a reliable positive indicator of the amount of competition between 2 species. I like this particularly because it matches our own experience with birds (Smith et al., 1978; Grant and Grant, 1980) and is more solidly based than our conclusions. Therefore the result is likely to be general in application. I emphasize this because lizards, like salamanders (Hairston, 1980a) but unlike most birds, may eat young of other species, so the interpretation of interspecific interactions between lizards in terms of competition is complicated by the possibility of simultaneous predation (Hairston, 1980a,b,c).

A Forward Look: Some Pointers for Future Research

Scientific activity ranges between the extremes of (1) normal science, in which progress is made within established conceptual frameworks, and (2) revolutionary science, in which those frameworks are challenged and occasionally replaced. All the chapters in this book seem to be toward the normal science end of the spectrum. Since lizard ecologists know best how to solve their own scientific problems, I am unlikely to be able to make any helpful remarks here. All I can do is to highlight one method of problem solving that is of general applicability, yet not always recognized for its value. I refer to the habit of turning a problem upside down, of approaching it from quite a different angle when the obvious approach has failed. There are a couple of interesting examples in this book, and I will mention one. Case (Chapter 14) was concerned with the question of whether morphological differences between coexisting species of lizards on islands represent some assortative process, such as competitive exclusion or displacement, or whether the observed differences are to be expected by chance. Faced with the problem of distinguishing between the two, he turned the problem around by specifying rules of assortment that should apply to a set of data and then testing to see whether such artificial data sets can be distinguished from randomly combined ones.

With regard to revolutionary science, it is impossible to predict where the next major breakthroughs in lizard ecology are going to be, because probably they will come from relatively unknown individuals in rebellion against received ideas. This is the conclusion I reach from thinking about some concepts of major importance in ecology that were not addressed in this book: evolutionary stable strategies (ESS), inclusive fitness, and group selection. The names associated with them are J. Maynard Smith, W. D. Hamilton, D. S. Wilson, and M. J. Wade. Admittedly Maynard Smith was not a young man when he published the first paper on ESS's, but Hamilton, whose preceding and similar work was on the concept of an unbeatable sex ratio, was in the early stages of his career (1967) when he published his sex ratio paper, and Wilson and Wade likewise were starting their careers with their initial papers on group selection.

Major advances might be expected to come from opposition to 3 current, powerful, but not entirely satisfactory theories: the equilibrium theory of island biogeography, optimal-foraging theory, and kin-selection theory. Roughgarden, Heckel, and Fuentes (Chapter 16) came closest in this book to a contribution in the vein of revolutionary science by proposing an alternative to equilibrium island-biogeography theory, although as yet it is not developed as a general model. Optimal-foraging (or diet) theory needs to be set in opposition to an equally well-developed alternative, such as one based on the principle of satisficing (Simon, 1956). More generally, the concept of single optimal solutions to biological problems needs to be evaluated in relation to the concept of alternative solutions, such as mixed stable strategies in behavior, variants maintained by frequency-dependent selection in ecological genetics, and multiple adaptive peaks in evolution. Finally, kin-selection theory needs to be set against individual-selection theory, difficult as that is, to see which best explains the numerous observations made of apparently cooperative behavior among animals. As I mentioned earlier, a major task in ecology is to distinguish clearly between alternative theories.

A final way in which I shall try to identify profitable directions for future research in lizard ecology is to summarize some of our work in bird ecology, and hope that an intellectual current may pass between these electrodes. It is my conviction that behavioral studies of individuals of known identity can contribute greatly to our understanding of population phenomena and their evolution. This is largely behavioral ecology, which can be defined as the study of populations through an analysis of the behavior of individuals of known identity. Marking individuals uniquely has been done for many years in order to estimate population sizes and other parameters of all types of vertebrates including lizards (for example, Tinkle, 1967), but in the process the value of knowing which individual is which has largely been ignored. With a rise in the importance of sociobiology, and in particular the need to know genetic relatedness, this is changing.

Since 1975 we have been color-banding and studying the 2 Darwin's finch species on Daphne Major Island, Galápagos; these are *Geospiza fortis,* the medium ground finch, and *G. scandens,* the cactus ground finch. Approximately 95 percent of each population is now color banded. Our research is not motivated by sociobiological concerns, although these are relevant; rather we ask: "Does it matter that an individual has a particular phenotype?" The phenotypic traits of greatest interest to us are body size, beak size, and beak shape. We have obtained a partial answer by studying survival and reproduction, food supply, and feeding behavior. In 1977 a drought occurred, and the populations of *G. fortis* and *G. scandens* fell from about 1,200 and 400 individuals to 180 and 120 respectively. Survival was not random with respect to phenotype, so our answer is that to a certain extent it does matter to an individual what phenotype it has. As food supply declined, *G. fortis,* a feeding generalist, was subjected to directional selection (Boag and Grant, 1981); large individuals survived better than small individuals because only they could deal with the largest and hardest foods (actually woody fruits). *Geospiza scandens,* a feeding specialist on different parts of *Opuntia* cactus, was subjected to stabilizing selection. Our previous studies, which showed high heritabilities of beak and body size (Boag and Grant, 1978), argue for genetic responses and not just nongenetic phenotypic responses to environmental stress.

Most of these results and all of the interpretation would not have been possible without knowing the identity of individuals. In lizard ecology there should be plenty of scope for answering behavioral and evolutionary questions by similar methods. Fox's work (Fox, 1975, 1978, Chapter 8) is the closest to ours, but it lacks the link between identified selection pressures and observed population responses. Ballinger's work (Chapter 11) on clutch sizes in lizards holds promise too, especially as heritabilities of clutch sizes in birds have been estimated (Perrins and Jones, 1974), although I interpret his genetic variance model to be more a stimulating metaphor than a literal description.

As an example of the scope for such studies in lizard ecology, I will mention one problem area: character displacement. Character displacement is an evolutionary explanation for the enhanced differences between some sympatric species. I believe the process of divergence could be experimentally reproduced by bringing together representatives of allopatric populations that exhibit similar niches, using Slatkin's (1980) theoretical work to identify the most likely conditions under which it will occur. Heritabilities of body size or jaw size would have to be determined to ensure that there is enough additive genetic variance for selection to act on, but the data from birds (Grant and Price, 1981) give encouragement that it might be present. The experimental protocols of Dunham (1980, Chapter 12) could be used, although fencing might be required (for example, Fox, Chapter 8) to prevent the introduced species from escaping into an unnatural area and establishing itself. Of course such experiments might fail, but a success would be very valuable for 2 reasons. First, it would establish the validity of ecological character displacement. Evidence for it so far is indirect, and although a good example is available from a study of lizards (Huey and Pianka, 1974), it is not without its problems (Dunham, Smith, and Taylor, 1979; but see Pianka, Huey, and Lawlor, 1979). Second, this type of experimental manipulation could be used to test other evolutionary ideas, such as those which attempt to account for the evolutionary and ecological events in the final stages of allopatric speciation.

Peter R. Grant

References

Acknowledgments

Contributors

Index

References

The following frequently cited symposia are indicated in the references by the editors' names only.

Burghardt, G. M., and A. S. Rand, eds. 1982. *Iguanas of the world: behavior, ecology, and evolution.* Park Ridge, New Jersey: Noyes.

Gans, Carl, and W. R. Dawson, eds. 1976. *Biology of the Reptilia.* London: Academic Press.

Gans, Carl, and F. H. Pough, eds. 1982. *Biology of the Reptilia,* vols. 12 and 13. London: Academic Press.

Gans, Carl, and D. W. Tinkle, eds. 1977. *Biology of the Reptilia.* London: Academic Press.

Greenberg, Neil, and P. D. MacLean, eds. 1978. *Behavior and neurology of lizards.* Rockville, Maryland: National Institute of Mental Health.

Milstead, W. W., ed. 1967. *Lizard ecology: a symposium.* Columbia: University of Missouri Press.

Introduction

Andrews, R. M. 1971. Structural habitat and time budget of *Anolis polylepis* (Iguanidae). *Ecology* 52:262–270.

——— 1979. Evolution of life histories: a comparison of *Anolis* from matched island and mainland habitats. *Breviora Mus. Comp. Zool.* 454:1–51.

Avery, R. A. 1982. Field studies of reptilian body temperatures and thermoregulation. In Carl Gans and F. H. Pough, vol. 12.

Ballinger, R. E. 1978. Variation in and evolution of clutch and litter size. In *The vertebrate ovary,* ed. R. E. Jones. New York: Plenum Press.

Bartholomew, G. A. 1982. Physiological control of body temperature. In Carl Gans and F. H. Pough, vol. 12.

Bennett, A. F. 1978. Activity metabolism of the lower vertebrates. *Annu. Rev. Physiol.* 40:447–469.

——— 1980. The metabolic foundations of vertebrate behavior. *Bioscience* 30:452–456.

Bennett, A. F., and K. A. Nagy. 1977. Energy expenditure in free-ranging lizards. *Ecology* 58:697–700.

Bennett, A. F., and J. A. Ruben. 1979. Endothermy and activity in vertebrates. *Science* (Wash., D.C.) 206:649–654.
Blair, W. F. 1960. *The rusty lizard: a population study.* Austin: University of Texas Press.
Carpenter, C. C. 1967. Aggression and social structure in iguanid lizards. In W. W. Milstead.
Carpenter, C. C., and G. W. Ferguson. 1977. Variation and evolution of stereotyped behavior in reptiles. In Carl Gans and D. W. Tinkle.
Case, T. J. 1975. Species numbers, density compensation, and colonizing ability of lizards on islands in the Gulf of California. *Ecology* 56:3–18.
——— 1978. On the evolution and adaptive significance of postnatal growth rates in the terrestrial vertebrates. *Q. Rev. Biol.* 53:243–282.
——— 1979. Character displacement and coevolution in some *Cnemidophorus* lizards. *Fortschr. Zool.* 25:235–282.
Collette, B. B. 1961. Correlations between ecology and morphology in anoline lizards from Havana, Cuba and southern Florida. *Bull. Mus. Comp. Zool.* 125:137–162.
Crews, David. 1975. Psychobiology of reptilian reproduction. *Science* (Wash., D.C.) 189:1059–65.
——— 1979. The hormonal control of behavior in a lizard. *Sci. Am.* 241:180–187.
Crews, David, and E. E. Williams. 1977. Hormones, reproductive behavior, and speciation. *Am. Zool.* 17:271–286.
Dawson, W. R. 1967. Interspecific variation in physiological responses of lizards to temperature. In W. W. Milstead.
——— 1975. On the physiological significance of the preferred body temperature of reptiles. In *Perspectives in biophysical ecology,* ed. D. M. Gates and R. B. Schmerl. New York: Springer-Verlag.
Dunham, A. E. 1980. An experimental study of interspecific competition between the iguanid lizards *Sceloporus merriami* and *Urosaurus ornatus. Ecol. Monogr.* 50:309–330.
Ferguson, G. W. 1971. Variation and evolution of the pushup displays of the side-blotched lizard genus *Uta* (Iguanidae). *Syst. Zool.* 20:79–101.
Fuentes, E. R. 1976. Ecological convergence of lizard communities in Chile and California. *Ecology* 57:3–17.
Gorman, G. C. 1969. Intermediate territorial display of a hybrid *Anolis* lizard (Sauria: Iguanidae). *Z. Tierpsychol.* 26:390–393.
Grant, P. R. 1967. Unusual feeding of lizards on an island. *Copeia* 1967:223–224.
Greenberg, Neil, and P. D. MacLean, eds. 1978. *Behavior and neurology of lizards.* Rockville, Maryland: National Institute of Mental Health.
Heatwole, Harold. 1976. *Reptile ecology.* St. Lucia: University of Queensland Press.
Hertz, P. E. 1974. Thermal passivity of a tropical forest lizard, *Anolis polylepis. J. Herpetol.* 8:323–327.
Hirshfield, M. F., and D. W. Tinkle. 1975. Natural selection and the evolution of reproductive effort. *Proc. Natl. Acad. Sci. U.S.A.* 72:2227–31.
Huey, R. B. 1979. Parapatry and niche complementarity of Peruvian desert

geckos (*Phyllodactylus*): the ambiguous role of competition. *Oecologia* (Berl.) 38:249–259.
——— 1982. Temperature, physiology, and the ecology of reptiles. In Carl Gans and F. H. Pough, vol. 12.
Huey, R. B., and E. R. Pianka. 1974. Ecological character displacement in a lizard. *Am. Zool.* 14:1127–36.
——— 1981. Ecological consequences of foraging mode. *Ecology* 62:991–999.
Huey, R. B., and Montgomery Slatkin. 1976. Costs and benefits of lizard thermoregulation. *Q. Rev. Biol.* 51:363–384.
Inger, R. F., and R. K. Colwell. 1977. Organization of contiguous communities of amphibians and reptiles in Thailand. *Ecol. Monogr.* 47:229–253.
Jenssen, T. A. 1973. Shift in the structural habitat of *Anolis opalinus* due to congeneric competition. *Ecology* 54:863–869.
——— 1977. Evolution of anoline lizard display behavior. *Am. Zool.* 17:203–215.
Kiester, A. R. 1977. Communication in amphibians and reptiles. In *How animals communicate,* ed. T. A. Sebeok. Bloomington: Indiana University Press.
Kiester, A. R., and Montgomery Slatkin. 1974. A strategy of movement and resource utilization. *Theoret. Popul. Biol.* 6:1–20.
Kluger, M. J. 1979. Fever in ectotherms: evolutionary implications. *Am. Zool.* 19:295–304.
Lister, B. C. 1976. The nature of niche expansion in West Indian *Anolis* lizards, I: ecological consequences of reduced competition. *Evolution* 30:659–676.
MacArthur, R. H., and E. R. Pianka. 1966. On optimal use of a patchy environment. *Am. Nat.* 100:603–609.
Milstead, W. W. 1961. Competitive relations in lizard populations. In *Vertebrate speciation,* ed. W. F. Blair. Austin: University of Texas Press.
———, ed. 1967. *Lizard ecology: a symposium.* Columbia: University of Missouri Press.
Moberly, W. R. 1968. The metabolic responses of the common iguana, *Iguana iguana,* to walking and diving. *Comp. Biochem. Physiol.* 27:21–32.
Moermond, T. C. 1979a. Habitat constraints on the behavior, morphology, and community structure of *Anolis* lizards. *Ecology* 60:152–164.
——— 1979b. The influence of habitat structure on *Anolis* foraging behavior. *Behaviour* 70:147–167.
Nagy, K. A. 1982. Energy requirements of free-living iguanid lizards. In G. M. Burghardt and A. S. Rand.
Nevo, Eviatar, G. C. Gorman, M. F. Soulé, S. Y. Yang, Robert Clover, and Vojislav Jovanović. 1972. Competitive exclusion between insular *Lacerta* species (Sauria, Lacertidae). *Oecologia* (Berl.) 10:183–190.
Norris, K. S. 1967. Color adaptation in desert reptiles and its thermal relationships. In W. W. Milstead.
Pacala, Stephen, and Jonathan Roughgarden. 1982. Resource partitioning and interspecific competition in two two-species insular *Anolis* lizard communities. *Science* 217:444–446.
Pianka, E. R. 1967. On lizard species diversity: North American flatland deserts. *Ecology* 48:333–351.

——— 1969. Sympatry of desert lizards (*Ctenotus*) in Western Australia. *Ecology* 50:1012–30.

——— 1973. The structure of lizard communities. *Annu. Rev. Ecol. Syst.* 4:53–74.

——— 1974. Niche overlap and diffuse competition. *Proc. Natl. Acad. Sci. U.S.A.* 71:2141–45.

——— 1975. Niche relations of desert lizards. In *Ecology and evolution of communities,* ed. M. Cody and J. Diamond. Cambridge, Mass.: Harvard University Press.

——— 1977. Reptilian species diversity. In Carl Gans and D. W. Tinkle.

Pianka, E. R., and W. S. Parker. 1975. Age-specific reproductive tactics. *Am. Nat.* 109:453–464.

Porter, W. P., J. W. Mitchell, W. A. Beckman, and C. B. DeWitt. 1973. Behavioral implications of mechanistic ecology. *Oecologia* (Berl.) 13:1–54.

Pough, F. H. 1980. The advantages of ectothermy for tetrapods. *Am. Nat.* 115:113–120.

Price, D. H. de Solla. 1963. *Little science, big science.* New York: Columbia University Press.

Rand, A. S. 1964. Ecological distribution in anoline lizards of Puerto Rico. *Ecology* 45:745–752.

——— 1967. The adaptive significance of territoriality in iguanid lizards. In W. W. Milstead.

Rand, W. M., and A. S. Rand. 1976. Agonistic behavior in nesting iguanas: a stochastic analysis of dispute settlement dominated by the minimization of energy cost. *Z. Tierpsychol.* 40:279–299.

Rand, A. S., and E. E. Williams. 1971. An estimation of redundancy and information content of anole dewlaps. *Am. Nat.* 104:99–103.

Regal, P. J. 1978. Behavioral differences between reptiles and mammals: an analysis of activity and mental capabilities. In Neil Greenberg and P. D. MacLean.

Roughgarden, Jonathan. 1972. Evolution of niche width. *Am. Nat.* 106:683–718.

——— 1974. Niche width: biogeographic patterns among *Anolis* lizard populations. *Am. Nat.* 108:429–442.

——— 1979. A local concept of structural homology for ecological communities with examples from simple communities in West Indian *Anolis* lizards. *Fortschr. Zool.* 25:149–157.

Roughgarden, Jonathan, and E. R. Fuentes. 1977. The environmental determinants of size in solitary populations of West Indian *Anolis* lizards. *Oikos* 29:44–51.

Roughgarden, Jonathan, W. P. Porter, and D. G. Heckel. 1981. Resource partitioning of space and its relationship to body temperature in *Anolis* lizard populations. *Oecologia* (Berl.) 50:256–264.

Roughgarden, Jonathan, J. Rummel, and Stephen Pacala. 1982. Experimental evidence of strong present-day competition between the *Anolis* population of the Anguilla Bank—a preliminary report. In *Advances in herpetology and evolutionary biology: essays in honor of Ernest E. Williams, Spec. Publ. Bull. Mus. Comp. Zool.*

Ruibal, Rodolfo. 1961. Thermal relations of five species of tropical lizards. *Evolution* 15:98–111.
——— 1967. Evolution and behavior in West Indian anoles. In W. W. Milstead.
Sage, R. D. 1973. Ecological convergence of the lizard faunas of the chaparral communities in Chile and California. In *Ecological studies,* vol. 7, ed. F. di-Castri and H. A. Mooney. New York: Springer-Verlag.
Schall, J. J., and E. R. Pianka. 1978. Geographical trends in numbers of species. *Science* (Wash., D.C.) 201:679–686.
——— 1980. Evolution of escape behavior diversity. *Am. Nat.* 115:551–566.
Schoener, T. W. 1968. The *Anolis* lizards of Bimini: resource partitioning in a complex fauna. *Ecology* 49:704–726.
——— 1969. Models of optimal size for solitary predators. *Am. Nat.* 103:277–313.
——— 1971. Theory of feeding strategies. *Annu. Rev. Ecol. Syst.* 2:369–404.
——— 1974a. Resource partitioning in ecological communities. *Science* (Wash., D.C.) 185:27–38.
——— 1974b. Competition and the form of habitat shift. *Theoret. Popul. Biol.* 6:265–307.
——— 1975. Presence and absence of habitat shift in some widespread lizard species. *Ecol. Monogr.* 45:233–258.
——— 1977. Competition and the niche. In Carl Gans and D. W. Tinkle.
——— 1979. Inferring the properties of predation and other injury-producing agents from injury frequencies. *Ecology* 60:1110–15.
Schoener, T. W., and Amy Schoener. 1980. Ecological and demographic correlates of injury rates in some Bahamian *Anolis* lizards. *Copeia* 1980:839–850.
——— 1982a. Intraspecific variation in home-range size in some *Anolis* lizards. *Ecology* 63:809–823.
——— 1982b. On the voluntary departure of lizards from very small islands. In *Advances in herpetology and evolutionary biology: essays in honor of Ernest E. Williams, Spec. Publ. Bull. Mus. Comp. Zool.*
Schoener, T. W., and G. C. Gorman. 1968. Some niche differences in three Lesser Antillean lizards of the genus *Anolis. Ecology* 49:819–830.
Schoener, T. W., R. B. Huey, and E. R. Pianka. 1978. A biogeographic extension of the compression hypothesis: competitors in narrow sympatry. *Am. Nat.* 113:295–298.
Scott, N. J., Jr. 1982. Chronological bibliography of herpetological community studies. In *Herpetological communities,* ed. N. J. Scott, Jr. Washington, D.C.: Wildlife Research Reports no. 13, U.S. Fish and Wildlife Service.
Sexton, O. J., and Harold Heatwole. 1968. An experimental investigation of habitat selection and water loss in some anoline lizards. *Ecology* 49:762–767.
Simon, C. A. 1975. The influence of food abundance on territory size in the iguanid lizard *Sceloporus jarrovi. Ecology* 56:993–998.
Smith, D. C. 1981. Competitive interactions of the striped plateau lizard (*Sceloporus virgatus*) and the tree lizard (*Urosaurus ornatus*). *Ecology* 62:679–687.
Stamps, J. A. 1977a. Social behavior and spacing patterns in lizards. In Carl Gans and D. W. Tinkle.

——— 1977b. The relationship between resource competition, risk, and aggression in a tropical territorial lizard. *Ecology* 58:344–358.

Stamps, J. A., and Sanford Tanaka. 1981. The relationship between food and social behavior in juvenile lizards (*Anolis aeneus*). *Copeia* 1981:422–434.

Stamps, J. A., Sanford Tanaka, and V. V. Krishnan. 1981. The relationship between selectivity and food abundance in a juvenile lizard. *Ecology* 62:1079–92.

Tinkle, D. W. 1967. Home range, density, dynamics, and structure of a Texas population of the lizard *Uta stansburiana.* In W. W. Milstead.

——— 1969. The concept of reproductive effort and its relation to the evolution of life histories of lizards. *Am. Nat.* 103:501–516.

Tinkle, D. W., H. M. Wilbur, and S. G. Tilley. 1970. Evolutionary strategies in lizard reproduction. *Evolution* 24:55–74.

Tracy, C. R. 1982. Biophysical modelling in reptilian physiology and zoology. In Carl Gans and F. H. Pough, vol. 12.

Trivers, R. L. 1972. Parental investment and sexual selection. In *Sexual selection and the descent of man,* ed. B. Campbell. Chicago: Aldine-Atherton.

Turner, F. B. 1977. The dynamics of populations of squamates, crocodilians, and rhynchocephalians. In Carl Gans and D. W. Tinkle.

Turner, F. B., P. A. Medica, and B. W. Kowalewsky. 1976. Energy utilization by a desert lizard (*Uta stansburiana*). *US/ IBP Desert Biome Monograph No. 1.* Logan: Utah State University Press.

Vitt, L. J., and J. D. Congdon. 1978. Body shape, reproductive effort, and relative clutch mass in lizards: resolution of a paradox. *Am. Nat.* 112:595–608.

Vitt, L. J., J. D. Congdon, and N. A. Dickson. 1977. Adaptive strategies and energetics of tail autotomy in lizards. *Ecology* 58:326–337.

Williams, E. E. 1969. The ecology of colonization as seen in the zoogeography of anoline lizards on small islands. *Q. Rev. Biol.* 44:345–389.

——— 1972. The origin of faunas: evolution of lizard congeners in a complex island fauna—a trial analysis. *Evolutionary Biology* 4:47–89.

Overview of Part I

Bartholomew, G. A., and V. A. Tucker. 1963. Control of changes in body temperature, metabolism, and circulation by the agamid lizard, *Amphibolurus barbatus. Physiol. Zool.* 36:199–218.

——— 1964. Size, body temperature, thermal conductance, oxygen consumption, and heart rate in Australian varanid lizards. *Physiol. Zool.* 37:341–354.

Bartholomew, G. A., V. A. Tucker, and A. K. Lee. 1965. Oxygen consumption, thermal conductance, and heart rate in the Australian skink, *Tiliqua scincoides. Copeia* 1965:169–173.

Cowles, R. B., and C. M. Bogert. 1944. A preliminary study of the thermal requirements of desert reptiles. *Bull. Am. Mus. Nat. Hist.* 83:261–296.

Kluger, M. J. 1979. Fever in ectotherms: evolutionary implications. *Am. Zool.* 19:295–304.

Milstead, W. W., ed. 1967. *Lizard ecology: a symposium.* Columbia: University of Missouri Press.

Moberly, W. R. 1966. The physiological correlates of activity in the common iguana, *Iguana iguana.* Ph.D. dissertation, University of Michigan.

——— 1968a. The metabolic responses of the common iguana, *Iguana iguana,* to activity under restraint. *Comp. Biochem. Physiol.* 27:1–20.

——— 1968b. The metabolic responses of the common iguana, *Iguana iguana,* to walking and diving. *Comp. Biochem. Physiol.* 27:21–32.

Pough, F. H. 1980. The advantages of ectothermy for tetrapods. *Am. Nat.* 115:92–112.

Regal, P. J. 1978. Behavioral differences between reptiles and mammals: an analysis of activity and mental capabilities. In Neil Greenberg and P. D. MacLean.

Turner, F. B. 1977. The dynamics of populations of squamates, crocodilians, and rhynchocephalians. In Carl Gans and D. W. Tinkle.

1. Ecological Consequences of Activity Metabolism

Asplund, K. K. 1970. Metabolic scope and body temperatures of whiptail lizards (*Cnemidophorus*). *Herpetologica* 26:403–411.

Bakker, R. T. 1972. Locomotor energetics of lizards and mammals compared. *Physiologist* 15:76.

Bartholomew, G. A., and V. A. Tucker. 1964. Size, body temperature, thermal conductance, oxygen consumption, and heart rate in Australian varanid lizards. *Physiol. Zool.* 37:341–354.

Bennett, A. F. 1972. The effect of activity on oxygen consumption, oxygen debt, and heart rate in the lizards *Varanus gouldii* and *Sauromalus hispidus. J. Comp. Physiol.* 79:259–280.

——— 1978. Activity metabolism of the lower vertebrates. *Annu. Rev. Physiol.* 40:447–469.

——— 1980. The thermal dependence of lizard behaviour. *Anim. Behav.* 28:752–763.

——— 1982. The energetics of reptilian activity. In Carl Gans and F. H. Pough, vol. 13.

Bennett, A. F., and W. R. Dawson. 1972. Aerobic and anaerobic metabolism during activity in the lizard *Dipsosaurus dorsalis. J. Comp. Physiol.* 81:289–299.

——— 1976. Metabolism. In Carl Gans and W. R. Dawson.

Bennett, A. F., and T. T. Gleeson. 1976. Activity metabolism in the lizard *Sceloporus occidentalis. Physiol. Zool.* 49:65–76.

——— 1979. Metabolic expenditure and the cost of foraging in the lizard *Cnemidophorus murinus. Copeia* 1979:573–577.

Bennett, A. F., T. T. Gleeson, and G. C. Gorman. 1981. Anaerobic metabolism in a lizard (*Anolis bonairensis*) under natural conditions. *Physiol. Zool.* 54:237–341.

Bennett, A. F., and G. C. Gorman. 1979. Population density and energetics of lizards on a tropical island. *Oecologia* (Berl.) 42:339–358.

Bennett, A. F., and Paul Licht. 1972. Anaerobic metabolism during activity in lizards. *J. Comp. Physiol.* 81:277–288.

Bennett, A. F., and J. A. Ruben. 1979. Endothermy and activity in vertebrates. *Science* (Wash., D.C.) 206:649–654.

Dawson, W. R. 1967. Interspecific variation in physiological responses of lizards to temperature. In W. W. Milstead.

Dmi'el, Razi, and Dyna Rappeport. 1976. Effect of temperature on metabolism during running in the lizard *Uromastix aegyptius. Physiol. Zool.* 45:77–84.

Fry, F. E. J. 1947. Effects of the environment on animal activity. *Univ. Toronto Stud. Biol. Ser.* 55:1–62.

Gleeson, T. T. 1979. Foraging and transport costs in the Galapagos marine iguana, *Amblyrhynchus cristatus. Physiol. Zool.* 52:549–557.

——— 1980. Lactic acid production during field activity in the Galapagos marine iguana, *Amblyrhynchus cristatus. Physiol. Zool.* 53:157–162.

Gleeson, T. T., G. S. Mitchell, and A. F. Bennett. 1980. Cardiovascular responses to graded activity in the lizards *Varanus* and *Iguana. Am. J. Physiol.* 239:R174–R179.

Hemmingsen, A. M. 1960. Energy metabolism as related to body size and respiratory surfaces, and its evolution. *Rep. Steno Hosp.* 9:1–110.

John-Alder, H. B., and A. F. Bennett. 1981. Thermal dependence of endurance, oxygen consumption, and cost of locomotion in a lizard. *Am. J. Physiol.* 241 (Reg. Integ. Comp. Physiol. *10*):R342–R349.

Moberly, W. R. 1968. The metabolic responses of the common iguana, *Iguana iguana,* to walking and diving. *Comp. Biochem. Physiol.* 27:21–32.

Pough, F. H. 1980. The advantages of ectothermy for tetrapods. *Am. Nat.* 115:92–112.

Regal, P. J. 1978. Behavioral differences between reptiles and mammals: an analysis of activity and mental capacities. In Neil Greenberg and P. D. MacLean.

Schmidt-Nielsen, Knut. 1972. Locomotion: energy cost of swimming, flying, and running. *Science* (Wash., D.C.) 177:222–228.

Taylor, C. R. 1973. Energy cost of animal locomotion. In *Comparative physiology: locomotion, respiration, transport and blood,* ed. L. Bolis, Knut Schmidt-Nielsen, and S. H. P. Maddrell. New York: American Elsevier.

Taylor, C. R., Knut Schmidt-Nielsen, and J. L. Raab. 1970. Scaling of energetic cost of running to body size in mammals. *Am. J. Physiol.* 219:1104–7.

Templeton, J. R. 1970. Reptiles. In *Comparative physiology of thermoregulation.* Vol. 1, *Invertebrates and nonmammalian vertebrates,* ed. G. C. Whittow. New York: Academic Press.

Tucker, V. A. 1967. The role of the cardiovascular system in oxygen transport and thermoregulation in lizards. In W. W. Milstead.

——— 1970. Energetic cost of locomotion in animals. *Comp. Biochem. Physiol.* 34:841–846.

Wilson, K. J. 1974. The relationships of maximum and resting oxygen consumption and heart rates to weight in reptiles of the order Squamata. *Copeia* 1974:781–785.

Wilson, K. J., and A. K. Lee. 1974. Energy expenditure of a large herbivorous lizard. *Copeia* 1974:338–348.

Wood, S. C., Kjell Johansen, M. L. Glass, and G. M. O. Maloiy. 1978. Aerobic metabolism of the lizard *Varanus exanthematicus:* effects of activity, temperature, and size. *J. Comp. Physiol.* 127:331–336.

2. Ecological Energetics

Alexander, C. E., and W. G. Whitford. 1968. Energy requirements of *Uta stansburiana. Copeia* 1968:678–683.

Anderson, R. A., and W. H. Karasov. 1981. Contrasts in energy intake and expenditure in sit-and-wait and widely foraging lizards. *Oecologia* (Berl.) 49:67–72.

Andrews, R. M. 1979. Reproductive effort of female *Anolis limifrons* (Sauria: Iguanidae). *Copeia* 1979:620–626.

Avery, R. A. 1970. Utilization of caudal fat by hibernating common lizards, *Lacerta vivipara. Comp. Biochem. Physiol.* 37:119–121.

——— 1974. Storage lipids in the lizard *Lacerta vivipara:* a quantitative study. *J. Zool.* (Lond.) 173:419–425.

Ballinger, R. E., and D. R. Clark. 1973. Energy content of lizard eggs and the measurement of reproductive effort. *J. Herpetol.* 7:129–132.

Banse, Karl, and Steven Mosher. 1980. Adult body mass and annual production/ biomass relationships of field populations. *Ecol. Monogr.* 50:355–379.

Beatley, J. C. 1976. *Vascular plants of the Nevada Test Site and central-southern Nevada: ecological and geographic distributions.* National Technical Information Center, U.S. Dept. of Commerce, Springfield, Virginia.

Bennett, A. F., and W. R. Dawson. 1976. Metabolism. In Carl Gans and W. R. Dawson.

Bennett, A. F., and K. A. Nagy. 1977. Energy expenditure in free-ranging lizards. *Ecology* 58:697–700.

Brody, Samuel. 1945. *Bioenergetics and growth.* New York: Hafner Press.

Bustard, R. H. 1967. Gekkonid lizards adapt fat storage to desert environments. *Science* (Wash., D.C.) 158:1197–98.

Carpenter, C. C. 1962. A comparison of the patterns of display of *Urosaurus, Uta,* and *Streptosaurus. Herpetologica* 18:145–152.

Charnov, E. L. 1976. Optimal foraging: attack strategy of a mantid. *Am. Nat.* 110:141–151.

Congdon, J. D., R. E. Ballinger, and K. A. Nagy. 1979. Energetics, temperature, and water relations in winter aggregated *Sceloporus jarrovi* (Sauria: Iguanidae). *Ecology* 60:30–35.

Derickson, W. K. 1974. Lipid deposition and utilization in the sagebrush lizard, *Sceloporus graciosus:* its significance for reproduction and maintenance. *Comp. Biochem. Physiol.* 49A:267–272.

——— 1976a. Lipid storage and utilization in reptiles. *Am. Zool.* 16:711–723.

——— 1976b. Ecological and physiological aspects of reproductive strategies in two lizards. *Ecology* 57:445–458.

Dixon, J. R. 1967. Aspects of the biology of the lizards of White Sands, New Mexico. *Los Ang. Cty. Mus. Contrib. Sci.* No. 129.

Dixon, J. R., and P. A. Medica. 1966. Summer food of four species of lizards from the vicinity of White Sands, New Mexico. *Los Ang. Cty. Mus. Contrib. Sci.* No. 121.

Dixon, W. S., and F. J. Massey. 1969. *Introduction to statistical analysis.* New York: McGraw-Hill.

Ferguson, G. W. 1966. Effect of follicle-stimulating hormone and testosterone propionate on the reproduction of the side-blotched lizard, *Uta stansburiana. Copeia* 1966:495–498.

Fitch, H. S. 1970. Reproductive cycles in lizards and snakes. *Univ. Kans. Mus. Nat. Hist. Misc. Publ.* 52:1–247.

Fox, S. F. 1978. Natural selection on behavioral phenotypes of the lizard *Uta stansburiana. Ecology* 59:834–847.

Goldberg, S. R. 1977. Reproduction in a mountain population of the side-blotched lizard, *Uta stansburiana* (Reptilia, Lacertilia, Iguanidae). *J. Herpetol.* 11:31–35.

Golley, F. B. 1968. Secondary productivity in terrestrial communities. *Am. Zool.* 8:53–59.

Hahn, W. E. 1967. Estradiol-induced vitellinogenesis and concomitant fat mobilization in the lizard *Uta stansburiana. Comp. Biochem. Physiol.* 23:83–93.

Hahn, W. E., and D. W. Tinkle. 1965. Fat body cycling and experimental evidence for its adaptive significance to ovarian follicle development in the lizard *Uta stansburiana. J. Exp. Zool.* 158:79–86.

Hirshfield, M. R., and D. W. Tinkle. 1975. Natural selection and the evolution of reproductive effort. *Proc. Natl. Acad. Sci. U.S.A.* 72:2227–31.

Hoff, C. L., and F. R. Kay. 1970. Herbivorous feeding in the lizard *Uta stansburiana stejnegeri. Southwest. Nat.* 15:137–138.

Huey, R. B., and Montgomery Slatkin. 1976. Costs and benefits of lizard thermoregulation. *Q. Rev. Biol.* 51:363–384.

Humphreys, W. F. 1979. Production and respiration in animal populations. *J. Anim. Ecol.* 48:427–453.

Irwin, L. N. 1965. Diel activity and social interaction of the lizard *Uta stansburiana stejnegeri. Copeia* 1965:99–101.

King, J. R. 1974. Seasonal allocation of time and energy resources in birds. In *Avian energetics,* ed. R. A. Paynter. Cambridge, Mass.: Nuttall Ornithol. Club.

Kitchell, J. F., and J. T. Windell. 1972. Energy budget for the lizard *Anolis carolinensis. Physiol. Zool.* 45:178–188.

Kleiber, Max. 1975. *The fire of life.* 2nd ed. Huntington, N.Y.: Krieger.

Licht, Paul. 1973. Environmental influences on the testis cycles of the lizards *Dipsosaurus dorsalis* and *Xantusia vigilis. Comp. Biochem. Physiol.* 45A:7–20.

——— 1979. Reproductive endocrinology of reptiles and amphibians: gonadotropins. *Annu. Rev. Physiol.* 41:337–351.

Licht, Paul, and G. C. Gorman. 1970. Reproductive and fat cycles in Caribbean *Anolis* lizards. *Univ. Calif. Publ. Zool.* 95:1–52.

——— 1975. Altitudinal effects on the seasonal testis cycles of tropical *Anolis* lizards. *Copeia* 1975:496–504.

Lifson, Nathan, and Ruth McClintock. 1966. Theory of use of the turnover rates of body water for measuring energy and material balance. *J. Theor. Biol.* 12:46–74.

Lindemann, R. L. 1942. The trophic dynamic aspect of ecology. *Ecology* 23:399–418.

Mayhew, W. W. 1967. Comparative reproduction in three species of the genus *Uma.* In W. W. Milstead.

McNeill, S., and J. H. Lawton. 1970. Annual production and respiration in animal populations. *Nature* (Lond.) 225:472–474.

Medica, P. A., and F. B. Turner. 1976. Reproduction by *Uta stansburiana* (Reptilia, Lacertilia, Iguanidae) in southern Nevada. *J. Herpetol.* 10:123–128.

Merker, G. P., and K. A. Nagy. 1982. Energy utilization in free-ranging lizards (*Sceloporus virgatus*). Manuscript.

Mood, A. M., F. A. Graybill, and D. C. Boes. 1974. *Introduction to the theory of statistics.* New York: McGraw-Hill.

Nagy, K. A. 1972. Water and electrolyte budgets of a free-living desert lizard, *Sauromalus obesus. J. Comp. Physiol.* 79:39–62.

——— 1973. Behavior, diet, and reproduction in a desert lizard, *Sauromalus obesus. Copeia* 1973:93–102.

——— 1975. Water and energy budgets of free-living animals: measurement using isotopically labeled water. In *Environmental physiology of desert organisms,* ed. N. F. Hadley. Stroudsburg, Pa.: Dowden, Hutchinson and Ross.

——— 1980. CO_2 production in animals: analysis of potential errors in the doubly labeled water method. *Am. J. Physiol.* 238:R466–R473.

——— 1982. Energy requirements of free-living iguanid lizards. In G. M. Burghardt and A. S. Rand.

Nagy, K. A., and D. P. Costa. 1980. Water flux in animals: analysis of potential errors in the tritiated water method. *Am. J. Physiol.* 238:R454–R465.

Nagy, K. A., and V. H. Shoemaker. 1975. Energy and nitrogen budgets of the free-living desert lizard *Sauromalus obesus. Physiol. Zool.* 48:252–262.

Norris, K. S. 1967. Color adaptation in desert reptiles and its thermal relationships. In W. W. Milstead.

Nussbaum, R. A. 1981. Seasonal shifts in clutch size and egg size in the side-blotched lizard, *Uta stansburiana* Baird and Girard. *Oecologia* (Berl.) 49:8–13.

Parker, W. S. 1974. Home range, growth, and population density of *Uta stansburiana* in Arizona. *J. Herpetol.* 8:135–139.

Parker, W. S., and E. R. Pianka. 1975. Comparative ecology of populations of the lizard *Uta stansburiana. Copeia* 1975:615–632.

Pearson, A. K., Paul Licht, K. A. Nagy, and P. A. Medica. 1978. Endocrine function and reproductive impairment in an irradiated population of the lizard *Uta stansburiana. Radiat. Res.* 76:610–623.

Pianka, E. R. 1976. Natural selection of optimal reproductive tactics. *Am. Zool.* 16:775–784.

Pianka, E. R., and W. S. Parker. 1975. Age-specific reproductive tactics. *Am. Nat.* 109:453–464.

Pough, F. H. 1973. Lizard energetics and diet. *Ecology* 54:837–844.

——— 1980. The advantages of ectothermy for tetrapods. *Am. Nat.* 115:92–112.

Pyke, G. H. 1978. Optimal foraging in hummingbirds: testing the marginal value theorem. *Am. Zool.* 18:739–754.

Pyke, G. H., H. R. Pulliam, and E. L. Charnov. 1977. Optimal foraging theory: a selective review of theory and tests. *Q. Rev. Biol.* 52:137–154.

Ricklefs, R. E. 1974. Energetics of reproduction in birds. In *Avian energetics,* ed. R. A. Paynter. Cambridge, Mass.: Nuttall Ornithol. Club.

Roberts, L. A. 1968. Oxygen consumption in the lizard *Uta stansburiana. Ecology* 49:809–819.

Sanborn, S. R. 1977. A comparison of arthropod abundance and food consumed by *Uta stansburiana* (Sauria: Iguanidae) at the Nevada Test Site. M.A. thesis, California State University, Long Beach.

Schmidt-Nielsen, Knut. 1979. *Animal physiology: adaptation and environment.* 2nd ed. Cambridge: Cambridge University Press.

Schoener, T. W. 1971. Theory of feeding strategies. *Annu. Rev. Ecol. Syst.* 2:369–404.

Shine, Richard. 1980. Costs of reproduction in reptiles. *Oecologia* (Berl.) 46:92–100.

Shoemaker, V. H., K. A. Nagy, and W. R. Costa. 1976. Energy utilization and temperature regulation by jackrabbits (*Lepus californicus*) in the Mojave Desert. *Physiol. Zool.* 49:364–375.

Stearns, S. C. 1976. Life-history tactics: a review of the ideas. *Q. Rev. Biol.* 51:3–47.

Tanner, W. W., and J. M. Hopkin. 1972. Ecology of *Sceloporus occidentalis longipes* Baird and *Uta stansburiana stansburiana* Baird and Girard on Rainier Mesa, Nevada Test Site, Nye County, Nevada. *Brig. Young Univ. Sci. Bull.,* Biol. Ser. 15, no. 4.

Thomas, R. D. K., and E. C. Olson, eds. 1980. *A cold look at the warm-blooded dinosaurs.* American Association for the Advancement of Science, Selected Symposium Series, 28. Boulder, Colorado: Westview Press.

Tinkle, D. W. 1967. The life and demography of the side-blotched lizard, *Uta stansburiana. Misc. Publ. Mus. Zool. Univ. Mich.* 132:1–182.

——— 1969. The concept of reproductive effort and its relation to the evolution of life histories of lizards. *Am. Nat.* 103:501–516.

Tinkle, D. W., and N. F. Hadley. 1975. Lizard reproductive effort; caloric estimates and comments on its evolution. *Ecology* 56:427–434.

Tinkle, D. W., Don McGregor, and Sumner Dana. 1962. Home range ecology of *Uta stansburiana stejnegeri. Ecology* 43:223–229.

Tinkle, D. W., H. M. Wilbur, and S. G. Tilley. 1970. Evolutionary strategies in lizard reproduction. *Evolution* 24:55–74.

Turner, F. B. 1970. The ecological efficiency of consumer populations. *Ecology* 51:471–472.

Turner, F. B., and R. M. Chew. 1981. Production by desert animals. In *Arid land ecosystems: structure, functioning, and management,* vol. 2, Internat. Biol. Prog. 17, ed. D. W. Goodall and R. E. Perry. Cambridge: Cambridge University Press.

Turner, F. B., G. A. Hoddenbach, P. A. Medica, and J. R. Lannom. 1970. The demography of the lizard, *Uta stansburiana* Baird and Girard, in southern Nevada. *J. Anim. Ecol.* 39:505–519.

Turner, F. B., and P. A. Medica. 1977. Sterility among female lizards (*Uta stansburiana*) exposed to continuous radiation. *Radiat. Res.* 70:154–163.

Turner, F. B., P. A. Medica, and B. W. Kowalewsky. 1976. Energy utilization by a desert lizard (*Uta stansburiana*). *US/IBP Desert Biome Monograph No. 1.* Logan: Utah State University Press.

Turner, F. B., P. A. Medica, J. R. Lannom, and G. A. Hoddenbach. 1969. A demographic analysis of continuously irradiated and nonirradiated populations of the lizard, *Uta stansburiana. Radiat. Res.* 38:349–356.

Turner, F. B., P. A. Medica, and D. D. Smith. 1974. Reproduction and survivorship of the lizard, *Uta stansburiana,* and the effects of winter rainfall, density, and predation on these processes. *US/IBP Desert Biome Res. Memo. 74-26.* Logan: Utah State University Press.

Vitt, L. J., and R. D. Ohmart. 1975. Ecology, reproduction, and reproductive effort of the iguanid lizard *Urosaurus graciosus* on the lower Colorado River. *Herpetologica* 31:56–65.
Vleck, C. M., David Vleck, and D. F. Hoyt. 1980. Patterns of metabolism and growth in avian embyros. *Am. Zool.* 20:405–416.
Williams, G. C. 1966a. Natural selection, the costs of reproduction, and a refinement of Lack's principle. *Am. Nat.* 100:687–690.
——— 1966b. *Adaptation and natural selection.* Princeton: Princeton University Press.
Wilson, K. J., and A. K. Lee. 1974. Energy expenditure of a large herbivorous lizard. *Copeia* 1974:338–348.
Wood, R. A., K. A. Nagy, N. S. MacDonald, S. T. Wakakuwa, R. J. Beckman, and Howard Kaaz. 1975. Determination of oxygen-18 in water contained in biological samples by charged particle activation. *Anal. Chem.* 47:646–650.

3. Biophysical Analyses of Energetics, Time-Space Utilization, and Distributional Limits

Anderson, R. V., C. R. Tracy, and Z. Abramsky. 1980. Habitat selection in two species of short-horned grasshopper. *Oecologia* (Berl.) 38:359–374.
Atsatt, S. R. 1939. Color changes as controlled by temperature and light in the lizards of the desert regions of southern California. *Publ. Univ. Calif. Los Angeles Biol. Sci.* 1:237–276.
Bakken, G. S. 1976. A heat transfer analysis of animals: unifying concepts and the application of metabolism data to field ecology. *J. Theor. Biol.* 60:337–384.
——— 1981. How many equivalent black body temperatures are there? *J. Therm. Biol.* 6:59–60.
Ballinger, R. E., Jon Hawker, and O. J. Sexton. 1970. The effect of photoperiod acclimation on the thermoregulation of the lizard, *Sceloporus undulatus. J. Exp. Zool.* 171:43–48.
Bartlett, P. N., and D. M. Gates. 1967. The energy budget of a lizard on a tree trunk. *Ecology* 48:315–322.
Beckman, W. A., J. W. Mitchell, and W. P. Porter. 1973. Thermal model for prediction of a desert iguana's daily and seasonal behavior. *J. Heat. Trans.* 257–262.
Bennett, A. F., and W. R. Dawson. 1972. Aerobic and anaerobic metabolism during activity in the lizard *Dipsosaurus dorsalis. J. Comp. Physiol.* 81:289–299.
——— 1976. Metabolism. In Carl Gans and W. R. Dawson.
Bird, R. B., W. E. Stewart, and E. N. Lightfoot. 1960. *Transport phenomena.* New York: Wiley and Sons.
Brattstrom, B. H. 1965. Body temperatures of reptiles. *Am. Midl. Nat.* 73:376–422.
Brody, Samuel. 1945. *Bioenergetics and growth.* New York: Hafner Press.
Brunt, D. 1932. Notes on radiation in the atmosphere. *Q. J. R. Meteorol. Soc.* 58:389–418.
Campbell, G. S. 1977. *An introduction to environmental biophysics.* New York: Springer-Verlag.

Carpenter, C. C. 1969. Behavioral and ecological notes on the Galapagos land iguanas. *Herpetologica* 25:155–164.

Chapman, B. M., and R. F. Chapman. 1964. Observations on the lizard *Agama agama* in Ghana. *Proc. Zool. Soc. Lond.* 143:121–132.

Christian, K. A., and C. R. Tracy. 1981. The effect of the thermal environment on the ability of hatchling Galapagos land iguanas to avoid predation during dispersal. *Oecologia* (Berl.) 49:218–223.

——— 1982. Reproductive behavior of Galapagos land iguanas (*Conolophus pallidus*) on Isla Santa Fe, Galapagos. In G. M. Burghardt and A. S. Rand.

Christian, K. A., C. R. Tracy, and W. P. Porter. 1982. Seasonal shifts in body temperature and use of microhabitats by Galapagos land iguanas (*Conolophus pallidus*). *Ecology* 62: in press.

Church, N. S. 1960. Heat loss and body temperatures of flying insects. II. Heat conduction within the body and its loss by radiation and convection. *J. Exp. Biol.* 37:186–212.

Cole, L. C. 1943. Experiments on toleration of high temperatures in lizards with reference to adaptive coloration. *Ecology* 24:94–108.

Cowles, R. B. 1958. Possible origin of dermal temperature regulation. *Evolution* 12:347–357.

Cowles, R. B., and C. M. Bogert. 1944. A preliminary study of the thermal requirements of desert reptiles. *Bull. Am. Mus. Nat. Hist.* 83:261–296.

Daniel, P. M. 1960. Growth and cyclic behavior in the West African lizard, *Agama agama africana. Copeia* 1960:94–97.

Dawson, W. R., and G. A. Bartholomew. 1958. Metabolic and cardiac responses to temperature in the lizard, *Dipsosaurus dorsalis. Physiol. Zool.* 31:100–111.

DeWitt, C. B. 1967. Precision of thermoregulation and its relation to environmental factors in the desert iguana, *Dipsosaurus dorsalis. Physiol. Zool.* 40:49–66.

Garratt, J. R., and B. B. Hicks. 1973. Momentum, heat and water vapour transfer to and from natural and artificial surfaces. *Q. J. R. Meteorol. Soc.* 99:680–687.

Gates, D. M. 1962. *Energy exchange in the biosphere.* New York: Harper and Row.

——— 1975. Introduction: biophysical ecology. In *Perspectives in biophysical ecology,* ed. D. M. Gates and R. B. Schmerl. New York: Springer-Verlag.

——— 1980. *Biophysical ecology.* New York: Springer-Verlag.

Harlow, H. J., S. S. Hillman, and M. Hoffman. 1976. The effects of temperature on digestive efficiency in the herbivorous lizard *Dipsosaurus dorsalis. J. Comp. Physiol.* 111:1–6.

Heath, J. E. 1962. Temperature regulation and diurnal activity in horned lizards. *Univ. Calif. Publ. Zool.* 64:97–129.

Huey, R. B., E. R. Pianka, and J. A. Hoffman. 1977. Seasonal variation in thermoregulatory behavior and body temperature of diurnal Kalahari lizards. *Ecology* 58:1066–75.

Hutchison, V. H., and J. L. Larimer. 1960. Reflectivity of the integuments of some lizards from different habitats. *Ecology* 41:199–209.

Idso, S. B., and R. D. Jackson. 1969. Thermal radiation from the atmosphere. *J. Geophys. Res.* 74:5397–5403.

James, F. C., and W. P. Porter. 1979. Behavior-microclimate relationships in the African rainbow lizard. *Copeia* 1979: 585–593.

Keen, Robert. 1979. Effects of fluctuating egg temperature on duration of egg development of *Chydorus sphaericus* (Cladocera, Crustacea). *J. Therm. Biol.* 4:5–8.

Kendeigh, S. C. 1949. Effect of temperature and season on the energy resources of the English sparrow. *Auk* 66:113–127.

Kingsolver, J. G. 1979. Thermal and hydric aspects of environmental heterogeneity in the pitcher plant mosquito. *Ecol. Monogr.* 49:357–376.

Klauber, L. M. 1939. Studies of reptile life in the arid southwest. II, Speculations on protective coloration and protective reflectivity. *Bull. Zool. Soc. San Diego, Calif.* 14:65–79.

Kleiber, Max. 1961. *The fire of life.* New York: Wiley.

Kowalski, G. J., and J. W. Mitchell. 1976. Heat transfer from spheres in the naturally turbulent, outdoor environment. *J. Heat Trans.* 98:649–653.

Kreith, Frank. 1968. *Principles of heat transfer.* Scranton, Penn.: International Textbook Co.

Marshall, A. J., and Raymond Hook. 1960. The breeding biology of equatorial vertebrates: reproduction of the lizard *Agama agama lionotus* Boulenger at lat. 0°01′N. *Proc. Zool. Soc. Lond.* 134:197–205.

Mayhew, W. W. 1965. Growth response to photoperiodic stimulation in the lizard *Dipsosaurus dorsalis. Comp. Biochem. Physiol.* 14:209–216.

McCullough, E. M., and W. P. Porter. 1971. Computing clear day solar spectra for the terrestrial ecological environment. *Ecology* 52:1008–15.

Minnich, J. E. 1970. Evaporative water loss from the desert iguana, *Dipsosaurus dorsalis. Copeia* 1970:575–578.

Minnich, J. E., and V. H. Shoemaker. 1970. Diet, behavior, and water turnover in the desert iguana, *Dipsosaurus dorsalis. Am. Midl. Nat.* 84:496–509.

Mitchell, J. W. 1976. Heat transfer from spheres and other animal forms. *Biophys. J.* 16:561–569.

Mitchell, J. W., W. A. Beckman, R. T. Bailey, and W. P. Porter. 1975. Microclimatic modeling of the desert. In *Heat and mass transfer in the biosphere. Part I, transfer processes in the plant environment,* ed. D. A. deVries and N. H. Afgan. Washington, D. C.: Scripta Book Co.

Moberly, W. R. 1962. Hibernation in the desert iguana, *Dipsosaurus dorsalis. Physiol. Zool.* 36:152–160.

Murrish, David, and Knut Schmidt-Nielsen. 1970. Exhaled air temperatures and water conservation in lizards. *Respir. Physiol.* 10:151–158.

Muth, F. A. 1980. Physiological ecology of desert iguana (*Dipsosaurus dorsalis*) eggs: temperature and water relations. *Ecology* 61:1335–43.

Nagy, K. A. 1977. Cellulose digestion and nutrient assimilation in *Sauromalus obesus,* a plant eating lizard. *Copeia* 1977: 355–362.

Norris, K. S. 1953. The ecology of the desert iguana *Dipsosaurus dorsalis. Ecology* 34:265–287.

——— 1967. Color adaptation in desert reptiles and its thermal relationships. In W. W. Milstead.

Norris, K. S., and W. R. Dawson. 1964. Observations on the water economy and electrolyte excretion of chuckwallas (Lacertilia, *Sauromalus*). *Copeia* 1964: 638–646.

Parker, W. S., and E. R. Pianka. 1975. Comparative ecology of populations of the lizard *Uta stansburiana*. *Copeia* 1975:615–632.

Parry, D. A. 1951. Factors determining the temperature of terrestrial arthropods in sunlight. *J. Exp. Biol.* 28:445–462.

Porter, W. P. 1967. Solar radiation through the living body walls of vertebrates with emphasis on desert reptiles. *Ecol. Monogr.* 37:273–296.

Porter, W. P., and R. L. Busch. 1978. Fractional factorial designs applied to growth and reproductive success in deer mice. *Science* (Wash., D.C.) 202:907–910.

Porter, W. P., and D. M. Gates. 1969. Thermodynamic equlilibria of animals with environment. *Ecol. Monogr.* 39:227–244.

Porter, W. P., and F. C. James. 1979. Behavioral implications of mechanistic ecology. II: The African rainbow lizard, *Agama agama*. *Copeia* 1979:594–619.

Porter, W. P., and P. A. McClure. 1982. Population implications of mechanistic ecology (in preparation).

Porter, W. P. and K. S. Norris. 1969. Lizard reflectivity change and its effect on light transmission through body wall. *Science* (Wash., D.C.) 163:482–484.

Porter, W. P., J. W. Mitchell, W. A. Beckman, and C. B. DeWitt. 1973. Behavioral implications of mechanistic ecology. *Oecologia* (Berl.) 13:1–54.

Regal, P. J. 1967. Voluntary hypothermia in reptiles. *Science* (Wash., D.C.) 155:1551–53.

Riechert, Susan, and C. R. Tracy. 1975. Thermal balance and prey availability: bases for a model relating web-site characteristics to spider reproductive success. *Ecology* 56:265–284.

Roughgarden, Jonathan. 1979. *Theory of population genetics and evolutionary ecology: An introduction.* New York: Macmillan.

Roughgarden, Jonathan, W. P. Porter, and David Heckel. 1981. Resource partitioning of space and its relationship to body temperature in *Anolis* lizard populations. *Oecologia* (Berl.) 50:256–264.

Schmidt-Nielsen, Knut. 1964. *Desert animals: physiological problems of heat and water.* Oxford: Clarendon Press.

Schoener, T. W. and G. C. Gorman. 1968. Some niche differences in three Lesser-Antillean lizards of the genus *Anolis*. *Ecology* 49:819–830.

Sellers, W. D. 1965. *Physical climatology.* Chicago: University of Chicago Press.

Sexton, O. J., and Lee Claypool. 1978. Nest sites of a northern population of an oviparous snake, *Opheodrys vernalis* (Serpentes, Colubridae). *J. Nat. Hist.* 12:365–370.

Skoczylas, Rafal. 1970. Influence of temperature on gastric digestion in the grass snake *Natrix natrix* L. *Comp. Biochem. Physiol.* 33:793–804.

Spotila, J. R., P. W. Lommen, G. S. Bakken, and D. M. Gates. 1973. A mathematical model for body temperatures of large reptiles: implications for dinosaur ecology. *Am. Nat.* 107:391–404.

Swinbank, W. C. 1963. Long-wave radiation from clear skies. *Q. J. R. Meteorol. Soc.* 89:339.

Templeton, J. R. 1960. Respiration and water loss at higher temperatures in the desert iguana, *Dipsosaurus dorsalis*. *Physiol. Zool.* 33:136–145.

Tinkle, D. W. 1967. The life and demography of the side-blotched lizard, *Uta stansburiana*. *Misc. Publ. Mus. Zool., Univ. Mich.* 132:1–182.

Tracy, C. R. 1976. A model of the dynamic exchanges of water and energy between a terrestrial amphibian and its environment. *Ecol. Monogr.* 46:293–326.

Turner, F. B., P. A. Medica, and B. W. Kowalewsky. 1976. Energy utilization by a desert lizard (*Uta stansburiana*). *US/ IBP Desert Biome Monograph No. 1.* Logan: Utah State University Press.

Van Wijk, W. R., ed. 1963. *Physics of plant environment.* Amsterdam: North Holland Publishing.

Waldschmidt, Steve. 1982. The spatial distributions in two populations of the lizard *Uta stansburiana.* In preparation.

Waldschmidt, Steve, and C. R. Tracy. 1982. Lizard-thermal environment interactions: implications for sprint performance and space utilization in the lizard *Uta stansburiana. Ecology* (in press).

Wathen, Patricia, J. W. Mitchell, and W. P. Porter. 1971. Theoretical and experimental studies of energy exchange from jackrabbit ears and cylindrically shaped appendages. *Biophys. J.* 11:1030–47.

——— 1973. Heat transfer from animal appendage-shapes—cylinders, arcs, and cones. *Trans. ASME,* Paper No. 73/WA/ Bio 10:1–8.

Welch, W. R. 1980. Evaporative water loss from endotherms in thermally and hygrically complex environments: an empirical approach for interspecific comparisons. *J. Comp. Physiol.* 139B:135–143.

Wiens, J. A., and G. S. Innis. 1974. Estimation of energy flow in bird communities: a population bioenergetics model. *Ecology* 55:730–746.

Winslow, C. E. A., L. P. Herrington, and A. P. Gagge. 1937. Physiological reactions of the human body to varying environmental temperatures. *Am. J. Physiol.* 120:1–22.

Zucker, Albert. 1980. Procedural and anatomical considerations of the determination of cutaneous water loss in squamate reptiles. *Copeia* 1980:425–439.

4. Lizard Malaria: Parasite-Host Ecology

Anderson, R. M., and R. M. May. 1979. Population biology of infectious diseases: part I. *Nature* (Lond.) 280:361–367.

Aragão, H. de B., and A. Neiva. 1909. A contribution to the study of the intraglobular parasites of lizards. Two new species of *Plasmodium, Pl. diploglossi* n. sp. and *Pl. tropiduri* n. sp. *Mem. Inst. Oswaldo Cruz* 1:44–50.

Ayala, S. C. 1970. Lizard malaria in California; description of a strain of *Plasmodium mexicanum,* and biogeography of lizard malaria in western North America. *J. Parasitol.* 56:417–425.

——— 1973. The phlebotomine sandfly-protozoan parasite community of central California grasslands. *Am. Midl. Nat.* 89:266–280.

——— 1977. Plasmodia of reptiles. In *Parasitic Protozoa, vol. 3, Gregarines, Haemogregarines, Coccidia, Plasmodia, and Haemoproteids,* ed. J. P. Kreier. New York: Academic Press.

——— 1978. Checklist, host index, and annotated bibliography of *Plasmodium* from reptiles. *J. Protozool.* 25:87–100.

Ayala, S. C., and Dwayne Lee. 1970. Saurian malaria: development of sporozoites in two species of phebotomine sandflies. *Science* (Wash., D.C.) 167:891–892.

Ayala, S. C., and J. L. Spain. 1976. A population of *Plasmodium colombiense* sp. n. in the iguanid lizard, *Anolis auratus*. *J. Parasitol.* 62:177–189.

Barbehenn, K. R. 1969. Host-parasite relationships and species diversity in mammals: an hypothesis. *Biotropica* 1:29–35.

Bennett, A. F. 1978. Activity metabolism of the lower vertebrates. *Annu. Rev. Physiol.* 40:447–469.

Bonorris, J. S., and G. H. Ball. 1955. *Schellackia occidentalis* n. sp., a blood-inhabiting coccidian found in lizards in southern California. *J. Protozool.* 2:31–34.

Burnet, F. M. 1962. *Natural history of infectious disease.* Cambridge: Cambridge University Press.

Cornell, Howard. 1974. Parasitism and distributional gaps between allopatric species. *Am. Nat.* 108:880–883.

Diggs, L. W., D. Sturm, and A. Bell. 1978. *The morphology of human blood cells.* No. Chicago: Abbott Laboratories.

Dogiel, V.A. 1966. *General parasitology.* New York: Academic Press.

Friedman, M. J., and William Trager. 1981. The biochemistry of resistance to malaria. *Sci. Am.* 244:154–164.

Gaffney, F. G., and L. C. Fitzpatrick. 1973. Energetics and lipid cycles in the lizard, *Cnemidophorus tigris*. *Copeia* 1973:446–452.

Guerrero, Stella, César Rodríguez, and S. C. Ayala. 1977. Prevalencia de hemoparasitos en lagartijas de la isla Barro Colorado, Panama. *Biotropica* 9:118–123.

Hahn, W. E., and D. W. Tinkle. 1964. Fat body cycling and experimental evidence for its adaptive significance to ovarian follicle development in the lizard *Uta stansburiana*. *J. Exp. Zool.* 158:79–86.

Harrison, Gordon. 1978. *Mosquitoes, malaria, and man.* New York: Dutton.

Jordan, H. B. and M. B. Friend. 1971. The occurrence of *Schellackia* and *Plasmodium* in two Georgia lizards. *J. Protozool.* 18:485–487.

Livingstone, F. B. 1971. Malaria and human polymorphisms. *Annu. Rev. Genet.* 5:33–64.

Manwell, R. D. 1955. Some evolutionary possibilities in the history of malaria parasites. *Indian J. Malariol.* 9:247–253.

Naumov, N. P. 1972. *The ecology of animals.* Urbana: University of Illinois Press.

Nawalinski, T., G. A. Shad, and A. B. Chowdhury. 1978. Population biology of hookworms in children in rural West Bengal. *Am. J. Trop. Med. Hyg.* 27:1152–61.

Price, P. W. 1977. General concepts on the evolutionary biology of parasites. *Evolution* 31:405–420.

——— 1980. *Evolutionary biology of parasites.* Princeton: Princeton University Press.

Russell, P. F., L. S. West, R. D. Manwell, and George MacDonald. 1963. *Practical malariology,* 2nd ed. Oxford: Oxford University Press.

Ruth, S. B. 1977. A comparison of the demography and female reproduction in sympatric western fence lizards (*Sceloporus occidentalis*) and sagebrush lizards (*Sceloporus graciosus*) on Mount Diablo, California. Ph.D. dissertation, University of California, Berkeley.

Schall, J. J. 1978. Reproductive strategies in sympatric whiptail lizards (*Cne-*

midophorus): two parthenogenetic and three bisexual species. *Copeia* 1978:108–116.
——— 1982. Lizard malaria: costs to vertebrate host's reproductive success. Submitted.
Schall, J. J., A. F. Bennett, and R. W. Putnam. 1982. Lizards infected with malaria: physiological and behavioral consequences. *Science* (Wash., D.C.) 217:1057–59.
Scorza, J. V. 1971. Anaemia in lizard malaria infections. *Parasitology* 13:391–405.
Telford, S. R. 1970. A comparative study of endoparasitism among some southern California lizard populations. *Am. Midl. Nat.* 83:516–554.
——— 1971. Parasitic diseases of reptiles. *J. Am. Vet. Med. Assoc.* 159:1644–52.
——— 1972. The course of infection of Japanese saurian malaria (*Plasmodium sasai,* Telford and Ball) in natural and experimental hosts. *Japan. J. Exp. Med.* 42:1–21.
——— 1977. The distribution, incidence, and general ecology of saurian malaria in Middle America. *Int. J. Parasitol.* 7:299–314.
——— 1978. The saurian malarias of Venezuela: haemosporidian parasites of gekkonid lizards. *Int. J. Parasitol.* 8:341–353.
Warner, R. E. 1968. The role of introduced diseases in the extinction of the endemic Hawaiian avifauna. *Condor* 70:101–120.
Wenyon, C. M. 1909. Report of traveling pathologist and protozoologist. In *Third report, Wellcome Research Laboratory, Khartoum,* ed. A. Balfour. London: Bailliere, Tindall, and Cox.

Overview of Part II

Greenberg, B., and G. K. Noble. 1944. Social behavior of the American chameleon (*Anolis carolinensis* Voigt). *Physiol. Zool.* 17:392–439.
Noble, G. K. 1931. *Biology of the Amphibia.* New York: McGraw-Hill.
Noble, G. K., and H. T. Bradley. 1933. The mating behavior of lizards: its bearing on the theory of natural selection. *Ann. N.Y. Acad. Sci.* 35:25–100.
Noble, G. K., and H. J. Clausen. 1936. The aggregation behavior of *Storeria dekayi* and other snakes with special reference to the sense organs involved. *Ecol. Monogr.* 6:269–316.
Noble, G. K., and A. Schmidt. 1937. The structure and function of the facial and labial pits of snakes. *Proc. Am. Philos. Soc.* 77:392–439.
Pough, F. H. 1980. The advantages of ectothermy for tetrapods. *Am. Nat.* 115:92–112.
Rose, Barbara. 1982. Lizard home ranges: methodology and functions. *J. Herpetol.* 16:253–270.
Waldschmidt, S. R. 1979. The effect of statistically based models of home-range size estimates in *Uta stansburiana. Am. Midl. Nat.* 101:236–240.

5. The Adaptive Zone and Behavior of Lizards

Anderson, R. A., and W. H. Karasov. 1981. Contrasts in energy intake and expenditure in sit-and-wait and widely foraging lizards. *Oecologia* (Berl.) 49:67–72.

Auffenberg, Walter. 1978. Social and feeding behavior in *Varanus komodoensis.* In Neil Greenberg and P. D. MacLean.

Avery, R. A. 1976. Thermoregulation, metabolism and social behavior in Lacertidae. In *Morphology and biology of reptiles,* ed. A. Bellairs and C. Cox. London: Academic Press.

Bartholomew, G. A., and J. Hudson. 1961. Desert ground squirrels. *Sci. Am.* 205:107–116.

Bennett, A. F. 1980. The thermal dependence of lizard behavior. *Anim. Behav.* 28:752–762.

Bennett, A. F., and J. A. Ruben. 1979. Endothermy and activity in vertebrates. *Science* (Wash., D.C.) 206:649–654.

Berman, D., and P. J. Regal. 1967. The loss of the ophidian middle ear. *Evolution* 21:641–643.

Bock, W. J. 1977. Adaptation and the comparative method. In *Major patterns in vertebrate evolution,* ed. M. K. Hecht, P. C. Goody, and B. M. Hecht. New York: Plenum Press.

Bock, W. J., and G. von Wahlert. 1965. Adaptation and the form-function complex. *Evolution* 19:269–299.

Brattstrom, B. H. 1974. The evolution of reptilian social behavior. *Am. Zool.* 14:35–49.

——— 1978. Learning studies in lizards. In Neil Greenberg and P. D. MacLean.

——— 1979. Amphibian temperature regulation studies in the field and laboratory. *Am. Zool.* 19:345–356.

Brown, G. D. 1974. The biology of marsupials of the Australian arid zone. *J. Aust. Mammal Soc.* 1:269–288.

Brown, J. H. 1978. The theory of insular biogeography and the distribution of boreal birds and mammals. *Great Basin Nat. Mem.* 2:209–227.

Burghardt, G. M. 1977. Learning processes in reptiles. In Carl Gans and D. W. Tinkle.

Carpenter, C. C. 1978. Ritualistic social behaviors in lizards. In Neil Greenberg and P. D. MacLean.

——— 1980. An ethological approach to reproductive success in reptiles. In *Reproductive biology and diseases of captive reptiles,* ed. J. B. Murphy and J. T. Collins. Society for the Study of Amphibians and Reptiles.

Carpenter, C. C., and G. W. Ferguson. 1977. Variation and evolution of stereotyped behavior in reptiles. In Carl Gans and D. W. Tinkle.

Coulson, R. A. 1979. Anaerobic glycolysis: the Smith and Wesson of the heterotherms. *Perspect. Biol. Med.* 22:465–479.

Cowles, R. B., and C. M. Bogert. 1944. A preliminary study of the thermal requirements of desert reptiles. *Bull. Am. Mus. Nat. Hist.* 83:261–296.

Crews, David, and L. D. Garrick. 1980. Methods of inducing reproduction in captive reptiles. In *Reproductive biology and diseases of captive reptiles,* ed. J. B. Murphy and J. T. Collins. Society for the Study of Amphibians and Reptiles.

Ferguson, G. W., and C. H. Bohlen. 1978. Demographic analysis: a tool for the study of natural selection of behavioral traits. In Neil Greenberg and P. D. MacLean.

Fitch, H. S. 1968. Temperature and behavior of some equatorial lizards. *Herpetologica* 24:35–38.

——— 1980. Reproductive strategies of reptiles. In *Reproductive biology and diseases of captive reptiles,* ed. J. Murphy and J. Collins. Society for the Study of Amphibians and Reptiles.

Freed, A. N. 1980. An adaptive advantage of basking behavior in an anuran amphibian. *Physiol. Zool.* 53:433–444.

Gans, Carl. 1974. *Biomechanics: an approach to vertebrate biology.* Philadelphia: Lippincott.

Garrick, L. D. 1979. Lizard thermoregulation: operant responses for heat at different thermal intensities. *Copeia* 1979:258–266.

Gates, D. M. 1975. Introduction: biophysical ecology. In *Perspectives in biophysical ecology,* ed. D. M. Gates and R. B. Schmerl. New York: Springer-Verlag.

Gould, S. J., and R. C. Lewontin. 1979. The spandrels of San Marco and the Panglossian paradigm: a critique of the adaptationist programme. *Proc. R. Soc. London B. Biol. Sci.* 205:581–598.

Greenberg, Neil. 1978. Ethological considerations in the experimental study of lizard behavior. In Neil Greenberg and P. D. MacLean.

Heath, J. E. 1965. Temperature regulation and diurnal activity in horned lizards. *Univ. Calif. Publ. Zool.* 64:97–136.

Heatwole, Harold. 1977. Habitat selection in reptiles. In Carl Gans and D. W. Tinkle.

Hinde, R. A. 1970. *Animal behavior: a synthesis of ethology and comparative psychology.* New York: McGraw-Hill.

Huey, R. B., and E. R. Pianka. 1981. Ecological consequences of foraging mode. *Ecology* 62:991–999.

Huey, R. B., and Montgomery Slatkin. 1976. Costs and benefits of lizard thermoregulation. *Q. Rev. Biol.* 51:363–384.

Huey, R. B., and R. D. Stevenson. 1979. Integrating thermal physiology and ecology of ectotherms: a discussion of approaches. *Am. Zool.* 19:357–366.

Jenssen, T. A. 1978. Display diversity in anoline lizards and problems of interpretation. In Neil Greenberg and P. D. MacLean.

MacArthur, R. H., and E. O. Wilson. 1967. *The theory of island biogeography.* Princeton: Princeton University Press.

Marcellini, D. L. 1978. The acoustic behavior of lizards. In Neil Greenberg and P. D. MacLean.

Moberly, W. R. 1968a. The metabolic responses of the common iguana, *Iguana iguana,* to activity under restraint. *Comp. Biochem. Physiol.* 27:1–20.

——— 1968b. The metabolic responses of the common iguana, *Iguana iguana,* to walking and diving. *Comp. Biochem. Physiol.* 27:21–32.

Moermond, T. C. 1979a. The influence of habitat structure on *Anolis* foraging behavior. *Behaviour* 70:147–167.

——— 1979b. Habitat constraints on the behavior, morphology and community structure of *Anolis* lizards. *Ecology* 60:152–164.

Newsome, A., and L. Corbett. 1975. Outbreaks of rodents in semi-arid and arid Australia: causes, preventions, and evolutionary considerations. In *Rodents in desert environments,* ed. I. Prakash and P. K. Ghosh. The Hague: W. Junk.

Northcutt, R. G. 1978. Forebrain and midbrain organization in lizards and its phylogenetic significance. In Neil Greenberg and P. D. MacLean.

Pearson, O. P., and D. Bradford. 1976. Thermoregulation of lizards and toads at high altitudes in Peru. *Copeia* 1976:155–170.

Pianka, E. R., and J. J. Schall. 1981. Species densities of Australian vertebrates. In *Ecological biogeography in Australia,* ed. A. Keast. The Hague: W. Junk.

Platel, R. 1979. Brain weight–body weight relationships. In *Biology of the Reptilia,* ed. Carl Gans, R. G. Northcutt, and P. Ulinski. New York: Academic Press.

Pough, F. H. 1980. The advantages of ectothermy for tetrapods. *Am. Nat.* 115:92–112.

Rand, A. S., and W. M. Rand. 1978. Display and dispute settlement in nesting iguanas. In Neil Greenberg and P. D. MacLean.

Regal, P. J. 1968. An analysis of heat-seeking in a lizard. Ph.D. dissertation, University of California, Los Angeles.

——— 1975. The evolutionary origin of feathers. *Q. Rev. Biol.* 50:35–66.

——— 1978. Behavioral differences between reptiles and mammals: an analysis of activity and mental capabilities. In Neil Greenberg and P. D. MacLean.

——— 1980. Temperature and light requirements of captive reptiles. In *Reproductive biology and diseases of captive reptiles,* ed. J. B. Murphy and J. T. Collins. Society for the Study of Amphibians and Reptiles.

——— 1982. Pollination by wind and by animals: ecology of the geographic trends. *Annu. Rev. Ecol. Syst.* 13:497–524.

Regal, P. J., and Carl Gans. 1980. The revolution in thermal physiology: implications for dinosaurs. In *A cold look at the warm-blooded dinosaurs,* ed. R. D. K. Thomas and E. C. Olson. American Association for the Advancement of Science, Selected Symposium Series, 28. Boulder, Colorado: Westview Press.

Schall, J. J., and E. R. Pianka. 1978. Geographical trends in numbers of species. *Science* (Wash., D.C.) 201:679–686.

——— 1980. Evolution of escape behavior diversity. *Am. Nat.* 115:551–556.

Schoener, T. W. 1971. Theory of feeding strategies. *Annu. Rev. Ecol. Syst.* 2:369–404.

Sherbrooke, W. C. 1975. Reproductive cycle of a tropical teiid lizard, *Neusticurus ecpleopus* Cope, in Peru. *Biotropica* 7:194–207.

Simpson, G. G. 1953. *The major features of evolution.* New York: Columbia University Press.

Stamps, J. A. 1976. Egg retention, rainfall, and egg laying in a tropical lizard *Anolis aeneus. Copeia* 1976:759–764.

——— 1977. Social behavior and spacing patterns in lizards. In Carl Gans and D. W. Tinkle.

Steenis, C. G. G. van. 1969. Plant speciation in Malaysia, with special reference to the theory of non-adaptive saltatory evolution. *Biol. J. Linn. Soc.* 1:97–133.

Waddington, C. H. 1969. The theory of evolution today. In *Beyond reductionism,* ed. A. Koestler and J. Smythies. New York: Macmillan.

Williams, G. C. 1966. *Adaptation and natural selection: a critique of some current evolutionary thought.* Princeton: Princeton University Press.
Wright, S. J. 1979. Competition between insectivorous lizards and birds in Central Panama. *Am. Zool.* 19:1145–56.

6. A Review of Lizard Chemoreception

Abel, Erich. 1951. Über das Geruchsvermögen der Eidechsen. *Österr. Zool. Zeitschr.* 3:84–125.
Auffenberg, Walter. 1978. Social and feeding behavior in *Varanus komodoensis.* In Neil Greenberg and P. D. MacLean.
Bellairs, Angus. 1970. *The life of reptiles,* vol. 2. New York: Universe Books.
Bellairs, Angus, and J. D. Boyd. 1950. The lachrymal apparatus in lizards and snakes. II. The anterior part of the lachrymal duct and its relationship with the palate and with the nasal and vomeronasal organs. *Proc. Zool. Soc. Lond.* 120:269–310.
Berry, K. H. 1974. The ecology and social behavior of the chuckwalla, *Sauromalus obesus obesus* Baird. *Univ. Calif. Publ. Zool.* 101:1–60.
Bissinger, B. E., and C. A. Simon. 1979. Comparison of tongue extrusions in representatives of six families of lizards. *J. Herpetol.* 13:133–139.
——— 1981. The chemical detection of conspecifics by juvenile Yarrow's spiny lizard, *Sceloporus jarrovi. J. Herpetol.* 15:77–81.
Bogert, C. M., and R. Martín del Campo. 1956. The gila monster and its allies. The relationships, habits, and behavior of the lizards of the family Helodermatidae. *Bull. Am. Mus. Nat. Hist.* 109:1–238.
Broman, I. 1920. Das Organon vomero-nasale Jacobsoni—ein Wassergeruchsorgan. *Anat. Hefte* 58:137–191.
Brooks, G. R. 1967. Population ecology of the ground skink, *Lygosoma laterale* (Say). *Ecol. Monogr.* 37:71–87.
Burghardt, G. M. 1970. Chemical perception in reptiles. In *Communication by chemical signals,* vol. 1, ed. J. W. Johnston, Jr., D. G. Moulton, and A. Turk. New York: Appleton-Century-Crofts.
——— 1973. Chemical release of prey attack: extension to naive newly hatched lizards, *Eumeces fasciatus. Copeia* 1973:178–181.
——— 1977. The ontogeny, evolution, and stimulus control of feeding in humans and reptiles. In *The chemical senses and nutrition,* ed. M. R. Kare and O. Maller. New York: Academic Press.
——— 1980. Behavioral and stimulus correlates of vomeronasal functioning in reptiles: feeding, grouping, sex, and tongue use. In *Chemical signals: vertebrates and aquatic invertebrates,* ed. D. Müller-Schwarze and R. M. Silverstein.
Burghardt, G. M., H. W. Greene, and A. S. Rand. 1977. Social behavior in hatchling green iguanas: life at a reptile rookery. *Science* (Wash., D.C.) 195:689–691.
Burghardt, G. M., and C. H. Pruitt. 1975. The role of the tongue and senses in feeding of naive and experienced garter snakes. *Physiol. Behav.* 14:185–194.
Burkholder, G. L., and W. T. Tanner. 1974. A new gland in *Sceloporus graciosus* males (Sauria: Iguanidae). *Herpetologica* 30:368–371.
Camp, C. L. 1923. Classification of the lizards. *Bull. Am. Mus. Nat. Hist.* 48:289–481.

Carpenter, C. C. 1962. Patterns of behavior in two Oklahoma lizards. *Am. Midl. Nat.* 67:132–151.

——— 1975. Reviews and comments. *Copeia* 1975: 388–389.

——— 1978. Tongue display by the common bluetongue (*Tiliqua scincoides:* Reptilia, Lacertilia, Scincidae). *J. Herpetol.* 12:428–429.

Chiu, K. W., B. Lofts, and H. W. Tsui. 1970. The effect of testosterone on the sloughing cycle and epidermal glands of the female gecko, *Gekko gecko* L. *Gen. Comp. Endocrinol.* 15:12–19.

Chiu, K. W., and P. F. A. Maderson. 1975. The microscopic anatomy of epidermal glands in two species of gekkonine lizards, with some observations of testicular activity. *J. Morphol.* 147:23–40.

Chiu, K. W., P. F. A. Maderson, S. A. Alexander, and K. L. Wong. 1975. Sex steroids and epidermal glands in two species of gekkonine lizards. *J. Morphol.* 147:9–22.

Cole, C. J. 1966a. Femoral glands of the lizard, *Crotaphytus collaris. J. Morphol.* 118:119–136.

——— 1966b. Femoral glands in lizards: a review. *Herpetologica* 22:199–206.

Cowles, R. B., and R. L. Phelan. 1958. Olfaction in rattlesnakes. *Copeia* 1958:73–83.

Czaplicki, J. A., and R. H. Porter. 1974. Visual cues mediating the selection of goldfish (*Carassius auratus*) by two species of *Natrix. J. Herpetol.* 8:129–134.

DeFazio, Antoinette, C. A. Simon., G. A. Middendorf, and Daniel Romano. 1977. Iguanid substrate licking: a response to novel situations in *Sceloporus jarrovi. Copeia* 1977:706–709.

Distel, Hanjürgen. 1978. Behavioral responses to the electrical stimulation of the brain in the green iguana. In Neil Greenberg and P. D. MacLean.

Ditmars, R. L. 1931. *Snakes of the world.* New York: Macmillan.

Duvall, David. 1979. Western fence lizard (*Sceloporus occidentalis*) chemical signals. I. Conspecific discriminations and release of a species-typical visual display. *J. Exp. Zool.* 210:321–326.

——— 1980. Pheromonal mechanisms in the social behavior and communication of the western fence lizard, *Sceloporus occidentalis biseriatus.* Ph.D. dissertation, University of Colorado.

Duvall, David, Renée Herskowitz, and Jeanne Trupiano-Duvall. 1980. Responses of five-lined skinks (*Eumeces fasciatus*) and ground skinks (*Scincella lateralis*) to conspecific and interspecific chemical cues. *J. Herpetol.* 14:121–127.

Duvall, David, Jeanne Trupiano, and H. M. Smith. 1979. An observation of maternal behavior in the Mexican desert spiny lizard, *Sceloporus rufidorsum. Trans. Kans. Acad. Sci.* 82:60–62.

Evans, L. T. 1959. A motion picture study of maternal behavior of the lizard *Eumeces obsoletus* Baird and Baird. *Copeia* 1959:103–110.

——— 1961. Structure as related to behavior in the organization of populations in reptiles. In *Vertebrate speciation,* ed. W. F. Blair. Austin: University of Texas Press.

——— 1967. Introduction. In W. W. Milstead.

Ferguson, G. W. 1966. Releasers of courtship and territorial behaviour in the side-blotched lizard *Uta stansburiana. Anim. Behav.* 14:89–92.

Fitch, H. S. 1954. Life history and ecology of the five-lined skink *Eumeces fasciatus. Univ. Kans. Mus. Nat. Hist. Misc. Publ.* 8:1–156.

——— 1967. Ecological studies of lizards on the University of Kansas Natural History Reservation. In W. W. Milstead.

FitzSimons, V. F. 1943. The lizards of South Africa. *Transvaal Mus. Mem.* 1:1–528.

Gabe, M., and Hubert Saint-Girons. 1965. Contribution à la morphologie comparée du cloaque et des glandes épidermoides de la région cloacale chez lépidosauriens. *Mem. Mus. Nat. Hist., Nat. Ser. A. Zool.* 33:150–292.

——— 1967. Données histologiques sur le tégument et les glandes épidermoides céphaliques des lépidosauriens. *Acta Anat.* 67:571–594.

Gelbach, Gretchen. 1979. Licking by neonatal lizards in response to novel stimuli. *Herpetol. Rev.* 10:56.

Gettkandt, Albert. 1931. Die Analyse des Funktionskreises der Nahrung bei der Kutscherpeitschenschlange *Zamenis flagelliformis* L. nebst Ergänzungsversuchen bei der Ringelnatter *Tropidonotus natrix* L. *Z. Vgl. Physiol.* 14:1–39.

Gillingham, J. C. 1976. Reproductive behavior of the rat snakes of eastern North America, genus *Elaphe.* Ph.D. dissertation, University of Oklahoma.

Gove, Doris. 1978. The form, variation, and evolution of tongue-flicking in reptiles. Ph.D. dissertation, University of Tennessee.

——— 1979. A comparative study of snake and lizard tongue-flicking with an evolutionary hypothesis. *Z. Tierpsychol.* 51:58–76.

Gravelle, Karen. 1981. Chemical communication in the iguanid lizard *Sceloporus jarrovi.* Ph.D. dissertation, Hunter College of the City University of New York.

Gravelle, Karen, and C. A. Simon. 1980. Field observations on the use of the tongue-Jacobson's organ system in two iguanids, *Sceloporus jarrovi* and *Anolis trinitatis. Copeia* 1980: 356–359.

Greenberg, B. 1943. Social behavior of the western banded gecko, *Coleonyx variegatus* Baird. *Physiol. Zool.* 16:110–122.

Greenberg, Neil. 1977. An ethogram of the blue spiny lizard *Sceloporus cyanogenys* (Reptilia, Lacertilia, Iguanidae). *J. Herpetol.* 11:177–195.

Hunsaker, Don. 1962. Ethological isolating mechanisms in the *Sceloporus torquatus* group of lizards. *Evolution* 16:62–74.

Kahmann, H. 1932. Sennesphysiologische Studien an Reptilien: I. Experimentelle Untersuchungen über das Jakobsonische Organ der Eidenschsen und Schlangen. *Zool. Jahrb. Abt. Allg. Zool. Physiol. Tiere* 51:173–238.

——— 1939. Über das Jakobsonische Organ der Echsen. *Z. Vgl. Physiol.* 26:669–695.

Klauber, L. M. 1956. *Rattlesnakes: their habits, life histories, and influence on mankind.* 2 vols. Berkeley and Los Angeles: University of California Press.

Kluge, A. G. 1967. Higher taxonomic categories of gekkonid lizards and their evolution. *Bull. Am. Mus. Nat. Hist.* 135:1–60.

Kratzing, J. E. 1975. The fine structure of the olfactory and vomeronasal organs of a lizard (*Tiliqua scincoides scincoides*). *Cell Tissue Res.* 165:239–252.

Kroll, J. C., and H. W. Reno. 1971. A re-examination of the cloacal sacs and gland of the blind snake, *Leptotyphlops dulcis* (Reptilia: Leptotyphlopidae). *J. Morphol.* 133:273–280.

Kubie, J. L. 1977. The role of the vomeronasal organ in garter snake prey trailing and courtship. Ph.D. dissertation, State University of New York, Downstate Medical Center.

Lewis, T. H. 1951. The biology of *Leiolopisma laterale* (Say). *Am. Midl. Nat.* 45:232–240.

Maderson, P. F. A. 1967. The history of the escutcheon scales of *Gonatodes* (Gekkonidae) with a comment on the squamate sloughing cycle. *Copeia* 1967: 743–752.

——— 1968a. The epidermal glands of *Lygodactylus* (Gekkonidae, Lacertilia). *Breviora Mus. Comp. Zool.* 288:1–35.

——— 1968b. On the presence of "escutcheon scales" in the eublepharine gekkonid *Coleonyx*. *Herpetologica* 24:99–103.

——— 1970. Lizard glands and lizard hands: models for evolutionary study. *Forma Functio* 3:179–204.

——— 1972. The structure and evolution of holocrine epidermal glands in sphaerodactyline and eublepharine gekkonid lizards. *Copeia* 1972: 559–571.

Maderson, P. F. A., and K. W. Chiu. 1970. Epidermal glands in gekkonid lizards: evolution and phylogeny. *Herpetologica* 26:233–238.

Madison, D. M. 1977. Chemical communication in amphibians and reptiles. In *Chemical signals in vertebrates,* ed. D. Müller-Schwarze and M. M. Mozell. New York: Plenum.

Malan, M. E. 1946. Contributions to the comparative anatomy of the nasal capsule and the organ of Jacobson of the Lacertilia. *Ann. Univ. Stellenbosch Ser. A* 24:69–137.

McDowell, S. B. 1972. The evolution of the tongue in snakes, and its bearing on snake origins. In *Evolutionary biology,* vol. 6, ed. T. Dobzhansky, M. K. Hecht, and W. C. Steere. New York: Appleton-Century-Crofts.

Menchel, S., and P. F. A. Maderson. 1975. The post-natal development of holocrine epidermal specializations in gekkonid lizards. *J. Morphol.* 147:1–8.

Meredith, M., and G. M. Burghardt. 1978. Electrophysiological studies of the tongue and accessory bulb in garter snakes. *Physiol. Behav.* 21:1001–8.

Mertens, Robert. 1955. Die Amphibien und Reptilien Südwestafrikas. *Abh. Senckenb. Naturforsch. Ges.* 490:1–172.

Milstead, W. W. 1961. Competitive relations in lizard populations. In *Vertebrate speciation,* ed. W. F. Blair. Austin: University of Texas Press.

Moncrieff, R. W. 1977. *The chemical senses.* 3rd ed. London: Hill.

Noble, G. K. 1937. The sense organs involved in the courtship of *Storeria, Thamnophis* and other snakes. *Bull. Am. Mus. Nat. Hist.* 73:673–725.

Noble, G. K., and K. P. Kumpf. 1936. The function of Jacobson's organ in lizards. *J. Genet. Psychol.* 48:371–382.

Noble, G. K., and E. R. Mason. 1933. Experiments on the brooding habits of the lizards *Eumeces* and *Ophisaurus*. *Am. Mus. Novit.* 619:1–29.

Nobel, G. K., and H. K. Teale. 1930. The courtship of some iguanid and teiid lizards. *Copeia* 1930:54–56.

Northcutt, R. G. 1978. Forebrain and midbrain organization in lizards and its phylogenetic significance. In Neil Greenberg and P. D. MacLean.

Parcher, S. R. 1974. Observations on the natural histories of six Malagasy Chamaeleontidae. *Z. Tierpsychol.* 34:500–523.

Parsons, T. S. 1959a. Nasal anatomy and the phylogeny of reptiles. *Evolution* 13:175–187.
——— 1959b. Studies on the comparative embryology of the reptilian nose. *Bull. Mus. Comp. Zool.* 120:101–277.
——— 1967. Evolution of the nasal structure in the lower tetrapods. *Am. Zool.* 7:397–413.
——— 1970. The nose and Jacobson's organ. In *Biology of the reptilia,* vol. 2, ed. Carl Gans and T. S. Parsons. New York: Academic Press.
Pitman, C. R. S. 1974. *A guide to the snakes of Uganda,* rev. ed. Codicote: Wheldon and Wesley.
Porter, K. R. 1972. *Herpetology.* Philadelphia: Saunders.
Pratt, C. W. McE. 1948. The morphology of the ethmoidal region of *Sphenodon* and lizards. *Proc. Zool. Soc. Lond.* 118:171–201.
Rensch, Bernhard, and M. Eisentraut. 1927. Experimentelle Untersuchungen über den Geschmackssinn der Reptilien. *Z. Vgl. Physiol.* 5:607–612.
Simon, C. A., Karen Gravelle, B. E. Bissinger, Israel Eiss, and Rodolfo Ruibal. 1981. The role of chemoreception in the iguanid lizard *Sceloporus jarrovi. Anim. Behav.* 29:46–54.
Smith, M. A. 1935. *The fauna of British India, including Ceylon and Burma. Reptilia and Amphibia.* Vol. 2, *Sauria.* London: Taylor and Francis.
Taylor, E. H., and A. B. Leonard. 1956. Concerning the relationship of certain neotropical gekkonid lizard genera, with comments on the microscopical structure of the glandular scales. *Univ. Kans. Sci. Bull.* 38:1019–29.
Tinkle, D. W. 1967. The life and history of the side-blotched lizard, *Uta stansburiana. Misc. Publ. Mus. Zool. Univ. Mich.* 132:1–182.
Underwood, Garth. 1951. Reptilian retinas. *Nature* (Lond.) 167:183–185.
——— 1970. The eye. In *Biology of the Reptilia,* vol. 2, ed. Carl Gans and T. S. Parsons. New York: Academic Press.
Wevers, J., Jr. 1910. Einige Beobachtungen und grosseren Terrarientieron. *Wochenschr. Aquar. Terrerienk.* 7 (Lacerta, no. 21): 82–83.
Whiting, A. M. 1967. Amphisbaenian cloacal glands. *Am. Zool.* 7:776.
——— 1969. Squamate cloacal glands: morphology, histology, and histochemistry. Ph.D. dissertation, Pennsylvania State University.
Wilde, W. S. 1938. The role of Jacobson's organ in the feeding reaction of the common garter snake, *Thamnophis sirtalis sirtalis* (Linn). *J. Exp. Zool.* 77:445–465.

7. Food Availability and Territorial Establishment of Juvenile *Sceloporus undulatus*

Brown, J. L. 1964. The evolution of diversity in avian territorial systems. *Wilson Bull.* 6:160–169.
——— 1975. *The evolution of behavior.* New York: Norton.
Brown, J. L., and G. H. Orians. 1970. Spacing patterns in mobile animals. *Annu. Rev. Ecol. Syst.* 1:239–262.
Carpenter, C. C. 1966. The marine iguana of the Galapagos Islands, its behavior and ecology. *Proc. Calif. Acad. Sci.* 34:329–376.
Croze, Harvey. 1970. Searching image in carrion crows. *Z. Tierpsychol. Beih.* V, 1–85.

Ferguson, G. W., and C. H. Bohlen. 1978. Demographic analysis: a technique for studying natural selection of behavioral traits. In Neil Greenberg and P. D. MacLean.

Ferguson, G. W., and Todd Brockman. 1980. Geographic differences of growth rate potential of *Sceloporus* lizards (Sauria: Iguanidae). *Copeia* 1980:259–264.

Ferguson, G. W., C. H. Bohlen, and H. P. Woolley. 1980. *Sceloporus undulatus:* comparative life history and regulation of a Kansas population. *Ecology* 61:312–322.

Fox, S. F. 1978. Natural selection on behavioral phenotypes of the lizard *Uta stansburiana. Ecology* 59:834–847.

Fox, S. F., Elizabeth Rose, and Ronald Myers. 1981. Dominance and acquisition of superior home-ranges in the lizard *Uta stansburiana. Ecology* 62:888–893.

Hinde, R. A. 1956. The biological significance of the territories of birds. *Ibis* 98:340–369.

Krekorian, C. O. 1976. Home-range size and overlap and their relationship to food abundance in the desert iguana *Dipsosaurus dorsalis. Herpetologica* 32:405–412.

Le Beouf, B. J., and R. S. Peterson. 1969. Social status and mating activity in elephant seals. *Science* (Wash., D.C.) 163:91–93.

Miller, G. R., and Adam Watson. 1978. Territories and the food plant of individual red grouse. 1. Territory size, number of mates and brood size compared with the abundance, production and diversity of heather. *J. Anim. Ecol.* 47:298–306.

Moss, R., Adam Watson, and Raymond Parr. 1975. Maternal nutrition and breeding success in red grouse (*Lagopus lagopus scoticus*). *J. Anim. Ecol.* 44:233–244.

Nice, M. M. 1941. The role of territory in bird life. *Am. Midl. Nat.* 26:441–487.

Rand, A. S. 1967. The adaptive significance of territoriality in iguanid lizards. In W. W. Milstead.

——— 1968. A nesting aggregation of iguanas. *Copeia* 1968:552–561.

Robel, R. J. 1966. Booming territory size and mating success of the greater prairie chicken (*Tympanarchus cupido pinnatus*). *Anim. Behav.* 14:328–331.

Schoener, T. W. 1981. An empirically based estimate of home range. *J. Theor. Biol.* 20:281–325.

Siegel, Sidney. 1956. *Nonparametric statistics for the behavioral sciences.* New York: John Wiley.

Simon, C. A. 1975. The influence of food abundance on territory size in the iguanid lizard *Sceloporus jarrovi. Ecology* 56:993–998.

Simon, C. A., and G. A. Middendorf. 1980. Spacing in juvenile lizards (*Sceloporus jarrovi*). *Copeia* 1980:141–146.

Smith, H. M. 1950. Handbook of reptiles and amphibians of Kansas. *Univ. Kans. Publ. Mus. Nat. Hist. Misc. Publ.* 2:1–336.

Sokal, R. R., and F. J. Rohlf. 1969. *Biometry.* San Francisco: W. H. Freeman.

Stamps, J. A. 1977. Social behavior and spacing in lizards. In Carl Gans and D. W. Tinkle.

Tinkle, D. W. 1967. Life and demography of the side-blotched lizard *Uta stansburiana. Misc. Publ. Mus. Zool., Univ. Mich.* 132:1–182.

Watson, Adam, and G. R. Miller. 1971. Territory size and aggression in a fluctuating red grouse population. *J. Anim. Ecol.* 40:367–383.

Yedlin, I. N., and G. W. Ferguson. 1973. Variations of aggressiveness of free-living male and female collared lizards, *Crotaphytus collaris. Herpetologica* 29:268–275.

8. Fitness, Home-Range Quality, and Aggression in *Uta stansburiana*

Beardmore, J. A., and Louis Levine. 1963. Fitness and environmental variation. I. A study of some polymorphic populations of *Drosophila pseudoobscura. Evolution* 17:121–129.

Blair, W. F. 1960. *The rusty lizard: a population study.* Austin: University of Texas Press.

Boag, D. A. 1973. Spatial relationships among members of a population of wall lizards. *Oecologia* (Berl.) 12:1–13.

Brackin, M. F. 1978. The relation of rank to physiological state in *Cnemidophorus sexlineatus* dominance hierarchies. *Herpetologica* 34:185–191.

Brattstrom, B. H. 1974. The evolution of reptilian social organization. *Am. Zool.* 14:35–49.

Brown, J. L., and G. H. Orians. 1970. Spacing patterns in mobile animals. *Annu. Rev. Ecol. Syst.* 1:239–257.

Bryant, E. H. 1974. On the adaptive significance of enzyme polymorphisms in relation to environmental variability. *Am. Nat.* 108:1–19.

——— 1976. A comment on the role of environmental variation in maintaining polymorphisms in natural populations. *Evolution* 30:188–190.

Crews, David. 1974. Castration and androgen replacement on male facilitation of ovarian activity in the lizard *Anolis carolinensis. J. Comp. Physiol. Psychol.* 87:963–969.

Dunham, A. E. 1978. Food availability as a proximate factor influencing individual growth rates in the iguanid lizard *Sceloporus merriami. Ecology* 59:770–778.

Felsenstein, Joseph. 1976. The theoretical population genetics of variable selection and migration. *Annu. Rev. Genet.* 10:253–280.

Ferguson, G. W., and C. H. Bohlen. 1978. Demographic analysis: a tool for the study of natural selection of behavioral traits. In Neil Greenberg and P. D. MacLean.

Ferguson, G. W., and S. F. Fox. Annual variation of survival advantage of large juvenile side-blotched lizards *Uta stansburiana stejnegeri.* In preparation.

Ferner, J. W. 1974. Home-range size and overlap in *Sceloporus undulatus erythrocheilus* (Reptilia: Iguanidae). *Copeia* 1974:332–337.

Fox, S. F. 1973. Natural selection in the lizard *Uta stansburiana.* Ph.D. dissertation, Yale University.

——— 1975. Natural selection on morphological phenotypes of the lizard *Uta stansburiana. Evolution* 29:95–107.

——— 1978. Natural selection on behavioral phenotypes of the lizard *Uta stansburiana. Ecology* 59:834–847.

——— Natural selection on scale characters in a lizard population reexamined after several years. In preparation.

Fox, S. F., and S. R. Mays. Home-range acquisition in juvenile lizards implanted with testosterone propionate. In preparation.

Fox, S. F., Elizabeth Rose, and Ronald Myers. 1981. Dominance and the acquisition of superior home ranges in the lizard *Uta stansburiana. Ecology* 62:888–893.

Gans, Carl, and W. R. Dawson. 1976. Reptilian physiology: an overview. In Carl Gans and W. R. Dawson.

Gillespie, J. H. 1974. The role of environmental grain in the maintenance of genetic variation. *Am. Nat.* 108:831–836.

——— 1975. The role of migration in the genetic structure of populations in temporally and spatially varying environments. I. Conditions for polymorphism. *Am. Nat.* 109:127–135.

Gillespie, J. H., and C. H. Langley. 1974. A general model to account for enzyme variation in natural populations. *Genetics* 76:837–884.

Haldane, J. B. S., and S. D. Jayakar. 1963. Polymorphism due to selection of varying direction. *J. Genet.* 58:237–242.

Hecht, M. K. 1952. Natural selection in the lizard genus *Aristelliger. Evolution* 6:112–124.

Hedrick, P. W. 1974. Genetic variation in a heterogeneous environment. I. Temporal heterogeneity and the absolute dominance model. *Genetics* 78:757–770.

——— 1976. Genetic variation in a heterogeneous environment. II. Temporal heterogeneity and directional selection. *Genetics* 84:145–157.

Hedrick, P. W., M. E. Ginevan, and E. P. Ewing. 1976. Genetic polymorphism in heterogeneous environments. *Annu. Rev. Ecol. Syst.* 7:1–32.

Hoddenbach, G. A., and F. B. Turner. 1968. Clutch size of the lizard *Uta stansburiana* in southern Nevada. *Am. Midl. Nat.* 80:262–265.

Huey, R. B., and E. R. Pianka. 1981. Ecological consequences of foraging mode. *Ecology* 62:991–999.

Hutchinson, G. E., and R. H. MacArthur. 1959. On the theoretical significance of aggressive neglect in interspecific competition. *Am. Nat.* 93:133–134.

Levinton, J. 1973. Genetic variation in a gradient of environmental variability: marine bivalvia (Mollusca). *Science* (Wash., D.C.) 180:75–76.

McDonald, J. F., and F. J. Ayala. 1974. Genetic response to environmental heterogeneity. *Nature* (Lond.) 250:572–574.

McKinney, C. O., R. K. Selander, W. E. Johnson, and S. Y. Yang. 1972. Genetic variation in the side-blotched lizard (*Uta stansburiana*). Studies in Genetics VII, University of Texas Publication no. 7213:307–318.

Nagy, K. A. 1973. Behavior, diet, and reproduction in a desert lizard, *Sauromalus obesus. Copeia* 1973:93–102.

Parker, W. S., and E. R. Pianka. 1975. Comparative ecology of populations of the lizard *Uta stansburiana. Copeia* 1975:615–632.

Pasteur, Georges. 1977. Endocyclic selection in reptiles. *Am. Nat.* 111: 1027–30.

Philibosian, Richard. 1975. Territorial behavior and population regulation in the lizards, *Anolis acutus* and *A. cristatellus. Copeia* 1975:428–444.

Pianka, E. R. 1967. On lizard species diversity: North American flatland deserts. *Ecology* 48:333–351.

——— 1970. Comparative autecology of the lizard *Cnemidophorus tigris* in different parts of its geographic range. *Ecology* 51:703–720.

——— 1973. The structure of lizard communities. *Annu. Rev. Ecol. Syst.* 4:53–73.

——— 1975. Niche relations of desert lizards. In *Ecology and evolution of communities,* ed. M. L. Cody and J. M. Diamond. Cambridge, Mass.: Harvard University Press.

Pough, F. H. 1973. Lizard energetics and diet. *Ecology* 54:837–844.

Powell, J. R. 1971. Genetic polymorphisms in varied environments. *Science* (Wash., D.C.) 174:1035–36.

Rand, A. S. 1967a. The adaptive significance of territoriality in iguanid lizards. In W. W. Milstead.

——— 1967b. Ecology and social organization in the iguanid lizard *Anolis lineatopus. Proc. U.S. Nat. Mus.* 122:1–79.

Ruby, D. E. 1978. Seasonal changes in the territorial behavior of the iguanid lizard *Sceloporus jarrovi. Copeia* 1978:430–438.

——— 1981. Phenotypic correlates of male reproductive success in the lizard, *Sceloporus jarrovi.* In *Natural selection and social behavior,* ed. R. D. Alexander and D. W. Tinkle. New York: Chiron Press.

Ruibal, Rodolfo, and Richard Philibosian. 1974. Aggression in the lizard *Anolis acutus. Copeia* 1974:349–357.

Schoener, T. W. 1971. Theory of feeding strategies. *Annu. Rev. Ecol. Syst.* 2:369–404.

Simon, C. A., and G. A. Middendorf. 1976. Resource partitioning by an iguanid lizard: temporal and microhabitat aspects. *Ecology* 57:1317–20.

——— 1980. Spacing in juvenile lizards (*Sceloporus jarrovi*). *Copeia* 1980:141–146.

Stamps, J. A. 1977. Social behavior and spacing patterns in lizards. In Carl Gans and D. W. Tinkle.

——— 1978. A field study of the ontogeny of social behavior in the lizard *Anolis aeneus. Behaviour* 66:1–31.

Stamps, J. A., and Sanford Tanaka. 1981. The relationship between food and social behavior in juvenile lizards (*Anolis aeneus*). *Copeia* 1981:422–434.

Tinkle, D. W. 1965. Population structure and effective size of a lizard population. *Evolution* 19:569–573.

——— 1967a. Home range, density, dynamics, and structure of a Texas population of the lizard *Uta stansburiana.* In W. W. Milstead.

——— 1967b. The life and demography of the side-blotched lizard, *Uta stansburiana. Misc. Publ. Mus. Zool., Univ. Mich.* no. 132.

——— 1969. Evolutionary implications of comparative population studies in the lizard *Uta stansburiana.* In Int. Conf. Syst. Biol., Univ. Mich., 1967, *Systematic Biology.* National Acad. Sci., Washington, D.C., no. 1692, pp. 133–154.

Tinkle, D. W., D. McGregor, and S. Dana. 1962. Home range ecology of *Uta stansburiana stejnegeri. Ecology* 43:223–229.

Tinkle, D. W., H. M. Wilbur, and S. G. Tilley. 1970. Evolutionary strategies in lizard reproduction. *Evolution* 24:55–74.

Trivers, R. L. 1976. Sexual selection and resource-accruing abilities in *Anolis garmani. Evolution* 30:253–269.

Tubbs, A. A., and G. W. Ferguson. 1976. Effects of artificial crowding on behavior, growth, and survival of juvenile spiny lizards. *Copeia* 1976:820–823.

Turner, F. B., G. A. Hoddenbach, P. A. Medica, and J. R. Lannom. 1970. The demography of the lizard, *Uta stansburiana* Baird and Girard, in southern Nevada. *J. Anim. Ecol.* 39:505–519.

Turner, F. B., J. R. Lannom, Jr., P. A. Medica, and G. A. Hoddenbach. 1969a. Density and composition of fenced populations of leopard lizards (*Crotaphytus wislizenii*) in southern Nevada. *Herpetologica* 25:247–257.

——— 1969b. A demographic analysis of fenced populations of the whiptail lizard, *Cnemidophorus tigris,* in southern Nevada. *Southwest. Nat.* 14:189–202.

Wiens, J. A. 1977. On competition and variable environments. *Am. Scien.* 65:590–597.

Wilson, E. O. 1975. *Sociobiology: the new synthesis.* Cambridge, Mass.: Harvard University Press.

Yedlin, I. N., and G. W. Ferguson. 1973. Variations in aggressiveness of free-living male and female collared lizards, *Crotaphytus collaris. Herpetologica* 29:268–275.

9. Sexual Selection, Sexual Dimorphism, and Territoriality

Alcala, A. C., and W. C. Brown. 1967. Population ecology of the tropical scincoid lizard *Emoia atrocostata* in the Philippines. *Copeia* 1967:596–604.

Andrews, R. M. 1971. Structural habitat and time budget of a tropical *Anolis* lizard. *Ecology* 52:262–271.

——— 1979. Evolution of life histories: a comparison of *Anolis* lizards from matched island and mainland habitats. *Breviora Mus. Comp. Zool.* 454:1–51.

Atchley, W. R., C. T. Gaskins, and Dwane Anderson. 1976. Statistical properties of ratios. I. Empirical results. *Syst. Zool.* 25:137–148.

Avery, D. F., and W. W. Tanner. 1971. Evolution of the iguanine lizards (Sauria, Iguanidae) as determined by osteological and myological characters. *Brig. Young Univ. Sci. Bull.* 12:1–79.

Avery, R. A. 1975. Age structure and longevity of common lizard *Lacerta vivipara* populations. *J. Zool.* (Lond.) 176:555–558.

Baharav, D. W. 1975. Movement of the horned lizard *Phrynosoma solare. Copeia* 1975:649–657.

Ballinger, R. D. 1973. Comparative demography of two viviparous iguanid lizards (*Sceloporus jarrovi* and *Sceloporus poinsetti*). *Ecology* 54:269–283.

Barbault, Robert. 1974. Structure et dynamique d'un peuplement de lezards: les scincides de la savane de Lamto (Cote-d'Ivoire). *Terre Vie* 28:352–428.

Bartholomew, G. A. 1970. A model for the evolution of pinniped polygyny. *Evolution* 24:546–559.

Bartholomew, G. A., and V. A. Tucker. 1964. Size, body temperature, thermal conductance, oxygen consumption, and heart rate in Australian varanid lizards. *Physiol. Zool.* 37:341–354.

Bennett, A. F., and W. R. Dawson. 1976. Metabolism. In Carl Gans and W. R. Dawson.

Bennett, A. F., and K. A. Nagy. 1977. Energy expenditure in free-ranging lizards. *Ecology* 58:697–700.

Berry, K. H. 1974. The ecology and social behavior of the chuckwalla, *Sauromalus obesus obesus* Baird. *Univ. Calif. Publ. Zool.* 101:1–44.
Blair, W. F. 1960. *The rusty lizard.* Austin: University of Texas Press.
Blanc, C. P. and C. C. Carpenter. 1969. Studies on the Iguanidae of Madagascar. III. Social and reproductive behavior of *Chalarodon madagascariensis. J. Herpetol.* 3:125–134.
Boag, D. A. 1973. Spatial relationships among members of a population of wall lizards. *Oecologia* (Berl.) 12:1–13.
Boersma, P. D. An ecological study of the Galapagos marine iguana. In *Galapagos symposium volume,* ed. R. Bowman. Washington, D.C.: American Association for the Advancement of Science. In press.
Bogdanov, O. P., and E. V. Vashetko. 1972. On ecology of lizard *Eremias persica. Zool. Zh.* 51:310–312.
Bostic, D. L. 1965. Home range of the teiid lizard, *Cnemidophorus hyperythrus beldingi. Southwest. Nat.* 10:278–281.
Bradbury, T. W., and S. L. Vehrencamp. 1977. Social organization and foraging in emballonurid bats. III. Mating systems. *Behav. Ecol. Sociobiol.* 2:1–17.
Brooks, G. R. 1967. Population ecology of the ground skink *Lygosoma laterale. Ecol. Monogr.* 37:71–87.
Brown, J. L. 1964. The evolution of diversity in avian territorial systems. *Wilson Bull.* 76:160–169.
Brown, J. L., and G. H. Orians. 1970. Spacing patterns in mobile animals. *Annu. Rev. Ecol. Syst.* 1:239–269.
Bruton, M. N. 1977. Feeding, social behaviour, and temperature preferences in *Agama atra. Zool. Afr.* 12:183–199.
Burghardt, G. M., H. W. Greene, and A. S. Rand. 1977. Social behavior in hatchling green iguanas: life at a reptile rookery. *Science* (Wash., D.C.) 195:689–691.
Burrage, R. B. 1973. Comparative ecology and behaviour of *Chamaeleo pumilus pumilus* (Gmelin) and *C. namaquensis* (Smith) (Sauria: Chamaeleonidae). *Ann. S. Afr. Mus.* 51:1–158.
—— 1974. Population structure in *Agama atra* and *Cordylus cordylus* in the vicinity of De Kelders, C. P. *Ann. S. Afr. Mus.* 66:1–23.
Bustard, H. R. 1970. A population study of the scincid lizard *Egernia striolata* in northern New South Wales I. *Proc. K. Ned. Akad. Wet. Ser. C Biol. Med. Sci.* 73:186–212.
Carothers, J. M. 1981. Dominance and competition in a herbivorous lizard. *Behav. Ecol. Sociobiol.* 8:261–266.
Carpenter, C. C. 1966. The marine iguana of the Galapagos Islands, its behavior and ecology. *Proc. Calif. Acad. Sci.* 34:329–376.
——— 1977. Variation and evolution of stereotyped behavior in reptiles. Part I. A survey of stereotyped reptilian behavioral patterns. In Carl Gans and D. W. Tinkle.
Case, Ted. 1982. Ecology and evolution of the insular gigantic chuckwallas *Sauromalus hispidus* and *Sauromalus varius.* In G. M. Burghardt and A. S. Rand.
Chayes, Felix. 1949. On ratio correlation in petrography. *J. Geol.* 57:239–254.

Clutton-Brock, T. H., and P. H. Harvey. 1978. Mammals, resources, and reproductive strategies. *Nature* (Lond.) 273:192–195.
Clutton-Brock, T. H., P. H. Harvey, and B. Rudder. 1977. Sexual dimorphism, socionomic sex ratio, and body weight in primates. *Nature* (Lond.) 269:797–800.
Crews, David. 1975. Effects of different components of male courtship behaviour on environmentally induced ovarian recrudescence and mating preferences in the lizard *Anolis carolinensis. Anim. Behav.* 23:349–356.
Cruce, Mihai. 1977. Structure et dynamique d'une population de *Lacerta t. taurica* Pallas. *Terre Vie* 31:611–636.
Danielyan, F. D. 1971. A comparative study of the population density and migrations of parthenogenetic and bisexual rock lizards in Armenia. *Zool. Zh.* 50:145–147.
Darwin, Charles. 1871. *The descent of man, and selection in relation to sex.* 2 vols. New York: Appleton.
Davies, N. B. 1978. Ecological questions about territorial behavior. In *Behavioural ecology, an evolutionary approach,* ed. J. R. Krebs and N. B. Davies. Sunderland, Mass.: Sinauer.
Dugan, Beverly. 1982. The mating behavior of the green iguana, *Iguana iguana.* In G. M. Burghardt and A. S. Rand.
Dunham, A. E. 1978. Food availability as a proximate factor influencing individual growth rates in the iguanid lizard *Sceloporus merriami. Ecology* 59:770–778.
Dunham, A. E., D. W. Tinkle, and J. W. Gibbons. 1978. Body size in island lizards: a cautionary tale. *Ecology* 59:1230–39.
Dutton, R. H., L. C. Fitzpatrick, and J. L Hughes. 1975. Energetics of the rusty lizard *Sceloporus olivaceus. Ecology* 56:1378–87.
Eibl-Eibesfeldt, Irenaus. 1955. Der Comment-Kampf der Meerechse (*Amblyrhynchus cristatus*). *Z. Tierpsychol.* 12:49–62.
Emlen, S. T., and L. W. Oring. 1977. Ecology, sexual selection, and the evolution of mating systems. *Science* (Wash., D.C.) 197:215–223.
Evans, L. T. 1951. Field study of the social behavior of the black lizard, *Ctenosaura pectinata. Am. Mus. Novit.* 1493:1–26.
Ferguson, G. W. 1971. Observations on the behavior and interactions of two sympatric *Sceloporus* in Utah. *Am. Midl. Nat.* 86:190–195.
Ferguson, G. W., C. H. Bohlen, and H. P. Woolley. 1980. *Sceloporus undulatus:* comparative life history and regulation of a Kansas population. *Ecology* 61:313–323.
Ferner, J. W. 1974. Home-range size and overlap in *Sceloporus undulatus erythrocheilus* (Reptilia: Iguanidae). *Copeia* 1974:332–337.
——— 1976. Notes on natural history and behavior of *Sceloporus undulatus erythrocheilus* in Colorado. *Am. Midl. Nat.* 96:291–302.
Fitch, H. S. 1940. A field study of the growth and behavior of the fence lizard. *Univ. Calif. Publ. Zool.* 44:151–172.
——— 1954. Life history and ecology of the five-lined skink *Eumeces fasciatus. Univ. Kans. Publ. Mus. Nat. Hist.* 8:1–156.
——— 1955. Habits and adaptations of the great plains skink (*Eumeces obsoletus*). *Ecol. Monogr.* 25:59–83.

——— 1956. An ecological study of the collared lizard (*Crotaphytus collaris*). *Univ. Kans. Publ. Mus. Nat. Hist.* 8:213–274.

——— 1958. Natural history of the six-lined racerunner (*Cnemidophorus sexlineatus*). *Univ. Kans. Publ. Mus. Nat. Hist.* 11:11–62.

——— 1959. A population of the six-lined racerunner *Cnemidophorus sexlineatus. Herpetologica* 15:81–87.

——— 1976. Sexual size differences in mainland anoles. *Occas. Pap. Mus. Nat. Hist. Univ. Kans.* 50:1–21.

Fitch, H. S., and P. L. von Achen. 1977. Spatial relationships and seasonality in the skinks *Eumeces fasciatus* and *Scincella laterale* in northeastern Kansas. *Herpetologica* 33:303–313.

Fleming, T. H., and R. S. Hooker. 1975. *Anolis cupreus:* the response of a lizard to tropical seasonality. *Ecology* 56:1243–61.

Fox, S. F. 1978. Natural selection on behavioral phenotypes of the lizard *Uta stansburiana. Ecology* 59:834–847.

Fox, S. F., Elizabeth Rose, and Ronald Myers. 1981. Dominance and the acquisition of superior home ranges in the lizard *Uta stansburiana. Ecology* 62:888–893.

Gennaro, A. L. 1972. Home range and movements of *Holbrookia maculata maculata* in eastern New Mexico. *Herpetologica* 28:165–168.

Ghiselin, M. T. 1974. *The economy of nature and the evolution of sex.* Berkeley: University of California Press.

Goss-Custard, J. D., R. I. Dunbar, and F. P. Aldrich-Blake. 1972. Survival, mating, and rearing strategies in the evolution of primate social structure. *Folia Primatol.* 17:1–19.

Hall, R. J. 1971. Ecology of a population of the Great Plains skink (*Eumeces obsoletus*). *Univ. Kans. Sci. Bull.* 49:357–388.

Harris, V. A. 1964. *The life of the rainbow lizard.* London: Hutchinson.

Heusner, A. A., and E. W. Jameson, Jr. 1981. Seasonal changes in oxygen consumption and body composition of *Sceloporus occidentalis. Comp. Biochem. Physiol.* 69A:363–372.

Hirth, E. F. 1963. Ecology of two lizards on a tropical beach. *Ecol. Monogr.* 33:83–112.

Iverson, J. B. 1977. Behavior and ecology of the rock iguana *Cyclura carinata.* Ph.D. dissertation, University of Florida.

Jameson, E. W., Jr., A. A. Heusner, and Donna Lem. 1980. Seasonal, sexual, and altitudinal variations in stomach content and ingested fat in *Sceloporus occidentalis. J. Herpetol.* 14:255–261.

Jenssen, T. A. 1970. The ethoecology of *Anolis nebulosus* Sauria, Iguanidae. *J. Herpetol.* 4:1–38.

Johnson, S. R. 1965. An ecological study of the chuckwalla, *Sauromalus obesus* Baird in the western Mojave Desert. *Am. Midl. Nat.* 73:1–29.

Jorgensen, C. O., and W. W. Tanner. 1963. The application of the probability density function to determine the home range of *Uta stansburiana* and *Cnemidophorus tigris. Herpetologica* 19:105–115.

Krekorian, C. O. 1976. Home-range size and overlap and their relationship to food abundance in the desert iguana, *Dipsosaurus dorsalis. Herpetologica* 32:405–412.

Mackay, W. P. 1975. The home range of the banded rock lizard *Petrosaurus mearnsi*. *Southwest. Nat.* 20:113–120.

Mather, C. M. 1970. Some aspects of the life history of the ground skink *Lygosoma laterale*. *Tex. J. Sci.* 21:429–438.

McCoy, C. J. 1965. Life history and ecology of *Cnemidophorus tigris septentrionalis*. Ph.D. dissertation, University of Colorado.

McNab, B. K. 1963. Bioenergetics and the determination of home-range size. *Am. Nat.* 97:133–140.

Milstead, W. W. 1970. Late summer behavior of the lizards *Sceloporus merriami* and *Urosaurus ornatus* in the field. *Herpetologica* 26:341–357.

Mitchell, F. J. 1973. Studies on the ecology of the agamid lizard, *Amphibolurus maculosus* (Mitchell). *Trans. R. Soc. S. Aust.* 97:47–76.

Mount, R. H. 1963. The natural history of the red-tailed skink, *Eumeces egregius* Baird. *Am. Midl. Nat.* 70:356–385.

Müller, Horst. 1971. Ökophysiologische und ökethologische Untersuchungen an *Cnemidophorus lemniscatus* L. (Reptilia: Teiidae) in Kolumbien. *Forma Functio* 4:189–224.

Nagy, K. A. 1973. Behavior, diet, and reproduction in a desert lizard *Sauromalus obesus*. *Copeia* 1973:93–103.

Norris, K. S. 1953. The ecology of the desert iguana, *Dipsosaurus dorsalis*. *Ecology* 34:265–287.

Nussbaum, R. A., and L. V. Diller. 1976. The life history of the side-blotched lizard, *Uta stansburiana* Baird and Girard, in north-central Oregon. *Northwest Sci.* 50:243–260.

Orians, G. H. 1961. The ecology of blackbird (*Agelaius*) social systems. *Ecol. Monogr.* 31:285–312.

——— 1969. On the evolution of mating systems in birds and mammals. *Am. Nat.* 103:589–603.

——— 1979. *Some adaptations of marsh-nesting blackbirds*. Princeton: Princeton University Press.

Owen-Smith, Norman. 1977. On territoriality in ungulates and an evolutionary model. *Q. Rev. Biol.* 52:1–38.

Parker, W. S. 1974. Home range, growth, and population density of *Uta stansburiana* in Arizona. *J. Herpetol.* 8:135–141.

Parker, W. S., and E. R. Pianka. 1975. Comparative ecology of populations of the lizard *Uta stansburiana*. *Copeia* 1975:615–632.

Pianka, E. R. 1971. Comparative ecology of two lizards. *Copeia* 1971:129–138.

Pianka, E. R., and W. S. Parker. 1975. Ecology of horned lizards: a review with special reference to *Phrynosoma platyrhinos*. *Copeia* 1975: 141–162.

Ralls, Katherine. 1976. Mammals in which females are larger than males. *Q. Rev. Biol.* 51:245–276.

——— 1977. Sexual dimorphism in mammals: avian models and unanswered questions. *Am. Nat.* 111:917–938.

Rand, A. S. 1967. Ecology and social organization in the iguanid lizard *Anolis lineatopus*. *Proc. U. S. Nat. Mus.* 122:1–79.

——— 1968. A nesting aggregation of iguanas. *Copeia* 1968:552–561.

Rand, W. M., and A. S. Rand. 1976. Agonistic behavior in nesting iguanas: a stochastic analysis of dispute settlement dominated by the minimization of energy cost. *Z. Tierpsychol.* 40:279–299.

Rensch, Bernard. 1959. *Evolution above the species level.* London: Methuen.

Rose, Barbara. 1981. Factors affecting activity in *Sceloporus virgatus. Ecology* 62:706–716.

——— 1982. Lizard home ranges: methodology and functions. *J. Herpetol.* 16:253–270.

Roughgarden, Jonathan, and E. R. Fuentes. 1977. The environmental determinants of size in solitary populations of West Indian *Anolis* lizards. *Oikos* 29:44–51.

Ruby, D. E. 1976. The behavioral ecology of the viviparous lizard, *Sceloporus jarrovi.* Ph.D. dissertation, University of Michigan.

——— 1978. Seasonal changes in the territorial behavior of the iguanid lizard *Sceloporus jarrovi. Copeia* 1978:430–438.

——— 1981. Phenotypic correlates of male reproductive success in the lizard *Sceloporus jarrovi.* In *Natural selection and social behavior,* ed. R. D. Alexander and D. W. Tinkle. New York: Chiron Press.

Ruibal, Rodolfo, and Richard Philibosian. 1974. The population ecology of the lizard *Anolis acutus. Ecology* 55:525–537.

Ruibal, Rodolfo, Richard Philibosian, and J. L. Adkins. 1972. Reproductive cycle and growth in the lizard *Anolis acutus. Copeia* 1972:509–519.

Schall, J. J. 1974. Population structure of the Aruban whiptail lizard, *Cnemidophorus arubensis,* in varied habitats. *Herpetologica* 30:38–44.

Schoener, T. W. 1966. The ecological significance of sexual dimorphism in size in the lizard *Anolis conspersus. Science* (Wash., D.C.) 155:474–476.

——— 1968. The *Anolis* lizards of Bimini: resource partitioning in a complex fauna. *Ecology* 49:704–727.

——— 1969a. Models of optimum size for solitary predators. *Am. Nat.* 103:227–313.

——— 1969b. Size patterns in West Indian *Anolis* lizards. I. Size and species diversity. *Syst. Zool.* 18:386–401.

——— 1970. Size patterns in West Indian *Anolis* lizards. II. Correlations with the sizes of particular sympatric species—displacement and convergence. *Am. Nat.* 104:155–174.

——— 1971. Theory of feeding strategies. *Annu. Rev. Ecol. Syst.* 2:369–404.

——— 1977. Competition and the niche. In Carl Gans and D. W. Tinkle.

——— 1981. An empirically based estimate of the home range. *Theoret. Popul. Biol.* 20:281–325.

Schoener, T. W., and Amy Schoener. 1980. Densities, sex ratios, and population structure in four species of Bahamian *Anolis* lizards. *J. Anim. Ecol.* 49:19–53.

Searcy, W. A. 1979. Sexual selection and body size in male red-winged blackbirds. *Evolution* 33:649–661.

Selander, R. K. 1966. Sexual dimorphism and differential niche utilization in birds. *Condor* 68:113–151.

——— 1972. Sexual selection and dimorphism in birds. In *Sexual selection and the descent of man, 1871–1971,* ed. B. Campbell. Chicago: Aldine.

Sexton, O. J., H. F. Heatwole, and E. H. Meseth. 1963. Seasonal population changes in the lizard *Anolis limifrons* in Panama. *Am. Midl. Nat.* 69:482–491.

Shine, Richard. 1978. Sexual size dimorphism and male combat in snakes. *Oecologia* (Berl.) 33:269–277.

——— 1979. Sexual selection and sexual dimorphism in the amphibia. *Copeia* 1979:297–307.

Simon, C. A. 1973. The effect of food abundance on territory size in the lizard *Sceloporus jarrovi.* Ph.D. dissertation, University of California, Riverside.

——— 1975. The influence of food abundance on territory size in the iguanid lizard *Sceloporus jarrovi. Ecology* 56:993–998.

Simon, C. A., and G. A. Middendorf. 1980. Spacing in juvenile lizards (*Sceloporus jarrovi*). *Copeia* 1980:141–147.

Stamps, J. A. 1973. Displays and social organization in female *Anolis aeneus. Copeia* 1973:264–272.

——— 1977a. The relationship between resource competition, risk and aggression in a tropical territorial lizard. *Ecology* 58:349–358.

——— 1977b. Social behavior and spacing patterns in lizards. In Carl Gans and D. W. Tinkle.

——— 1978. A field study of the ontogeny of social behavior in the lizard *Anolis aeneus. Behaviour* 66:1–31.

Stamps, J. A., and D. P. Crews. 1976. Seasonal changes in reproduction and social behavior in the lizard *Anolis aeneus. Copeia* 1976:467–476.

Stamps, J. A., S. K. Tanaka, and V. V. Krishnan. 1981. The relationship between selectivity and food abundance in a juvenile lizard. *Ecology* 62:1079–92.

Stebbins, R. C. 1944. Field notes on a lizard, the mountain swift, with special reference to territorial behavior. *Ecology* 25:233–245.

Stebbins, R. C., J. M. Lowenstein, and N. W. Cohen. 1967. A field study of the lava lizard (*Tropidurus albemarlensis*) in the Galapagos Islands. *Ecology* 48:839–851.

Tanaka, L. K., and S. K. Tanaka. 1982. Rainfall and seasonal changes in arthropod abundance on a tropical oceanic island. *Biotropica* 14:114–123.

Tanner, W. W. 1957. A taxonomic and ecological study of the western skink (*Eumeces skiltonianus*). *Great Basin Nat.* 17:59–94.

——— 1965. A comparative population study of small vertebrates in the uranium areas of the Upper Colorado River Basin of Utah. *Brig. Young Univ. Sci. Bull.* 7:1–31.

Tanner, W. W., and J. M. Hopkin. 1972. Ecology of *Sceloporus occidentalis longipes* Baird and *Uta stansburiana stansburiana* Baird and Girard on Rainier Mesa, Nevada Test Site, Nye County, Nevada. *Brig. Young Univ. Bull. Biol. Ser.* 15:1–39.

Tanner, W. W., and J. E. Krogh. 1974. Ecology of the leopard lizard, *Crotaphytus wislizenii* at the Nevada Test Site, Nye County, Nevada. *Herpetologica* 30:63–72.

Tertyshnikov, M. F. 1970. Home ranges of the sand lizard (*Lacerta agilis*) and the motley lizard (*Eremias arguta*) and peculiarities of their utilization. *Zool. Zh.* 49:1377–85.

Tinkle, D. W. 1967. The life and demography of the side-blotched lizard, *Uta stansburiana. Misc. Publ. Mus. Zool., Univ. Mich.* 132:1–182.

——— 1973. A population analysis of the sagebrush lizard *Sceloporus graciosus* in southern Utah. *Copeia* 1973:284–296.

——— 1976. Comparative data on the population ecology of the desert spiny lizard, *Sceloporus magister. Herpetologica* 32:1–6.

Tinkle, D. W., and D. W. Woodard. 1967. Relative movements of lizards in natural populations as determined from recapture radii. *Ecology* 48:166–168.

Tinkle, D. W., Dan McGregor, and Sumner Dana. 1962. Home-range ecology of *Uta stansburiana stejnegeri. Ecology* 43:223–229.

Tollestrup, Kristine. 1979. The ecology, social structure, and foraging behavior of two closely related species of leopard lizards, *Gambelia silus* and *Gambelia wislizenii.* Ph.D. dissertation, University of California, Berkeley.

Trivers, R. L. 1972. Parental investment and sexual selection. In *Sexual selection and the descent of man, 1871–1971,* ed. B. Campbell. Chicago: Aldine.

——— 1976. Sexual selection and resource-accruing abilities in *Anolis garmani. Evolution* 30:253–269.

Turner, F. B., R. I. Jennrich, and J. D. Weintraub. 1969. Home ranges and body size of lizards. *Ecology* 50:1076–81.

Verner, Jared. 1964. Evolution of polygamy in the long-billed marsh wren. *Evolution* 18:252–261.

Wade, M. J. 1979. Sexual selection and variance in reproductive success. *Am. Nat.* 114:742–746.

Werner, D. I. 1978. On the biology of *Tropidurus delanonis* Baur (Iguanidae). *Z. Tierpsychol.* 47:337–395.

——— 1982. Social organization and ecology of land iguanas, *Conolophus subcristatus* on Fernandina Island, Galápagos. In G. M. Burghardt and A. S. Rand.

Whitford, W. G., and Martha Bryant. 1979. Behavior of a predator and its prey: the horned lizard (*Phrynosoma cornutum*) and harvester ants (*Pogonomyrmex* spp.). *Ecology* 60:686–695.

Wiewandt, T. A. 1977. Ecology, behavior, and management of the Mona Island ground iguana, *Cyclura stejnegeri.* Ph.D. dissertation, Cornell University.

Wilson, E. O. 1975. *Sociobiology: the new synthesis.* Cambridge, Mass.: Harvard University Press.

Wittenberger, J. F. 1976. The ecological factors selecting for polygyny in altricial birds. *Am. Nat.* 109:779–799.

——— 1980. Group size and polygamy in social mammals. *Am. Nat.* 115:197–222.

Worthington, R. O., and E. R. Arvizo. 1973. Density, growth, and home range of the lizard *Uta stansburiana stejnegeri* in southern Dona Ana County, New Mexico. *Great Basin Nat.* 33:124–129.

Yadgarov, T. 1973. On the ecology of the long-legged skink (*Eumeces schneideri princeps*). *Zool. Zh.* 49:1377–85.

10. Psychobiology of Parthenogenesis

Adkins, E. K. 1980. Non-mammalian psychosexual differentiation. In *Handbook of behavioral neurobiology,* ed. R. W. Goy and D. W. Pfaff. New York: Plenum Press.

Adler, N. T. 1974. The behavioral control of reproductive physiology. In *Reproductive behavior,* ed. W. Montagna and W. A. Sadler. New York: Plenum Press.

Arslan, M., P. Zaidi, J. Lobo, A. A. Zaidi, and M. H. Qazi. 1978. Steroid levels in preovulatory and gravid lizards (*Uromastix hardwicki*). *Gen. Comp. Endocrinol.* 34:300–303.

Beach, F. A. 1968. Factors involved in the control of mounting behavior by female mammals. In *Perspectives in reproduction and sexual behavior,* ed. M. Diamond. Bloomington: Indiana University Press.

——— 1979. Animal models for human sexuality. In *Sex, hormones, and behaviour,* Ciba Foundation Symposium 62. Amsterdam: Excerpta Medica.

Bogart, J. P. 1980. Evolutionary implications of polyploidy in amphibians and reptiles. In *Polyploidy: biological relevance,* ed. W. H. Lewis. New York: Plenum Press.

Brattstrom, B. H. 1974. The evolution of reptilian social behavior. *Am. Zool.* 14:35–50.

Brown, W. M., and J. W. Wright. 1979. Mitochondrial DNA analysis and the origin and relative age of parthenogenetic lizards (genus *Cnemidophorus*). *Science* (Wash., D.C.) 203:1247–49.

Bull, J. J. 1978. Sex chromosome differentiation: an intermediate stage in a lizard. *Can. J. Genet. Cytol.* 20:205–209.

——— 1980. Sex determination in reptiles. *Q. Rev. Biol.* 55:3–21.

Bull, J. J., and R. C. Voigt. 1979. Temperature-dependent sex determination in turtles. *Science* (Wash., D.C.) 206:1186–88.

Callard, I. P., G. V. Callard, V. Lance, J. L. Bolaffi, and J. S. Rosset. 1978. Testicular regulation in nonmammalian vertebrates. *Biol. Reprod.* 18:16–43.

Carpenter, C. C. 1960. Aggressive behavior and social dominance in the six-lined racerunner (*Cnemidophorus sexlineatus*). *Anim. Behav.* 8:61–66.

——— 1962. Patterns of behavior in two Oklahoma lizards. *Am. Midl. Nat.* 67:132–151.

Carpenter, C. C., and G. W. Ferguson. 1977. A survey of stereotyped reptilian behavior patterns. In Carl Gans and D. W. Tinkle.

Charnier, M. 1966. Action de la température sur la sex-ratio chez l'embryon d'*Agama agama. C. R. Seances Soc. Biol. Fil.* 160:620–622.

Cherfas, N. B. 1966. Natural triploidy in females of the unisexual form of silver carp (goldfish) (*Carassius auratus gibelio*). *Genetika* 5:16–24.

Cole, C. J. 1966. Femoral glands in lizards: a review. *Herpetologica* 22:199–206.

——— 1975. Evolution of parthenogenetic species of reptiles. In *Intersexuality in the animal kingdom,* ed. R. Reinboth. New York: Springer-Verlag.

——— 1978. The value of virgin birth. *Nat. Hist.* 57:56–63.

Cole, C. J., and C. R. Townsend. 1977. Parthenogenetic reptiles: new subjects for laboratory research. *Experientia* (Basel) 33:285–288.

Cole, C. J., C. H. Lowe, and J. W. Wright. 1969. Sex chromosomes in teiid whiptail lizards (genus *Cnemidophorus*). *Am. Mus. Novit.* 2395:1–14.

Crews, David. 1974. Effects of group stability and male aggressive and sexual behavior on environmentally induced ovarian recrudescence in the lizard, *Anolis carolinensis. J. Zool.* (Lond.) 172:419–441.

——— 1975. Psychobiology of reptilian reproduction. *Science* (Wash., D.C.) 189:1059–65.

——— 1979. Control of reptilian reproduction cycles. In *Endocrine control of sexual behavior,* ed. C. Beyer. New York: Raven Press.

——— 1980. Interrelationships among ecological, behavioral, and neuroendocrine processes in the reproductive cycle of *Anolis carolinensis* and other reptiles. In *Advances in the study of behavior,* vol. 11, ed. J. S. Rosenblatt, R. A. Hinde, C. G. Beer, and M. C. Busnel. London: Academic Press.

Crews, David, and K. T. Fitzgerald. 1980. Sexual behavior in parthenogenetic lizards (*Cnemidophorus*). *Proc. Nat. Acad. Sci. U.S.A.* 77:499–502.

Cuellar, Orlando. 1971. Reproduction and the mechanism of meiotic restitution in the parthenogenetic lizard, *Cnemidophorus uniparens. J. Morphol.* 133:139–166.

——— 1974. On the origin of parthenogenesis in vertebrates: the cytological factors. *Am. Nat.* 108:625.

——— 1976. Intraclonal histocompatibility in a parthenogenetic lizard: evidence of genetic homogeneity. *Science* (Wash., D.C.) 193:150–153.

——— 1977a. Genetic homogeneity and speciation in the parthenogenetic lizards, *Cnemidophorus velox* and *C. neomexicanus:* evidence from intraspecific histocompatibility. *Evolution* 31:24–31.

——— 1977b. Animal parthenogenesis: a new evolutionary-ecological model is needed. *Science* (Wash., D.C.) 197:837–843.

——— 1979. On the ecology of coexistence in parthenogenetic and bisexual lizards in the genus *Cnemidophorus. Am. Zool.* 19:773–786.

——— 1981. Long-term analysis of reproductive periodicity in the lizard *Cnemidophorus uniparens. Am. Midl. Nat.* 105:93–101.

Cuellar, Orlando, and C. O. McKinney. 1976. Natural hybridization between parthenogenetic and bisexual lizards: detection of uniparental source by skin grafting. *J. Exp. Zool.* 196:341–350.

Darevsky, J. S. 1966. Natural parthenogenesis in a polymorphic group of Caucasian rock lizards related to *Lacerta saxicola* Eversmann. *J. Ohio Herpetol. Soc.* 5:115–152.

Duvall, David. 1979. Western fence lizard (*Sceloporus occidentalis*) chemical signals. I. Conspecific discrimination and release of a species-typical visual display. *J. Exp. Zool.* 210:321–326.

Fitch, H. S. 1958. Natural history of the six-lined racerunner. *Univ. Kans. Publ. Mus. Nat. Hist.* 11:11–62.

Forbes, T. R. 1961. Endocrinology of reproduction in cold-blooded vertebrates. In *Sex and internal secretions,* ed. W. C. Young. Baltimore: Williams and Wilkins.

Fox, Harold. 1977. The urogenital system of reptiles. In *Biology of the reptilia,* vol. 6, ed. Carl Gans and T. S. Parsons. London: Academic Press.

Gorman, G. C. 1973. The chromosomes of the Reptilia, a cytotaxonomic interpretation. In *Cytotaxonomy and vertebrate evolution,* ed. A. B. Chiarelli and E. Capanna. London: Academic Press.

Gorski, R. A. 1979. The neuroendocrinology of reproduction: an overview. *Biol. Reprod.* 20:11–127.

Greenough, W. T. 1976. Enduring brain effects of differential experience and training. In *Neural mechanisms of learning and memory,* ed. M. R. Rosenzweig and E. L. Bennet. Cambridge, Mass.: MIT Press.

Gustafson, Jill, and David Crews. 1981. Effects of group size and physiological state of a cagemate on reproductive effort in the parthenogenetic lizard, *Cnemidophorus uniparens* (*Teiidae*). *Behav. Ecol. Sociobiol.* 8:267–272.

Hardy, D. 1962. Ecology and behavior of the six-lined racerunner, *Cnemidophorus sexlineatus. Univ. Kans. Sci. Bull.* 43:1–73.

Hubbs, Clark. 1964. Interactions between a bisexual fish species and its gynogenetic sexual parasite. *Bull. Texas Mem. Mus.* 8:1–72.

Hubel, D. H., T. N. Wiesel, and S. LeVay. 1977. Plasticity of ocular dominance columns in monkeys. *Phil. Trans. R. Soc. Lond.* (B) 278:377–409.

Leshner, A. I. 1978. *An introduction to behavioral endocrinology.* Oxford: Oxford University Press.

Leuck, B. E. 1980. Life with and without sex: comparative behavior of three species of whiptail lizards (*Cnemidophorus:* Teiidae). Ph.D. dissertation, University of Oklahoma.

Licht, Paul, J. Wood, D. W. Owens, and F. Wood. 1979. Serum gonadotropins and steroids associated with breeding activities in the green sea turtle, *Chelonia mydas.* I. Captive animals. *Gen. Comp. Endocr.* 39:274–289.

Lowe, C. H., and J. W. Wright. 1964. Species of the *Cnemidophorus exsanguis* subgroup of whiptail lizards. *J. Ariz. Acad. Sci.* 3:77–80.

——— 1966. Evolution of parthenogenetic species of *Cnemidophorus* (whiptail lizards) in western North America. *J. Ariz. Acad. Sci.* 4:81–87.

Lowe, C. H., J. W. Wright, C. J. Cole, and R. L. Bezy. 1970. Chromosomes and evolution of the species group *Cnemidophorus* (Reptilia: Teiidae) *Syst. Zool.* 19:128–141.

McBride, Glen. 1976. The study of social organizations. *Behaviour* 59:96–115.

MacGregor, H. C., and T. M. Uzzell. 1964. Gynogenesis in salamanders related to *Ambystoma jeffersonianum. Science* (Wash., D.C.) 143:1043–45.

Maslin, T. P. 1962. All-female species of the lizard genus *Cnemidophorus,* Teiidae. *Science* (Wash., D.C.) 135:212–213.

——— 1967. Skin grafting in the bisexual teiid lizard *Cnemidophorus sexlineatus* and in the unisexual *C. tesselatus. J. Exp. Zool.* 166:137–150.

——— 1971. Conclusive evidence of parthenogenesis in three species of *Cnemidophorus* (Teiidae). *Copeia* 1971:156–158.

Mitchell, J. C. 1979. Ecology of southwestern Arizona whiptail lizards (*Cnemidophorus:* Teiidae): population densities, resource partitioning, and niche overlap. *Can. J. Zool.* 57:1487–99.

Neaves, W. B. 1969. Gene dosage at the lactate dehydrogenase B locus in the triploid and diploid teiid lizards. *Science* (Wash., D.C.) 164:557–558.

Noble, G. K., and H. T. Bradley. 1933. The mating behavior of lizards: its bearing on the theory of sexual selection. *Ann. N.Y. Acad. Sci.* 35:25–100.

Ohno, Susumu. 1979. *Major sex-determining genes.* Monographs on Endocrinology, vol. 2, ed. F. Gross, A. Labhard, M. B. Lipsett, T. Mann, L. T. Samuels, and J. Zander. New York: Springer-Verlag.

Olsen, M. W. 1960. Nine-year summary of parthenogenesis in turkeys. *Proc. Soc. Exp. Biol. Med.* 105:279–281.

Raynaud, Albert, and Claude Pieau. 1971. Evolution des canaux de Müller et activité enzymatique Δ-3β hydroxsteroide deshydrogenasique dans les glandes génitales, chez les embryons de lézard vert (*Lacerta viridis* Laur.). *C. R. Acad. Sci.* 273:2335.

Schultz, R. J. 1969. Hybridization, unisexuality, and polyploidy in the teleost *Poeciliopsis* (Poeciliidae) and other vertebrates. *Am. Nat.* 103:605–619.

——— 1971. Special adaptive problems associated with unisexual fishes. *Am. Zool.* 11:351–360.
——— 1979. Role of polyploidy in the evolution of fishes. In *Polyploidy: biological relevance,* ed. W. H. Lewis. New York: Plenum Press.
Serena, M. 1980. Why is there so much sex? Ph.D. dissertation, University of Colorado.
Uzzell, T. M. 1964. Relations of diploid and triploid species of the *Ambystoma jeffersonium* complex. *Copeia* 1964:257–300.
——— 1970. Meiotic mechanisms of naturally occurring unisexual vertebrates. *Am. Nat.* 104:433–435.
Uzzell, T. M., and S. Goldblatt. 1967. Serum proteins of salamanders of the *Ambystoma jeffersonium* complex, and the origin of the triploid species of this group. *Evolution* 21:345–354.
Vandenbergh, J. G. 1975. Hormones, pheromones, and behavior. In *Hormonal correlates of behavior,* ed. B. E. Eleftherioud and R. L. Sprott. New York: Plenum Press.
Wagner, Ernie. 1980. Gecko husbandry and reproduction. In *Reproductive biology and diseases of captive reptiles,* ed. J. Murphy and J. T. Collins. Society for the Study of Amphibians and Reptiles.
Werner, Y. L. 1980. Apparent homosexual behavior in an all-female population of a lizard, *Lepidodactylus lugubris* and its probable interpretation. *Z. Tierpsychol.* 54:144–150.
White, M. J. D. 1973. *Animal cytology and evolution.* Cambridge: Cambridge University Press.
Wright, J. W., and C. H. Lowe. 1968. Weeds, polyploids, parthenogenesis, and the geographical and ecological distribution of all-female species of *Cnemidophorus. Copeia* 1968:128–138.
Yntema, C. L. 1976. Effects of incubation temperatures on sexual differentiation in the turtle *Chelydra serpentina. J. Morphol.* 150:453–462.
——— 1979. Temperature levels and period of sex determination during incubation of eggs of *Chelydra serpentina. J. Morphol.* 159:17–27.

Overview of Part III

Andrews, R. M. 1979. Evolution of life histories: a comparison of *Anolis* from matched island and mainland habitats. *Breviora Mus. Comp. Zool.* 454:1–51.
Blair, W. F. 1960. *The rusty lizard.* Austin: University of Texas Press.
Christiansen, F. B., and T. M. Fenchel. 1979. Evolution of marine invertebrate reproductive patterns. *Theoret. Popul. Biol.* 16:267–282.
Cole, L. C. 1954. The population consequences of life-history phenomena. *Q. Rev. Biol.* 29:103–137.
Collette, B. B. 1961. Correlations between ecology and morphology in anoline lizards from Havana, Cuba, and southern Florida. *Bull. Mus. Comp. Zool.* 125:137–162.
Connor, E. F., and Daniel Simberloff. 1979. The assembly of species communities: chance or competition? *Ecology* 60:1132–40.
Dunham, A. E. 1980. An experimental study of interspecific competition between the iguanid lizards *Sceloporus merriami* and *Urosaurus ornatus. Ecol. Monogr.* 50:309–330.

Haigh, J., and John Maynard Smith. 1972. Can there be more predators than prey? *Theoret. Popul. Biol.* 3:290–299.

Heatwole, Harold. 1967. In W. W. Milstead, p. 62.

Lack, David. 1946. Competition for food by birds of prey. *J. Anim. Ecol.* 15:123–129.

MacArthur, R. H. 1965. Patterns of species diversity. *Biol. Rev.* 40:510–533.

Milstead, W. W. 1957. Some aspects of competition in natural populations of whiptail lizards (genus *Cnemidophorus*). *Tex. J. Sci.* 9:410–447.

——— 1961. Competitive relations in lizard populations. In *Vertebrate speciation,* ed. W. F. Blair. Austin: University of Texas Press.

Orians, G. H. 1979. *Some adaptations of marsh-nesting blackbirds.* Princeton: Princeton University Press.

Pacala, Stephen, and Jonathan Roughgarden. 1982. Resource partitioning and interspecific competition in two two-species insular *Anolis* lizard communities. *Science* 217:444–446.

Pianka, E. R. 1967. On lizard species diversity: North American flatland deserts. *Ecology* 48:333–351.

——— 1971. Comparative ecology of two lizards. *Copeia* 1971:129–138.

——— 1974. Niche overlap and diffuse competition. *Proc. Nat. Acad. Sci. U.S.A.* 71:2141–45.

——— 1977. Reptilian species diversity. In Carl Gans and D. W. Tinkle.

Pianka, E. R., R. B. Huey, and L. R. Lawlor. 1979. Niche segregation in desert lizards. In *Analysis of ecological systems,* ed. D. J. Horn, R. Mitchell, and G. R. Stairs. Columbus: Ohio State University Press.

Rand, A. S. 1964. Ecological distribution in anoline lizards of Puerto Rico. *Ecology* 45:745–752.

Rand, A. S., and E. E. Williams. 1969. The anoles of La Palma: aspects of their ecological relationships. *Breviora Mus. Comp. Zool.* 327:1–18.

Roughgarden, Jonathan, John Rummel, and Stephen Pacala. 1982. Experimental evidence of strong present-day competition between the *Anolis* population of the Anguilla Bank—a preliminary report. In *Advances in herpetology and evolutionary biology: essays in honor of Ernest E. Williams, Spec. Publ. Mus. Comp. Zool.*

Royama, T. 1970. Factors governing the hunting behavior and selection of food by the great tit (*Parus major* L.) *J. Anim. Ecol.* 39:619–668.

Ruibal, Rodolfo. 1961. Thermal relations of five species of tropical lizards. *Evolution* 15:98–111.

Schoener, T. W. 1969. Size patterns in West Indian *Anolis* lizards. I. Size and species diversity. *Syst. Zool.* 18:386–401.

——— 1970. Size patterns in West Indian *Anolis* lizards. II. Correlations with the sizes of particular sympatric species—displacement and convergence. *Am. Nat.* 104:155–174.

——— 1974a. Resource partitioning in ecological communities. *Science* (Wash., D.C.) 185:27–39.

——— 1974b. The compression hypothesis and temporal resource partitioning. *Proc. Nat. Acad. Sci. U.S.A.* 71:4169–72.

——— 1977. Competition and the niche. In Carl Gans and D. W. Tinkle.

Schoener, T. W., and Amy Schoener. 1980. Densities, sex ratios, and population structure in four species of Bahamian *Anolis* lizards. *J. Anim. Ecol.* 49:19–53.

Simberloff, Daniel, and William Boecklen. 1981. Santa Rosalia reconsidered. *Evolution* 35:1206–28.

Smith, J. M. N., P. R. Grant, B. R. Grant, I. J. Abbott, and L. K. Abbott. 1978. Seasonal variation in feeding habits of Darwin's finches. *Ecology* 59:1137–50.

Strong, D. R., Jr., Daniel Simberloff, and L. G. Abele. 1982. *Ecological communities: conceptual issues and the evidence.* Princeton: Princeton University Press.

Strong, D. R., Jr., L. A. Szyska, and Daniel Simberloff. 1979. Tests of communitywide character displacement against null hypotheses. *Evolution* 33:897–913.

Svärdson, Gunnar. 1949. Competition and habitat selection in birds. *Oikos* 1:157–174.

Tinkle, D. W. 1967. The life and demography of the side-blotched lizard, *Uta stansburiana. Misc. Publ. Mus. Zool., Univ. Mich.* 132:1–182.

——— 1969. The concept of reproductive effort and its relation to the evolution of life histories of lizards. *Am. Nat.* 153:501–516.

Tinkle, D. W., H. M. Wilbur, and S. G. Tilley. 1970. Evolutionary strategies in lizard reproduction. *Evolution* 24:55–74.

Turner, F. B. 1977. The dynamics of populations of squamates, crocodilians, and rhynchocephalians. In Carl Gans and D. W. Tinkle.

Whittaker, R. H. 1965. Dominance and diversity in land plant communities. *Science* (Wash., D.C.) 147:250–260.

Wiens, J. A. 1977. On competition and variable environments. *Am. Sci.* 65:590–597.

Williams, E. E. 1972. The origin of faunas: evolution of lizard congeners in a complex island fauna—a trial analysis. *Evolutionary Biology* 4:47–89.

11. Life-History Variations

Andrews, R. M. 1976. Growth rate in island and mainland anoline lizards. *Copeia* 1976:477–482.

Andrews, R. M., and A. S. Rand. 1974. Reproductive effort in anoline lizards. *Ecology* 55:1317–27.

Ayala, S. C., and J. L. Spain. 1975. Annual oögenesis in the lizard *Anolis auratus* determined by a blood-smear technique. *Copeia* 1975:138–141.

Ballinger, R. E. 1973. Comparative demography of two viviparous iguanid lizards (*Sceloporus jarrovi* and *Sceloporus poinsetti*). *Ecology* 54:269–283.

——— 1977. Reproductive strategies: food availability as a source of proximal variation in a lizard. *Ecology* 58:628–635.

——— 1978. Variation in and evolution of clutch and litter size. In *The vertebrate ovary: comparative biology and evolution,* ed. R. E. Jones. New York: Plenum Press.

——— 1979. Intraspecific variation in demography and life history of the lizard, *Sceloporus jarrovi,* along an altitudinal gradient in southeastern Arizona. *Ecology* 60:901–909.

——— 1980. Food limiting effects in populations of *Sceloporus jarrovi* (Iguanidae). *Southwest. Nat.* 25:554–557.

Ballinger, R. E., and R. A. Ballinger. 1979. Food resource utilization during periods of low and high food availability in *Sceloporus jarrovi* (Sauria: Iguanidae). *Southwest. Nat.* 24:347–363.

Ballinger, R. E., and J. D. Congdon. 1981. Population ecology and life-history strategy of a montane lizard (*Sceloporus scalaris*) in southeastern Arizona. *J. Nat. Hist.* 15:213–222.

Ballinger, R. E., Jon Hawker, and O. J. Sexton. 1969. The effect of photoperiod acclimation on the thermoregulation of the lizard, *Sceloporus undulatus*. *J. Exp. Zool.* 171:43–48.

Barbault, Robert. 1974. Ecologie comparée des lézards *Mabuya blandingi* (Hallowell) et *Panaspis kitsoni* (Boulenger) dans les forêts de Lamto (Côte-d'Ivoire). *Terre Vie* 28:272–295.

——— 1976. Population dynamics and reproductive patterns of three African skinks. *Copeia* 1976:483–490.

Bartholomew, G. A. 1950. The effects of artificially controlled temperature and day length on gonadal development in a lizard, *Xantusia vigilis*. *Anat. Rec.* 106:49–60.

——— 1953. The modification by temperature of the photoperiodic control of gonadal development in the lizard *Xantusia vigilis*. *Copeia* 1953:45–50.

——— 1959. Photoperiodism in reptiles. In *Photoperiodism and related phenomena in plants and animals,* ed. R. B. Withrow. Washington, D.C.: American Association for the Advancement of Science.

Bellairs, Angus. 1970. *The life of reptiles,* vol. 2. New York: Universe Books.

Blair, W. F. 1960. *The rusty lizard.* Austin: University of Texas Press.

Bogart, J. P. 1980. Evolutionary implications of polyploidy in amphibians and reptiles. In *Polyploidy: biological relevance,* ed. W. H. Lewis. New York: Plenum Press.

Bowker, R. G., and D. W. Johnson. 1980. Thermoregulatory precision in three species of whiptail lizards (Lacertidae: Teiidae). *Physiol. Zool.* 53:176–185.

Bradshaw, S. D. 1971. Growth and mortality in a field population of *Amphibolurus* lizards exposed to seasonal cold and aridity. *J. Zool.* (Lond.) 165:1–25.

Brattstrom, B. H. 1965. Body temperature of reptiles. *Am. Midl. Nat.* 73:376–422.

Brown, K. M., and O. J. Sexton. 1973. Stimulation of reproductive activity of female *Anolis sagrei* by moisture. *Physiol. Zool.* 46:168–172.

Bustard, H. R. 1968. The ecology of the Australian gecko, *Gehyra variegata,* in northern New South Wales. *J. Zool.* (Lond.) 154:113–138.

Carpenter, C. C. 1963. Patterns of behavior in three forms of the fringe-toed lizards (*Uma*-Iguanidae). *Copeia* 1963:406–412.

——— 1967. Aggression and social structure of iguanid lizards. In W. W. Milstead.

Carpenter, C. C., and G. W. Ferguson. 1977. Variation and evolution of stereotyped behavior in reptiles. In Carl Gans and D. W. Tinkle.

Cole, C. J. 1975. Evolution of parthenogenetic species of reptiles. In *Intersexuality in the animal kingdom,* ed. R. Reinboth. New York: Springer-Verlag.

——— 1979. Chromosome inheritance in parthenogenetic lizards and evolution of allopolyploidy in reptiles. *J. Hered.* 70:95–102.

Cole, L. C. 1954. The population consequences of life-history phenomena. *Q. Rev. Biol.* 29:103–137.

Congdon, J. D., R. E. Ballinger, and K. A. Nagy. 1979. Energetics, temperature and water relations in winter-aggregated *Sceloporus jarrovi* (Sauria: Iguanidae). *Ecology* 60:30–35.

Cowles, R. B., and C. M. Bogert. 1944. A preliminary study of the thermal requirements of desert reptiles. *Bull. Am. Mus. Nat. Hist.* 83:261–296.

Crews, David. 1975a. Psychobiology of reptilian reproduction. *Science* (Wash., D.C.) 189:1059–65.

——— 1975b. Effects of different components of male courtship behavior on environmentally induced ovarian recrudescence and mating preferences in the lizard, *Anolis carolinensis. Anim. Behav.* 21:349–356.

——— 1977. The annotated anole: studies on the control of lizard reproduction. *Am. Sci.* 65:428–434.

——— 1978. Integration of internal and external stimuli in the regulation of lizard reproduction. In Neil Greenberg and P. D. MacLean.

——— 1979. Neuroendocrinology of lizard reproduction. *Biol. Reprod.* 20:51–73.

Cuellar, Orlando. 1977. Animal parthenogenesis. *Science* (Wash., D.C.) 197:837–843.

Darevsky, I. S. 1958. Natural parthenogenesis in certain subspecies of rocky lizard, *Lacerta saxicola* Eversmann. *Dokl. Biol. Sci.* 122:877–879.

Davis, W. B., and J. R. Dixon. 1961. Reptiles (exclusive of snakes) of the Chilpancingo region, Mexico. *Proc. Biol. Soc. Wash.* 74:37–56.

Dawson, W. R. 1975. On the physiological significance of the preferred body temperatures of reptiles. In *Perspectives of biophysical ecology,* vol. 12, *Ecological studies,* ed. D. M. Gates and R. B. Schmerl. New York: Springer-Verlag.

Derickson, W. K. 1974. Lipid deposition and utilization in the sagebrush lizard, *Sceloporus graciosus:* its significance for reproduction and maintenance. *Comp. Biochem. Physiol.* 49A:267–272.

——— 1976a. Lipid storage and utilization in reptiles. *Am. Zool.* 16:711–723.

——— 1976b. Ecological and physiological aspects of reproductive strategies in two lizards. *Ecology* 57:445–458.

Dessauer, H. C. 1955. Seasonal changes in the gross organ composition of the lizard *Anolis carolinensis. J. Exp. Zool.* 128:1–12.

Dunham, A. E. 1978. Food availability as a proximate factor influencing individual growth rates in the iguanid lizard *Sceloporus merriami. Ecology* 59:770–778.

Evans, L. T. 1951. Field study of the social behavior of the black lizard, *Ctenosaura pectinata. Am. Mus. Novit.* 1943:1–26.

Falconer, D. S. 1960. *Introduction to quantitative genetics.* New York: Ronald Press.

Ferguson, G. W. 1966. Releasers of courtship and territorial behavior in the side-blotched lizard *Uta stansburiana. Anim. Behav.* 14:89–92.

——— 1971. Variation and evolution of the pushup displays of the side-blotched lizard genus *Uta* (Iguanidae). *Syst. Zool.* 29:79–101.

——— 1976. Color change and reproductive cycling in female collared lizards (*Crotaphytus collaris*). *Copeia* 1976:491–494.

——— 1977a. Display and communication in reptiles: an historical perspective. *Am. Zool.* 17:167–176.

——— 1977b. Variation and evolution of stereotyped behavior in reptiles. Part II. Social displays of reptiles: communications value, ultimate causes of variation, taxonomic significance, and heritability of population differences. In Carl Gans and D. W. Tinkle.

Ferguson, G. W., and C. H. Bohlen. 1978. Demographic analysis: a tool for the study of natural selection of behavioral traits. In Neil Greenberg and P. D. MacLean.

Ferguson, G. W., and T. Brockman. 1980. Geographic differences of growth rate of *Sceloporus* lizards (Sauria: Iguanidae). *Copeia* 1980:259–264.

Fisher, R. A. 1930. *The genetical theory of natural selection.* Oxford: Clarendon Press.

Fitch, A. V. 1964. Temperature tolerances of embryonic *Eumeces. Herpetologica* 20:184–187.

Fitch, H. S. 1970. Reproductive cycles in lizards and snakes. *Univ. Kans. Mus. Nat. Hist. Misc. Publ.* 52:1–247.

——— 1978. Sexual size differences in the genus *Sceloporus. Univ. Kans. Sci. Bull.* 51:441–461.

Fox, Wade, and H. C. Dessauer. 1957. Photoperiodic stimulation of appetite and growth in the male lizard, *Anolis carolinensis. J. Exp. Zool.* 134:557–575.

Gaffney, F. C., and L. C. Fitzpatrick. 1973. Energetics and lipid cycles in the lizard, *Cnemidophorus tigris. Copeia* 1973:446–452.

Goldberg, S. R. 1977. Reproduction in a mountain population of the side-blotched lizard, *Uta stansburiana* (Reptilia, Lacertilia, Iguanidae). *J. Herpetol.* 11:31–35.

Gorman, G. C. 1969. Intermediate territorial display of a hybrid *Anolis* lizard (Sauria, Iguanidae). *Z. Tierpsychol.* 26:390–393.

Gorman, G. C., and Paul Licht. 1974. Seasonality in ovarian cycles among tropical *Anolis* lizards. *Ecology* 55:360–369.

Gould, S. J., and R. C. Lewontin. 1979. The spandrels of San Marco and the Panglossian paradigm: a critique of the adaptationist programme. *Proc. R. Soc. Lond.* (B) 205:581–598.

Guillette, L. J., Jr. 1981. On the occurrence of oviparous and viviparous forms of the Mexican lizard *Sceloporus aeneus. Herpetologica* 37:11–15.

Guillette, L. J., Jr., R. E. Jones, K. T. Fitzgerald, and H. M. Smith. 1980. Evolution of viviparity in the lizard genus *Sceloporus. Herpetologica* 36:201–215.

Gunther, W. C. 1964. *Analysis of variance.* Englewood Cliffs, N.J.: Prentice-Hall.

Hahn, W. E., and D. W. Tinkle. 1965. Fat-body cycling and experimental evidence for its adaptive significance to ovarian follicle development in the lizard *Uta stansburiana. J. Exp. Zool.* 158:79–86.

Heatwole, Harold. 1976. *Reptile ecology.* St. Lucia: University of Queensland Press.

Heatwole, Harold, and O. J. Sexton. 1966. Herpetofaunal comparisons between two climatic zones in Panama. *Am. Midl. Nat.* 75:45–60.

Hellmich, W. G. 1951. On ecotypic and autotypic characters, a contribution to the knowledge of the evolution of the genus *Liolaemus* (Iguanidae). *Evolution* 5:359–369.

Hipp, T. G. 1977. Reproductive cycle and correlated hematological characteristics in *Crotaphytus collaris* in west central Texas. Master's thesis, Angelo State University, San Angelo, Texas.

Hoddenbach, G. A., and F. B. Turner. 1968. Clutch size of the lizard *Uta stansburiana* in southern Nevada. *Am. Midl. Nat.* 66:262–265.

Huey, R. B. 1974. Behavior thermoregulation in lizards: importance of associated costs. *Science* (Wash., D.C.) 184:1001–3.

——— 1977. Egg retention in some high altitude *Anolis* lizards. *Copeia* 1977:373–375.

Huey, R. B., and Montgomery Slatkin. 1976. Costs and benefits of lizard thermoregulation. *Q. Rev. Biol.* 51:363–384.

Huey, R. B., and R. D. Stevenson. 1979. Integrating thermal physiology and ecology of ectotherms: a discussion of approaches. *Am. Zool.* 19:357–366.

Hunsaker, Don II. 1959. Birth and litter sizes of the blue spiny lizard *Sceloporus cyanogenys*. *Copeia* 1959:260–261.

——— 1962. Ethological isolating mechanisms in the *Sceloporus torquatus* group of lizards. *Evolution* 16:62–74.

Hurtubia, Jaime, and Francesco di Castri. 1973. Segregation of lizard niches in the Mediterranean region of Chile. In *Mediterranean type ecosystems: origin and structure,* ed. F. di Castri and H. A. Mooney. New York: Springer-Verlag.

Inger, R. F., and Bernard Greenberg. 1966. Annual reproductive patterns of lizards from a Bornean rain forest. *Ecology* 47:1007–21.

Iverson, J. B. 1979. Behavior and ecology of the rock iguana *Cyclura carinata. Bull. Fla. State Mus. Biol. Sci.* 24:175–358.

Jaksić, F. M., and Herman Núñez. 1979. Escaping behavior and morphological correlates in two *Liolaemus* species of central Chile (Lacertilia: Iguanidae). *Oecologia* (Berl.) 42:119–122.

Jaksić, F. M., Herman Núñez, and F. P. Ojeda. 1980. Body proportions, microhabitat selection, and adaptive radiation of *Liolaemus* lizards in central Chile. *Oecologia* (Berl.) 45:178–181.

Jenssen, T. A. 1971. Display analysis of *Anolis nebulosus* (Sauria, Iguanidae). *Copeia* 1971:197–209.

Kay, F. R., B. W. Miller, and C. L. Miller. 1970. Food habits and reproduction of *Callisaurus draconoides* in Death Valley, California. *Herpetologica* 26:431–436.

Kluge, A. G. 1967. Higher taxonomic categories of gekkonid lizards and their evolution. *Bull. Am. Mus. Nat. Hist.* 135:1–59.

Lack, David. 1947. The significance of clutch size. Parts I–II. *Ibis* 89:302–352.

——— 1950. Family-size in titmice of the genus *Parus*. *Evolution* 4:279–290.

——— 1965. Evolutionary ecology. *J. Anim. Ecol.* 34:223–231.

——— 1968. *Ecological adaptations for breeding birds.* London: Methuen.

Licht, Paul. 1967. Environmental control of annual testicular cycles in the lizard *Anolis carolinensis*. II. Seasonal variations in the effects of photoperiod and temperature on testicular recrudescence. *J. Exp. Zool.* 166:243–253.

——— 1969. Illumination thresholds and spectral sensitivity of photosexual responses in the male lizard (*Anolis carolinensis*). *Comp. Biochem. Physiol.* 30:233–246.

——— 1971. Regulation of the annual testis cycle by photoperiod and temperature in the lizard *Anolis carolinensis. Ecology* 52:240–252.

——— 1973a. Influence of temperature and photoperiod on the annual ovarian cycle in the lizard *Anolis carolinensis. Copeia* 1973:465–472.

——— 1973b. Environmental influences on the testis cycles of the lizards *Dipsosaurus dorsalis* and *Xantusia vigilis. Comp. Biochem. Physiol.* 45A:7–20.

——— 1974. Response of *Anolis* lizards to food supplementation in nature. *Copeia* 1974:215–221.

Licht, Paul, and G. C. Gorman. 1970. Reproductive and fat cycles in Caribbean *Anolis* lizards. *Univ. Calif. Publ. Zool.* 95:1–52.

Licht, Paul, and W. R. Moberly. 1965. Thermal requirements for embryonic development in the tropical lizard *Iguana iguana. Copeia* 1965:515–517.

Licht, Paul, H. E. Hoyer, and P. G. W. J. van Dordt. 1969. Influence of photoperiod and temperature on testicular recrudescence and body growth in the lizards, *Lacerta sicula* and *Lacerta muralis. J. Zool.* (Lond.) 157:469–501.

Lundelius, E. L. 1957. Skeletal adaptations in two species of *Sceloporus. Evolution* 11:65–83.

MacArthur, R. H., and E. O. Wilson. 1967. *The theory of island biogeography.* Princeton: Princeton University Press.

Marion, K. R. 1970. Temperature as the reproductive cue for the female fence lizard *Sceloporus undulatus. Copeia* 1970:562–564.

Marshall, A. J., and Raymond Hook. 1960. The breeding biology of equatorial vertebrates: reproduction of the lizard *Agama agama lionotus* Boulenger at Lat. 0°01′N. *Proc. Zool. Soc. Lond.* 134:197–205.

Martin, R. F. 1973. Reproduction in the tree lizard (*Urosaurus ornatus*) in central Texas: drought conditions. *Herpetologica* 29:27–32.

——— 1977. Variation in reproductive productivity of range margin tree lizards (*Urosaurus ornatus*). *Copeia* 1977:83–92.

Maslin, T. P. 1971. Parthenogenesis in reptiles. *Am. Zool.* 11:361–380.

Mayhew, W. W. 1961. Photoperiodic response of female fringe-toed lizards. *Science* (Wash., D.C.) 134:2104–5.

——— 1963. Reproduction in the granite spiny lizard, *Sceloporus orcutti. Copeia* 1963:144–152.

——— 1964. Photoperiodic responses in three species of the lizard genus *Uma. Herpetologica* 20:95–113.

——— 1965. Growth response to photoperiodic stimulation in the lizard *Dipsosaurus dorsalis. Comp. Biochem. Physiol.* 14:209–216.

——— 1967. Comparative reproduction in three species of the genus *Uma.* In W. W. Milstead.

Mayr, Ernst. 1961. Cause and effect in biology. *Science* (Wash., D.C.) 134:1501–6.

McCoy, C. J., and G. A. Hoddenbach. 1966. Geographic variation in ovarian cycles and clutch size in *Cnemidophorus tigris* (Teiidae). *Science* (Wash., D.C.) 154:1671–72.

McKinney, C. O. 1971a. Individual and intrapopulation variations in the pushup display of *Uta stansburiana. Copeia* 1971:159–160.

——— 1971b. An analysis of zones of intergradation in the side-blotched lizard, *Uta stansburiana.* (Sauria: Iguanidae). *Copeia* 1971:596–613.

Mellish, C. H. 1936. The effects of anterior pituitary extract and certain environmental conditions on the genital system of the horned lizard (*Phrynosoma cornutum* Harlan). *Anat. Rec.* 67:23–33.

Michael, E. D. 1972. Growth rates in *Anolis carolinensis. Copeia* 1972:575–577.

Michel, Larry. 1976. Reproduction in a southwest New Mexican population of *Urosaurus ornatus. Southwest. Nat.* 21:281–299.

Miller, M. R. 1948. The seasonal histological changes occurring in the ovary, corpus luteum, and testis of the viviparous lizard, *Xantusia vigilis. Univ. Calif. Publ. Zool.* 47:197–224.

Mittleman, M. B. 1942. A summary of the iguanid genus *Urosaurus. Bull. Mus. Comp. Zool.* 91:103–181.

Neill, W. T. 1964. Viviparity in snakes: some ecological and zoogeographical considerations. *Am. Nat.* 98:35–55.

Newlin, M. E. 1976. Reproduction in the bunch grass lizard, *Sceloporus scalaris. Herpetologica* 32:171–184.

Orians, G. H. 1962. Natural selection and ecological theory. *Am. Nat.* 96:257–263.

Packard, G. C. 1966. The influence of ambient temperature and aridity on modes of reproduction and excretion of amniote vertebrates. *Am. Nat.* 100:667–682.

Packard, G. C., C. R. Tracy, and J. J. Roth. 1977. The physiological ecology of reptilian eggs and embryos, and the evolution of viviparity within the class Reptilia. *Bio. Rev.* 52:71–105.

Parker, W. S., and E. R. Pianka. 1973. Notes on the ecology of the iguanid lizard, *Sceloporus magister. Herpetologica* 29:143–152.

——— 1975. Comparative ecology of populations of the lizard *Uta stansburiana. Copeia* 1975:615–632.

Pianka, E. R. 1970a. On r- and K-selection. *Am. Nat.* 104:592–597.

——— 1970b. Comparative autecology of the lizard *Cnemidophorus tigris* in different parts of its geographic range. *Ecology* 51:703–720.

Pianka, E. R., and W. S. Parker. 1975. Ecology of horned lizards: a review with special reference to *Phrynosoma platyrhinos. Copeia* 1975:141–162.

Rand, A. S., and E. E. Williams. 1970. An estimation of redundancy and information content of anole dewlaps. *Am. Nat.* 104:99–103.

Ruby, Douglas. 1977. The function of shudder displays in the lizard *Sceloporus jarrovi. Copeia* 1977:110–114.

Ruibal, Rodolfo. 1967. Evolution and behavior of West Indian anoles. In W. W. Milstead.

Ruibal, Rodolfo, Richard Philibosian, and J. L. Adkins. 1972. Reproductive cycle and growth in the lizard *Anolis acutus. Copeia* 1972:509–518.

Sage, R. D. 1973. Ecological convergence of the lizard faunas of the chaparral communities in Chile and California. In *Mediterranean type ecosystems: origin and structure,* ed. H. A. Mooney. New York: Springer-Verlag.

Scheffé, Henry. 1959. *The analysis of variance.* New York: John Wiley.

Sergeev, A. M. 1940. Researches on the viviparity of reptiles. *135th Anniv. Publ. Moscow Soc. Nat.,* pp. 1–34. Cited in Tinkle and Gibbons, 1977.

Sexton, O. J., and K. R. Marion. 1974. Duration of incubation of *Sceloporus undulatus* eggs at constant temperature. *Physiol. Zool.* 47:91–98.

Sexton, O. J., and Olga Turner. 1971. The reproductive cycle of a neotropical lizard. *Ecology* 52:159–164.

Sexton, O. J., Joan Bauman, and Edward Ortleb. 1972. Seasonal food habits of *Anolis limifrons. Ecology* 53:182–186.

Sexton, O. J., E. P. Ortleb, L. M. Hathaway, R. E. Ballinger, and Paul Licht. 1971. Reproductive cycles of three species of anoline lizards from the isthmus of Panama. *Ecology* 52:201–215.

Shine, Richard. 1980. Costs of reproduction in reptiles. *Oecologia* (Berl.) 46:92–100.

Simon, Carol. 1975. The influence of food abundance on territory size in the iguanid lizard *Sceloporus jarrovi. Ecology* 56:993–998.

Smith, D. C. 1977. Interspecific competition and the demography of two lizards. Ph.D. dissertation, University of Michigan.

Smith, H. M., G. Sinelnik, J. D. Fawcett, and R. E. Jones. 1973. A survey of the chronology of ovulation in anoline lizard genera. *Trans. Kans. Acad. Sci.* 75:107–120.

Smith, R. E. 1968. Experimental evidence for a gonadal-fat body relationship in two teiid lizards (*Ameiva festiva, Ameiva quadrilineata*). *Biol. Bull.* 134:325–331.

Stamps, J. A. 1976. Egg retention, rainfall, and egg laying in a tropical lizard *Anolis aeneus. Copeia* 1976:759–764.

——— 1977. The relationship between resource competition, risk, and aggression in a tropical territorial lizard. *Ecology* 58:349–358.

Stamps, J. A., and D. P. Crews. 1976. Seasonal changes in reproduction and social behavior in the lizard *Anolis aeneus. Copeia* 1976:467–476.

Stearns, S. C. 1976. Life-history tactics: a review of the ideas. *Q. Rev. Biol.* 51:3–47.

——— 1977. The evolution of life-history traits: a critique of the theory and a review of the data. *Annu. Rev. Ecol. Syst.* 8:145–171.

——— 1980. A new view of life-history evolution. *Oikos* 35:266–281.

——— 1982. The role of development in the evolution of life histories. In *Evolution and development,* ed. J. T. Bonner. New York: Springer-Verlag.

Templeton, J. R. 1970. Reptiles. In *Comparative physiology of thermoregulation,* vol. 1, ed. G. C. Whittow. New York: Academic Press.

Tinkle, D. W. 1967. The life and demography of the side-blotched lizard, *Uta stansburiana. Misc. Publ. Mus. Zool., Univ. Mich.* 132:1–182.

——— 1969. The concept of reproductive effort and its relation to the evolution of life histories of lizards. *Am. Nat.* 103:501–516.

——— 1976. Comparative data on the population ecology of the desert spiny lizard, *Sceloporus magister. Herpetologica* 32:1–6.

Tinkle, D. W., and R. E. Ballinger. 1972. *Sceloporus undulatus:* a study of the intraspecific comparative demography of a lizard. *Ecology* 53:570–584.

Tinkle, D. W., and J. W. Gibbons. 1977. The distribution and evolution of viviparity in reptiles. *Misc. Pub. Mus. Zool., Univ. Mich.* 154:1–55.

Tinkle, D. W., and L. N. Irwin. 1965. Lizard reproduction: refractory period and response to warmth in *Uta stansburiana* females. *Science* (Wash., D.C.) 148:1613–14.

Tinkle, D. W., H. M. Wilbur, and S. G. Tilley. 1970. Evolutionary strategies in lizard reproduction. *Evolution* 24:55–74.
Turner, F. B. 1977. The dynamics of populations of squamates, crocodilians and rhynchocephalians. In Carl Gans and D. W. Tinkle.
Turner, F. B., G. A. Hoddenbach, P. A. Medica, and J. R. Lannom. 1970. The demography of the lizard, *Uta stansburiana* Baird and Girard, in southern Nevada. *J. Anim. Ecol.* 39:505–519.
Vinegar, M. B. 1975. Demography of the striped plateau lizard, *Sceloporus virgatus. Ecology* 56:172–175.
Vitt, L. J. 1981. Lizard reproduction: habitat specificity and constraints on relative clutch mass. *Am. Nat.* 117:506–514.
Vitt, L. J., and J. D. Congdon. 1978. Body shape, reproductive effort, and relative clutch mass in lizards: resolution of a paradox. *Am. Nat.* 112:595–608.
Weekes, H. C. 1935. A review of placentation among reptiles with particular regard to the function and evolution of the placenta. *Proc. Zool. Soc. Lond.* 1935:625–645.
Werler, J. E. 1949. Reproduction of captive Texas and Mexican lizards. *Herpetologica* 5:67–70.
Wilbur, H. M., D. W. Tinkle, and J. P. Collins. 1974. Environmental certainty, trophic level, and resource availability in life-history evolution. *Am. Nat.* 108:805–817.
Williams, G. C. 1966. *Adaptation and natural selection.* Princeton: Princeton University Press.

12. Realized Niche Overlap, Resource Abundance, and Intensity of Interspecific Competition

Abrams, Peter. 1976. Limiting similarity and the form of the competition coefficient. *Theor. Popul. Biol.* 8:356–375.
——— 1980. Some comments on measuring niche overlap. *Ecology* 61:44–49.
Brown, J. H., and D. W. Davidson. 1977. Competition between seed-eating rodents and ants in desert ecosystems. *Science* (Wash., D.C.) 196:880–882.
Clover, R. C. 1975. Morphological variations in populations of *Lacerta* from islands in the Adriatic Sea. Ph.D. dissertation, Oregon State University, Corvallis.
Colwell, R. K., and D. J. Futuyma. 1971. On the measurement of niche breadth and overlap. *Ecology* 52:567–576.
Connell, J. H. 1974. Field experiments in marine ecology. In *Experimental marine biology,* ed. R. Mariscal. New York: Academic Press.
——— 1975. Some mechanisms producing structure in natural communities: a model and evidence from field experiments. In *Ecology and evolution of communities,* ed. M. L. Cody and J. M. Diamond. Cambridge, Mass.: Harvard University Press.
Dayton, P. K. 1971. Competition, disturbance, and community organization: the provision and subsequent utilization of space in a rocky intertidal community. *Ecol. Monogr.* 41:351–389.
Diamond, J. M. 1978. Niche shifts and the rediscovery of interspecific competition. *Am. Sci.* 66:322–331.
Dunham, A. E. 1978. Food availability as a proximate factor influencing indi-

vidual growth rates in the iguanid lizard *Sceloporus merriami. Ecology* 59:770–778.

——— 1980. An experimental study of interspecific competition between the iguanid lizards *Sceloporus merriami* and *Urosaurus ornatus. Ecol. Monogr.* 50:309–330.

——— 1981. Populations in a fluctuating environment: the comparative population ecology of *Sceloporus merriami* and *Urosaurus ornatus. Misc. Publ. Mus. Zool., Univ. Mich.* 158:1–62.

Grant, P. R. 1969. Experimental studies of competitive interaction in a two-species system. I. *Microtus* and *Clethrionomys* species in enclosures. *Can. J. Zool.* 47:1059–82.

——— 1970. Experimental studies of competitive interaction in a two-species system. II. The behavior of *Microtus, Peromyscus,* and *Clethrionomys* species. *Anim. Behav.* 18:411–426.

——— 1971. Experimental studies of competitive interaction in a two-species system. III. *Microtus* and *Peromyscus* species in enclosures. *J. Anim. Ecol.* 40:323–350.

——— 1972. Interspecific competition between rodents. *Annu. Rev. Ecol. Syst.* 3:79–106.

Hairston, N. G. 1980. The experimental test of an analysis of field distributions: competition in terrestrial salamanders. *Ecology* 61:817–826.

Hall, J. J., W. E. Cooper, and E. E. Werner. 1970. An experimental approach to the production dynamics and structure of freshwater animal communities. *Limnol. Oceanogr.* 15:839–928.

Horn, H. S. 1966. Measurement of "overlap" in comparative ecological studies. *Am. Nat.* 100:419–424.

Huey, R. B. and E. R. Pianka. 1974. Ecological character displacement in a lizard. *Am. Zool.* 14:1127–36.

——— 1977. Patterns of niche overlap among broadly sympatric versus narrowly sympatric Kalahari lizards (Scincidae: Mabuya). *Ecology* 58:119–128.

Huey, R. B., E. R. Pianka, M. E. Egan, and L. W. Coons. 1974. Ecological shifts in sympatry: Kalahari fossorial lizards (*Typhlosaurus*). *Ecology* 55:304–316.

Hurlbert, S. H. 1978. The measurement of niche overlap and some relatives. *Ecology* 59:67–77.

Hutchinson, G. E. 1957. Concluding remarks. *Cold Spring Harbor Symp. Quant. Biol.* 22:415–427.

——— 1975. Variations on a theme by Robert MacArthur. In *Ecology and evolution of communities,* ed. M. L. Cody and J. M. Diamond. Cambridge, Mass.: Harvard University Press.

Inger, R. F., and Bernard Greenberg. 1966. Ecological and competitive relations among three species of frogs (genus *Rana*). *Ecology* 47:746–759.

Jaeger, R. G. 1970. Potential extinction through competition between two species of terrestrial salamanders. *Evolution* 24:632–642.

——— 1971. Competitive exclusion as a factor influencing the distribution of two species of terrestrial salamanders. *Ecology* 52:632–637.

——— 1972. Food as a limited resource in competition between two species of terrestrial salamanders. *Ecology* 53:535–546.

Johnson, D. H. 1980. The comparison of usage and availability measurements for evaluating resource preference. *Ecology* 61:65–71.
Lack, David. 1946. Competition for food by birds of prey. *J. Anim. Ecol.* 15:123–129.
——— 1947. *Darwin's finches.* Cambridge: Cambridge University Press.
——— 1971. *Ecological isolation in birds.* Cambridge, Mass.: Harvard University Press.
Lawlor, L. R. 1980. Overlap, similarity, and competition coefficients. *Ecology* 61:245–251.
Levine, S. H. 1976. Competitive interactions in ecosystems. *Am. Nat.* 110:903–910.
Levins, Richard. 1968. *Evolution in changing environments.* Princeton: Princeton University Press.
Lister, B. C. 1976a. The nature of niche expansion in West Indian *Anolis* lizards. I. Ecological consequences of reduced competition. *Evolution* 30:659–676.
——— 1976b. The nature of niche expansion in West Indian *Anolis* lizards. II. Evolutionary components. *Evolution* 30:677–692.
MacArthur, R. H., and Richard Levins. 1967. The limiting similarity, convergence, and divergence of coexisting species. *Am. Nat.* 101:377–385.
MacArthur, R. H., and E. R. Pianka. 1966. On optimal use of a patchy environment. *Am. Nat.* 100:603–609.
May, R. M. 1973. *Stability and complexity in model systems.* Princeton: Princeton University Press.
May, R. M., and R. H. MacArthur. 1972. Niche overlap as a function of environmental variability. *Proc. Nat. Acad. Sci. U.S.A.* 69:1109–13.
McClure, M. S., and P. W. Price. 1975. Competition among sympatric *Erythroneura* leafhoppers (Homoptera, Cicadellidae) on American sycamore. *Ecology* 56:1388–97.
Menge, B. A. 1972. Competition for food between two intertidal starfish species and its effects on body size and feeding. *Ecology* 53:635–644.
Milstead, W. W. 1957. Some aspects of competition in natural populations of whiptail lizards (genus *Cnemidophorus*). *Tex. J. Sci.* 9:410–447.
——— 1961. Competitive relations in lizard populations. In *Vertebrate speciation,* ed. W. F. Blair. Austin: University of Texas Press.
——— 1965. Changes in competing populations of whiptail lizards (*Cnemidophorus*) in southwestern Texas. *Am. Midl. Nat.* 73:75–80.
Morisita, M. 1959. Measuring of interspecific association and similarity between communities. *Mem. Fac. Sci. Kyushu Univ., Ser. E* (*Biol.*) 3:64–80.
Müller, C. H. 1966. The role of chemical inhibition (allelopathy) in vegetational composition. *Bull. Torrey Bot. Club* 93:332–351.
Nevo, Eviatar, G. C. Gorman, M. F. Soulé, S. Y. Yang, Robert Clover, and Vojislav Radovanović. 1972. Competitive exclusion between insular *Lacerta* species (Sauria, Lacertidae). *Oecologia* (Berl.) 10:183–190.
Petraitis, P. S. 1979. Likelihood measures of niche breadth and overlap. *Ecology* 60:703–710.
Pianka, E. R. 1969. Sympatry of desert lizards (*Ctenotus*) in Western Australia. *Ecology* 50:1012–30.

——— 1970. Comparative autecology of the lizard *Cnemidophorus tigris* in different parts of its geographic range. *Ecology* 51:703–720.

——— 1973. The structure of lizard communities. *Annu. Rev. Ecol. Syst.* 4:53–74.

——— 1974. Niche overlap and diffuse competition. *Proc. Nat. Acad. Sci. U.S.A.* 71:2141–45.

——— 1975. Niche relations of desert lizards. In *Ecology and evolution of communities,* ed. M. L. Cody and J. M. Diamond. Cambridge, Mass.: Harvard University Press.

——— 1976. Competition and niche theory. In *Theoretical ecology: principles and applications,* ed R. May. Philadelphia: Saunders.

Pianka, E. R., R. B. Huey, and L. R. Lawlor. 1979. Niche segregation in desert lizards. In *Analysis of ecological systems,* ed. D. J. Horn, R. Mitchell, and G. R. Stairs. Columbus: Ohio State University Press.

Pielou, E. C. 1972. Niche width and overlap: a method for measuring them. *Ecology* 53:687–692.

——— 1975. *Ecological diversity.* New York: John Wiley.

Pyke, G. H., H. R. Pulliam, and E. L. Charnov. 1977. Optimal foraging: a selective review of theory and tests. *Q. Rev. Biol.* 52:137–154.

Reynoldson, T. B. 1964. Evidence for intraspecific competition in field population of triclads. *J. Anim. Ecol.* 33:187–201.

Reynoldson, T. B., and L. F. Bellamy. 1971. The establishment of interspecific competition in field populations with an example of competition action between *Polycelis nigra* (Mull.) and *P. tenuis* (Ijima) (Turbellaria, Tricladida). In *Dynamics of populations,* ed. P. J. den Boer and G. Gradwell. Wageningen, The Netherlands: Centre for Agricultural Publishing and Documentation.

Roughgarden, Jonathan. 1972. The evolution of niche width. *Am. Nat.* 106:683–718.

——— 1974a. Species packing and the competition function with illustrations from coral reef fish. *Theoret. Popul. Biol.* 5:163–186.

——— 1974b. Niche width: biogeographic patterns among *Anolis* lizards. *Am. Nat.* 108:429–442.

Sale, P. R. 1974. Overlap in resource use, and interspecific competition. *Oecologica* (Berl.) 17:245–256.

Schoener, T. W. 1968. The *Anolis* lizards of Bimini: resource partitioning in a complex fauna. *Ecology* 49:704–726.

——— 1969. Size patterns in West Indian *Anolis* lizards. I. Size and species diversity. *Syst. Zool.* 18:386–401.

——— 1970. Nonsynchronous spatial overlap of lizards in patchy habitats. *Ecology* 51:408–418.

——— 1974a. Resource partitioning in ecological communities. *Science* (Wash., D.C.) 185:27–39.

——— 1974b. Some methods of calculating competition equations from resource utilization spectra. *Am. Nat.* 108:332–340.

——— 1975. Presence and absence of habitat shift in some widespread lizard species. *Ecol. Monogr.* 45:233–258.

——— 1977. Competition and the niche. In Carl Gans and D. W. Tinkle.

Schoener, T. W., and G. Gorman. 1968. Some niche differences in three Lesser Antillean lizards of the genus *Anolis*. *Ecology* 49:819–830.

Schroder, G. D., and M. L. Rosenzweig. 1975. Perturbation analysis of competition in habitat utilization between *Dipodomys ordii* and *Dipodomys merriami*. *Oecologica* (Berl.) 19:9–28.

Selby, S. M. 1965. *Standard math tables,* 14th ed. Cleveland: Chemical Rubber Co.

Simpson, E. H. 1949. Measurement of diversity. *Nature* (Lond.) 163:688.

Smith, D. C. 1977. Interspecific competition and the demography of two lizards. Ph.D. dissertation, University of Michigan.

Smith, J. N. M., P. R. Grant, B. R. Grant, I. J. Abbott, and L. K. Abbott. 1978. Seasonal variation in feeding habits of Darwin's finches. *Ecology* 59:1137–50.

Svärdson, Gunnar. 1949. Competition and habitat selection in birds. *Oikos* 1:157–174.

Tinkle, D. W. 1982. Results of experimental density manipulation in an Arizona lizard community. *Ecology* 63:57–65.

Vandermeer, J. H. 1972. Niche theory. *Annu. Rev. Ecol. Syst.* 3:107–132.

Werner, E. E., and D. J. Hall. 1976. Niche shifts in sunfishes: experimental evidence and significance. *Science* (Wash., D.C.) 191:404–406.

Wiens, J. A. 1977. On competition and variable environments. *Am. Scien.* 65:590–597.

Wilbur, H. M. 1972. Competition, predation, and the structure of the *Ambystoma-Rana sylvatica* community. *Ecology* 53:3–21.

13. Temporal Separation of Activity and Interspecific Dietary Overlap

Baumgartner, A. M., and F. M. Baumgartner. 1944. Hawks and owls in Oklahoma 1939–1942: food habits and population changes. *Wilson Bull.* 56:209–215.

Bogert, C. M. 1949. Thermoregulation in reptiles—a factor in evolution. *Evolution* 3:195–211.

Case, T. J., and M. E. Gilpin. 1974. Interference competition and niche theory. *Proc. Nat. Acad. Sci. U.S.A.* 71:3073–77.

Cody, M. L. 1974. *Competition and the structure of bird communities.* Princeton: Princeton University Press.

Connor, E. F., and Daniel Simberloff. 1979. The assembly of species communities: chance or competition? *Ecology* 60:1132–40.

Craighead, J. J., and F. C. Craighead. 1956. *Hawks, owls and wildlife.* Harrisburg, Penn.: Stackpole Company and Wildlife Management Institute. (Reprinted 1969 by Dover Publications.)

Fitch, H. S. 1947. Predation by owls in the Sierran foothills of California. *Condor* 49:137–151.

Hirshfield, M. F., C. R. Feldmeth, and D. L. Soltz. 1980. Genetic differences in physiological tolerances of Amargosa pupfish (*Cyprinodon nevadensis*) populations. *Science* (Wash., D.C.) 207:999–1001.

Huey, R. B. 1982. Temperature, physiology, and the ecology of reptiles. In Carl Gans and F. H. Pough, vol. 12.

Huey, R. B., and E. R. Pianka. 1981. Ecological consequences of foraging mode. *Ecology* 62:991–999.
Huey, R. B., and Montgomery Slatkin. 1976. Costs and benefits of lizard thermoregulation. *Q. Rev. Biol.* 51:363–384.
Huey, R. B., E. R. Pianka, and J. A. Hoffman. 1977. Seasonal patterns of thermoregulatory behavior and body temperature of diurnal Kalahari lizards. *Ecology* 58:1066–75.
Inger, R. F., and R. K. Colwell. 1977. Organization of contiguous communities of amphibians and reptiles in Thailand. *Ecol. Monogr.* 47:229–253.
Jaksić, F. M. 1982. Inadequacy of activity time as a niche difference: the case of diurnal and nocturnal raptors. *Oecologia* (Berl.) 52:171–175.
Jaksić, F. M., H. W. Greene, and J. L. Yañez. 1981. The guild structure of a community of predatory vertebrates in central Chile. *Oecologia* (Berl.) 49:21–28.
Korschgen, L. J., and H. B. Stuart. 1972. Twenty years of avian predator-small mammal relationships in Missouri. *J. Wildl. Manage.* 36:269–282.
Lack, David. 1946. Competition for food by birds of prey. *J. Anim. Ecol.* 15:123–129.
Levins, Richard. 1968. *Evolution in changing environments.* Princeton: Princeton University Press.
Lewis, T., and L. R. Taylor. 1964. Diurnal periodicity of flight by insects. *Trans. R. Entomol. Soc. Lond.* 116:393–476.
MacArthur, R. H., and Richard Levins. 1967. The limiting similarity, convergence and divergence of coexisting species. *Am. Nat.* 101:377–385.
Meagher, T. R., and D. S. Burdick. 1980. The use of nearest neighbor frequency analyses in studies of associations. *Ecology* 61:1253–55.
Munro, J. A. 1929. Notes on the food habits of certain raptors in British Columbia and Alberta. *Condor* 31:112–116.
Mushinsky, H. R., and J. J. Hebrard. 1977a. The use of time by sympatric water snakes. *Can. J. Zool.* 55:1545–50.
——— 1977b. Food partitioning by five species of water snakes in Louisiana. *Herpetologica* 33:162–166.
Norris, K. S., and J. L. Kavanau. 1966. The burrowing of the western shovel-nosed snake, *Chionactis occipitalis,* and the undersand environment. *Copeia* 1966:650–664.
Orians, G. H., and F. Kuhlman. 1956. Red-tailed hawk and horned owl populations in Wisconsin. *Condor* 58:371–385.
Pianka, E. R. 1969a. Habitat specificity, speciation, and species density in Australian desert lizards. *Ecology* 50:498–502.
——— 1969b. Sympatry of desert lizards (*Ctenotus*) in Western Australia. *Ecology* 50:1012–30.
——— 1971. Lizard species density in the Kalahari desert. *Ecology* 52:1024–29.
——— 1973. The structure of lizard communities. *Annu. Rev. Ecol. Syst.* 4:53–74.
——— 1974. Niche overlap and diffuse competition. *Proc. Nat. Acad. Sci. U.S.A.* 71:2141–45.
Pianka, E. R., and R. B. Huey. 1978. Comparative ecology, resource utiliza-

tion, and niche segregation among gekkonid lizards in the southern Kalahari. *Copeia* 1978:691–701.
Pianka, E. R., R. B. Huey, and L. R. Lawlor. 1979. Niche segregation in desert lizards. In *Analysis of ecological systems,* ed. D. J. Horn, R. Mitchell, and G. R. Stairs. Columbus: Ohio State University Press.
Pimm, S. L. 1980. Properties of food webs. *Ecology* 61:219–225.
Pimm, S. L., and J. H. Lawton. 1981. Are food webs compartmented? *J. Anim. Ecol.* 49:879–898.
Porter, W. P., J. W. Mitchell, W. A. Beckman, and C. B. DeWitt. 1973. Behavioural implications of mechanistic ecology (thermal and behavioural modeling of desert ectotherms and their microenvironment). *Oecologia* (Berl.) 13:1–54.
Schoener, T. W. 1970. Nonsynchronous spatial overlap of lizards in patchy habitats. *Ecology* 51:408–418.
——— 1974. Resource partitioning in ecological communities. *Science* (Wash., D.C.) 185:27–39.
——— 1977. Competition and the niche. In Carl Gans and D. W. Tinkle.
——— 1982. Size differences among sympatric bird-eating hawks: a worldwide survey. In *Ecological communities: conceptual issues and the evidence,* ed. D. R. Strong, Jr., Daniel Simberloff, and L. G. Abele. Princeton: Princeton University Press.
Simpson, E. H. 1949. Measurement of diversity. *Nature* (Lond.) 163:688.
Wilson, D. S., and A. B. Clark. 1977. Above ground predator defence in the harvester termite, *Hodotermes mossambicus* (Hagen). *J. Entomol. Soc. South Afr.* 40:271–282.

14. Sympatry and Size Similarity in *Cnemidophorus*

Abbott, Ian, L. K. Abbott, and P. R. Grant. 1977. Comparative ecology of Galapagos ground finches (*Geospiza* Gould): evaluation of the importance of floristic diversity and interspecific competition. *Ecol. Monogr.* 47:151–184.
Brown, W. M., and J. W. Wright. 1979. Mitochondrial DNA analyses and the origin and relative age of parthenogenetic lizards (genus *Cnemidophorus*). *Science* (Wash., D. C.) 203:1247–49.
Case, T. J. 1975. Species number, density compensation, and colonizing ability of lizards on islands in the Gulf of California. *Ecology* 56:3–18.
——— 1979. Character displacement and coevolution in some *Cnemidophorus* lizards. *Fortschr. Zool.* 25:235–282.
Case, T. J., and Ron Sidell. 1982. Distinguishing pattern from chance in the structure of natural communities. *Evolution,* in press.
Caswell, Hal. 1976. Community structure: a neutral model analysis. *Ecol. Monogr.* 46:327–354.
Connor, E. F., and D. S. Simberloff. 1979. The assembly of species communities: chance or competition? *Ecology* 60:1132–40.
Cuellar, Orlando. 1979. On the ecology of coexistence in parthenogenetic and bisexual lizards of the genus *Cnemidophorus*. *Am. Zool.* 19:773–786.
Degenhardt, W. G. 1966. A method of counting some diurnal ground lizards of the genera *Holbrookia* and *Cnemidophorus* with results from the Big Bend National Park. *Am. Midl. Nat.* 75:61–100.

Dixon, J. R., and P. A. Medica. 1966. Summer food of four species of lizards from the vicinity of White Sands, New Mexico. *Los Ang. Cty. Mus. Contrib. Sci.* no. 121:1–6.

Grant, P. R., and Ian Abbott. 1980. Interspecific competition, island biogeography, and null hypotheses. *Evolution* 34:332–341.

Hastings, J. R., and R. M. Turner. 1969. Climatological data and statistics for Baja California. In *Meteorology and climatology of arid regions. Univ. Arizona Inst. Atmos. Phys. Tech. Rep.* 18.

Hastings, J. R., R. M. Turner, and D. K. Warren. 1972. An atlas of some plant distributions in the Sonoran Desert. In *Meteorology and climatology of arid regions. Univ. Arizona Inst. Atmos. Phys. Tech. Rep.* 21.

Johnson, C. R. 1969. Observations on northern California populations of *Cnemidophorus tigris* (Sauria, Teiidae). *Herpetologica* 25:316–318.

Johnson, D. H., M. D. Bryant, and A. H. Miller. 1948. Vertebrate animals of the Providence Mountains area of California. *Univ. Calif. Publ. Zool.* 48:221–376.

Lack, David. 1947. *Darwin's finches.* Cambridge: Cambridge University Press.

Lawlor, L. R. 1980. Structure and stability in natural and randomly constructed competitive communities. *Am. Nat.* 116:394–408.

MacArthur, R. H. 1972. *Geographical ecology: patterns in the distribution of species.* New York: Harper & Row.

Medica, P. A. 1967. Food habits, habitat preference, reproduction, and diurnal activity in four sympatric species of whiptail lizards (*Cnemidophorus*) in South Central New Mexico. *Bull. South. Calif. Acad. Sci.* 66:251–276.

Miller, A. H., and R. C. Stebbins. 1964. *The lives of desert animals in Joshua Tree National Monument.* Berkeley: University of California Press.

Milstead, W. W. 1957a. Some aspects of competition in natural populations of whiptail lizards (Genus *Cnemidophorus*). *Tex. J. Sci.* 9:410–447.

——— 1957b. Observations on the natural history of four species of whiptail lizard, *Cnemidophorus* (Sauria, Teiidae) in Trans-Pecos Texas. *Southwest. Nat.* 2:105–121.

Pianka, E. R., R. B. Huey, and L. R. Lawlor. 1979. Niche segregation in desert lizards. In *Analysis of ecological systems,* ed. D. J. Horn, R. Mitchell, and G. R. Stairs. Columbus: Ohio State University Press.

Poole, R. W., and Rathcke, B. J. 1979. Regularity, randomness, and aggregation in flowering phenologies. *Science* (Wash., D.C.) 203:470–471.

Root, R. B. 1967. The niche exploitation pattern of the blue-grey gnatcatcher. *Ecol. Monogr.* 37:317–350.

Sale, P. F. 1974. Overlap in resource use and interspecific competition. *Oecologia* (Berl.) 17:245–256.

Schall, J. J. 1976. Comparative ecology of sympatric parthenogenetic and bisexual species of *Cnemidophorus.* Ph.D. dissertation, University of Texas.

Shreve, Forest. 1951. Vegetation and flora of the Sonoran Desert. Vol. 1, Vegetation. *Carnegie Inst. Wash. Publ.* 591:1–192.

Simberloff, Daniel. 1978. Using island biogeographic distributions to determine if colonization is stochastic. *Am. Nat.* 112:713–726.

Strong, D. R., L. A. Szyska, and D. S. Simberloff. 1979. Tests of community-wide character displacement against null hypotheses. *Evolution* 33:897–913.

Wright, J. W. 1968. Variation in three sympatric sibling species of whiptail lizards, genus *Cnemidophorus. J. Herpetol.* 1:1–19.
Wright, J. W., and C. H. Lowe. 1968. Weeds, polyploids, parthenogenesis, and the geographical and ecological distribution of all-female species of *Cnemidophorus. Copeia* 1968:129–138.

15. Ecomorphs, Faunas, Island Size, and Diverse End Points in Island Radiations of *Anolis*

Amadon, Dean. 1966. The superspecies concept. *Syst. Zool.* 15:246–249.
Arnold, D. L. 1980. Geographic variation in *Anolis brevirostris* (Sauria: Iguanidae) in Hispaniola. *Breviora Mus. Comp. Zool.* 461:1–31.
Cochran, D. M. 1935. New reptiles and amphibians collected in Haiti by P. J. Darlington. *Proc. Boston Nat. Hist. Soc.* 40:367–376.
Diamond, J. M. 1975. Assembly of species communities. In *Ecology and evolution of communities,* ed. M. L. Cody and J. M. Diamond. Cambridge, Mass.: Harvard University Press.
——— 1977. Continental and insular speciation in Pacific land birds. *Syst. Zool.* 26:263–268.
Etheridge, R. E. 1960. The relationships of the anoles (Reptilia: Sauria; Iguanidae): an interpretation based on skeletal morphology. University Microfilms, Ann Arbor, Michigan.
Fretwell, S. D. 1981. Reply to Sih and Dixon. *Amer. Nat.* 117:560.
Gorman, G. C., D. G. Buth, and J. S. Wyles. 1980. *Anolis* lizards of the eastern Caribbean: a case study of evolution. III. A cladistic analysis of albumin immunological data and the definition of species groups. *Syst. Zool.* 29:143–158.
Hertz, P. E. 1976. *Anolis alumina,* new species of grass anole from the Barahona Peninsula of Hispaniola. *Breviora Mus. Comp. Zool.* 437:1–19.
Hicks, Robert. 1973. New studies on a montane lizard of Jamaica. *Breviora Mus. Comp. Zool.* 404:1–23.
Huey, R. B., and T. P. Webster. 1976. Thermal biology of *Anolis* lizards in a complex fauna: the *cristatellus* group of Puerto Rico. *Ecology* 57:985–994.
Jenssen, T. A. 1973. Shift in the structural habitat of *Anolis opalinus* due to congeneric competition. *Ecology* 54:863–869.
——— 1981. Unusual display behavior by *Anolis grahami* from western Jamaica. *Copeia* 1981:728–733.
Karr, J. R., and F. C. James. 1975. Eco-morphological configurations and convergent evolution in species and communities. In *Ecology and evolution of communities,* ed. M. L. Cody and J. M. Diamond. Cambridge, Mass: Harvard University Press.
Lack, David. 1947. *Darwin's finches.* Cambridge: Cambridge University Press.
——— 1976. *Island biology illustrated by the land birds of Jamaica.* Berkeley: University of California Press.
Lazell, J. D. Jr. 1962. The anoles of the eastern Caribbean (Sauria, Iguanidae). V. Geographic differentiation in *Anolis oculatus* on Dominica. *Bull. Mus. Comp. Zool.* 127:466–476.
——— 1964. The anoles (Sauria, Iguanidae) of the Guadeloupéen Archipelago. *Bull. Mus. Comp. Zool.* 131:359–401.

——— 1966. Contributions to the herpetology of Jamaica: studies on *Anolis reconditus* Underwood and Williams. *Bull. Inst. Jamaica, Sci. Ser.* 18:1–15.

——— 1972. The anoles (Sauria: Iguanidae) of the Lesser Antilles. *Bull. Mus. Comp. Zool.* 143:1–115.

Lister, B. C. 1976. The nature of niche expansion in West Indian *Anolis* lizards. I. Ecological consequences of reduced competition. *Evolution* 30:659–676.

MacArthur, R. H., and E. O. Wilson. 1967. *The theory of island biogeography.* Princeton: Princeton University Press.

Mayr, Ernst. 1963. *Animal species and evolution.* Cambridge, Mass.: Harvard University Press.

Mayr, Ernst, and L. L. Short. 1970. Species taxa of North American birds: a contribution to comparative systematics. *Publ. Nuttall Ornithological Club* 9:1–127.

Moermond, T. C. 1979. Habitat constraints on the behavior, morphology, and community structure of *Anolis* lizards. *Ecology* 60:152–164.

——— 1981. Prey-attack behavior of *Anolis* lizards. *Z. Tierpsychol.* 56:128–136.

Rand, A. S. 1964. Ecological distribution in anoline lizards of Puerto Rico. *Ecology* 45:745–752.

——— 1967. The ecological distribution of the anoline lizards around Kingston, Jamaica. *Breviora Mus. Comp. Zool.* 272:1–18.

——— 1969. Competitive exclusion among anoles (Sauria: Iguanidae) on small islands in the West Indies. *Breviora Mus. Comp. Zool.* 319:1–16.

Rand, A. S., and E. E. Williams. 1969. The anoles of La Palma: aspects of their ecological relationships. *Breviora Mus. Comp. Zool.* 327:1–19.

Roughgarden, Jonathan. 1974. Niche width: biogeographic patterns among *Anolis* lizard populations. *Am. Nat.* 108:429–442.

Ruibal, Rodolfo. 1964. An annotated checklist and key to the anoline lizards of Cuba. *Bull. Mus. Comp. Zool.* 130:473–520.

Schoener, T. W. 1968. The *Anolis* lizards of Bimini: resource partitioning in a complex fauna. *Ecology* 49:704–726.

——— 1970. Nonsynchronous spatial overlap of lizards in patchy habitats. *Ecology* 51:408–418.

Schoener, T. W., and G. C. Gorman. 1968. Some niche differences in three Lesser Antillean lizards of the genus *Anolis. Ecology* 49:819–830.

Schoener, T. W., and Amy Schoener. 1971a. Structural habitats of West Indian *Anolis* lizards. I. Lowland Jamaica. *Breviora Mus. Comp. Zool.* 368: 1–53.

——— 1971b. Structural habitats of West Indian *Anolis* lizards. II. Puerto Rican uplands. *Breviora Mus. Comp. Zool.* 375:1–39.

Schwartz, Albert. 1968. Geographic variation in *Anolis distichus* Cope (Lacertilia, Iguanidae) in the Bahama Islands and Hispaniola. *Bull. Mus. Comp. Zool.* 137:255–310.

——— 1973. A new species of montane anole (Sauria, Iguanidae) from Hispaniola. *Ann. Carnegie Mus.* 44:183–195.

——— 1974a. A new species of primitive *Anolis* (Sauria, Iguanidae) from the Sierra de Baoruco, Hispaniola. *Breviora Mus. Comp. Zool.* 423:1–19.

——— 1974b. An analysis of variation in the Hispaniolan giant anole, *Anolis ricordii* Dumeril and Bibron. *Bull. Mus. Comp. Zool.* 146:86–146.

——— 1977. The Hispaniolan *Anolis* (Reptilia, Lacertilia, Iguanidae) of the *hendersoni* complex. *J. Herpetol.* 12:355–370.
——— 1978. A new species of aquatic *Anolis* (Sauria, Iguanidae) from Hispaniola. *Ann. Carnegie Mus.* 47:261–279.
Schwartz, Albert, and Richard Thomas. 1975. A checklist of West Indian amphibians and reptiles. *Spec. Publ. Carnegie Mus. Nat. Hist.* 1:1–216.
——— 1977. Two new species of *Sphaerodactylus* (Reptilia, Lacertilia, Gekkonidae) from Hispaniola. *J. Herpetol.* 11:61–66.
Schwartz, Albert, Richard Thomas, and L. D. Ober. 1978. First supplement to a checklist of West Indian amphibians and reptiles. *Spec. Publ. Carnegie Mus. Nat. Hist.* 5:1–35.
Shochat, Daniel, and H. C. Dessauer. 1981. Comparative immunological study of albumins of *Anolis* lizards of the Caribbean islands. *Comp. Biochem. Physiol.* 68A:67–73.
Sih, Andrew, and John Dixon. 1981. Tests of some predictions from the MacArthur-Levins competition models: a critique. *Am. Nat.* 117:550–559.
Thomas, Richard, and Albert Schwartz. 1967. The *monticola* group of the lizard genus *Anolis* in Hispaniola. *Breviora Mus. Comp. Zool.* 261:1–27.
Thomas, Richard, and K. R. Thomas. 1977. Distributional records of amphibians and reptiles from Puerto Rico. *Herpetol. Rev.* 8(2):1–40.
Turesson, G. W. 1922. The genotypic response of the plant species to the habitat. *Hereditas* 3:211–350.
Underwood, Garth, and E. E. Williams. 1959. The anoline lizards of Jamaica. *Bull. Inst. Jamaica, Sci. Ser.* 9:1–48.
Vanzolini, P. E., and E. E. Williams. 1981. The vanishing refuge: a mechanism for ecogeographic speciation. *Papeis Avuls. Zool. S. Paulo* 34:251–255.
Webster, T. P., W. P. Hall, and E. E. Williams. 1972. Fission in the evolution of a lizard karyotype. *Science* (Wash., D.C.) 177:611–613.
Williams, E. E. 1961. Notes on Hispaniolan herpetology. 3. The evolution and relationships of the *Anolis semilineatus* group. *Breviora Mus. Comp. Zool.* 136:1–8.
——— 1963a. Notes on Hispaniolan herpetology. 8. The forms related to *Anolis hendersoni* Cochran. *Breviora Mus. Comp. Zool.* 186:1–13.
——— 1963b. *Anolis whitemani,* new species from Hispaniola (Sauria, Iguanidae). *Breviora Mus. Comp. Zool.* 197:1–8.
——— 1965. The species of Hispaniolan green anoles (Sauria, Iguanidae). *Breviora Mus. Comp. Zool.* 227:1–16.
——— 1969. The ecology of colonization as seen in the zoogeography of anoline lizards on small islands. *Q. Rev. Biol.* 44:345–389.
——— 1972. The origin of faunas: evolution of lizard congeners in a complex island fauna—a trial analysis. *Evolutionary Biology* 6:47–89.
——— 1975. *Anolis marcanoi,* new species, sibling to *Anolis cybotes:* description and field evidence. *Breviora Mus. Comp. Zool.* 430:1–9.
——— 1976. West Indian anoles: a taxonomic and evolutionary summary. I. Introduction and a species list. *Breviora Mus. Comp. Zool.* 440:1–21.
——— 1977. Anoles out of place. *Third Anolis Newsletter:* 110–118.
Williams, E. E., and A. S. Rand. 1969. *Anolis insolitus,* a new dwarf anole of zoogeographic importance from the mountains of the Dominican Republic. *Breviora Mus. Comp. Zool.* 326:1–21.

16. Coevolutionary Theory and the Biogeography and Community Structure of *Anolis*

Bond, James. 1971. *Birds of the West Indies.* Boston: Houghton Mifflin.

Diamond, J. M. 1973. Distributional ecology of New Guinea birds. *Science* (Wash., D.C.) 179:759–769.

Diamond, J. M., M. E. Gilpin, and Ernst Mayr. 1976. Species-distance relation for birds of the Solomon Archipelago, and the paradox of the great speciators. *Proc. Nat. Acad. Sci. U.S.A.* 73:2160–64.

Heckel, D. G. and Jonathan Roughgarden. 1979. A technique for estimating the size of lizard populations. *Ecology* 60:966–975.

Lazell, J. D., Jr. 1972. The anoles (Sauria, Iguanidae) of the Lesser Antilles. *Bull. Mus. Comp. Zool.* 143:1–115.

MacArthur, R. H., and E. O. Wilson. 1967. *The theory of island biogeography.* Princeton: Princeton University Press.

Rand, A. S. 1964. Ecological distribution of the anoline lizards of Puerto Rico. *Ecology* 45:745–752.

Roughgarden, Jonathan. 1974. Niche width: biogeographic patterns among *Anolis* lizard populations. *Am. Nat.* 108:429–442.

——— 1979. *Theory of population genetics and evolutionary ecology, an introduction.* New York: Macmillan.

Roughgarden, Jonathan, and E. R. Fuentes. 1977. The environmental determinants of size in solitary populations of West Indian *Anolis* lizards. *Oikos* 29:44–51.

Roughgarden, Jonathan, W. P. Porter, and D. G. Heckel. 1981. Resource partitioning of space and its relationship to body temperature in *Anolis* lizard populations. *Oecologia* (Berl.) 50:256–264.

Roughgarden, Jonathan, John Rummel, and Stephen Pacala. 1982. Experimental evidence of strong present-day competition between *Anolis* populations of the Anguilla Bank. In *Advances in herpetology and evolutionary biology: essays in honor of Ernest E. Williams, Spec. Publ. Mus. Comp. Zool.*

Schoener, T. W. 1969. Size patterns in West Indian *Anolis* lizards. I. Size and species diversity. *Syst. Zool.* 18:386–401.

——— 1974. Resource partitioning in ecological communities. *Science* (Wash., D.C.) 185:27–39.

Schoener, T. W., and G. C. Gorman. 1968. Some niche differences in three Lesser Antillean lizards of the genus *Anolis. Ecology* 49:819–830.

Simberloff, D. S., and E. O. Wilson. 1970. Experimental zoogeography of islands: a two-year record of colonization. *Ecology* 51:934–937.

Slatkin, Montgomery, and John Maynard Smith. 1979. Models of coevolution. *Q. Rev. Biol.* 54:233–263.

Voous, Kilt. 1957. The birds of Aruba, Curaçao, and Bonaire. *Stud. Fauna Curaçao.* 7:1–260.

Williams, E. E. 1972. Origin of faunas: evolution of lizard congeners in a complex island fauna—a trial analysis. *Evolutionary Biology* 6:47–89.

Conclusion: Lizard Ecology, Viewed at a Short Distance

Boag, P. T., and P. R. Grant. 1978. Heritability of external morphology in Darwin's finches. *Nature* (Lond.) 274:793–794.

——— 1981. Intense natural selection in a population of Darwin's finches (Geospizinae) in the Galápagos. *Science* (Wash., D.C.) 214:82–85.
Dunham, A. E. 1978. An experimental study of interspecific competition between the iguanid lizards *Sceloporus merriami* and *Urosaurus ornatus*. Ph.D. dissertation, University of Michigan.
——— 1980. An experimental study of interspecific competition between the iguanid lizards *Sceloporus merriami* and *Urosaurus ornatus*. *Ecol. Monogr.* 50:309–330.
Dunham, A. E., G. R. Smith, and J. N. Taylor. 1979. Evidence for ecological character displacement in western American catastomid fishes. *Evolution* 33:877–896.
Fox, S. F. 1975. Natural selection on morphological phenotypes of the lizard *Uta stansburiana*. *Evolution* 29:95–107.
——— 1978. Natural selection on behavioral phenotypes of the lizard *Uta stansburiana*. *Ecology* 59:834–847.
Grant, P. R. 1965. The adaptative significance of some island size trends in birds. *Evolution* 19:355–367.
——— 1966. Ecological compatibility of bird species on islands. *Am. Nat.* 100:451–462.
——— 1967. Unusual feeding of lizards on an island. *Copeia* 1967:223–224.
——— 1968. Bill size, body size, and the ecological adaptations of bird species to competitive situations on islands. *Syst. Zool.* 17:319–333.
——— 1972. Interspecific competition among rodents. *Annu. Rev. Ecol. Syst.* 3:79–106.
Grant, P. R., and B. R. Grant. 1980. The breeding and feeding characteristics of Darwin's finches on Isla Genovesa, Galápagos. *Ecol. Monogr.* 50:381–410.
Grant, P. R., and T. D. Price. 1981. Population variation in continuously varying traits as an ecological genetics problem. *Am. Zool.* 21:795–811.
Hairston, N. G. 1980a. Species packing in the salamander genus *Desmognathus:* what are the interspecific interactions involved? *Am. Nat.* 115:354–366.
——— 1980b. Evolution under interspecific competition: field experiments on terrestrial salamanders. *Evolution* 34:409–420.
——— 1980c. The experimental test of an analysis of field distributions: competition in terrestrial salamanders. *Ecology* 61:817–826.
Hamilton, W. D. 1967. Extraordinary sex ratios. *Science* (Wash., D.C.) 156:477–488.
Huey, R. B., and E. R. Pianka. 1974. Ecological character displacement in a lizard. *Am. Zool.* 14:1127–36.
Milstead, W. W. 1961. Competitive relations in lizard populations. In *Vertebrate speciation,* ed. W. F. Blair. Austin: University of Texas Press.
Perrins, C. M., and P. J. Jones. 1974. The inheritance of clutch size in the Great Tit (*Parus major* L.). *Condor* 76:225–229.
Pianka, E. R. 1978. *Evolutionary ecology,* 2nd ed. New York: Harper and Row.
Pianka, E. R., R. B. Huey, and L. R. Lawlor. 1979. Niche segregation in desert lizards. In *Analysis of ecological systems,* ed. D. J. Horn, R. D. Mitchell, and G. R. Stairs. Columbus: Ohio State University Press.
Simon, H. A. 1956. Rational choice and the structure of the environment. *Psych. Rev.* 63:129–138.

Slatkin, Montgomery. 1980. Ecological character displacement. *Ecology* 61:163–177.

Smith, D. C. 1977. Interspecific competition and the demography of two lizards. Ph.D. dissertation, University of Michigan.

Smith, J. N. M., P. R. Grant, B. R. Grant, I. J. Abbott, and L. K. Abbott. 1978. Seasonal variation in feeding habits of Darwin's ground finches. *Ecology* 59:1137–50.

Tinkle, D. W. 1967. The life and demography of the side-blotched lizard, *Uta stansburiana. Misc. Publ. Mus. Zool., Univ. Mich.* 132:1–182.

Acknowledgments

1. Ecological Consequences of Activity Metabolism

Financial support was provided by NSF grants PCM 77-24208 and 81-02331 and NIH grant K04 AM00351.

2. Ecological Energetics

This work was funded through Contract DE-AM03-76-SF00012 between the United States Department of Energy and the University of California, Los Angeles, and by an intramural grant from UCLA for computing services. I am very grateful to Philip Medica for his outstanding field work on this project. I thank Bernardo Maza, Richard Castetter, and Arthur Vollmer for assistance in the field, Sarah Rockhold, Alan Beck, and Stan Wakakuwa for laboratory assistance, Fred Turner for calculating the demography data and for assistance in preparing the manuscript, David Bradford for help with statistics, and Justin Congdon, Raymond Huey, and Fred Turner for reviewing the manuscript. The cooperation and support of Auda Morrow and the Civil Effects Test Organization at the Nevada Test Site is gratefully acknowledged.

3. Biophysical Analyses of Energetics, Time-Space Utilization, and Distributional Limits

Much of the work presented here came from publications with our students and colleagues. We especially thank Keith A. Christian, Allan F. Muth, and Frances C. James for sharing data and ideas. We thank David Socha for field assistance in the Galapagos. Raymond Huey and James Spotila provided helpful critical reviews of this manuscript. William R. Welch, Steve Waldschmidt, Ray Ozanne and Mike Wrabetz contributed to the development of the crossed equation figure. We thank Cheryle Hughes and Scott Turner for illustrations and Don Chandler for photography. Professor Porter would like to thank Kenneth S.

Norris, who through his contagious curiosity, excitement, and enthusiasm for things not intuitively obvious, started him on this path, and David M. Gates, John W. Mitchell, William A. Beckman and Glen E. Myers, who guided him over the rocks and helped him when he stumbled. This work was supported by Guggenheim Fellowships to both authors, by a Romnes Faculty Fellowship, by Earthwatch, and by DOE grants DEAC 02-76EV02270 and EY-76-S-02-2270.

4. Lizard Malaria: Parasite-Host Ecology

This study was conducted while I was an NIH National Research Service Award Postdoctoral Fellow at the University of California, Berkeley. John E. Simmons was my sponsor at Berkeley and provided encouragement, space, and equipment. Rene W. Schall played a major role in this study, often assisting me in the field as well as drawing the figures. The superintendent of the Hopland Field Station, A. H. Murphy, granted permission for my work at that excellent facility. Armand Kuris and Raymond Huey read the manuscript and made many helpful suggestions. Many of my friends and colleagues offered advice, criticism, and encouragement throughout the project; their help will always be remembered and appreciated.

5. The Adaptive Zone and Behavior of Lizards

I thank Roger Anderson and Daryl Karns for discussions and comments on the manuscript, and I particularly thank R. Anderson for permission to cite his observations.

6. A Review of Lizard Chemoreception

This chapter could not have been written without the help and encouragement of two of my students, Barbara Bissinger and Karen Gravelle, who provided background information and portions of the research presented within and contributed ideas to our many discussions. David Duvall commented on an early version of the manuscript and contributed data, figures, and ideas, all of which made this a stronger paper. I also thank Gordon Burghardt, Charles J. Cole, Raymond Huey, and Eric Pianka for their valuable comments, as well as Wynne Brown for her drawings. This paper was written with the support of PSC-BHE grant no. 13060 from the Research Foundation of the City University of New York.

7. Food Availability and Territorial Establishment of Juvenile *Sceloporus undulatus*

We thank Doyle Wilson and Eugene Suiter for permission to conduct the study on land under their control. We thank the Wayne Oak family, especially Ira, for understanding and cooperation that strongly facilitated the study. Steve Hammack was especially instrumental in conducting aspects of the study. Charles Bohlen, Sam Brush, Perri Eason,

Jill Glanville, Matt Harden, Dave Hudson, John Sims, Lynne Smith, and Shayne Tidwell all provided valuable aid. Several citizens of the towns of Macksville, St. John, and Larned, Kansas, too numerous to mention, provided help and support for which we are thankful. Support for the study was provided by National Science Foundation grants to the senior author, DEB-77-01208 and DEB-79-05123, and grants from the Texas Christian University Research Foundation (5-24906 and B-7892).

8. Fitness, Home-Range Quality, and Aggression in *Uta stansburiana*

Donald W. Tinkle's fine work on lizard ecology inspired my own work and that of many others. His research furnished an excellent theoretical and empirical base from which this study has grown. Uppermost, I thank him. I thank too all those who have helped gather data and discuss interpretations: Karen Fox, Howard Fox, Elizabeth Rose, Gary Slizgi, Ronald Myers, Stanley Mays, Lurinda Burge, John Allee, Charles Boydstun, Charles Bloom, and Judith Rosenthal. I thank the gracious people of Winkler County, and especially the cantaloupe-growing region around Pecos, Texas, for logistical and other aid. The use of the Texas Tech field station at Wink and the cooperation of the Vest-Pigmon ranch made this study possible. I also value discussions of these data with Philip Ashmole, Tom Uzzell, Ed Goldstein, Fred Hotchkiss, Mike Baker, Bob Tamarin, and Tom Gavin. Financial support was provided by the Yale University Biology Department, the Boston University Graduate School, the School of Biological Sciences of Oklahoma State University, a Sigma Xi grant, and NSF grants GB-31577 and DEB-78-07156.

9. Sexual Selection, Sexual Dimorphism, and Territoriality

Research for this chapter was funded by NSF grant BNS-79-07241.

10. Psychobiology of Parthenogenesis

We wish to thank D. Duvall, N. Greenberg, R. Huey, P. Mason, and an anonymous reviewer for their many helpful comments on the manuscript. The assistance of Brian Camazine and Kevin Fitzgerald in the collection of *Cnemidophorus uniparens* is gratefully acknowledged. We also thank Janice Riley for her able technical assistance in the histological preparation of embryonic material. This research was supported in part by the President's Fund of Harvard College, a Sloan Fellowship in Basic Neuroscience, and a NIMH Research Scientist Development Award (MH00135) to David Crews.

11. Life-History Variations

This research was supported by NSF grants GB-35490, BMS-72-02292A01, and DEB-77-24477. I thank Dr. Robert Martin for allowing me access to his original data on central Texas *Urosaurus,* L. Michel and

M. Newlin for field assistance, Dr. Vince LaRiccia for statistical advice, and Drs. K. Keeler, A. Joern, J. Lynch, R. Jones and D. Clark for critical comments on an earlier draft of the manuscript and idea. I appreciate constructive comments by Drs. J. Roughgarden and S. Stearns. The late Donald W. Tinkle's friendly interest in and intellectual stimulation of my research can never be repaid.

12. Realized Niche Overlap, Resource Abundance, and Intensity of Interspecific Competition

The late Donald Tinkle's friendship, encouragement, and support for this work will never be forgotten. Drs. Justin Congdon, Donald Tinkle, Thomas W. Schoener, J. N. M. Smith, Laurie Vitt, and an anonymous reviewer provided valuable criticism of earlier versions of this chapter. I am greatly indebted to the many people who have helped me in the fieldwork upon which this report is based. The assistance of the National Park Service in Big Bend National Park is gratefully acknowledged. Without their help this study would have been impossible. I am especially grateful to my wife, Marsha, for without her continual encouragement, hard work, patience, and attention to logistic minutia I would not have completed this project. Finally, I am indebted to Karen Overall for her support and tireless assistance in the preparation of the manuscript.

13. Temporal Separation of Activity and Interspecific Dietary Overlap

We thank Robert K. Colwell, Fabian M. Jaksić, Gordon H. Orians, and Thomas W. Schoener for numerous insightful suggestions. Dr. Schoener forced us to confront difficult statistical problems. He, Robert Colwell, Joseph Felsenstein, and Stuart Pimm clarified many statistical complications and generously helped us with the analysis and programming. The National Geographic Society and the National Science Foundation provided financial support for our research.

14. Sympatry and Size Similarity in *Cnemidophorus*

I am very grateful to Rob Colwell, Mike Gilpin, Peter Grant, Eric Pianka, Dan Sulzbach, and Mark Taper for their comments and suggestions on the manuscript. Mark Taper, Todd Case, and Benita Epstein provided assistance in the field.

15. Ecomorphs, Faunas, Island Size, and Diverse End Points in Island Radiations of *Anolis*

I am deeply indebted to too many coworkers and colleagues who have shared field experience and ideas with me to mention them all by name. Raymond Huey, Timothy Moermond, Thomas Schoener, Jared Diamond, Kurt Fristrup, and Anna Haynes have improved the present chapter by their comments and criticisms. The major part of the studies reported were supported by grants from the National Science Foundation. Laszlo Meszoely has done the illustrations.

16. Coevolutionary Theory and the Biogeography and Community Structure of *Anolis*

We thank Ronald Burton, Stephen Pacala, Warren Porter, and John Rummel for their invaluable assistance during this work. We thank the Fleming family of St. Maarten for allowing our long-term studies on their property. We also thank Peter Abrams, Thomas Schoener, and two anonymous reviewers for their comments on the manuscript. Finally, we thank the National Science Foundation for its support.

Contributors

Royce E. Ballinger, School of Life Sciences, University of Nebraska-Lincoln, Lincoln

Albert F. Bennett, School of Biological Sciences, University of California, Irvine

Kent L. Brown, Department of Biology, Texas Christian University, Fort Worth

Ted J. Case, Department of Biology, University of California at San Diego, La Jolla

David Crews, Museum of Comparative Zoology, Harvard University, Cambridge, Massachusetts (present address: Department of Zoology, University of Texas, Austin)

William R. Dawson, Division of Biological Sciences, University of Michigan, Ann Arbor

Arthur E. Dunham, Department of Biology, University of Pennsylvania, Philadelphia

Gary W. Ferguson, Department of Biology, Texas Christian University, Fort Worth

Stanley F. Fox, Department of Zoology, Oklahoma State University, Stillwater

Eduardo R. Fuentes, Instituto de Ciencias Biológicas, Pontificia Universidad Católíca de Chile, Santiago

Peter R. Grant, Division of Biological Sciences, University of Michigan, Ann Arbor

Jill E. Gustafson, Museum of Comparative Zoology, Harvard University, Cambridge, Massachusetts (present address: Boston University School of Medicine, Boston, Massachusetts)

David Heckel, Department of Zoology, Clemson University, Clemson, South Carolina

Raymond B. Huey, Department of Zoology, University of Washington, Seattle

John L. Hughes, Department of Biological Sciences, Texas Christian University, Fort Worth

Kenneth A. Nagy, Laboratory of Biomedical and Environmental Sciences, University of California, Los Angeles
Eric R. Pianka, Department of Zoology, University of Texas, Austin
Stuart L. Pimm, Department of Zoology and Graduate Program in Ecology, University of Tennessee, Knoxville
Warren P. Porter, Department of Zoology, University of Wisconsin, Madison
Philip J. Regal, Museum of Natural History and Department of Ecology and Behavioral Biology, University of Minnesota, Minneapolis
Jonathan Roughgarden, Department of Biological Sciences, Stanford University, Stanford, California
Rodolfo Ruibal, Department of Biology, University of California, Riverside
Jos. J. Schall, Department of Zoology, University of Vermont, Burlington
Thomas W. Schoener, Department of Zoology, University of California, Davis
Carol A. Simon, Department of Biology, City College of the City University of New York, New York, and Department of Herpetology, American Museum of Natural History, New York
Judy A. Stamps, Department of Zoology, University of California, Davis
Richard R. Tokarz, Museum of Comparative Zoology, Harvard University, Cambridge, Massachusetts (present address: Ramopo College, Mahwah, New Jersey)
C. Richard Tracy, Department of Zoology and Entomology, Colorado State University, Fort Collins
Ernest E. Williams, Museum of Comparative Zoology, Harvard University, Cambridge, Massachusetts

Index

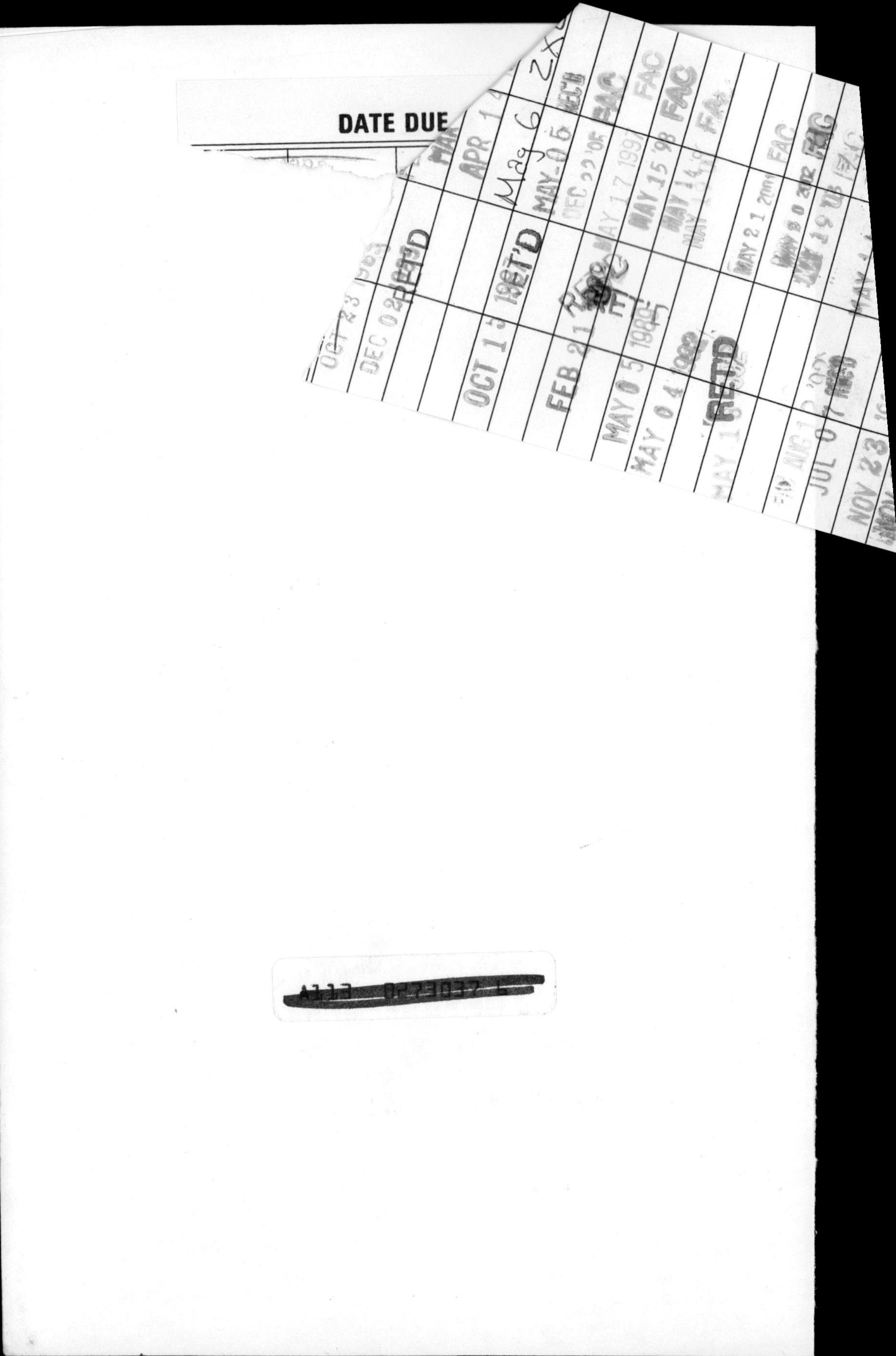